TERRESTRIAL
PLANT
ECOLOGY

The Benjamin/Cummings Series in the Life Sciences

F. J. Ayala and J. A. Kiger, Jr.
Modern Genetics (1980)

F. J. Ayala and J. W. Valentine
Evolving: The Theory and Processes of Organic Evolution (1979)

M. G. Barbour, J. H. Burk, and W. D. Pitts
Terrestrial Plant Ecology (1980)

L. L. Cavalli-Sforza
Elements of Human Genetics, second edition (1977)

R. E. Dickerson and I. Geis
The Structure and Action of Proteins, second edition (1981)

L. E. Hood, I. L. Weissman, and W. B. Wood
Immunology (1978)

L. E. Hood, J. H. Wilson, and W. B. Wood
Molecular Biology of Eucaryotic Cells (1975)

A. L. Lehninger
Bioenergetics: The Molecular Basis of Biological Energy Transformations,
second edition (1971)

S. E. Luria, S. J. Gould, and S. Singer
Life: Molecules to Organisms (1981)

A. P. Spence and E. B. Mason
Human Anatomy and Physiology (1979)

J. D. Watson
Molecular Biology of the Gene, third edition (1976)

I. L. Weissman, L. E. Hood, and W. B. Wood
Essential Concepts in Immunology (1978)

N. K. Wessells
Tissue Interactions and Development (1977)

W. B. Wood, J. H. Wilson, R. M. Benbow, and L. E. Hood
Biochemistry: A Problems Approach, second edition (1981)

TERRESTRIAL PLANT ECOLOGY

Michael G. Barbour
University of California, Davis

Jack H. Burk
California State University, Fullerton

Wanna D. Pitts
San Jose State University

The Benjamin/Cummings Publishing Company, Inc.
Menlo Park, California • Reading, Massachusetts
London • Amsterdam • Don Mills, Ontario • Sydney

Sponsoring Editors: *Mary Forkner and Jim Behnke*
Copy Editor: *Pat Broughton Burner*
Production Editor: *Pat Broughton Burner*
Cover Designer: *John Edeen*
Book Designer: *John Edeen*
Artists: *Pat Odell and Ginny Mickelson*

The art for the part opening pages is from the following sources: Parts I, III, and IV: Brine, M. D. 1886. *From Gold to Grey: Being Poems and Pictures of Life and Nature.* New York: Cassell & Company, Limited. Part II: Frye, A. E. 1899. *Introductory Geography.* Boston, MA: Ginn & Company.

Library of Congress Cataloging in Publication Data

Barbour, Michael G
    Terrestrial plant ecology.

    Bibliography: p.
    Includes index.
    1. Botany—Ecology   2. Botany—North America—
Ecology.   I. Burk, Jack H., 1942–   joint author.
II. Pitts, Wanna D., 1932–   joint author.
III. Title.
QK901.B345   581.5′264   79–23591
ISBN 0-8053-0540-8
    EFGHIJ-MA-83

The Benjamin/Cummings Publishing Company, Inc.
2727 Sand Hill Road
Menlo Park, California 94025

PREFACE

Our purpose in writing this book has been to condense the literature pertaining to the ecology of wild plants into a digestible survey for the undergraduate student. Most readers will be in a one quarter or one semester course at the upper division level and will have had previous courses in basic biology, botany, and possibly in general ecology or environmental science.

This is not a detailed, thorough, unbiased reference work, nor does it take us to a new level of understanding. Rather, it is a basic synthesis which we hope will be a useful starting point for instructors and students. The present science of plant ecology is so broad as to encourage specialization and to narrow one's perspective of the entire field. We believe it is essential to maintain a broad outlook, and to examine the important aspects of all the specialized branches of plant ecology, if we are to grasp the essentials of the interactions between plants and their environment.

This book is divided into four parts. The first part (Chapters 1 and 2) introduces the broad aspects of plant ecology and gives an historical overview of the development of the science. Part II (Chapters 3–6) discusses autecology, the responses of populations to their environment, and Part III (Chapters 7–12) focuses on synecology, that is, community attributes. Part IV (Chapters 13–19) describes the role of five environmental factors (light, temperature, fire, soil, and water) at both the autecological and the synecological levels. The concluding chapter, Chapter 19, describes the major vegetation types of North America, attempting to show how the many factors and responses covered individually in earlier sections interact to produce the vegetation of this continent. Except for this last chapter, which has its own list of references, the references cited are listed in one section at the end of the book. We hope these references offer a good starting point for the reader who wishes to go further with some topic of interest. Most chapters conclude with a relatively detailed summary.

We believe that the study of plant ecology may provide lessons that take us beyond the conventional, narrow limits of "science." It is possible that our social and psychological evolution has been influenced by vegetation. Natural settings have generated negative feelings in western (Judeo–Christian) societies, as well described by Roderick Nash:*

*Nash, R. 1973. *Wilderness and the American Mind.* Yale University Press.

... appreciation of the wilderness must be seen as recent, revolutionary, and incomplete. . . . Ambivalence, a blend of attraction and repulsion, is still the most accurate way to characterize the present feeling toward wilderness. . . . Wilderness has risen far on the scale of man's priorities. But the depth and intensity of previous antipathy suggests that it still has a long way to go.

The poet–anthropologist Loren Eisely reminds us how important our development in natural, wild vegetation will continue to be to us despite these negative feelings. "Man was born and took shape among earth's leafy shadows," and now he is to embark upon a new and difficult journey:*

> At the climactic moment of his journey into space he has met himself at the doorway of the stars. And the looming shadow before him has pointed backward into the tangled gloom of a forest from which it has been his purpose to escape. Man has crossed, in his history, two worlds. He must now enter another and forgotten one, but with the knowledge gained on the pathway to the moon. He must learn that, whatever his powers as a magician, he lies under the spell of a greater and a green enchantment which, try as he will, he can never avoid, however far he travels. The spell has been laid on him since the beginning of time—the spell of the natural world from which he sprang.

Many persons, some anonymous, have read all or part of the manuscript, providing valuable insights and corrections and helping maintain the intended focus of the book. We thank these reviewers, who include: Jerry M. Baskin, Gary L. Cunningham, Andrew M. Greller, H. Thomas Harvey, Gordon L. Huntington, Subodh K. Jain, Robert P. McIntosh, Jack Major, Bruce M. Pavlik, Robert W. Pearcy, Otto T. Solbrig, Robin Smith, James P. Syvertsen, Irwin A. Ungar, John L. Vankat, and Lynn D. Whittig. Final decisions concerning approach, presentation, and content were our own and we take full responsibility for any oversights or inaccuracies contained herein.

Several other individuals have contributed long hours in manuscript preparation and have tolerated and understood the sometimes intolerant authors. Special thanks go to Heather Stout, who worked many long hours organizing the Literature Cited, securing permissions, and preparing the index. We also thank Patricia Burk, Maggie Hummer, Jerry Pitts, and the many ecology students who "tried out" sections of this book as a text in courses during the past two years. We thank you for your help and patience.

M.G.B., J.H.B., W.D.P.

*Eisely, L. 1970. *The Invisible Pyramid.* Scribners.

CONTENTS

PART I

BACKGROUND AND BASIC CONCEPTS

The origins of informal plant ecology go back hundreds of years to plant geographers, taxonomists, and naturalists who were people of energy and insight, such as Humboldt, De Candolle, and Darwin. In contrast, the science of ecology has had a very brief, intensive history which began less than a century ago. In that brief period, certain energetic men and women have had such a profound effect on the developing science that many current research projects and viewpoints are simply extensions of their work.

In this part, we will examine some of that history and some of those personalities, and we will present a common language to use in our study of plant ecology. There are many approaches to plant ecology, but each one attempts to answer the same basic question: how do plants cope with their environment? Some of the approaches we will discuss are paleoecology, phytosociology, community dynamics, systems ecology, and autecology.

CHAPTER 1

INTRODUCTION

Plant ecologists try to discover an underlying order to vegetation. They do this at finer and finer levels for the same reasons that biologists, chemists, and physicists pursue their worlds to the level of DNA, hydrogen bonds, and subatomic particles: there seems to be a human need to know the complete story, to explain the past, to predict the future. What threads link plants to each other and to their environment? How flexible are these threads, and how intermeshed? How do plants "solve" the problems of dispersal, germination in a suitable site, competition, and the acquisition of energy and nutrients? How can they withstand unfavorable periods of fire, flood, or storm?

What can plants tell us by their presence, vigor, or abundance about the past, present, and future course of their habitat? Can plants be used as a scientific tool to analyze the intricacies of the environment or to test hypotheses about evolution?

What can plants tell us about our management hopes for the land? Once a forest is cut, what will replace it, how long will the process take, and how can we most efficiently manipulate this process? Once domestic livestock graze at a given density for a given time, what will the vegetation look like and how many animals will it then be able to support? Once topsoil is removed by strip mining in semiarid regions, what plants should be introduced to stabilize the remodeled landscape? Once brushfields have been sprayed, burned, and replanted as grasslands, what will happen to watershed quality, soil nutrient levels, and the rate of siltation behind nearby dams? What is the residence time of herbicides in soils and are there side effects on nontargeted organisms? How many backpackers can use an alpine trail without altering adjacent vegetation? If fire or flood is a natural catastrophe which must recur with a certain frequency to maintain particular vegetation types, how can we incorporate such regular disasters into state and federal park management plans?

All of these questions, and more, are being investigated by plant ecologists. Some researchers are more interested in generating basic information that has to do with the description of vegetation or the biology of component species. Others are more interested in applying that basic information to management problems. Applied plant ecologists may be called range managers, foresters, or agronomists, but they are all plant ecologists and they all share the same pleasure in discovering the subtle ways in which plants are adapted to their environment. Their objective is very close to a formal definition of **ecology**: the study of organisms in relation to their natural environment.

ECOLOGY, ENVIRONMENT, AND VEGETATION

The word ecology was coined little more than 100 years ago by the German zoologist Ernst Haeckel. He spelled the word "oekologie," but ecologists soon dropped the first o, to the annoyance and confusion of some purists. The word comes from the Greek roots *oikos*, home, and *logos*, the study of. Thus, oekologie translates loosely as: the study of organisms in their home, their environment.

Environment* is the summation of all **biotic** (living) and **abiotic** (nonliving) factors which surround and potentially influence an organism. Examples of biotic factors include competition, mutualism, allelopathy, and other interactions between organisms, which are described in Chapters 5 and 6. Abiotic factors include all chemical and physical aspects of the environment which influence a plant's growth and distribution.

The environment can be divided into two parts: the macroenvironment and the microenvironment. The **macroenvironment** is the prevailing, regional environment and the **microenvironment** is the environment close enough to an object to be influenced by it (Figure 1–1). The microenvironment may be quite different from the macroenvironment. For example, the microenvironment beneath a forest canopy is different from the macroenvironment above it in such traits as humidity, wind speed, and light intensity; the microenvironment beneath a rock in desert soil may be cooler and moister than other parts of the macroenvironment; the microenvironment just 1 mm above a leaf surface may differ in wind speed, humidity, and temperature from the macroenvironment 10 mm away. Each organ or part of a plant is exposed to a different microenvironment, as shown in Figure 1–2. Obviously, the microenvironment is what a plant responds to, and so the microenvironment will be emphasized in this book.

Plant ecology is concerned not only with individual plants and plant species, but also with vegetation. **Vegetation** consists of all the

---

*Environment can be synonymous with habitat. A habitat is homogeneous; that is, it cannot be broken up into distinctly different subunits. Some (Whittaker et al. 1973) restrict habitat to mean the environment around an individual organism or species, and they use the term biotope to mean the environment around a community. We will not adopt this distinction, however.

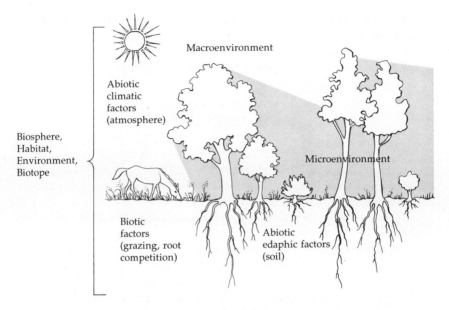

Figure 1-1. Components of the environment.

plant species in a region (the flora) and it is concerned with the pattern of how all those species are spatially or temporally distributed. If the region is large, its vegetation will consist of several prominent plant communities. Each vegetation type is characterized by the **life form*** of its dominant plants (the largest, most abundant, characteristic plants). Examples of life forms include annual herbs, broadleaf evergreen trees, drought-deciduous shrubs, plants with bulbs or rhizomes, needleleaf evergreen trees, perennial bunchgrass, and dwarf shrubs (Figure 1-3). Vegetation is also characterized by the architecture of its canopy layers, as shown in Figure 8-9. Different forest types have one, two, three, or four tree canopy layers.

Architecture and life form both contribute to the **physiognomy** (outer appearance) of vegetation, and each vegetation type has its own characteristic physiognomy. A vegetation type which extends over a

---

*Life form may include any or all of the following, depending on the context: (a) the size, life span, and woodiness of a taxon, for example, herb, annual, perennial, herbaceous perennial, woody perennial, shrub, tree, or vine; (b) the degree of independence of a taxon, for example green and rooted to the ground, parasitic, saprophytic, or epiphytic; (c) the morphology of a taxon, for example stem succulent, leaf succulent, rosette form, spinescent, or pubescent; (d) the leaf traits of a taxon, for example large, small, sclerophyllous, evergreen, winter-deciduous, drought-deciduous, needleleaf, or broadleaf; (e) the location of perennating buds, as defined by Raunkiaer (1934); and (f) **phenology**, the timing of life cycle events in relation to environmental cues.

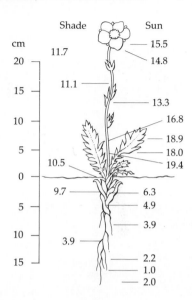

Figure 1–2. Temperature of the microenvironment (°C) near different parts of an alpine plant in sun and shade on a July day when the temperature of the macro-environment was 11.7°C. (Redrawn from Tikhomirov 1963. Also cited in Larcher 1975.)

large region is called a **formation**. For example, a tropical rain forest is a formation dominated by broadleaf evergreen trees, and it is characteristic of thousands of square kilometers in humid tropical regions on several continents. Part of the formation classification scheme adopted by UNESCO (1973) appears in Table 1–1.

Formations may be subdivided into **associations**. An association is the collection of all plant **populations*** coexisting in a given habitat. It has the following attributes: (a) it has a relatively fixed floristic composition, (b) it exhibits a relatively uniform physiognomy, and (c) it occurs in a relatively consistent type of habitat. The same species tend to recur together wherever a particular habitat repeats itself. Associations are usually named by their dominant or most characteristic taxa; thus the red fir forest in California's Sierra Nevada, which occurs on well-drained, undisturbed sites between 1800 and 2400 m, is a red fir *(Abies magnifica)* association in a conifer forest formation.

Typically, several to many associations may belong to the same formation, all sharing a similar physiognomy, but each differing qualitatively or quantitatively in species composition. A fascinating ecological phenomenon is the similarity of vegetation types in similar macro-environments scattered around the world. It is as though a particular

---

*A population is a group of individuals of the same species occupying a habitat small enough to permit interbreeding among all members of the group. Some populations do not interbreed, being self-pollinators or reproducing only asexually, but they exist in a habitat small enough to permit the *potential* exchange of genes. Very restricted species, endemic to a small area, may consist of only a single population, but most species include many populations. Consequently, the terms species and population are not usually synonymous.

(a)

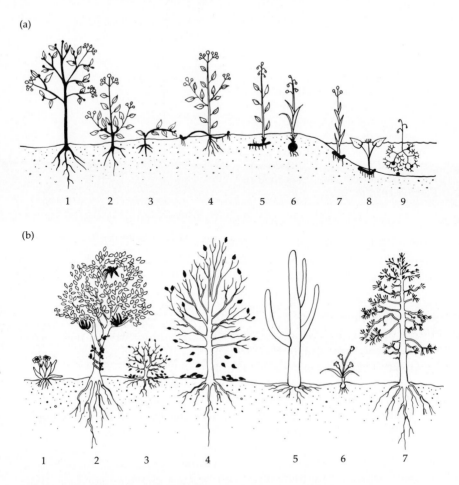

(b)

Figure 1–3. Examples of life forms (growth forms). (a) A scheme of classification by Raunkiaer (1934) based on the location of overwintering, perennating parts, such as buds, bulbs, or seeds, shown in dark, solid areas: (1) phanerophyte, (2–3) chamaephytes, (4) hemicryptophyte, (5–9) cryptophytes (also called geophytes). (b) Life forms based on other criteria, such as length of life, succulence, and leaf traits: (1) annual herb, (2) broadleaf evergreen tree with a liana (a woody vine) and an epiphyte (a plant which grows on another plant but uses it only for mechanical support, not as a source of nutrients), (3) drought-deciduous shrub, (4) broadleaf deciduous tree, (5) stem succulent, (6) bulbous herbaceous perennial, (7) needleleaf evergreen. ((a) redrawn from C. Raunkiaer. 1934. *The Life Forms of Plants and Statistical Plant Geography.* Courtesy of Clarendon Press, Oxford.)

physiognomy has been selected for in similar but isolated habitats. Evidently there has been convergent evolution among vegetation types (see Chapter 7).

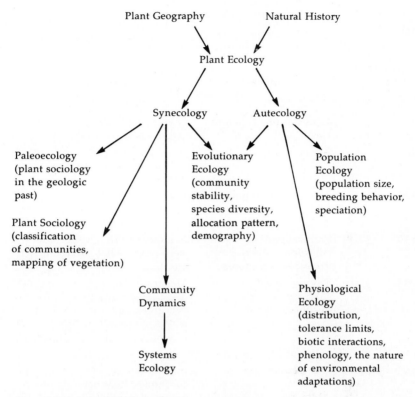

Figure 1–4. The relationship between and origins of specializations within the field of plant ecology.

SPECIALIZATIONS WITHIN PLANT ECOLOGY

Synecology

One large segment of plant ecology has followed directly from plant geography (Figure 1–4). This is **synecology**, and some of its many synonyms include community ecology, phytosociology, geobotany, vegetation science, and vegetation ecology.

One phase of synecology is **plant sociology**, the description and mapping of vegetation types and communities.* In the past 50 years there has been a proliferation of standard methods for sampling vegetation and treating and analyzing sampling data. With these standard methods valid conclusions can be drawn and vegetation from all

---

*Community (phytocoenose) is a general term which can be applied to any vegetation unit, from the regional to the very local.

Table 1–1. Part of the formation classification scheme adopted by UNESCO (1973).

Classification number	Formation	Description
I	Closed forest formation class	Dominants 5^+ m tall, crowns interlocking
IA	Mainly evergreen forest	Individual trees may shed leaves, but canopy as a whole remains green
IA1	Tropical ombrophilous forest (= tropical rain forest)	Dominants mainly broadleaved, evergreen, with drip tips; neither cold nor drought resistant
IA2	Tropical and subtropical evergreen seasonal forest	Number of drought-deciduous trees intermediate between IA1 and IA3
IA3	Tropical and subtropical semi-deciduous forest	Most upper canopy trees drought-deciduous; many understory trees evergreen and sclerophyllous; leaves without drip tips
IA4	Subtropical ombrophilous forest	Local variant, grading into IA1
IA5	Mangrove forest	Intertidal location of tropics and subtropics; dominated by ever-green sclerophyllous broadleaved trees with stilt roots or pneumatophores; vascular epiphytes rare
IA6	Temperate and subpolar evergreen ombrophilous forest	Occurs in extremely oceanic, frostfree climates of Southern Hemisphere, as in *Nothofagus* or *Podocarpus* forests of New Zealand
IA7	Temperate evergreen seasonal broadleaved forest	Dominated by hemisclerophyllous evergreen trees, rich in herbaceous undergrowth, but few epiphytes and lianas; grades into IA4, IA6, or IA8
IA8	Winter-rain evergreen broadleaved sclerophyllous forest	Dominated by sclerophyllous, ever-green trees with little understory but some lianas
IA9	Tropical and subtropical evergreen needleleaved forest	Dominated by needleleaved or scaleleaved evergreen trees; vascular epiphytes and lianas absent
IA10	Temperate and subpolar evergreen needleleaved forest	As IA9, but to the north

Classification number	Formation	Description
IB	Mainly deciduous forest	Majority of trees shed foliage simultaneously in connection with an unfavorable growing season
IB1	Tropical and subtropical drought-deciduous forest	Foliage shed during dry season (usually winter)
IB2	Cold-deciduous forests with evergreen trees	Foliage shed during frost season; deciduous trees dominant but evergreens present as in hemlock-hardwood forest
IB3	Cold-deciduous forests without evergreen trees	Deciduous trees absolutely dominant, vascular epiphytes absent
IC	Extremely xeromorphic forest	Dense stands of xeromorphic trees, with succulents and xeromorphic shrubs, often grading into woodland (II)
II	Woodland formation class	Dominants 5^+ m tall, crowns not usually touching, but canopy cover 40^+%, herbaceous layer may be present
IIA	Mainly evergreen woodland	Dominants evergreen as defined in IA
IIB	Mainly deciduous woodland	Dominants variously deciduous, as in IB
IIC	Extremely xeromorphic woodland	Similar to IC, but trees less dense
III	Scrub formation class	Dominants shrubs or dwarf trees 0.5–5 m tall
IIIA	Mainly evergreen scrub	Evergreen as defined in IA, includes chaparral
IIIB	Mainly deciduous scrub	Deciduous as defined in IB, includes riparian thickets
IIIC	Extremely xeromorphic (subdesert) shrubland	Very open stands of shrubs with xerophytic adaptations, some plants with thorns
IV	Dwarf-scrub and related communities	Dominants less than 0.5 m tall, includes arctic-alpine tundra, bogs, heaths

Table 1–1, continued

Classification number	Formation	Description
V	Herbaceous vegetation	Dominated by graminoids or forbs, more or less continuous cover, woody synusia less than 40% cover
VA	Tall graminoid vegetation	Dominant graminoids 2^+ m tall when in flower, forb cover less than 50%
VA1	Tall grassland with tree synusia 10–40% cover	An open woodland with graminoid cover greater than 50%
VA2	Tall grassland with tree synusia less than 10% cover	Savannas, sometimes with shrubs
VB	Medium tall grassland	Dominant graminoids 0.5–2 m tall when in flower, forb cover less than 50%
VC	Short grassland	Dominant graminoids less than 0.5 m tall when in flower, forb cover less than 50%, includes meadows, some types of tundra
VD	Forb vegetation	Forb cover greater than 50%, graminoid cover less than 50%

over the world can be compared on an equal basis. (The description of *past* vegetation types and associations, as they have existed through geologic time, is part of the field called **paleoecology**.)

Another phase of synecology is the examination of **community dynamics**, which includes processes such as the transfer of nutrients and energy between members, the antagonistic or symbiotic relationships between members, and the process and causes of **succession** (community change over time).* The study of community dynamics can be abstracted to a mathematical level, where complex formulae and computer programs summarize, simulate, or model the particular dynamic system being examined. This type of research is called **systems ecology**.

---

*Vegetation which does not change with time is in equilibrium with its environment and is said to be mature, stable, or in a **climax** state. In contrast, vegetation which is changing is said to be **seral**. In the process of succession, one seral community replaces another until a climax community is established and perpetuates itself. This path of succession through time is called a **sere**.

A third phase of community ecology tries to discover evolutionary themes that determine the fundamental nature of communities: what determines the number of species which may coexist in a habitat; how may communities be described in terms of function rather than taxa; what determines the stability or fragility of a community; how is physiognomy selected for; how may species be described in terms of adaption patterns; and how have plants and animals coevolved in the complex, gradual formation of the communities which exist today? This phase is called **evolutionary ecology**, and it overlaps with autecology and population ecology.

Autecology

Another large segment of plant ecology deals with the adaptations and behavior of individual species or populations in relation to their environment. This is **autecology**, and synonyms include physiological ecology and ecophysiology. Subdivisions of autecology include demecology (speciation), population ecology (the regulation of population size), demography (the allocation of energy to different activities), and genecology (genetics). Autecologists try to explain the *why* of a given species' distribution: what phenological, physiological, morphological, behavioral, or genetic traits seem most important to the species' continued success in a given habitat? They try to illustrate the pervasive influence of the environment at the population, organismic, and suborganismic levels. There is some overlap with evolutionary ecology when autecologists attempt to summarize all this as a species' adaptive pattern for survival. Autecology flows easily into other specialties outside the field of ecology proper, such as physiology, genetics, evolution, biophysics, and biosystematics (a part of taxonomy).

A Final Word on Specialization

Plant ecology itself may be thought of as a specialization within ecology. Some scientists and educators criticize the division of ecology into plant and animal ecology, arguing that the division is artificial and damaging to an understanding of the interdependence that permeates ecosystems.* Plant ecology as a discrete science is partly an artificial creation, but no more so than any other science. There are many ways in which chemistry is linked to physiology, or physiology to behavior, or soils to geology, or mathematics to economics, yet these fields are accepted as reasonable specializations. We all specialize, and in this

*An **ecosystem** is the sum of the plant community, animal community, and environment in a particular region or habitat.

Figure 1–5. "Although humans make sounds with their mouths and occasionally look at each other, there is no solid evidence that they actually communicate among themselves." (By Sidney Harris: reprinted by permission of *American Scientist,* journal of Sigma Xi, the Scientific Research Society.)

manner progress is made. One person cannot master all of plant ecology, let alone plant and animal ecology both. In addition, we think a case could be made that the inherent differences in structure, behavior, and function between plants and animals are so profound that many principles of plant ecology cannot be translated into principles of animal ecology, and vice versa.

The answer to the charge that we artificially fragment ecology is not that we should specialize less, but that we should communicate more. All of us must read, listen, and debate extensively (Figure 1–5). The excellent reviews of the history of general ecology in North America by Egerton (1976) and McIntosh (1976) are a good beginning in this direction.

CHAPTER 2

A BRIEF HISTORY
OF PLANT ECOLOGY

The route a science takes in its development is determined by the personal traits, interests, culture, and social surroundings of certain people along the way, perhaps as much as it is determined by the hypotheses, facts, or approaches that each person has contributed. Good science may be impersonal, dispassionate, and unbiased, but most scientists are not. Consequently, this brief account of the history of plant ecology is flavored not only by people's ideas, but also by the people themselves.

FOUNDATIONS IN PLANT GEOGRAPHY

Prior to the Nineteenth Century

Plant ecology has been studied informally and pragmatically since the beginning of the human race. Gatherers and hunters mastered a knowledge of the distribution of wild food and forage plants which sustained them and the prey they hunted. Shamans (medicine men) learned the narrow habitat requirements of rare species which had healing, narcotic, or hallucinogenic qualities; no doubt being able to find such plants was as important as knowing how to use them. Aristotle and Theophrastus, ca. 300 B.C., may have been the first to write about plant geography and plant ecology, but no doubt illiterate peoples throughout the world had an understanding of these topics well before that time.

Some of the earliest formal ecology papers, dating back to the seventeenth century, were concerned with the succession of communities surrounding gradually filling lakes and bogs, and the term succession was used in its modern context by the beginning of the nineteenth century (Clements 1916). However, the real development of plant ecology came through books on plant geography written by people trained as taxonomists or general botanists. World exploration, especially in the nineteenth century, molded an increasingly rigorous ecological viewpoint. Carl Ludwig Willdenow (1765–1812) was the pioneer of this line of thought. He was an early plant geographer who noted that similar climates produce similar vegetation types, even in regions thousands of kilometers apart, such as southern Africa and Australia.

The Nineteenth Century

Willdenow's teaching at the University of Göttingen, in Germany, greatly influenced a wealthy, young Prussian student, Friedrich Heinrich Alexander von Humboldt (1769–1859) (Figure 2–1), who took botany to round out his education in higher mathematics, natural sciences, and chemistry. Not long after graduating, Humboldt met Johann Forster, who had accompanied James Cook on his world voyage of discovery. Forster's stories made Humboldt determined to visit the new world tropics. A decade later Humboldt met a young French botanist named Aimé Bonpland who had a similar desire. Plans crystallized, and they received the permission and protection of King Carlos IV of Spain to travel in what is now Latin America. They took with them the best equipment of the day for measuring latitude, elevation, temperature, humidity, and other physical factors, and for the next five years they traveled from steamy lowland rain forest to cold alpine paramo, and from arid desert to thorn scrub. They explored Cuba, Venezuela, Peru, Mexico, and the Orinoco and Amazon Rivers; they climbed to the top of Mt. Chimborazo (6072 m); they collected 60,000 plant specimens; and they even influenced President Jefferson in his decision to send Lewis and Clark west.

Humboldt returned to France and began to write his monumental 30-volume work, *Voyage aux regions equinoxiales*. The first 14 volumes were devoted to botany, and in those he coined the term association, he described vegetation in terms of physiognomy, he correlated the distribution of vegetation types with environmental factors, and he described the synergistic effects of some physical factors (for example, elevation, latitude, and temperature). His statement, "In the great chain of causes and effects no thing and no activity should be regarded in isolation," is a striking preview of our modern view of interdependence within communities and ecosystems. Near the end of his life he wrote a five-volume encyclopedia, *Kosmos*, which attempted to describe and explain the entire universe. Humboldt was one of the last Renaissance people, attempting to master all the knowledge of his time.

Humboldt's study of plant geography was furthered by Schouw, De Candolle, Kerner, and Grisebach, among others. J. F. Schouw (1789–1852), a professor at the University of Copenhagen, methodically described the effects of major environmental factors on plant distribution in an 1823 book, emphasizing the role of temperature. This search for single, most important, factors is still with us today but more and more we understand the interdependence of all factors. Schouw popularized the procedure of naming associations by combining the dominant genus with the suffix -etum. Thus, Quercetum is an oak woodland association (Quercus + etum) and Pinetum is a pine forest association (Pinus + etum). Some modern schemes of association nomenclature still use this concept.

Anton Kerner von Marilaun (1831–1898) studied medicine at the University of Vienna but gave up practice after he experienced a cholera epidemic as an intern. He turned to botany as a less traumatic career, became Professor of Botany at the University of Innsbruck, and then was commissioned by the Hungarian government to describe the vegetation in parts of eastern Hungary and Transylvania. The book that resulted, *Plant Life of the Danube Basin*, has fortunately been translated into English, so a larger audience can now appreciate the beauty of his vegetation descriptions and his clear understanding of succession. Later, Kerner became one of the first experimental ecologists. He established several transplant gardens at various elevations in the Tyrolean Alps, from Vienna at 180 m to his villa at 1200 m, and up to the alpine zone at 2200 m. In each garden he grew alpine and lowland forms of more than 300 species together. He found that some variations within a species were fixed and heritable, whereas other variations were nonheritable modifications induced by the environment.

August Grisebach (1814–1879) traveled widely and described more than 50 major vegetation types in very modern physiognomic terms, relating their distribution to various climatic factors. He succeeded Willdenow as Professor of Botany at Göttingen.

Alphonse Luis Pierre Pyramus De Candolle (1806–1893) was an herbarium taxonomist, an armchair plant geographer, but his access to vast plant collections led him to try to "... discern the laws of plant distribution." Like Schouw, he chose to study temperature. He summed temperatures according to a formula that was so useful his 1874 data later became the basis for Köppen's famous classification of climate, published half a century later (see Gates 1972 and Ackerman 1941).

Other biologists who contributed to the development of plant ecology in the mid- to late-nineteenth century include Oscar Drude, Adolf Engler, George Marsh, Asa Gray, and Charles Darwin. The theory of evolution and the evidence which Darwin marshalled to support it in his book, *Origin of Species*, are inherently ecological (see Harper 1967). It is fitting that Darwin's explorations and accounts were inspired by Humboldt's plant ecology work.

THE ESTABLISHMENT OF
PLANT ECOLOGY APART FROM
PLANT GEOGRAPHY

The plant geography period culminated with the publication of several extremely important, innovative books between 1895 and 1916.

At the end of those 20 years, plant ecology was firmly established as a science in its own right, and most of our modern research directions had been initiated. The major contributors were Warming, Schimper, and Paczoski in Europe, and Merriam, Cowles, and Clements in America.

Johannes Eugenius Bülow Warming (1841–1924)

As a young man in Holland, Johannes Warming (Figure 2–2) was offered the chance to help a paleontologist conduct a field study near Lagoa Santa, Brazil. It was an area of tropical woodland-savanna, 42 days travel northwest of Rio de Janeiro. He spent three years there, and 30 years later, when the last of his 2600 plant specimens had finally been identified or described, he wrote a book on the vegetation, a book whose classical ecological organization would serve as a model for any vegetation study done today. Introductory chapters on geology, soils,

Figure 2–1. Friedrich Heinrich Alexander von Humboldt, 1769–1859. (Courtesy of the Hunt Institute for Botanical Documentation.)

Figure 2–2. Johannes Eugenius Bülow Warming, 1841–1924. (Courtesy of the Hunt Institute for Botanical Documentation.)

and climate are followed by sections on each of the major vegetation types and communities. He discussed dominants and subdominants, the adaptive value of various life-forms, the effect of fire on community composition and succession, and the phenology of communities and taxa.

In the meantime, he had returned to teach at the University of Copenhagen as a successor to Schouw. He organized what may have been the world's first ecology course and he was recognized as an outstanding teacher. He published his lecture notes in 1895, in Danish, as the world's first plant ecology text. The book had an immediate impact on many botanists throughout the Western world. It was later translated into German, Polish, Russian, and finally English. Warming synthesized plant morphology, physiology, taxonomy, and biogeography into a coherent science for the first time. He concluded that soil has more of an effect on vegetation than climate, and he emphasized both moisture and temperature as prime climatic factors. He coined such useful, and still commonly used, terms as **halo-**, **hydro-**, **meso-**, and **xerophyte**, meaning plants of saline, wet, moist, and dry habitats, respectively.

Andreas Franz Wilhelm Schimper (1856–1901)

Andreas Schimper was born into a celebrated family of German botanists. He studied geology and botany at the University of Strassburg, and later taught plant histology, physiology, ecology, and geography at the University of Bonn. He traveled extensively in the tropics and concluded that the basic work of plant geography and taxonomy would soon be completed; he was not quite correct. Based on his conclusion, however, the goal he set for himself was to explain the *causes* of regional differences in floras and vegetation. Near the end of his short life he published *Plant Geography on a Physiological Basis*. The book's title is somewhat misleading because he stressed morphological features of presumed adaptive value and presented little of plant physiology in the modern sense (although many of his semi-physiological conclusions are still accepted today).

Like Warming, Schimper gave weight to both climatic and edaphic (soil) factors, and among climatic factors he emphasized temperature and moisture. He appeared to borrow heavily from Warming's text and figures (Warming's book had been translated into German two years before), but nowhere did he give Warming so much as a footnote of credit. It is possible that Schimper came to Warming's conclusions independently, but some of Warming's supporters still suspect Schimper of plagiarism (Goodland 1975).

Jozef Paczoski (1864–1941)

The spread of Jozef Paczoski's (pronounced pach · ós · ky) ideas suffered because they were published in a Slavic rather than a western European language, and they were not quickly translated (see Maycock 1967). He has only belatedly been credited in the U.S.S.R. and North America as the father of **phytosociology** (which he defined in his 1891 and 1896 papers as all the sociological relationships of plants). He showed how plants modify the habitat, creating their own microenvironment, he discussed the role of competition, the causes of succession, the role of fire, the interdependence of species in a community, the continuum nature of community boundaries, and such physiological adaptations as shade tolerance. He later published a text on, and founded the world's first department of, phytosociology at the University of Poznan, Poland. Many current phytosociology practices, terms, and concepts in Europe were conceived by him long before their current popularity.

Clinton Hart Merriam (1855–1942)

Clinton Merriam received an M.D. degree from Columbia University, but later devoted himself to biology, especially zoology, and served as Chief of the U.S. Biological Survey from 1885 to 1910. During that time he visited many parts of the western United States, writing with a naturalist's eye about new species of mammals and the distribution of vegetation types, and with an anthropologist's eye about Indian groups in California. He was a founder of the National Geographic Society.

As his large expeditions with long supply trains moved slowly through Arizona and the West, he was impressed with the similarities of elevational zones of vegetation from one mountain range to another. He believed that temperature, especially warmth in the growing season, was the determining environmental factor, and he developed formulae to sum degrees of warmth much as Grisebach had done. He then correlated each vegetation type to values of warmth, enlarged the vegetation types into life zones (which he named Boreal, Transitional, Canadian, Hudsonian, Tropical, etc.), and extrapolated the distribution of his life zones across the entire North American continent (Figure 2–3).

Merriam had a strong measure of self-confidence, for he later wrote, ". . . in its broader aspects the study of the geographic distribution of life in North America is completed. The primary regions and

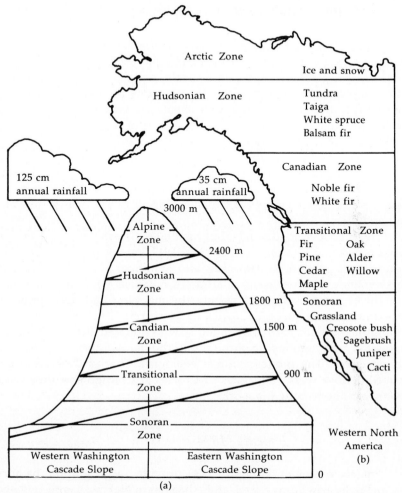

Figure 2–3. Example of Merriam's life zones (a) in the Cascade Mountains of Washington, and (b) diagrammatically extended across the western states. (Adapted from *Living Systems* by J. Ford and J. Monroe. 1971. Courtesy Canfield Press, New York.)

their subdivisions have been defined and mapped, the problems involved in the control of distribution have been solved, and the laws themselves have been formulated." Although some naturalists in the U.S. National Park Service and some biology books still use his life zone terminology and concept, many of his assumptions, calculations, and conclusions have been rightfully challenged (see, for example, Daubenmire 1938), and his latitudinal extrapolations are not widely accepted. Nevertheless, Merriam had a strong impact on American plant ecology.

Henry Chandler Cowles (1869–1939)

Henry Cowles was a geologist-turned-botanist who investigated plant succession on sand dunes around Lake Michigan from the perspective of a geologist. He had a gentle, amiable personality which attracted many students to him. Gleason wrote in 1940, ". . . he was a man of infectious gaiety and high spirits, of infinite humanity and humor. . . ." The University of Chicago, where he taught, became a center of plant ecology during the first 20 years of this century. He was an organizer of the Ecological Society of America in 1915 and served as its president three years later. Most professional ecologists in the United States today belong to this society. It publishes a research journal and sponsors scientific meetings at which ecologists can exchange information and ideas.

He published moderately, but his Ph.D. dissertation and later papers on dune succession from 1898 to 1911 emphasized the dynamic nature of vegetation, as Warming and Schimper somehow had not. This dynamic aspect of ecology attracted botanists who until then had thought of plant ecology as merely the mapping and description of static pieces of vegetation. Cowles had an impact on American plant ecology mainly through his students, who adopted his emphasis on succession and applied it throughout the midwestern and eastern United States. The pioneer animal ecologists Charles C. Adams and Victor Shelford were colleagues of Cowles at Chicago and they were strongly influenced by his work.

Frederick Edward Clements (1874–1945)

Frederick Clements (Figure 2–4) provided a geographic balance to Cowles; he was born, raised, and educated in Nebraska, and traveled widely in the western United States. During his student days at the University of Nebraska, he was prodded by Professor of Botany C. E. Bessey to expand his classical undergraduate education by adding a large dose of field botany. Together with some other students, Clements attempted to collect, identify, and describe the distribution of every plant species in the state, from algae to oaks. He collected hundreds of plants and identified many new species of fungi, ultimately co-authoring a book in 1898, *The Phytogeography of Nebraska*, which received wide recognition.

He married a university student, Edith Schwartz, the next year, and she switched her Ph.D. studies from German to botany. Long after (1960), she wrote a charming book about her life with Clements, and it is clear that she was a great help to him in his career, functioning alternately as driver, secretary, photographer, foreign language translator, and sometimes co-author.

Clements described much of the vegetation of North America, naming regional formations and associations, local variants, and seral stages with great authority, if not always with great precision. He wrote about the causes of succession, the use of certain species as environmental indicators, and the methods one might use in documenting succession or in identifying associations. His classical background meshed with his philosophical, precise, rigid personality to produce large, comprehensive books filled with many new terms and with conclusions that have a dogmatic tinge. He defined a plant association in terms of an organism in order to illustrate the interdependent nature of an association's component species and compare the association's development to the life cycle of an organism (see Chapters 7 and 10). The wealth of information, the new terms, the philosophical sweep of completeness, and the assertive conclusions confused some readers and offended others, and thus many of his ideas were criticized unjustly. Much of his plant community work retains validity today.

In 1917 he was hired by the Carnegie Institution of Washington, D.C. to direct research at a coastal laboratory at Santa Barbara, California, during the winter and an alpine laboratory at Pikes Peak, Colorado, during the summer. He devoted the rest of his life to full time research on the causes of plant distribution, experimenting with transplant gardens much as Kerner had done before him, yet arriving at completely different conclusions which are not considered valid today. He performed valuable public service work with the Soil Conservation Service during the Dust Bowl of the mid-1930s, and he wrote one of the first American plant ecology textbooks with John E. Weaver in 1929. He was actively engaged in research until a few weeks before his death at the age of 70.

His personality was quite different from that of Cowles. It was "powerful" according to Tansley (1947), "... decidedly puritan, even ascetic ... and his manner was apt to be tinged with a certain arrogance.... [Nevertheless] he had that best of all senses of humour which enables a man to laugh at himself.... [He was] by far the greatest individual creator of the modern science of vegetation."

THE PAST 60 YEARS

By 1930 there were many plant ecologists in Europe and America, contemporaries of Cowles, Clements, and Paczoski, each asking his or her own particular questions and expanding the field of plant ecology in a particular direction.

William Cooper, Edgar Transeau, and Emma Lucy Braun, all students of Cowles, described forests and bogs and paths of succession in the Midwest. Cooper also traveled long distances, for those days, to examine succession behind retreating glaciers in Alaska, on sand dunes along the Pacific coast, and in California chaparral. Transeau introduced ecology into general botany texts, and foreshadowed our current interest in ecosystem productivity and energy transfer by working out an energy budget of a cornfield in 1926. E. Lucy Braun became famous for her descriptions of the virgin deciduous forests of eastern North America. She and her sister traveled extensively through the Appalachian Mountains during Depression and Prohibition times, gaining acceptance by the local people when male strangers might have been rebutted.

In 1926 Henry Gleason wrote a widely read paper which attacked Clements' basic assumptions about the nature of associations and succession (see Chapter 10). The result was, as he wrote in 1953, "... to ecologists I was anathema. Not one believed my ideas; not one would even argue the matter. . . . For ten years, or thereabout, I was an ecological outlaw, sometimes referred to as 'a good man gone wrong'." Gleason's ideas have since become standard parts of current ecology books, now often taken with as much faith as Clements' ideas had been taken before.

Forrest Shreve and William Cannon labored in the deserts, Cannon relating morphology and anatomy to habitat in the tradition of Schimper and Warming, and Shreve describing the major communities of the warm deserts of North America as a research associate of the Carnegie Institution from 1908 to 1945. Shreve drew one of the first published vegetation maps of North America in 1917, and he was a member of the small group Cowles brought together to found the Ecological Society of America. His wife, Edith Shreve, wrote some excellent articles on plant-water relations of xerophytes.

In Europe, Christen Raunkiaer succeeded Warming as Professor of Botany at Copenhagen, and developed a life form classification that is still widely used (see Figure 1–3(a)) and a quantitative method of sampling vegetation whereby data could be treated statistically without bias.

Sir Arthur Tansley investigated the vegetation of Britain, founded the British Ecological Society, coined the term ecosystem, called for more physiological investigations in field studies, and later led a conservation movement in England decades before a movement equally strong developed in the United States. In his first presidential address to the British Ecological Society (1914), he was not hesitant to put the importance of the new science of ecology above the older, traditional fields of study. Tansley contributed a great deal to the philosophy and process of ecological research, and he called for a more experimental approach even though he himself was not adept at such an approach.

In Switzerland and later in France, Josias Braun-Blanquet (Figure 2–5), following the path laid down by Kerner, developed his methods of community sampling, data reduction, and association nomenclature which dominate much of plant ecology today. Eduard Rübel made many other early contributions in Switzerland, and cooperation between Rübel and Braun-Blanquet led to the development of an approach to plant synecology called the Zurich-Montpellier School of Phytosociology (see Chapter 8). As described in detail by Whittaker (1962), other schools or approaches to synecology were developing in Czechoslovakia, Germany, the Baltic countries, and the U.S.S.R.

Where to Now?

The amount of information and data which has accumulated on plant ecology, the diversity of questions that have interest and importance, and the sophistication of modern research methods mean that

Figure 2–4. Frederick Edward Clements, 1874–1945. (Courtesy of the Hunt Institute for Botanical Documentation.)

Figure 2–5. Josias Braun-Blanquet, 1884–present. (Courtesy of the Hunt Institute for Botanical Documentation.)

professional ecologists today typically concentrate their efforts in a single area of plant ecology. The Ecological Society of America ended its first year with 300 members; today there are more than 5000. Just as we lost our Renaissance people in the nineteenth century, we lost our general plant ecologists in the twentieth.

New research approaches or alliances have been created to deal with specialization. The International Biological Program (IBP) was organized in 1960 to generate teamwork investigations of certain ecosystems. Dozens of biologists—animal ecologists, plant ecologists, mathematicians, biochemists, climatologists, taxonomists—studied different aspects of a large project with the hope that the pieces could be put together and large questions answered. The IBP ended in 1974 but the results are only now being published in synthesis volumes.

The **biome***** view of IBP work has stimulated research on the role of plants in energy transfer, nutrient cycling, and ecosystem stability. To some extent it has also stimulated research in evolutionary ecology, plant demography, and ecosystem modeling. We expect that these areas will continue to receive a great deal of deserved attention.

Another area that will continue to be accented is ecology's contribution to land use planning. In 1969, passage of the National Environmental Policy Act (NEPA) by the U.S. Congress required that future federal activities which would have a significant effect on the environment be postponed until an adequate Environmental Impact Statement could be written and commented on by the public and other agencies. Since then, many states have adopted similar legislation for nonfederal projects. Some impact statements are written by federal, state, or local agencies, but many others are written by private consulting companies who hire ecologists and other specialists. New magazines, technical journals, and many books have been published for this applied, professional audience.

The response and tolerance of vegetation to pollution and to other disturbances is actively being investigated. In 1971, the Ecological Society of America helped to create The Institute of Ecology (TIE). TIE was created to deal with large environmental policy questions which are beyond the scope of individuals or even of single institutions. Its objective is to educate the public and government on environmental issues and to generate research that will improve the ecological soundness of regional decisions about land, air, and water use. The National Academy of Sciences has participated in similar activities.

---

*The biome is the sum of the plant community and animal community coexisting in the same region. It generally applies to the formation level, rather than the association level.

Basic ecological research will also continue and should be encouraged. This is research into areas, species, or concepts which has no immediate known practical objective. The objective is simply to increase our understanding of the biological world. Applied ecology depends on continuing work in basic ecology, however, and often basic research does lead to some practical applications (McIntosh 1974).

PART II

THE SPECIES AS AN ECOLOGICAL UNIT

An autecological question is asked in this part: how do the members of just one species in a community cope with their environment? An incidental, secondary question is also asked: what is an ecological species? A taxonomic species is made up of individuals and populations which may be genetically heterogeneous. An ecological species, however, is a genetically homogeneous collection of plants adapted to one particular set of microenvironmental conditions.

Part of an organism's environment is composed of adjacent plants and animals, which may or may not be members of the same species. Interactions between pairs of organisms lie along every possible part of a continuum between obligate and incidental, and between mutually beneficial and mutually detrimental. These interactions, whether mediated by chemical or physical factors, can affect the spatial distribution of individuals; in fact, a striking distribution pattern may be an ecologist's first clue to the existence of an interaction.

CHAPTER 3

THE SPECIES IN
THE ENVIRONMENTAL
COMPLEX

The surface of the earth is a network of environmental factors which vary in both space and time. These environmental factors determine the direction of evolution of plant species and may be correlated with the patterns of plant life on the planet. We will consider the ways that environmental extremes and the gradients between them are related to the physiological tolerances of species, and how plant evolution reflects variations in and predictability of the environmental complex. We will see the plant species as a dynamic set of reactions able to respond to an ever changing environment. We will consider examples of how these sedentary organisms are able to adjust to variations in the environment not only as individuals but also on the level of organs, such as leaves, which react physiologically and developmentally as individuals to variations in their immediate environment.

ENVIRONMENTAL FACTORS
AND PLANT DISTRIBUTION

Later chapters will review the importance of certain environmental factors one by one, but there are some general principles that we will discuss in this chapter: the law of the minimum, the theory of tolerance, and the holocoenotic concept of the environment.

The "Law" of the Minimum

In 1840, the agriculturalist and physiologist Justus von Liebig wrote that the yield of any crop depended on the soil nutrient most limited in amount. This conclusion has come to be called the **law of the minimum** and the factors covered have been expanded, so that a loose definition of the law would now be: the growth and/or distribution of a species is dependent on the one environmental factor most critically in demand.

28

The validity of the law has been shown in many parts of the world. The poor growth of many clover pastures in parts of Australia, for example, was shown to be a result of deficiencies in the micro-nutrients copper, zinc, or molybdenum. Addition of only 6–8 kg ha^{-1} of copper or zinc sulfate every 4 to 10 years increased plant growth 300% and increased the wool harvest from sheep grazing the vegetation even more. As little as 140 g ha^{-1} of sodium molybdate, applied every 5 to 10 years, increased pasture yield six to sevenfold (Moore 1970). In England, the range of certain calcicoles (plants on calcium-rich soil with basic pH) abruptly ends when soil pH drops below pH 5 (Grubb et al. 1969; Rorison 1969), and in California the abundance of typical chaparral shrubs declines (sometimes to zero) when the substrate changes to serpentinite, which has an exceptionally low level of calcium (Walker 1954).

Two limitations to the law of the minimum have become evident, however. First, organisms have an upper tolerance limit to every factor as well as a lower limit. Second, most factors act in concert rather than in isolation; a low level of one factor can sometimes be partially compensated for by appropriate levels of other factors, or the influence of one factor may be magnified as other factors reach their maximum or minimum limits.

The Theory of Tolerance

The pioneer American animal ecologist Victor Shelford (1913) noted weaknesses in Liebig's concept and proposed a modification which has come to be called the **theory of tolerance**. Ronald Good, a plant geographer, later elaborated on it (1931, 1953); his version is as follows:

1. Each and every plant species is able to exist and reproduce successfully only within a definite range of environmental conditions (Figure 3–1).
2. In order of importance, the environmental conditions are climatic, edaphic, and biotic.
3. Tolerance ranges may be broad for some factors and narrow for others, and they may change in their relative width with the phenological stage of the species.
4. Tolerance ranges cannot be determined from an examination of morphological features; instead, they are related to physiological features which can only be measured experimentally.
5. Tolerance ranges may change in the process of evolution; however this process is so slow that environmental change is typically accompanied by plant migration, rather than a change in tolerance.
6. The relative distribution of species with similar tolerance ranges for physical factors is determined finally by the result of competition (or other biotic interactions) between the species.

Tolerance ranges for:

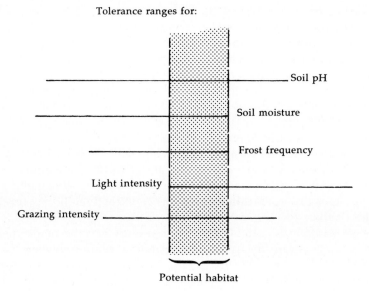

Potential habitat

Figure 3-1. According to the theory of tolerance, the range of habitats for a given species is the sum of tolerance limits for each environmental factor. The extent of each horizontal line represents the tolerance range for that factor. The stippled area is the region of overlap for all tolerance ranges, which represents the potential habitat.

Some ecologists disagree strongly with Good's second conclusion, believing that edaphic or biotic factors are more important than climatic ones, depending on the species under discussion. There have been a few experiments which dramatically show how the tolerance range or optimum for a physical factor is modified by competition (Harper 1964; Ellenberg 1958). For example, when the common weedy annuals wild radish *(Raphanus raphanistrum)* and spurrey *(Spergula arvensis)* are grown in separate pots in controlled conditions, their growth curves exhibit similar pH tolerance ranges and optima. Radish exhibits optimum growth at pH 5, and spurrey shows an optimum at pH 6 (Figure 3-2). When grown together, however, the optimum of spurrey shifts to pH 4 and its range for good growth becomes narrow, while the optimum for radish shifts slightly towards pH 6 and its tolerance range remains much as when grown alone.

The conditions under which a species can exist and grow best in isolation are its **potential range** and **potential optimum**, respectively. These may differ from its observed **ecological range** and observed **ecological optimum** in nature, where the species grows in competition with other species (Figure 3-3). The role of competition in plant distribution is very important. Perhaps Good's second and sixth statements could be combined and rewritten as, "The natural limit of distribution of a particular species is reached when, as a result of changing physical environmental factors, its ability to compete . . . is so much reduced that it can be ousted by other species" (Walter 1973). Undoubtedly, other biotic interactions, such as herbivory and pollination, also affect the distribution of species.

The Holocoenotic Concept of the Environment

Roughly 100 years after Humboldt wrote that everything was somehow interconnected and interdependent, Karl Friederich gave that ecosystem attribute a name: **holocoenotic** (Billings 1952) (holocoen is a synonym for ecosystem). The holocoenotic concept is a natural climax to the modifications others have applied to Liebig's law. The holocoenotic concept states that it is impossible to isolate the importance of single environmental factors to the distribution or abundance of a species, because the factors are interdependent and synergistic. Therefore, single factor ecology (the search for one all-important environmental factor which best determines plant distribution) is short-sighted and naive according to this concept.

This concept does not mean that all factors are necessarily equal, only that they are interactive. Certain factors in any ecosystem are of overriding importance, such as moisture in desert scrub, fire in chaparral scrub, or moving sand in coastal dune scrub. Billings (1970) calls these important factors **trigger factors**.

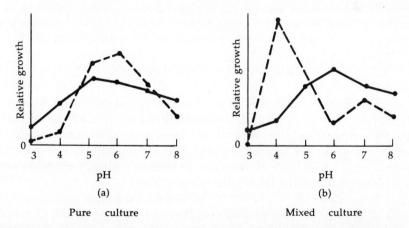

Figure 3–2. The relative growth of wild radish *(Raphanus raphanistrum)* (solid line) and spurry *(Spergula arvensis)* (dashed line) as a function of substrate pH, (a) when grown alone and (b) when grown in competition with each other. (From Ellenberg, H. 1958. Bodenreaktion (einschieplich kalfrage). In W. Ruhland (editor), *Handbuch der pflanzenphysiologie,* vol. 4. By permission of Springer-Verlag.)

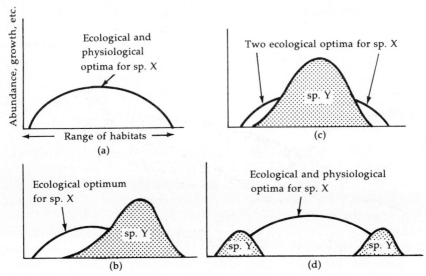

Figure 3–3. Examples of how the ecological optimum may be different from the physiological optimum due to competition. In (a), species X grows alone and laboratory experiments determine that its physiological optimum lies in the center of the curve, as shown. In nature, it faces competition from other species which displace it from habitats it could grow in alone, consequently its ecological optimum is shifted far to the left of its physiological optimum, as shown in (b). In (c), species X competes poorly and is displaced from the middle of its range; it appears to have two ecological optima. In (d), species X competes poorly at the range extremes, with a result that the ecological optimum still coincides with the physiological optimum, but the range is shortened. (From Walter, H. 1973. *Vegetation of the Earth in Relation to Climate and the Eco-physiological Conditions.* By permission of Springer-Verlag.)

THE TAXONOMIC SPECIES

Although the species is at the heart of our classification scheme, there is no unanimity on a working definition for species. We will use the following definition, synthesized from several sources: a **species** consists of groups of morphologically and ecologically similar natural populations which may or may not be interbreeding, but which are reproductively isolated from other such groups.

Three aspects of classification are combined in this definition: (1) external appearance (morphology), (2) breeding behavior, and (3) habitat distinctiveness. The last aspect is clearly third in importance to most taxonomists, and morphology or reproductive isolation receives the most weight in deciding what a species is. Traditional taxonomists weigh morphology heavily, but biosystematists give more weight to reproductive isolation.

The Traditional Concept of Species

Many biologists believe that living organisms do not vary continuously over the whole range, but they fall into more or less well-defined groups, which are commonly called species. Traditional taxonomists adhere to this philosophy of discrete species. A traditional taxonomist examines primarily plant morphology, searching for a few conservative, genetically controlled traits which consistently allow the separation of plants into well-defined groups. Traditional taxonomists do not limit themselves to the examination of only a few traits at the start of a study. They examine many morphological, anatomical, and chemical characters from many specimens, but eventually select only a few morphological characters to serve in defining the various species. The characters selected are those which show discontinuities and thus are most helpful in separating species. However, the selection process is somewhat subjective.

The Biosystematic (Biological) Concept of Species

Biosystematists are interested in determining natural biotic units: populations of plants which maintain their distinctiveness because of biological barriers that genetically isolate them from other populations. These isolating barriers may be due to breeding behavior (time of flowering, type of pollinator), habitat or geographic isolation, or inability to form fertile hybrids with closely related groups (perhaps because of chromosomal differences or pollen incompatibility with the stigma and style).

Conflicts arise between biosystematists and traditional taxonomists because the natural biotic units do not always correspond to well-defined groups. Two populations of the same traditional species may prove, upon crossing in the greenhouse, to yield no offspring, or infertile offspring, and thus belong to two different biosystematic species. The traditionalist argues that such non-visible traits as crossability are theoretically important, but they cannot be interpreted in the field with a hand lens, and thus are of no practical importance. Also, greenhouse crosses may not be a valid imitation of crossing frequency in nature. One further difficulty with the biosystematic approach is that crossability is seldom all or nothing, so subjective decisions still have to be made. For example, if populations A and B are 78% interfertile, are A and B in the same species?

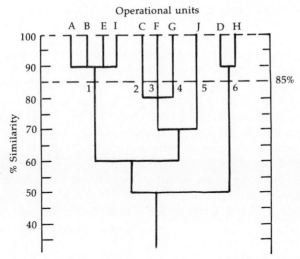

Figure 3–4. A dendrogram that could result from numerical taxonomy procedures applied to 10 operational units, A–H. The percentage similarity is shown for pairs and groups of operational units. If members of the same species were defined as showing at least 85% similarity (dashed line), then the 10 operational units would be lumped into six species, as indicated by the intersections of lines 1–6 with the dashed line. (From *Botany.* Weier, Stocking, and Barbour. Copyright 1974 John Wiley and Sons, Inc. Reprinted by permission of John Wiley and Sons, Inc.)

The Numerical Concept of Species

Another approach to taxonomy is to consider all forms of evidence with equal emphasis. This statistical approach is called numerical taxonomy. A basic tenet for this method is that a great deal of evidence is required to separate taxa. Some 50 to 300 characteristics are used for a given study, and they range from morphological to biochemical characteristics.

To study the number and relationship of species within one genus, the numerical taxonomist begins by arbitrarily dividing the genus into a large number of operational units. If this number is sufficiently large, the division will have no effect on the outcome of the study. The operational units are examined for each of the traits and the data are recorded in relative terms (+ or − for present or absent, and 1, 2, 3, or 4 for amount). All possible pairings between operational units are then made and their similarity to each other is calculated by a standard formula. Similarity can be converted to a 0–100 scale, 100 meaning complete similarity (identical), and 0 meaning no similarity (opposites).

The pairings are then ranked in order of decreasing similarity in a dendrogram (Figure 3–4). In our example, one taxonomist might select the range of 85–100 similarity as appropriate for members of the same species; if a line is drawn across the graph at the 85 level, there would be six species. Another taxonomist might choose 65 as a lower limit, and then there would be three species, and so on.

The numerical method shows that when one considers many traits instead of a few, one will see a continuum of variation. The boundaries numerical taxonomists erect between species are recognized as arbitrary, but they do have some degree of mathematical objectivity.

Conclusions

First, the process of identifying and defining species differs from taxonomist to taxonomist, but there is one facet which all the diverse approaches share to some degree: the result is arbitrary. Species are partly natural and partly human artifacts.

Second, habitat characteristics are seldom important taxonomic criteria. Consequently, some species have such enormous ranges that it is doubtful they are genetically homogeneous entities. Quaking aspen (*Populus tremuloides*), for example, has the widest distribution of any North American tree, covering 110° of longitude and 47° of latitude. Red maple (*Acer rubrum*) not only occurs throughout the eastern deciduous forest, but within that area it can be as abundant on sandy, dry sites as it is on wet, bottomland sites. Subalpine fir (*Abies lasiocarpa*) extends south from sea level in the Yukon Territory to 3600 m at timberline near the Mexican border, and east from the maritime climate of coastal British Columbia to the harsh continental climate along the front range of the Rocky Mountains. Clearly, such taxonomic species are not ecological tools; that is, they are not very precise indicators of a relatively narrow set of environmental ranges. Some wide-ranging species have been subdivided into regional variants (subspecies or varieties), but even then the ranges often cover heterogeneous areas.

Other taxonomic species, with restricted ranges, might serve as excellent indicators of a certain environment, but these are seldom common or dominant species. Coast redwood (*Sequoia sempervirens*) is restricted to a relatively narrow coastal strip in California, perhaps 90 km^2 in extent. Within that area it is a dominant species and it serves as an indicator of summer fog and a certain moderate range of annual and diurnal temperature fluctuation, among other factors. At an extreme, the cypress *Cupressus stephensonii* is restricted to a single grove in southern California about 0.15 ha in size. This species may completely characterize its unique site, but the relationship has no practical, predictive value because of the plant's absence from the rest of the world.

Can the average, somewhat arbitrary, taxonomic species be rede-fined or subdivided to make it a better ecological tool? The answer is yes in theory, and often no in practice, as discussed in the next section.

THE ECOLOGICAL SPECIES

The ultimate objective of any science is to be able to make accu-rate predictions or inferences about a given system, be it chemical, physical, or biological. Plant ecologists would like to use species as deductive tools with which to understand ecosystems. If the ecological requirements of species A are known, and the resource allocation pat-tern is understood, then the presence and degree of vigor of species A anywhere permits us to make many inferences about the environment, such as soil depth, soil nutrient levels, frequency of frost, light inten-sity, length of growing season, frequency of disturbance, and the pres-ence or absence of other plants and animals which may interact with that species. This type of analysis is our objective. In this section we ask: is it theoretically possible to use plant species as deductive tools?

Ecotypes

Linnaeus and taxonomists after him recognized that some taxo-nomic species were not homogeneous: their member plants varied in height, leaf size, flowering time, or other attributes, with change in light intensity, latitude, elevation, or other site characteristics. It was thought that such differences within a species were plastic, not herita-ble responses. The early transplant gardens of Kerner (1895) in the Tirol supported this view.

This conclusion was challenged by the experiments of the bota-nist Göte Turesson in the early twentieth century. He hypothesized that many variations within a species were heritable and were of adap-tive value to particular habitats within the species' range. During the 1920s he attempted to arrive at, as he put it, ". . . an ecological under-standing of the Linnaean species." At first he examined plants from Sweden only, but later he studied species which ranged all over Europe. For each species (usually a perennial), he brought back vegeta-tive material or seeds from different habitats or regions and grew the plants in a test garden near his home at Åkarp, Sweden. He reasoned that if the morphological or phenological differences noted in the field were retained in the garden, then the traits were heritable and geneti-cally based.

Table 3–1 summarizes a typical result, using the herbaceous pe-rennial hawkweed *(Hieraceum umbellatum)*, which in southern Sweden grows on coastal sand dunes, on rocky headlands, and in inland fields

Table 3-1. Some morphological and phenological traits of hawkweed (*Hieraceum umbellatum*) ecotypes, as revealed in Turesson's uniform garden. (Reprinted by permission of McGraw-Hill Book Co. from *Plant Variation and Evolution* by Briggs and Walters, 1969.)

Traits	Ecotypes		
	Woodland	Field	Dune
habit	erect	prostrate	intermediate
leaves	broad	intermediate	narrow
pubescence	absent	present	absent
autumn dormancy	present	present	absent

and woodlands. The field differences were retained in the test garden. Were these genetically distinct types of plants different species? Turesson made all the crosses and found all the types to be interfertile. Therefore, the types were technically part of only a single species.

Turesson called these entities ecotypes. An **ecotype** is the product of genetic response of a population to a habitat. It is a population or group of populations distinguished by morphological and/or physiological characters, interfertile with other ecotypes of the same species, but usually prevented from naturally interbreeding by ecological barriers. (Although Turesson wrote in English, it is a very dense form of English and is best read in the "translation" written by Turrill in 1946.) Turesson also coined the terms ecospecies (approximately the equivalent of the biosystematist's natural biotic unit) and coenospecies (the equivalent of a genus with few species, or a section of a larger genus), but these terms and concepts have not been widely used since.

It is important to outline all the elements that were part of ecotypes as Turesson saw them: (a) they were genetically based, (b) their distinctiveness could be morphological, physiological, phenological, or all three, (c) they occurred in distinctive habitat types, (d) the genetic differences were adaptations to the different habitats, (e) they were potentially interfertile with other ecotypes of the same species, and (f) they were discrete entities, with clear differences separating one ecotype from another. Our modern concept of the ecotype does not agree with all these elements, as will be shown later in the chapter.

At the same time that Turesson was describing ecotypes in over 50 common European species, three biologists were reporting similar results with western North American perennial plants. In 1922, Jens Clausen, a geneticist and cytologist, David Keck, a taxonomist, and William Hiesey, a physiological ecologist, established a 323 km long study transect in California, supported by the Carnegie Institution of

Washington, D.C. The transect extended from near sea level at Stanford University (just south of San Francisco), across the Coast Ranges, through California's Central Valley, up the gradual west face of the Sierra Nevada to the timberline at 3000 m elevation, and partly down the steep, relatively arid east face of the range (Figure 3–5). Despite great environmental diversity along this transect, Clausen, Keck, and Hiesey were able to find about 180 species whose ranges extended over much or all of the transect.

Each species was collected at a variety of locations along the transect, brought back to greenhouses at Stanford, cloned (literally torn into parts and each part induced to root separately so that many genetically identical plants were produced), grown for 6 months, then transplanted to test gardens along the transect. Initially there were 11 gardens, but the number was soon reduced to three: Stanford, near sea level, Mather, in the mid-elevation Sierra Nevada, and Timberline. Table 3–2 summarizes the garden environments. About 60 species were rugged enough to survive this initial handling, and their growth, phenology, and mortality were followed for as long as 16 years.

The herbaceous perennial cinquefoil *(Potentilla glandulosa)* can serve as an example of their garden results (Table 3–3). Based on morphology, phenology, physiology, and habitat, there appeared to be four ecotypes in the species (Clausen, Keck, and Hiesey chose the more conservative, taxonomic term subspecies, but ecotype is synonymous in this case). Ecotype *typica* was a lowland form. It grew best at Stanford, survived at Mather, but did not last a year at Timberline. Its herbage was relatively frost tolerant, and at Stanford it was able to grow all year round, but at Timberline it could not withstand the hard winter frosts

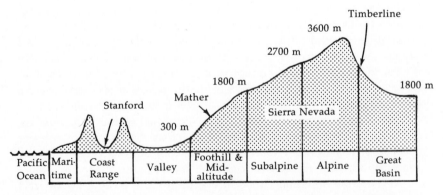

Figure 3–5. Profile, from west to east, of California at approximately 37°30′N latitude, showing the location of the three principal transplant gardens used by Clausen, Keck, and Hiesey. (From Clausen et al. 1940. Courtesy of the Carnegie Institution of Washington.)

Table 3–2. The environments of the three principal transplant gardens used by Clausen, Keck, and Hiesey. (From Clausen et al. 1940. Courtesy of the Carnegie Institution of Washington.)

Garden name	Elevation (m)	Surrounding vegetation	Growing season (mo)	Annual ppt. (cm)	Ppt. as snow (cm)
Stanford	30	oak woodland, chaparral	12.0	30	none
Mather	1400	mixed conifer forest	5.5	95	90
Timberline	3050	subalpine forest	2.0	73	485

Table 3–2, *continued*

Garden name	Mean max. temp. (°C)	Mean min. temp. (°C)
Stanford	36	− 3
Mather	36	−11
Timberline	23	−22

(Figure 3–6). Ecotypes *reflexa* and *hanseni* were mid-elevation forms, distinctively taller than *typica*. One occurred on dry sites, the other on wet meadow sites, and there were subtle morphological differences between them. Both flowered late in summer, so late that when grown at Timberline they often were hit by frost before flowering or seed set could be completed. Their herbage was not frost tolerant, and cold temperatures induced them to enter dormancy. They survived at all locations, but grew best at Mather. Ecotype *nevadensis*, a timberline form, was the shortest of the four. It flowered early in the season and its herbage was very frost tolerant. Its winter dormancy may have been induced by short daylength rather than temperature, for it was winter dormant even in the mild climate at Stanford. A period of winter cold may still have been physiologically necessary, however, for growth at Stanford declined in subsequent years. The herbage seemed more susceptible to disease at lower elevations than at Timberline. Crossing experiments showed that all the ecotypes were interfertile.

Table 3–3. Ecotypes (subspecies) of cinquefoil *(Potentilla glandulosa)* as revealed by the research of Clausen, Keck, and Hiesey. (From *Plants and Environment.* Daubenmire. Copyright 1974 John Wiley and Sons, Inc. Reprinted by permission of John Wiley and Sons, Inc.)

Coenospecies	Ecospecies	Ecotypes	Environment
		nevadensis	alpine and subalpine, 1600–3500 m
		hanseni	mid-elevation wet meadows, 250–2200 m
Potentilla, section Drymocallis	*glandulosa*	*reflexa*	mid-elevation dry slopes, 250–2200 m
		typica	Coast Range

Clausen, Keck, and Hiesey concluded that most species are composed of an assemblage of ecotypes, each ranging in size from a single population to a regional group of many populations; the wider the species' range is, the more ecotypes there are within the species. Ecotype research by many other investigators, on many different species throughout the world, has corroborated this conclusion. The terms race, genecotype, and ecological race are sometimes used as synonyms for ecotype. Random genetic variants (individuals or groups of individuals) within ecotypes are called biotypes. Populations whose uniqueness in nature is due to nongenetic plasticity are called ecophenes or phenecotypes, to distinguish them from ecotypes.

Ecoclines

The ecotype concept of Turesson may seem to give us the ecological species tool we seek, but more recent research has shown its limited practicality.

In Scotland, J. W. Gregor (1946) closely examined what at first appeared to be two ecotypes of the coastal plantain, *Plantago maritima.* One ecotype inhabited salt marshes regularly flooded by high tide with a soil salinity of approximately 2.5%. These plants had short leaves, small seeds, and thick, short, somewhat decumbent* flowering stalks. The other ecotype inhabited nonsaline meadows further inland, and these plants had longer leaves, larger seeds, and thinner, taller, more upright flowering stalks. Gregor collected seeds of each and sowed them in a test garden. In addition, he collected and sowed seeds from plants growing in an **ecotone**, an intermediate habitat. The resulting plants showed that field differences were genetically fixed, but

*Lying or growing along the ground but erect at or near the apex.

more importantly, they showed a continuous gradation from one extreme, one ecotype, to the other. There were no discrete boundaries between ecotypes or even between ecotypes and plants from the ecotone (Table 3–4). For example, 3.9% of the salt marsh ecotype plants had a habit grade score of 3, which is well within the range of meadow ecotype plants, and 2% of the meadow ecotype plants had a habit grade score of 2, which is well within the range of salt marsh ecotype plants.

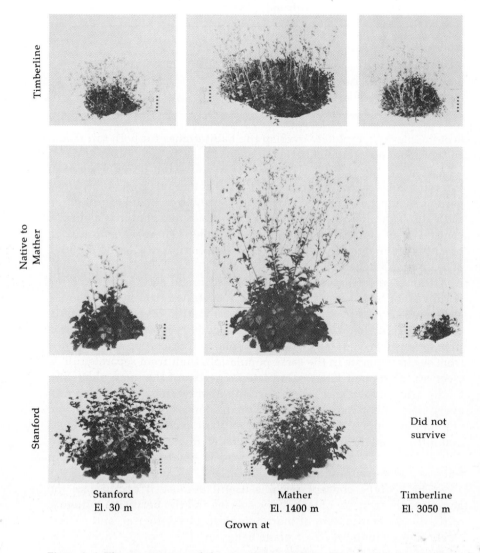

Timberline

Native to Mather

Stanford

Stanford
El. 30 m

Mather
El. 1400 m

Timberline
El. 3050 m

Did not survive

Grown at

Figure 3–6. The appearance of three ecotypes of *Potentilla glandulosa* as grown at three different locations. The top row shows ecotype *nevadensis*, the middle row is ecotype *reflexa*, and the bottom row shows ecotype *typica*. (Courtesy of Dr. William M. Hiesey.)

Table 3–4. Results of Gregor's cultivation experiments with plantain *(Plantago maritima)* ecotypes. Habit grade 1 includes the most prostrate, succulent plants, with shortest leaves and smallest seeds, habit grade 5 is the other extreme, and habit grades 2–4 are intermediate. (Reprinted by permission of McGraw-Hill Book Co. from *Plant Variation and Evolution* by Briggs and Walters, 1969.)

Habitat	Mean scape length (cm)	% of plants in habit grades 1–5				
		1	2	3	4	5
salt marsh, soil salinity 2.5%	23.0	74.5	21.6	3.9		
upper marsh edge, ecotone	38.6	10.8	20.6	66.7	2.0	
nonsaline meadow above marsh, soil salinity 0.25%	48.9		2.0	61.6	35.4	1.0

Ecotone plants completely overlap the habit grades for both salt marsh and meadow ecotype plants.

Olaf Langlet (1959), a compatriot of Turesson, brought seeds of the pine *Pinus sylvestris* from 580 sites throughout Sweden to a test garden. When he examined saplings for growth rate and morphological features, he found the extremes to be quite different, but a **cline**, a continuum of variation, connected the extremes. There were no sudden breaks in the range of variation, where one could say ecotype A ended and ecotype B began.

Cavers and Harper (1967*a* and *b*) also found significant variation within what had been called a discrete ecotype of the weedy annual curly dock *(Rumex crispus)*. They used germination response as an indicator of genetic heterogeneity. They first collected seeds from separate populations (seeds pooled from many plants in each population), then from separate plants in the same population, then from separate inflorescences (clusters of flowers and later of seeds) of the same plant, and finally from upper and lower portions of the same inflorescence. In every case, the range of germination response was very large from population to population, from plant to plant, etc. The implication was that other genetic, adaptive traits could fluctuate equally widely. Again, here was an ecotype which was as heterogeneous in certain traits as a wide-ranging species.

The ecotype concept remains useful because it emphasizes the genetic heterogeneity of taxonomic species and the pervasive influence of the local environment at morphological, physiological, and successively more subtle levels of plant behavior. The ecotype concept has practical importance to reforestation or revegetation projects in general. Seed or vegetative material from a wide-ranging species must be taken from that portion of the species' range most similar to the environment of the revegetation site, or the success of the project may be reduced.

However, the ecotype concept typically does not permit one to recognize homogeneous populations or groups of populations in nature to use as ecological tools. Ecotypes are no more discrete and distinct from each other than are species. Turesson's stairstep concept of ecotypes (Figure 3–7) must be replaced with an ecocline concept. An **ecocline** is a gradation in the attributes of a species (or community or ecosystem) associated with an environmental gradient. Sometimes ecocline is used to refer to the environmental gradient itself (Hanson 1962). Turesson, and to some extent Clausen, Keck, and Hiesey, thought of ecotypes as discrete because their sampling method prejudiced their results: plant material was selected from widely separate places and ecotones were ignored.

Sometimes habitats are discrete, with sudden changes from one habitat to the other (usually because of steep terrain or edaphic factors), but typically habitats intergrade, and plant populations vary with the same subtlety. An ecotype, then, is only an arbitrary segment of an ecocline, which may be convenient to recognize for reference purposes.

Genoecoclinodemes

Since the work of Turesson and Clausen, Keck, and Hiesey, ecotypes and ecoclines have been described in many species, but the rigor of ecotype definition has been lost. Many investigations of ecotypes

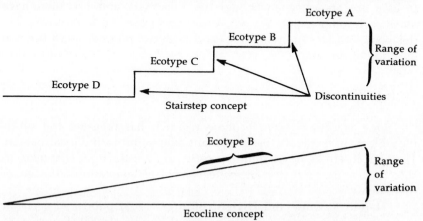

Figure 3–7. Diagrams that contrast two concepts of ecotypes. In the stairstep view, ecotypes are distinct and separated from each other by discontinuities in their morphology or other attributes (the vertical lines indicate discontinuities). In the ecocline view, there are no discontinuities, there is just a continuum of variation.

have not tested breeding behavior, some have not grown ecotypes together to test for genetic differentiation, and a few have implied habitat uniqueness without quantifying it. All of these missing items are important parts of Turesson's five criteria for identifying ecotypes.

Gilmour devised an entire system of nomenclature which eliminates the need to use the term ecotype and which gives precision to the results of autecological and biosystematic research (Gilmour and Heslop-Harrison 1954). All the terms are prefixes to a neutral suffix, -deme, which never stands alone but can be defined as a group of closely related individuals. As more information becomes known about any group, a series of single or multiple prefixes can be added, and these prefixes convey exactly what is known about the group, no more, no less.

For example, a topodeme is a group of individuals coexisting in a given locale; a gamodeme is a group of naturally interbreeding individuals (equivalent to the term population); an ecodeme is a group in a specific, unique habitat; a genodeme is a group which is genetically distinct from others; a phenodeme is a group whose differences are not yet known to be genetic; a plastodeme is a group whose differences are known to be nongenetic. A genoecodeme is a group whose differences are genetically fixed and which exists in a unique habitat; a genoecoclinodeme is a group whose differences are genetically fixed, which exists in a unique habitat, and which is part of a continuous cline of gradation. In short, a genoecoclinodeme is the equivalent of an ecotype.

Not surprisingly, this system of multiple prefixes has been slow to gain acceptance, even though its objectivity and precision have much to recommend it. We will continue to use the term ecotype in this book, but its synonym, genoecoclinodeme, reminds us of the real meaning of ecotype.

Ecotype Research at the Physiological Level

One aspect of recent ecotype research has revealed the subtle physiological, metabolic basis of plant adaptation to the local habitat. That aspect will be illustrated by a series of investigations that takes us successively closer to the ultimate control of adaptations, the genes themselves.

Harold Mooney and Dwight Billings (1961) published a classic study on the perennial herb alpine sorrel (Oxyria digyna) (Figure 3–8), for which they received the 1962 George Mercer award from the Ecological Society of America for the most outstanding ecological paper published by young ecologists in the previous 2 years. Oxyria has a circum-boreal distribution in the treeless arctic tundra, and extends

Figure 3–8. Alpine sorrel *(Oxyria digyna)* above the timberline in Wyoming. Note the heart shaped leaves and the profusion of flowers, both traits of the alpine ecotype. (Courtesy of W. D. Billings.)

south in alpine tundra along several mountain chains. In the conterminous United States it is found at high elevations in the Sierra Nevada and Rocky Mountains.

Arctic tundra vegetation and alpine tundra vegetation share many similarities, but there are important environmental differences. Alpine areas in the temperate zone experience higher light intensity and greater extremes of temperature in summer than do arctic areas.

Just as there are two environmental extremes, so there are two ecotypic extremes of *Oxyria*. Mooney and Billings showed that the arctic and alpine ecotypes differed morphologically and phenologically even when grown together from seed in controlled chambers which simulate a natural, uniform environment (the modern test garden) (Table 3–5). Mooney and Billings also showed that the metabolism of the two ecotypes differed.

By placing potted plants in a small plexiglass chamber whose microenvironment could be controlled, then measuring the CO_2 content of air passing into and out of the chamber, the rate of photosynthesis per unit leaf area or weight could be determined. As shown in Figure 3–9(a), the alpine ecotype had a significantly higher temperature optimum for photosynthesis. The amount of light required to saturate the photosynthesis system, beyond which increasing light fails to increase the rate of photosynthesis very much, was greater for the alpine ecotype (Figure 3–9(b)), again correlating well with the environmental differences in nature. These physiological differences most likely correlated with enzymatic and other biochemical factors, but Mooney and Billings did not pursue their investigation to that level.

One species of cattail, *Typha latifolia*, is widely distributed in the northern hemisphere. McNaughton (1966) collected dormant rhizomes from such disparate habitats as the cool, maritime, foggy Pacific coast at Point Reyes, California and the relatively hot, arid Sacramento Valley

Table 3–5. Some morphological, biochemical, and phenological traits of arctic and alpine ecotype extremes of alpine sorrel *(Oxyria digyna).* (Adapted from Mooney and Billings 1961. Copyright 1961 by the Ecological Society of America.)

Trait	Arctic	Alpine
rhizomes	present	absent
intensity of flowering	low	high
leaf anthocyanin level (red color)	low	high
critical photoperiod for flowering	24 hr at 70°N	15–17 hr at 38–48°N
leaf shape	ovate	heart shaped

near Red Bluff, California, more than 100 km inland. He potted the rhizomes and placed them in a greenhouse regulated at 30/25°C day/night, where they broke dormancy, produced shoots and grew for three months. He then took samples of leaf tissue, extracted their enzymes, and subjected the extract to a heat stress of 50°C for periods of time up to 30 minutes. This simulated the leaf temperature that Red Bluff plants might experience in nature. Point Reyes plants probably do not experience a leaf temperature in nature higher than 30°C.

After each period of heat stress, he tested the activity of three important respiratory enzymes: malate dehydrogenase, glutamate-oxaloacetate transaminase, and aldolase. He reasoned that the heat tolerance of the Red Bluff ecotype must inherently lie in the heat stability of some or all of its enzymes. As shown in Figure 3–10, malate dehydrogenase was significantly more heat stable in the Red Bluff ecotype than in the Point Reyes ecotype; the other two enzymes showed no difference. Enzymes are complex macromolecules with tertiary or quarternary structure and the reaction site is a relatively small part of their total architecture and chemistry. Thus, malate dehydrogenase could differ in the two ecotypes in many ways which might increase stability or activity (McNaughton 1972). Why only one of the enzymes, rather than all three, showed ecotypic differentiation is not clear.

In another paper, McNaughton (1967) showed that cattail ecotypes from Point Reyes at sea level and Wyoming at 1980 m elevation differed in their photosynthetic efficiency. The high elevation, short growing season ecotype exhibited about twice the photosynthetic rate of the Point Reyes ecotype. He was able to trace the difference to higher reducing activity (an important process in the light reactions of photosynthesis) in the chloroplasts of the Wyoming ecotype.

Respiration is also attuned to elevation at the enzymatic level, as Klikoff (1966) showed for populations of the grass squirrel tail *(Sitanion*

hystrix) from different elevations of the Sierra Nevada. Isolated mitochondria showed higher oxidative rates at lower temperatures with
increasing elevation of the parent plant.

Some species are differentiated into sun ecotypes (those which
germinate and develop in the open) and shade ecotypes (those which
develop beneath the canopy of other plants). A European species of
goldenrod *(Solidago virgaurea)*, a perennial herb, has such ecotypes. Olle

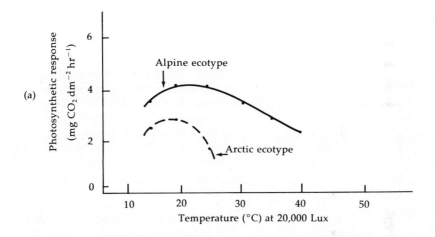

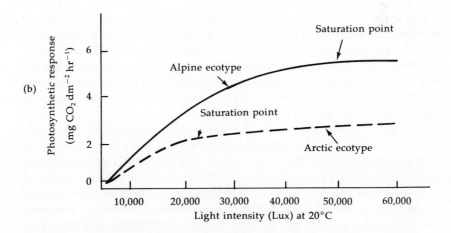

Figure 3-9. (a) Photosynthetic response of arctic (dashed line) and alpine
(solid line) ecotypes of *Oxyria digyna* to temperature at 20,000 lux. The arctic
ecotype temperature optimum for photosynthesis is 20°C; the alpine ecotype
optimum is 38°C. (b) Photosynthetic response to light at 20°C. (From
Mooney and Billings 1961. Copyright 1961 by the Ecological Society of
America.)

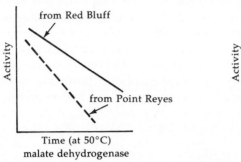

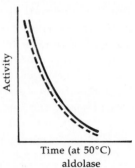

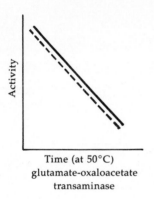

Figure 3–10. Enzyme response to incubation at 50°C. Enzymes were extracted from coastal (Point Reyes, California) (dashed line) and inland (Red Bluff, California) (solid line) ecotypes of *Typha latifolia.* (From McNaughton 1966. Thermal inactivation properties of enzymes from *Typha latifolia* L. ecotypes. *Plant Physiology* 41:1736–1738. By permission of the American Society of Plant Physiologists.)

Björkman, in a series of papers culminating in 1968, examined the differences between the two goldenrod ecotypes at successively more subtle levels. He collected vegetative material from plants growing in exposed heath vegetation in Norway and from plants growing beneath oak forests in Sweden, then cloned the material and grew plants in controlled environments.

Among other differences, he showed that the light saturation curve of the sun ecotype was different from the light saturation curve of the shade ecotype. The sun ecotype had a higher light saturation point and it showed a higher rate of photosynthesis at that saturation point. To find the reason for the difference, he first searched at the morphological level, asking if sun ecotype leaves perhaps absorbed more light; the answer was no. He then searched at the cellular level, asking if the chlorophyll concentration was higher in sun leaves; the answer was no. Finally he examined the enzyme level. The enzyme responsible for fixation of CO_2 into the dark reaction pathway of photosynthesis is ribulose biphosphate carboxylase (also called carboxydismutase). When he assayed for the concentration (activity) of this important enzyme, he found that it was two to five times greater in the sun ecotype, which was sufficient to account for the five times higher rate of photosynthesis of the sun ecotype at high light intensity.

Ultimately, this line of research should be able to show us which enzymes are most important to basic types of climatic or edaphic ecotypes, how these enzymes differ (in amount, as Björkman showed for one enzyme, or in structure, as McNaughton implied for another), and what the genetic basis for the difference is (how many loci and how many genes at each locus). When we get to that point, the information will have great practical importance for the breeding of ecotypes for use in reclaiming terrain such as nutritionally poor road cuts, toxic mine spoil, saline or coarse dredge spoil, strip mine areas, dunes which are mobile and low in nitrogen, eroded clear-cuts, etc. At present, we have yet to reach this level.

Acclimation

Acclimation (also called acclimatization) is a plastic, temporary change in an organism caused by an environment to which it has been exposed *in the past*. Matthaei (1905) may have been the first to document such a phenomenon in plants, and the effect of past temperatures on photosynthesis and respiration rates has been reviewed by Semikhatova (1960).

Billings et al. (1971) conducted an experiment which provides a good example of acclimation, and once again alpine sorrel proved to be a good study plant. Alpine sorrel *(Oxyria digyna)* seeds were collected from a range of habitats, germinated and grown in a uniform greenhouse for 4 months, then subdivided into three growth chamber environments: warm (32/21°C day/night), medium (21/10°C), and cold (12/4°C). After 5 to 6 months in the chambers, replicates of each collection were measured for net photosynthesis at a range of temperatures, from 10 to 43°C, and the optimum temperature for photosynthesis was noted. Table 3–6 shows that representatives of arctic and alpine

Table 3–6. The effect of acclimation temperature on the optimum temperature for photosynthesis of alpine and arctic ecotype extremes of alpine sorrel *(Oxyria digyna)*. (Adapted from Billings et al. Reproduced with permission of the Regents of the University of Colorado from *Arctic and Alpine Research* 3:277–289, 1971.)

		Optimum temperature for photosynthesis at each acclimation regime (°C)			
Population site	Ecotype	Warm (32/21°C)	Medium (21/10°C)	Cold (12/4°C)	Range
Sonora Pass, California	alpine	28	21.5	17	11
Pitmegea River, Alaska	arctic	21	20.5	20	1

ecotypes possessed different acclimation capacities. Optimum temperatures for alpine plants ranged up to 11°C, depending on the temperatures they had been growing at before the photosynthesis measurement, but the optimum temperatures for arctic plants ranged only 1°C.

Similar effects of preconditioning, or acclimation, have been shown for plants as diverse as pine trees (Rook 1969) and desert shrubs (Mooney and West 1964; Mooney and Harrison 1970; Strain and Chase 1966).

The relationship between plant and environment can be written:

$$\text{phenotype} = \text{genotype} + \frac{\text{prevailing}}{\text{environment}} + \frac{\text{past}}{\text{environment}}$$

How distant can past environment be to influence phenotype? The influential past environment may have been uncomfortably and unaccountably long ago, perhaps reaching back to parent generations, according to Rowe (1964). Seeds of groundsel (Senecio vulgaris) were germinated at different temperatures. The seedlings were immediately transferred to a common environment, were allowed to grow for 80 days, and the shoots were weighed. The sevenfold difference in shoot weights shown in Table 3–7 is hard to explain except as a result of temperature differences at the time of germination, 80 days earlier. Plants of the weedy annual prickly lettuce (Lactuca scariola), subjected to different daylengths or applications of growth regulators, produced progeny which differed in germination, seedling growth, and time of flowering (Gutterman et al. 1975).

Highkin (1958) grew pure-line peas under two sets of conditions: 24/14°C day/night and 26°C constant. Pollen from plants in either regime was transferred to the stigmas of other pure-line peas which had been kept separate in uniform conditions. The seeds were harvested, kept separate according to the temperature regime of the pollen donor, germinated, and grown in uniform conditions. The two types of

Table 3–7. The effect of germination temperature on subsequent growth of groundsel (Senecio vulgaris). (Adapted from Rowe 1964. Copyright 1964 by the Ecological Society of America.)

Germination temperature (°C)	Subsequent growing conditions, next 80 days	Final plant weight (mg)
10		147
14	all grown together at 17°C,	775
23	16 hr photoperiod	1078
30		390

progeny differed significantly in height and number of nodes, and there was a diminishing but still measureable carry-over effect for several generations.

A non-Lamarckian, genetic explanation for acclimation is possible, but our objective has only been to show that acclimation can occur, and the genetic basis for it need not concern us here. The significance of past environments on plant behavior is insufficiently recognized by plant autecologists, even though it appears that acclimation is important to an understanding of plant distribution.

SUMMARY

Plant ecologists would like to use species as deductive tools, as rather precise indicators of certain levels of environmental factors. This may not be a realistic objective for two reasons. First, plants respond to a complex of climatic, edaphic, and biotic factors, and the impact of single factors is difficult to isolate. The tolerance range of a species to factor X may be modified by factors Y or Z. Good's version of the theory of tolerance recognizes the special confounding effect of competition on tolerance ranges. Second, taxonomic species, whether recognized on morphological, biological, or statistical grounds, are partially artifacts of the human desire to classify. Wide-ranging species, which occur in many different habitats, are not genetically homogeneous and they cannot serve as ecological indicators.

Turesson searched for "... an ecological understanding of the Linnaean species." He discovered that taxonomic species were comprised of ecologically important subunits, which he called ecotypes. He defined ecotype as the genetic response of a population (or group of populations) to a habitat, distinguished by morphological and/or physiological characters, yet interfertile with other ecotypes of the same species. Most wide-ranging species are now known to be made up of many ecotypes, but the practical utility of the concept was diluted when it was discovered that ecotypes are just as heterogeneous, with boundaries just as vague, as species. Turesson's stairstep concept of ecotypes must be replaced with an ecocline concept, and the equivalent of the term ecotype in Gilmour's -deme terminology, genoecoclinodeme, properly emphasizes that fact.

The ecotype concept is still important from the standpoint of basic science, because it has led to research that shows the pervasive influence of the environment at all levels of plant behavior, from morphology and phenology to subtle levels of physiology, metabolism, and genetics. Recent ecological research has emphasized these subtle levels, but it has not yet reached the point where we know which enzymes are most important to basic types of climatic or edaphic ecotypes, nor how these enzymes differ from ecotype to ecotype, nor how many genes are involved, nor how we may breed for certain ecotypic features as we currently breed for seed yield or disease resistance. A confounding factor in ecotype research is acclimation: an environment to which an organism has been exposed in the past (sometimes a generation or more in the past) may cause a physiological change in the organism. Indeed, the capacity for acclimation may itself be an ecotypic trait.

CHAPTER 4

POPULATION DYNAMICS AND RESOURCE ALLOCATION

Plant populations have attributes which permit us to use them as tools to assess the environment. These features include the arrangement of individuals in space within a given community, the arrangement of individuals in time, that is, the age structure and growth rate of a population, and the resource allocation patterns of individuals, which characterize the mode of survival of the population in a particular environment.

Plant species are not evenly distributed over the geographic range within which they occur and success, as measured by the number or size of individuals, changes from site to site. Microenvironmental conditions also change through time as climates change, as organisms of different habit or tolerance occupy the site following a disturbance, as individuals die and leave space within the community, etc. The relative success of a species and its microenvironmental requirements can be deduced by measuring variations in distributional patterns and changes in population numbers in space and time. Wide variations in microenvironment and the enormous number of genetic variations possible imply that many different growth and reproduction patterns may be equally successful in the same geographic area.

Our goals in this chapter are to examine the importance of variations in spatial arrangements of individuals within a population, to examine the methods of measuring spatial variations, to consider the changes in populations with time as reflected in the age structure and life span of the individuals, and to consider various patterns of time and resource allocation, which are vitally important to the success of a population.

THE ARRANGEMENT OF INDIVIDUALS IN SPACE: DENSITY AND PATTERN

Density: Definitions and Methods

Density is the number of individuals per unit area, such as 300 sugar maples (*Acer saccharum*) per hectare in a Michigan deciduous forest, or 3000 creosote bushes (*Larrea tridentata*) per hectare in a New

Mexico desert scrub. It is not necessary to actually count every individual in a large area to arrive at a density value. Random sampling with quadrats whose combined area is perhaps only 1% of the overall area will give a close estimate.

A **quadrat** is a frame of any shape that can be placed over vegetation so that cover can be estimated, plants counted, or species listed. Its area is usually small enough that one person, standing at one point along its edge, can easily survey its extent, but quadrats for tree samples may be 10–50 m on a side, and so censusing can require more than one person.

Quadrats can be located randomly by constructing two imaginary axes along the edges of the area, dividing the axes into units, and picking pairs of units from a random numbers table. For the example shown in Figure 4–1, the random pair (5,15) meant that the investigator placed the center of the circular quadrat at a point 5 m up and 15 m over from the origin of the axes. Altogether, 12 quadrats, each 2 m², were placed in the 2400 m² area, so that 1% of the total area was included in the quadrats.

By chance, of course, all the random quadrats might be clustered in one section of the area. To avoid this possibility, the area can be subdivided into roughly equal areas and each section randomly sampled with fewer quadrats. This is called stratified random sampling, or restricted random sampling (Figure 4–1(b)).

Density can also be estimated by distance methods, which do not use quadrats. Random spots are picked from which the investigator begins to walk through the area (in some versions the path is a straight line, in others it is not), and at intervals the distance to the nearest individual is measured. Based on certain geometrical and statistical assumptions, density can be calculated, knowing only the average distance from point to plant. The assumptions and statistical machinations are detailed in several books such as Cox (1976), Greig-Smith (1964), Kershaw (1973), Mueller-Dombois and Ellenberg (1974), and Phillips (1959); Pielou (1969) requires a good mathematical background.

Pattern: Definitions and Methods

Density alone is a static measure. It does not reveal the dynamic interactions that may exist among members of the same species. The **pattern**, or spatial distribution, of the 300 sugar maples or the 3000 creosote bushes gives additional information about the species.

The same number of plants in an area can be arranged in basically three patterns: random, clumped, or regular (Figure 4–2). In a **random** pattern, the location of any one plant has no bearing on the location of

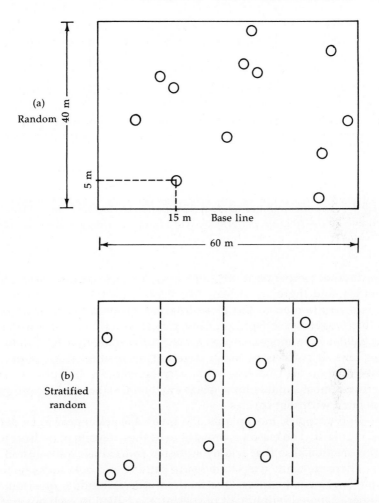

Figure 4–1. Placement of twelve 2 m² circular quadrats by (a) a random method, and (b) a stratified random method (three quadrats in each section). In each case, 1% of the total area has been included in the quadrats. One of the quadrats in (a) has been located by the random numbers 5 and 15, indicating that the investigator should locate the quadrat center 5 paces up from the base line and 15 paces down the base line. The other quadrat centers were located by using different pairs of random numbers.

another of the same species. In a **clumped** pattern (also called aggregated or underdispersed), the presence of one plant means there is a high probability of finding another of the same species nearby. A **regular**, or overdispersed, pattern is similar to the pattern of trees in an orchard. The members of a species appear to repel one another so that where one individual is found, there is a lower probability of finding another than would be expected by the assumption of randomness.

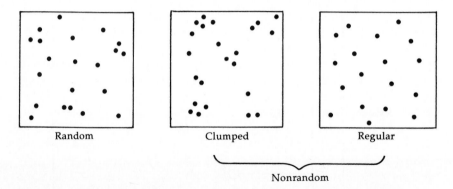

Figure 4–2. Random and nonrandom distribution patterns of individuals. Nonrandom distribution patterns may be clumped or regular. Dots represent individual plants as viewed from above.

In the temperate zone, members of most species seem to be clumped, and there are at least two reasons for this: (1) reproduction: seeds or fruits tend to fall close to a parent, or runners or rhizomes produce vegetative offspring near a parent, and (2) microenvironment: the habitat is homogeneous at a macroenvironmental level, but at a finer level it consists of many different microsites which permit the establishment of a species with varying degrees of success. Those microsites most suitable for a species will tend to become more densely populated with that species.

In the tropics, most plants of a given species appear to be distributed regularly. The reason is thought to be the intensity of pathogen and herbivore attack, so that widely scattered individuals are selected for.

There are many ways to measure pattern, and details can be found in the books cited earlier. One method utilizes random quadrats. The number of rooted individuals of species A is tallied in each quadrat and summarized in table form (Table 4–1). These are the *observed* data; the *expected* data, if members of species A were distributed at random, is generated by a rather simple formula, the Poisson distribution, which only requires that we know the average number of plants per quadrat. Discrepancy between the observed data and the expected data is evaluated by a chi-square calculation. In the example shown in Table 4–1, the chi-square value is greater than what one would expect by chance, so the conclusion is that members of species A are not distributed at random. Are they, then, clumped or regular? Inspection of the table shows that many fewer quadrats with zero individuals were encountered than expected, and many more quadrats with one individual were encountered than expected. This is the sort of result one expected from a regularly distributed population. If the population had been clumped, many quadrats would have had either zero or more than one plant, and very few would have had one plant. By deduction, then, we can say that the members of species A were distributed regularly. Other, shorter, treatments of the data are possible which permit the type of nonrandom distribution to be calculated rather than deduced.

One problem with this random quadrat method is that the quadrat size is critical to the results. If the quadrat size is too large or too small, the pattern is not revealed (Figure 4–3). A method that avoids this problem uses contiguous quadrats. The entire area to be sampled is subdivided into N quadrats and each one is visited and tallied. Plants may be counted or the percent of the ground covered by plants estimated. The variance (mean square) of all N quadrat values is then computed. Pairs of adjacent quadrats are then lumped together on paper and the variance of the $N/2$ values computed. The condensation continues until only two large blocks of quadrats remain. If the plants are distributed randomly, variance should double as block size doubles. If they are clumped, then when block size approaches the average clump size, variance will be much higher than expected. If the plants are regularly distributed, variance will be much lower than expected. The results can be more dramatically expressed in a graph. Figure 4–4

Table 4–1. Poisson analysis of quadrat data for a species with a nonrandom plant distribution. In the formulas, m is the average number of plants per quadrat, 1.56 in this case, and e^{-m} is 0.21. To employ this test, each category must have an expected value of $\geq$ 5% of the total quadrats; to achieve this, category 5 had to be lumped together with category 4. The χ^2 sum is thus based on 5 numbers, and the degrees of freedom is $5 - 2 = 3$. At the 99% level of significance for 3 degrees of freedom, any $\Sigma\chi^2$ value > 11.34 suggests that the null hypothesis be rejected. The null hypothesis here is that the plants are distributed randomly. Therefore, the plants are distributed nonrandomly.

No. of plants per quadrat (x)	Observed no. of quadrats with x plants	Expected no. of quadrats with x plants = $(e^{-m})\left(\dfrac{m^x}{x!}\right)(100)$	$\chi^2 = \dfrac{(\text{observed} - \text{expected})^2}{\text{expected}}$
0	13	21.0	3.0
1	51	32.8	10.1
2	23	25.6	0.3
3	3	13.3	8.0
4	0 } 10	5.2 } 6.8	1.5
5	10	1.6	
totals	100	99.5	$\Sigma\chi^2 = 22.9$

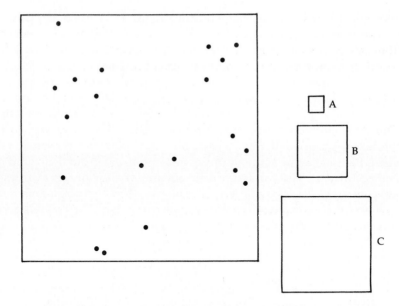

Figure 4-3. The relation of the quadrat size to the detection of the pattern. Only quadrats of size B will reveal a clumped distribution pattern; quadrats A and C will reveal a random plant distribution pattern.

shows a species that has levels of clumping at block sizes 8 and 64. If the basic quadrat size is 2 m^2 (= block size 1), then clumping occurs in areal units of average size 8 x 2 m^2 and 64 x 2 m^2, or 16 m^2 and 128 m^2, respectively.

Distance methods (nonquadrat methods) can also be used to detect distribution patterns. In this case, the distance between adjacent members of the same species is tallied (see Chapter 8). Frequency is another measure used to assess pattern. Frequency is the fraction of all quadrats which contains a given species. If 50 quadrats are placed and species A is noted as present in 10 of them, the frequency of A is 0.20, or 20%. If density is high but frequency is low, one could conclude that species A is clumped; if the reverse is true, then one could conclude that species A is regular. Density and frequency are usually independent measures; knowing one does not help predict the other unless the plants are distributed at random. In that case, 100 − % frequency = e^{-m}, where e^{-m} is the expected number of quadrats with no plants, the first entry in the Poisson distribution (Table 4-1) (Blackman 1935).

Frequency is usually not a reliable figure because it is so dependent on quadrat size. If the quadrats are too large, most species will have 100% frequency; if they are too small, many species will have close to 0% frequency. Cain and Castro (1959) recommended the quadrat sizes listed in Table 4-2 for different life forms in temperate zone vegetation. Daubenmire (1968a) suggested that the appropriate quadrat size for sampling a given life form be small enough that only one or two species show 100% frequency, but Blackman (1935) suggested that the maximum frequency should be 80%.

THE ARRANGEMENT OF INDIVIDUALS IN TIME: DEMOGRAPHY

Demography is the science or study of vital statistics: births, deaths, reproductive rates, and ages of individuals in populations. Demographers attempt to explain the dynamics of the number of individuals in a population. Every plant population has its own particular birth (germination) rate, age class distribution, and average life span. Unlike animals, which cease growing when mature, perennial plants possess primary and secondary meristems which theoretically permit continued growth in length and girth forever. Some, in addition, have the ability to reproduce asexually by runners, rhizomes, or layering so that the same genetic "individual" can also live endlessly, theoretically. Indeed, the longest-lived organisms on earth are plants, whether they be simple lichens 4500 years old (Thomson 1974), clones of shrubs 3000–4000 years old (Vasek et al. 1975; Cottam et al. 1959), or conifer trees nearly 5000 years old (Currey 1965). In addition, the seeds of some species may lie dormant for as long as 1000–10,000 years (Ødum 1965; Porsild et al. 1967; Harper and White 1974). However, even the

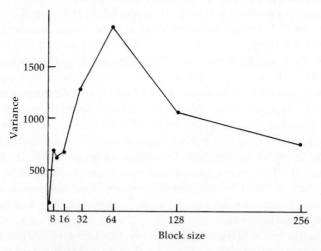

Figure 4–4. Variance/block size graph with a double peak, at block sizes 8 and 64, representing two scales of pattern. These blocks represent clusters of quadrats that equal 16 m² and 128 m², respectively. (From Kershaw, K. L. *Quantitative and Dynamic Plant Ecology,* Edward Arnold, London, 1973.)

Table 4–2. Quadrat sizes suggested by Cain and Castro (1959) to sample different life forms of vegetation in the northern temperate zone. (From Cain and Castro, *Manual of Vegetation Analysis*, Harper and Row, Publishers, Inc., New York, 1959.)

Life form	Quadrat area (m^2)
mosses	0.01–0.1
low herbs	1–2
tall herbs	4
low shrubs	4
tall shrubs	16
trees	100

record holders do eventually die, falling to disease, physical disasters, excessive herbivore damage, or a changing climate.

Plants present some unique problems to demographic studies, compared to animals. The concept of the individual is strained in perennials that reproduce vegetatively by sending out runners or rhizomes, by splitting at the stem base, or by producing arching canes that take root where they touch the ground. In this way some "individuals" can extend out over large areas, given sufficient time. The behavior of such interconnected units, in terms of mortality and other demographic events, may be different than for true individuals (Hutchings and Barkham 1976).

Another problem is that the timing of germination may be unrelated to the timing of reproduction. The Middle East desert plant *Blepharis persica* retains its seeds in the fruit for 10 or more years until a heavy rain triggers their release, at which time they germinate en masse within 3 hours (Gutterman 1972). The chaparral species of *Ceanothus* produces seeds with a hard coat which delays germination for many years, so that the small number of seeds which do germinate any given year do not describe the enormous seed bank in the soil. Scarification, as by fire, may release them from dormancy en masse. Some species produce several kinds of seeds, each with different germination requirements.

Finally, the degree of phenotypic plasticity exhibited by plants can be enormous (Bradshaw 1965), so that demographic events can vary for the same species over time or space. Growth rate, age at onset of reproduction, plant size, and life span can all be modified by the environment.

Life Spans

There are five characteristic plant life spans, and each has a basic corresponding life form. **Annual plants** live for 1 year or less. Their average life span is 1–8 months, depending on the species and then on the environment (the same desert species may complete its life cycle in 8 months one year, or in 1 month the next, depending on rainfall (Beatley 1967). Annuals with extremely short life cycles, such as *Boerrhavia repens* of the Sahara Desert, which goes from seed to seed in as few as 10 days (Cloudsley-Thompson and Chadwick 1964), deserve the special designation ephemeral. Annuals are **herbaceous**, that is, they lack a secondary meristem which would produce lateral, woody tissue. They die following seed production for a number of reasons: nutrient depletion, hormone changes, or inability of the nonwoody tissue to withstand unfavorable environmental conditions following the growing season (Leopold and Kriedemann 1975; Harris 1967). If the members of some species are in particularly favorable microhabitats, they may persist for more than 1 year (Barbour 1970*a*).

Biennial plants are also herbaceous, but typically live for 2 years. The first year is spent in vegetative growth, which is generally more below ground than above ground. Reproduction occurs in the second year, followed by plant death. Under poor growing conditions, or by experimental manipulation of the overwintering temperatures, the vegetative stage can be prolonged for more than 1 year.

Herbaceous perennials typically live for 20–30 years (Rabotnov 1969), though exceptional species are estimated to live for 400–800 years (Harper and White 1974). These plants die back to the root system and root crown at the end of each growing season. The root system becomes woody, but the above-ground system is herbaceous. They have a juvenile, vegetative stage for the first 2–8 years, then bloom and reproduce yearly, periodically every 2–3 years, or sometimes only once, at the end of their life span. Because they lack growth rings, few herbaceous perennials have been aged, and the methods used to age them depend on counting leaf scars or estimating the rate of spread in tussock (clumped) forms.

Suffrutescent shrubs (synonyms include sub-shrubs, half-shrubs, and hemixyles (see DuRietz 1931)) are half way between herbaceous perennials and true shrubs. They develop perennial, woody tissue only near the base of their stems, and the rest of the shoot system is herbaceous and dies back each year. They are small, perhaps 25 cm tall, and short-lived compared to true shrubs.

Woody perennials (trees and shrubs) have the longest life spans: shrubs average 30–50 years, angiosperm (broadleaf) trees average 200–300 years, and conifer (needleleaf) trees average 500–1000 years

(Harper and White 1974; see also Table 4–3). Woody perennials spend roughly the first 10% of their life in a juvenile, totally vegetative state, then enter a combined reproductive and vegetative state, reaching a peak of reproduction several years before death.

Age Distributions

Regardless of the life span, annual to perennial, one can recognize about eight important **age states** in an individual plant or a population (Rabotnov 1969): (1) viable seed, (2) seedling, (3) juvenile, (4) immature, vegetative, (5) mature, vegetative, (6) initial reproductive, (7) maximum vigor (reproductive and vegetative), and (8) senescent. If a population shows only the first four or five states, it is obviously invading and is part of a seral community. If a population shows all eight states, it is stable and is most likely part of a climax community. If it shows only the last four states, it may not maintain itself and may be part of a seral community.

Table 4–3. Average longevity of selected broadleaf and conifer trees common in North America. (Based on data in Fowells 1965.)

Broadleaf species	Average longevity (years)	Conifer species	Average longevity (years)
alder, red	100	cedar, Alaska	3500
aspen, bigtooth	60	cedar, eastern red	300
birch, sweet	265	Douglas fir	700
elm, rock	275	fir, noble	650
elm, slippery	200	fir, subalpine	250
hickory, several spp.	250	fir, white	420
maple, red	150	hemlock, eastern	1000
maple, sugar	350	juniper, western	900
tanbark oak	180	pine, jack	200
walnut, black	250	pine, lodgepole	525
willow, black	70	pine, ponderosa	600
oak, bur	250	redwood	2000
oak, Oregon white	500	sequoia, giant	2500
oak, southern red	150	spruce, black	250
oak, swamp white	300	spruce, Sitka	750
average	~200	average	~1000

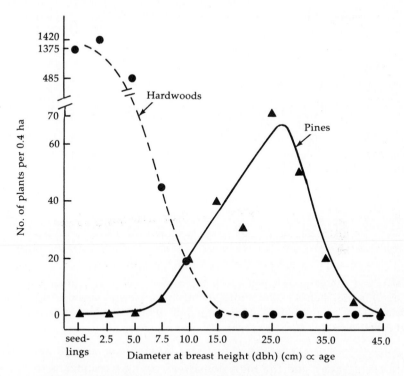

Figure 4–5. Summary of the ages of all trees in a forest near Gainesville, Florida. Actually, only diameter at breast height (dbh) was measured, not age, but dbh is roughly proportional to age in years. Succession is progressing from a pine community (solid line) to a hardwood community (mainly oak and wax myrtle) (dashed line). (Based on data in Daubenmire 1968a (*Plant Communities,* Harper and Row, Publishers, Inc., New York) and Heyward 1939 (by permission).)

Knowing the age distribution of a population may permit us to use demography as a predictive tool in community ecology. Figure 4–5 shows an age distribution of longleaf pine *(Pinus palustris)* in a northern Florida forest (actually, trunk diameter at breast height (dbh) is shown, not age, but in this case age correlates closely with dbh). The pine population has many mature trees, but no seedlings or saplings; it appears to be a senescent population which is not reproducing itself and is part of a seral community. What will replace it? The same graph shows two hardwood populations combined, water oak *(Quercus nigra)* and wax myrtle *(Myrica cerifera)*, which have many seedlings and saplings but no large (mature) individuals. The hardwoods are the invading populations, part of a community that will replace the pine forest.

In contrast, Figure 4–6 shows the age distribution of red spruce *(Picea rubens)* in virgin forests of the White Mountains of New Hampshire. Here we have a reverse J-shaped curve, with many seedlings and

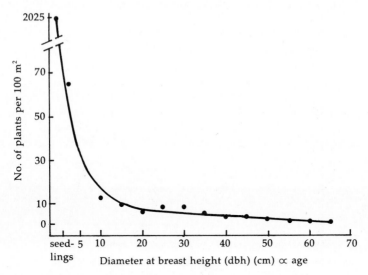

Figure 4–6. A reverse J-shaped age distribution curve for red spruce *(Picea rubens)* in four stands in the White Mountains of New Hampshire. (Based on data in Oosting and Billings 1951. Copyright 1951 by the Ecological Society of America.)

saplings, and fewer larger, older trees. Many ages, and all age states, are represented, and the high density of young trees means there is a high probability of the population maintaining itself as part of a climax community.

Not all stable, climax populations will show this same kind of age distribution curve, however. Woody perennials are long-lived and so can successfully maintain themselves even if their seedlings become established only sporadically in time. For instance, creosote bush *(Larrea tridentata)* is an evergreen shrub which dominates the warm deserts of North America. It seems to have all the attributes of a climax species which has maintained itself in the region for thousands of years (Vasek et al. 1975; Laudermilk and Munz 1938). However, the species has remarkably mesic requirements for good germination and seedling development (abundant moisture, neutral pH, low salinity, and moderate temperature) (Barbour, Cunningham, Oechel, and Bamberg 1977), and such conditions are unlikely to be met in nature perhaps more frequently than once in a decade or more. When a suitable year for establishment does arrive, we can expect an age distribution curve to show a peak. Robert and Alice Chew (1965) aged a *Larrea* stand in southeastern Arizona and found most shrubs to be 15–20 years old at the time (Figure 4–7(a)). No reproduction had taken place in the past 4 years, and the oldest shrub was 65; this is a relatively young stand. Barbour, MacMahon, Bamberg, and Ludwig (1977) sampled older stands, relating age to height, and stands on more arid sites often showed several age peaks, with no recent reproduction (Figure 4–7(b)). Evidently, mass germination and establishment in this species is a periodic phenomenon. There is no reverse J-shaped age curve, and only a few age classes are represented, yet the populations still maintain themselves.

In the mid-elevations of the Sierra Nevada Mountains, some groves of giant sequoia *(Sequoiadendron giganteum)* have many young trees and a reverse J-shaped age curve, while others do not (Figure 4–8). By correlating trunk diameter at breast height to age, Rundel (1971) showed that reproduction had not occurred for the past 500 years at sites such as Ponderosa Grove, and he predicted that these

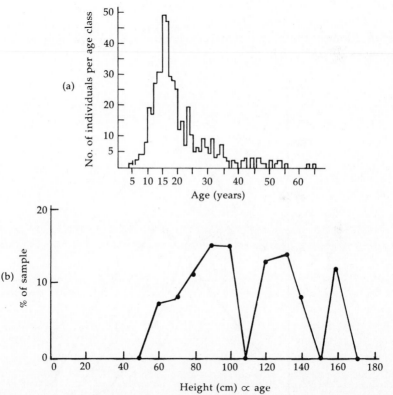

Figure 4-7. Age distribution curves for a long-lived woody perennial, creosote bush *(Larrea tridentata)*. (a) A relatively young stand which shows a peak of establishment 15–20 years ago and an absence of reproduction during the past 4 years. (b) An older stand in a more xeric (arid) environment, showing several past peaks of reproduction and an absence of reproduction for several years. ((a) From Chew and Chew 1965. Copyright 1965 by the Ecological Society of America. (b) from Barbour, MacMahon, Bamberg, and Ludwig 1977.)

groves would become extinct in the next 1000 years as the old trees died and were not replaced with their own kind. Yet, the life span of the giant sequoia is 2000–3000 years (Hartesveldt et al. 1975), so if a good year for establishment comes only once a millenium that should be sufficient to maintain the population.

Knobcone pine *(Pinus attenuata)* is a Californian closed cone conifer which typically occurs in single aged (even aged) stands that date back to periodic destructive fires (Vogl 1973). The reproduction of this species is dependent on fire melting a resin that otherwise seals cones shut and prevents seed dispersal. Apparently, conifers in the Canadian boreal forest also show single age populations, and these date back to severe storms, fires, or other natural disasters (Jones 1945). The frequency of these disasters is sufficient to maintain populations even though their age structure is not typical for climax species (Loucks 1970).

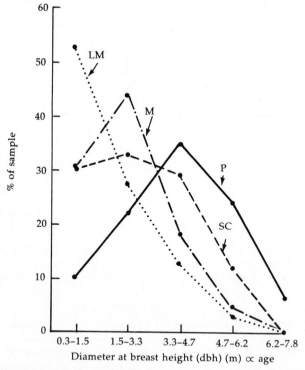

Figure 4-8. Age distribution curves for several stands of giant sequoia *(Sequoiadendron giganteum)* in the Sierra Nevada Mountains. Grove names are: LM, Long Meadow; M, Muir; SC, South Calaveras; and P, Ponderosa. Only LM shows a reverse J-shaped curve. (From Rundel et al. 1977. In *Terrestrial Vegetation of California.* Edited by Barbour and Major. Copyright 1977 John Wiley and Sons, Inc. Reprinted by permission of John Wiley and Sons, Inc.)

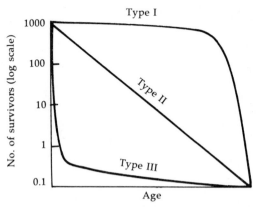

Figure 4-9. Hypothetical survivorship curves. (Reprinted by permission from Deevey 1947.)

Life Tables

If we follow a group of individuals from birth to death, the results can be summarized as a **survivorship curve**, which shows the fraction of the original group still alive at each age increment. There are three basic types of survivorship curves (Figure 4-9). Type I populations have low mortality when young. Most individuals die at maturity, within a relatively narrow age span. Type II populations have constant mortality at all ages. Type III populations have high mortality when young. Those few individuals that reach adulthood have a low death risk and continue living for a long time. Some plants show characteristics of all three curves, depending on the age of the population.

Janice Beatley (1967, 1969) carefully censused winter annuals in the northern Mojave Desert of southern Nevada for many years, recording mortality, growth, biomass, and reproduction of more than 50 species. Given sufficient rainfall (generally above 25 mm), they typically germinate in September (Figure 4-10). There is considerable mortality at this early establishment stage (Type III survivorship curve), then the population enters a period of slow vegetative growth during the winter with low, but constant, mortality (Type II curve). Accelerated rates of mortality accompany the beginning of rapid vegetative growth in April and the beginning of flowering in May (Type III curve). Finally, a Type I truncated survivorship curve marks the end of the population's cycle in July.

Another combination of survivorship curves is exhibited by the palm *Euterpe globosa* from tropical forests in Puerto Rico (Figure 4-11). High mortality of seeds and seedlings (Type III) is followed by linear mortality with age in the immature subcanopy stage (Type II), then by a plateau of zero mortality followed by a truncated decline (Type I) for canopy trees as they reach the end of their average life span.

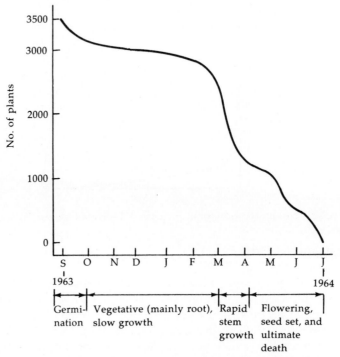

Figure 4–10. Actual survivorship curve for all winter annual species in a portion of the Mojave Desert, southern Nevada, from September of 1963, the beginning of winter rains, to July of 1964, when all mature plants died. Phenological stages are shown along the bottom axis. (Based on data in Beatley 1967 (copyright 1967 by the Ecological Society of America) and Beatley 1969 (by permission).)

The data from survivorship curves can be manipulated and transformed into a **life table** format. A life table (such as the one for *Euterpe*, Table 4–4) shows the number of deaths, survivors remaining, and expectation of future life for every age interval of a real or imaginary group. Life tables are useful summaries of population vigor and dynamics. They have been constructed for many animal species (see, for example, Krebs 1972 and Collier et al. 1973), but for very few plant species. An outstanding example of a plant demography study, complete with details on methods and formulae, is a paper on three short-lived herbaceous perennials by Sarukhan and Harper (1973).

THE BEHAVIOR OF INDIVIDUALS: RESOURCE ALLOCATION PATTERNS

The term "strategy" has been applied to the behavior of plants and animals, which is somewhat unfortunate because the term implies that these organisms are able to think and plan. A plant species has a pattern of resource allocation which minimizes its chances of extinction. Such patterns have been retained and refined through the process of natural selection. The resource allocation pattern of a species determines, in part, its niche—its functional address in a community. Winning in this existential game is simply being able to continue playing for as long as possible (Slobodkin and Rapoport 1974).

Organisms have a limited amount of time and energy with which to complete a life cycle. Time itself is not allocated, but is important in the gain of photosynthetic energy and in the utilization of energy for maintenance. A fraction of the total energy available is budgeted to each activity in the life cycle: the amount of time spent in a dormant state, in a juvenile, vegetative stage, or in a mature, reproductive stage; the amount of energy used for roots, shoots, leaves, flowers, seeds, or fruits; and the amount of energy used for growth, for maintenance, or for herbivore defense.

Organisms seem to lie on a continuum between two extremes of resource allocation: r and K. Extreme r-selection leads to short-lived plants which mature rapidly, occupy an open habitat in a seral community, and devote a large fraction of their photosynthate (their stored energy from photosynthesis) to producing flowers, fruits, and seeds. Their population size is density-independent; that is, population size is regulated by physical factors such as fire, flood, frost, drought, etc. Weeds and other species of pioneer communities are examples of r-selected populations.

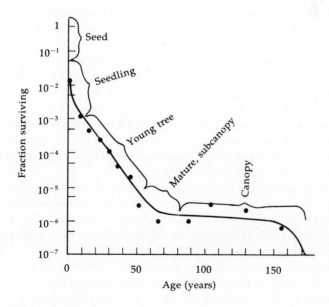

Figure 4-11. Actual survivorship curve for the tropical palm *Euterpe globosa.* (Reprinted by permission from Van Valen 1975.)

Table 4-4. A portion of a life table for the tropical palm *Euterpe globosa*. (Reprinted by permission from Van Valen 1975.)

Stage of life cycle	Height (m)	Age (yr)	Number per ~6000 m²	Probability of death before next stage	Expectation of future life (yr)
seed on tree	0.0	0	170,000	0.99⁺	0.2
seed on ground	0.0	0	46,000	0.96	0.4
germinated seed	0.0	1	2200	0.26	3.1
seedling	0.5	9	205	0.15	9
immature tree	1.0	15	80	0.07	9
	2.0	24	40	0.12	9
	3.0	30	18	0.15	9
	4.0	36	7	0.09	10
	5.5	45	3	0.25	17
reproductive	6.5	51	0.5	0.03	25
	9.0	66	0.2	0.00	67
overstory	12.0	88	0.2	0.00	63
	14.0	104	0.5	0.00	50
	16.0	130	0.3	0.02	28
senescent	18.0	156	0.1	0.10	9
	20.0	182	0	1.00	0

Extreme K-selection leads to long-lived plants which have a prolonged vegetative stage, occupy a closed, late seral, or climax community, and devote a small fraction of their photosynthate to reproduction. Their population size is density-dependent; that is, population size is regulated by biotic interactions such as competition. Population size is near the carrying capacity of the habitat. Forest trees are examples of K-selected populations. Table 4–5 summarizes general traits of r-selected and K-selected populations.

It is important to remember that r and K are relative terms. For example, two annual weeds may be compared and one considered to be in more of an extreme r mode than the other (Gadgil and Solbrig 1972; Hickman 1977). Many species lie in the middle of a continuum between r and K extremes (Grime 1977 and 1979).

The concept of r- and K-selection was first proposed by the very innovative ecologist Robert MacArthur (1962), who had a great impact on community and population ecology despite his tragically short life. The letters r and K come from the logistic equation for population growth rate:

$$\text{change in population size with time} = rN\left(\frac{K-N}{K}\right)$$

where r is the innate capacity for a population to increase; it is a constant for a given species, being the birth rate minus the death rate in a setting without any resource limitations. N is the population size. K is the highest population density that can be maintained in a constant, but real environment; it is the carrying capacity of the habitat. Ideal, limitless habitats would permit a geometric population increase, but real habitats have limits to growth, and as the population reaches K, the growth rate declines (Figure 4–12).

Reproductive Effort

A convenient way to place a population on the r–K continuum is to measure its **reproductive effort**, that is, the percentage of its total energy that is spent on reproduction (Harper and Ogden 1970). K-selected plants show relatively low reproductive effort, no matter what

Table 4–5. Some traits which correlate with r- and K-selection. (From Pianka, E. R. 1970. On r- and K-selection. *American Naturalist* 104:592–597. Copyright © 1970 by The University of Chicago.)

Trait	r-selection	K-selection
climate	variable and/or unpredictable; uncertain	fairly constant and/or predictable; more certain
mortality	often catastrophic; density-independent	density-dependent
survivorship	usually types I and II (see Figure 4–9)	often type III (see Figure 4–9)
population size	variable in time; not in equilibrium; usually well below carrying capacity of the habitat; recolonization each year	fairly constant in time; in equilibrium; at or near carrying capacity of the habitat; no recolonization necessary
intraspecific and interspecific competition	variable; often lax	usually keen
life span	short, usually less than 1 year	longer, usually more than 1 year
selection favors	rapid development; early reproduction; small body size; single reproduction period in life span	slower development; greater competitive ability; delayed reproduction; larger body size; repeated reproduction periods in life span
overall result	productivity	efficiency

year of their life is measured, while r-selected plants show a relatively high reproductive effort. Reproductive effort is partly a fixed, genetic trait, but it can be modified by the plant's environment.

For some species, reproductive effort is relatively constant regardless of the environment. Harper and Ogden (1970) grew common groundsel *(Senecio vulgaris)*, an annual weed, in the greenhouse under three levels of rooting stress (pot volume), but found reproductive effort to be roughly 20% regardless of stress. They pointed out that this value was comparable to that of other annuals, including cultivated crop species which had been bred for high yields.

Reproductive effort may be very plastic, changing with stress within the same species. Hickman (1975) observed the annual buckwheat *Polygonum cascadense* at five different sites in the Cascade Mountains of Oregon. The sites ranged from relatively xeric to relatively mesic. In the field, he found reproductive effort to decrease from 58% for a population on the driest site to 38% for a population on the most mesic site (Table 4–6). Were these ecotypic responses? Hickman collected seedlings from a xeric and a mesic site early in the season, transplanted them to a uniform greenhouse environment, and measured their reproductive effort (Table 4–6). There was no difference in reproductive effort between greenhouse-grown populations; hence the field differences were not genetic, ecotypic differences, but are examples of

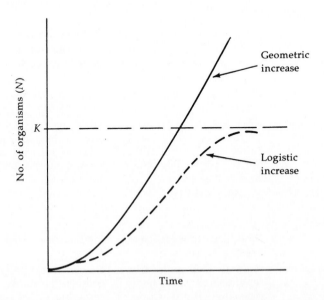

Figure 4–12. Geometric (solid line) versus logistic (dashed line) population growth over time. K is the carrying capacity of the habitat for a population showing density-dependent logistic growth (dashed line).

Table 4-6. Vegetative growth and reproductive effort in five populations of the annual buckwheat *Polygonum cascadense*. In the field, population 1 was on the most xeric site and population 5 was on the most mesic site. Young plants from two of the sites, 1 and 4, were also grown in a uniform (greenhouse) garden. (From Hickman 1975. By permission of the British Ecological Society.)

Trait	Field					Greenhouse	
	1	2	3	4	5	1	4
vegetative weight (mg)	13.6	16.0	43.2	49.8	325.3	48.4	73.6
reproductive weight (mg)	18.9	22.4	46.1	40.1	207.9	37.3	57.4
reproductive effort (%)	58.4	57.8	51.6	43.2	38.4	43.8	45.4

phenotypic plasticity. Consequently, reproductive effort may indicate something about the severity of a habitat as well as about a species' resource allocation pattern, and the two components might not be easy to separate without trials under uniform conditions.

As the environment becomes more severe and less predictable as to its suitability for plant growth over the course of a season or over the course of years, r-selected plants should predominate. One of the best studies to show this correlation was conducted by Abrahamson and Gadgil (1973) on four species of goldenrod (*Solidago*), all herbaceous perennials growing near each other in Massachusetts (Table 4-7). Field biomass samples revealed that reproductive effort was lower for species or populations in stable, shaded, hardwood forests than in seral, open meadows or in dry, disturbed, pioneer sites (Figure 4-13). When the four species were grown in greenhouses, they showed the same basic pattern, which indicated that the differences were genetically fixed.

Another example of the shift in reproductive effort with environment involves entire communities of herbs, rather than individual species. Herbaceous species were periodically harvested in replicate ¼ m² plots in three nearby communities: a 1 year old field (pioneer stage), a 10 year old field (later successional stage), and an oak forest (climax stage). All three communities were within 10 km of each other and were on the same soil type in northern Ohio. The plant matter was divided into shoots, roots, and reproductive parts. Figure 4-14 shows that the reproductive effort was greatest in the 1 year old field, which was dominated by r-selected species. The more predictable microenvironment, beneath a closed overstory tree canopy in the forest, was dominated by more K-selected perennial herbs with a high root:shoot ratio and a low reproductive effort.

Table 4-7. Species and habitats of goldenrod *(Solidago)*, whose reproductive efforts are summarized in Figure 4-13. (From Abrahamson, W. G. and M. Gadgil. 1973. Growth form and reproductive effort in goldenrods *(Solidago, Compositae). American Naturalist* 107:651-661. Copyright © 1973 by The University of Chicago.)

Trait	Dry, weedy sites (D)	Wet meadows (W)	Hardwood forest (H)
habitat	dry, disturbed	moist meadow	relatively undisturbed forest
associated species	dock *(Rumex),* yarrow *(Achillea)*	willow *(Salix),* elderberry *(Sambucus), Spiraea*	birch *(Betula),* oak *(Quercus),* maple *(Acer)*
canopy cover	open	closed	closed
successional stage	pioneer, start of succession	later successional stage	climax, end of succession
species of *Solidago* present	*S. nemoralis* *S. speciosa*	*S. rugosa* *S. canadensis*	*S. speciosa* *S. rugosa*

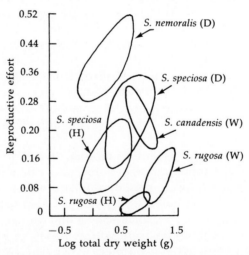

Figure 4-13. Reproductive effort of species and populations of *Solidago* in dry, weedy sites (D), wet meadows (W), and hardwood forest (H). Additional information is in Table 4-7. Reproductive effort here is the ratio of the dry weight of reproductive tissue to the total dry weight of aboveground tissue. Each closed curve embraces all points representing the individuals of a given group. (From Abrahamson, W. G. and M. Gadgil. 1973. Growth form and reproductive effort in goldenrods *(Solidago,* Compositae). *American Naturalist* 107:651-661. Copyright © 1973 by The University of Chicago.)

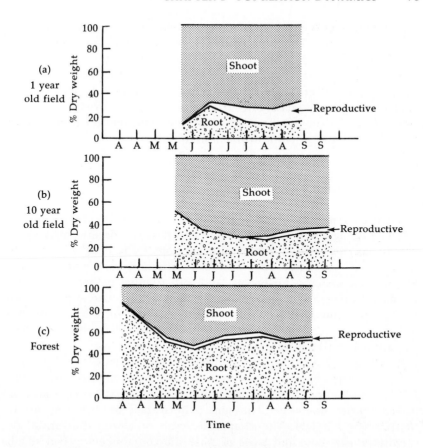

Figure 4-14. Dry weight of shoots, roots, and reproductive parts (as per cent of total weight) of herbs in (a) a 1 year old field, (b) a 10 year old field, and (c) a forest in northern Ohio. Samples were collected every 3 weeks between April and September of one growing season. (From Newell and Tramer 1978. Copyright 1978 by the Ecological Society of America.)

Reproductive effort can differ from one perennial to another over the cumulative life span of the plants. Some perennial species reproduce only once (semelparity), invest a considerable amount of stored energy into the event and become senescent during or following reproduction, for example, century plants (*Agave* species). There are both costs and benefits to postponing reproduction in this way. Other perennial species reproduce many times (iteroparity), though not necessarily every year nor with equal intensity each time. Each reproductive event uses only a small fraction of the plant's energy store.

Some authors (Harper and Ogden 1970; Hickman and Pitekla 1975) claim that reproductive effort can be measured by dividing the biomass (weight) of reproductive tissue by the biomass of vegetative tissue. However, reproductive effort involves the allocation of energy, not biomass, and vegetative matter does not contain the same caloric energy as reproductive matter. Stems and leaves of a typical species may average 4.2 kcal g dry wt^{-1}, roots may average 4.7 kcal g dry wt^{-1}, and seeds may average 5.1 kcal g dry wt^{-1} (Golley 1961). Also, by considering just the biomass, only the net yield is measured in reproductive effort; the energy used in respiration during the time flowers, fruits, and seeds are being formed is not measured.

Other Allocation Patterns

Species also differ in their allocation of time and energy within the vegetative stage. In a relatively small area of warm desert, for example, several species with different allocation patterns coexist: evergreen shrubs with extensive root systems; succulent cacti with shallow, fibrous root systems, no leaves, and a distinctive pathway of photosynthesis; drought-deciduous shrubs, dormant in late summer heat; winter-deciduous phreatophytes (trees or shrubs with deep roots that tap ground water); herbaceous perennials; annuals active in winter; and annuals active in summer. Several species within the same habitat may use the same pattern; consequently it is convenient and possibly even important to think of communities as being comprised of a handful of energy allocation pattern-types, rather than as being comprised of many different species. The same patterns may be shared by species in different habitats. The chaparral scrub of Chile, California, the Mediterranean area, and parts of Australia contains evergreen shrubs and drought-deciduous shrubs whose strategies seem identical to desert shrubs, even though rainfall is higher, soils are shallower, and the importance of fire and allelopathy is different.

Each allocation pattern has its own costs and benefits. The evergreen desert shrub creosote bush *(Larrea tridentata)*, for example, is able to photosynthesize year-round, which is a benefit. However, the leaves must be constructed so that water loss through transpiration is low, wilting is prevented, heat loss is high, and herbivore damage is minimal, for the leaves are present in the hottest part of the year as well as during cooler, mesic periods. These requirements are the costs of the creosote bush allocation pattern. It is energetically expensive to produce the many small, sclerophyllous, drought-resistant, nonpalatable leaves which remain on the plant 1–2 years. A further cost is that these leaf modifications inherently result in a low rate of gas exchange, and hence a low rate of photosynthesis (Mooney and Dunn 1970a, and b; Harrison et al. 1971).

Drought-deciduous desert shrubs, such as burrow bush *(Ambrosia dumosa)* or brittle bush *(Encelia farinosa)* are able to photosynthesize only part of the year. They drop their leaves and are dormant for several months, which is a cost. They must also produce a new crop of leaves (sometimes more than one) each year; this is another cost. But the leaves that are produced are inexpensive ones which have a high rate of photosynthesis, three to four times that of creosote bush (Figure 4–15) (Bamberg et al. 1975; Odening et al. 1974). Obviously, this is a benefit. Similar costs and benefits can be described and measured for other allocation patterns, and some will be discussed in more detail in later chapters.

SUMMARY

Some ecological characteristics of species or populations pertain to their distribution in space and time. Within a habitat, members of a population can be distributed in a random pattern, a clumped pattern, or a regular pattern. The type of distribution may reflect the type of reproduction, irregularity in the microenvironment, the degree of competitiveness, and the stage of succession. Certain sampling methods (random quadrats, contiguous quadrats, distance methods) and mathematical manipulations (Poisson distribution, variance-block size graphs, density, and frequency) can be employed to detect spatial pattern.

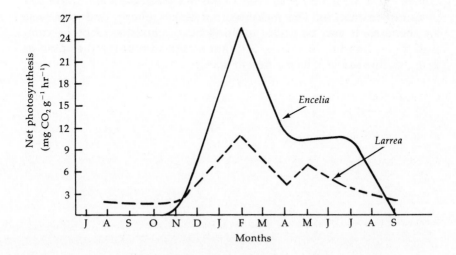

Figure 4–15. Seasonal patterns in maximum net photosynthesis rates for two desert shrubs: evergreen creosote bush *(Larrea tridentata)* (dashed line) and drought-deciduous brittle bush *(Encelia farinosa)* (solid line). (From Odening et al. 1974. Copyright 1974 by the Ecological Society of America.)

Demography, the study of vital statistics of populations, reveals ecological characteristics of populations over time. Annual, biennial, herbaceous perennial, suffrutescent, and woody perennial plant species each have characteristic life spans. Invading, successional populations will have early age states but no mature plants; regressive, senescent, successional populations will have late age states but no reproductive states. Climax populations may show all age states, with many more young plants than old (a reverse J-shaped curve), but there are important exceptions. Survivorship curves, and consequently life tables, of plant species may be quite complex, showing different patterns of mortality at different age states.

Organisms have a limited amount of time and energy with which to complete a life cycle. This means that a fraction of the total time and energy available is budgeted to each activity. An organism's resource allocation pattern involves the allocation of time to vegetative growth, seasonal dormancy, and reproduction, and the allocation of energy to growth, maintenance, reproduction, roots, shoots, and leaves. Organisms seem to lie on a continuum between two extremes of allocation patterns, r and K. r-selected species exhibit high reproductive effort, and K-selected species exhibit low reproductive effort, but reproductive effort can be influenced by environmental stress and it is not necessarily a constant for a given species. r and K are relative terms, and so are used in making comparisons. For example, an annual weed species is considered an r-selected species in comparison to a tree species in a climax forest, which is considered K-selected species. All allocation patterns can be analyzed and measured in terms of energy and time costs and benefits. It may be useful to think of communities as being comprised of a handful of energy allocation pattern-types, rather than as being comprised of many different species.

CHAPTER 5

SPECIES INTERACTIONS:
Competition and Amensalism

For the most part, Chapters 3 and 4 dealt with plant species and populations as though they existed in isolation. In nature, most communities consist of more than one plant population. In addition they show the influence of nonplant populations, such as decomposers (bacteria and fungi) in the soil, parasitic pathogens, and herbivorous animals. Interactions between these diverse populations modify the genetic potential of each species (its physiological optimum and range) to yield the communities around us (based on ecological optima and ranges). Harper (1964) has written a powerful review that demonstrates how the biology of organisms grown in isolation does not compare to their biology when grown in mixtures.

Many ecologists believe that the associated organisms in a community are somehow interdependent, that they are not associated by chance, and that disturbance of one organism will have consequences to all organisms. Clements (1916), and especially his followers, took this view to an extreme, equating climax communities with superorganisms, their component populations being as interdependent as the cells, tissues, or organs of a single organism. The objectives of this chapter and Chapter 6 are to survey the kinds of interactions which may occur between members of a community. This will lead, in Chapter 7, to an assessment of Clementsian views, and more moderate views, of community interdependence and integrity.

Table 5–1 catalogs all possible interactions according to a symbolic scheme developed by Burkholder (1952) (see also Odum 1971 and Malcolm 1966). Each interaction is described by its effect on two populations or organisms, A and B, when they are in contact (the interaction is "on") and when they are apart (the interaction is "off"). As an example, consider herbivory: when a herbivore and its food plant are together, the herbivore is stimulated (its growth, reproduction, or general success is improved) and the plant is depressed (its growth, reserves, reproduction, or general success declines); when the two are apart, the herbivore is depressed and the plant remains stable. In Table 5–1, herbivory, parasitism, and predation are identical, but the subtle, important differences for other interactions are apparent. Mathematically, there are 81 possible interactions with this symbolism, but

Table 5–1. A complete list of all biologically possible types of interactions according to Burkholder (1952). When organisms *A* and *B* are close enough to participate in the interaction, the interaction is "on"; otherwise it is "off." Stimulation is symbolized as +, no effect as 0, and depression as −. (From Burkholder 1952. Reprinted by permission of *American Scientist,* journal of Sigma Xi, The Scientific Research Society.)

Name of interaction	On		Off	
	A	*B*	*A*	*B*
neutralism	0	0	0	0
competition	−	−	0	0
mutualism	+	+	−	−
unnamed	+	+	0	−
protocooperation	+	+	0	0
commensalism	+	0	−	0
unnamed	+	0	0	0
amensalism	0	−	0	0
parasitism, predation, herbivory	+	−	−	0
unnamed	+	−	0	0

Burkholder concluded that only the 10 shown in Table 5–1 are logically possible. Among those 10, three are sufficiently rare or at least unexamined that they have no names. Neutralism is included for purposes of comparison and completeness, but it too may be a rarity in nature. This chapter will survey two of the remaining six interactions: competition and amensalism. Protocooperation, commensalism, mutualism, and herbivory will be discussed in Chapter 6.

Some interactions symbolized in Table 5–1 are negative (one or another partner is inhibited, as in competition or amensalism), and others are positive (one or another partner is stimulated, as in commensalism or mutualism). The existence of the interactions can only be conclusively demonstrated by elaborate experimentation. Rather simple field sampling, however, can provide the initial evidence for an interaction.

The sampling is based on the premise that positive interactions will produce positive spatial relationships between the partners; where one partner is found, the probability is high that the other will be found nearby. The two populations attract one another and they exist in a nonrandom, clumped pattern (Figure 5–1(a)). Similarly, negative interactions will produce negative spatial relationships; the two populations appear to repel one another and exist in a nonrandom, regular pattern (Figure 5–1(b)). If there is no interaction between the populations, then the location of one individual has no influence on the location of others; the two populations are said to be randomly distributed with respect to each other (Figure 5–1(c)).

The spatial relationship is revealed by sampling vegetation with random quadrats of an appropriate size. In each quadrat, the presence or absence of any two (or all) species is noted, then summarized in a contingency table (Table 5–2). This constitutes the observed data. Expected data, assuming a completely random distribution of the two taxa

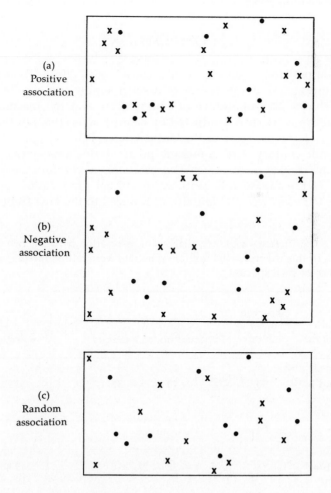

(a) Positive association

(b) Negative association

(c) Random association

Figure 5–1. An overhead, diagrammatic view of two species (dots and x's) which are (a) positively associated, (b) negatively associated, and (c) randomly associated.

to be examined, can be compared to observed data by a chi-square formula. The example in Table 5–2 yielded a chi-square value higher than expected by chance, so the two species (A and B) are not distributed randomly with respect to each other. Inspection of the table shows that there were more quadrats with only A or B than expected by chance, and there were fewer quadrats with both species present than expected by chance. The conclusion is that A and B are negatively associated. The reason for the association would then have to be determined by ecological observations and experimentation; the statistical treatment merely is the first step and does not constitute proof of a biological interaction.

COMPETITION

Competition results in mutually adverse effects to organisms which utilize a common resource in short supply. An excellent study which shows the importance of competition, and the mechanism by which it can work, was conducted by Grant Harris (1967) in the northern intermountain region of the United States. Before the middle of the nineteenth century, the dominant plant of this grassland area was bluebunch wheatgrass *(Agropyron spicatum)*, a perennial. About the middle of the nineteenth century, an annual grass called cheatgrass *(Bromus tectorum)* was accidentally introduced to the area from Europe.

Table 5–2. Contingency table for analysis of association between two species, A and B. In this example, 100 random quadrats were tallied for the presence or absence of each species.

Symbol and description	Observed no. of quadrats	Expected no. of quadrats	$\chi^2 =$ $\dfrac{(\text{observed} - \text{expected})^2}{\text{expected}}$
$a = A$ and B present	30	$\dfrac{(a + b)}{100} \times (a + c) = 38$	1.7
$b = A$ present, B absent	29	$(a + b) - 38 = 21$	3.0
$c = A$ absent, B present	35	$(a + c) - 38 = 27$	2.4
$d = A$ and B absent	6	$100 - (38 + 21 + 27) = 14$	4.6
totals	100		$\Sigma \chi^2 = 11.7$

From that time to the present, ranchers have noticed an enormous increase in the abundance of cheatgrass and an equally impressive decrease in the abundance of bluebunch wheatgrass. What caused the shift?

Both species have a similar life cycle. They germinate (or break dormancy if perennial) in fall, grow slowly during winter, grow rapidly in spring, form flowers in early summer, and die in June (or begin dormancy in mid-July if perennial). Harris studied growth and survival of these two species during a year, starting from seed for both. He found that the presence of cheatgrass greatly reduced growth and survival of bluebunch wheatgrass. In one field trial, bluebunch wheatgrass was sown in October in plots with different densities and cover of cheatgrass (0–100 plants m^{-2}, 0–100% cover). Seedlings were tallied the following June and again after summer drought in October, 7 and 12 months after sowing, respectively. Cheatgrass density had little effect on the number of wheatgrass seedlings still present in June, but it had a great effect on survival and biomass of those seedlings through the summer drought. Harris pointed out that his maximum experimental cheatgrass density of 100 plants m^{-2} was low compared to the entire region, where it could reach 3000–10,000 plants m^{-2}, and there is no establishment of bunchgrass where cheatgrass density is greater than 1000 plants m^{-2}. Clearly, cheatgrass is an important competitor, but how is it competing, and what happens during the summer months to cause the competitive effect?

Harris planted seeds of each species in long glass tubes filled with soil. The tubes were inserted in the field, level with surrounding soil. Every month the tubes were lifted and the depth of rooting was measured. Root growth during winter was much greater for cheatgrass than for wheatgrass (Figure 5–2). At the start of rapid spring growth in April, cheatgrass roots had penetrated 90 cm, in contrast to 20 cm for wheatgrass. This difference in depth allowed cheatgrass to absorb water from a much greater part of the soil profile than wheatgrass could. Consequently, when the upper soil became dry in early summer, with both species drawing water from it, only wheatgrass suffered, for water still remained available at greater depths, where only cheatgrass roots had penetrated. Harris measured soil water availability in the two soil regions at this time and found water potential was below −15 bars in the upper region (near the wilting point of many species), but was only −1 bars in the lower region (essentially still saturated, see pp. 432–3 and 443–7).

The recent success of cheatgrass, then, seems due to its winter root growth, which gives it a spring and early summer advantage over wheatgrass that started from seed at the same time. Each year cheatgrass increases in density, and each year this results in greater competition for moisture in the upper soil, and greater stress on wheatgrass.

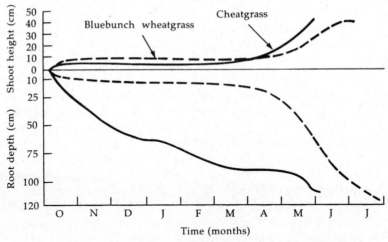

Figure 5–2. Elongation of shoots (above the 0 line) and roots (below the 0 line) of cheatgrass *(Bromus tectorum)* (solid lines) and of bluebunch wheatgrass *(Agropyron spicatum)* (dashed lines) in a field experiment. Plants were grown from the time of germination, in October, until the time of maturity of the annual cheatgrass. (From Harris 1967. Copyright 1967 by the Ecological Society of America.)

Competition and Habitat

In general, it appears that many plants endemic to severe habitats are restricted to them because they are poor competitors on less extreme sites. Severe habitats have very low plant cover, and so there is little competition. Plants in severe habitats tolerate the sort of environmental extremes that most other plants cannot, but they would grow better elsewhere, in the absence of competition. Research with serpentine endemics and halophytes illustrates this conclusion.

Serpentine endemics

Serpentine is a metamorphic, magnesium silicate rock, often green in color and slippery to the touch, which has a number of traits inimical to plant growth. It is low in such essential nutrients as N, Ca, K, and P; its pH may be far from neutrality (either acidic or basic); and it is high in such toxic elements as Ni and Cr. Soils which are derived from serpentine rock are sterile, support unusual, endemic floras, and are covered with vegetation whose physiognomy differs from that of surrounding vegetation on nonserpentine soil. Serpentine outcrops are found throughout the world (Europe, Rhodesia, Japan, Southeast Asia, Canada, and the United States; Whittaker 1954), but they are especially common in the Pacific Coast states.

Arthur Kruckeberg, of the University of Washington, experimented with serpentine and nonserpentine ecotypes and species (1954). He reported that herbaceous serpentine endemics became established from seed better and grew faster on nonserpentine soil, providing they were free from interspecific competition. When sown with typical nonserpentine species on nonserpentine soil, they became etiolated and did not survive. On serpentine soil, only the serpentine endemics survived, but there was considerable bare ground and the plants grew slowly.

James Griffin (1965) experimented with digger pine (*Pinus sabiniana*), a tree that is not endemic to serpentine but which increases in density on serpentine, especially at low elevations in the California foothills. He found that as elevation increased (and with it, increased rainfall, moderating temperatures, and greater competition from other species), the range of soil types on which digger pine was found narrowed and centered on serpentine. Serpentine outcrops, with low plant density and low interspecific competition, seemed to serve as a refuge for digger pine at the upper limit of its range.

This is not to say that all serpentine taxa have ranges limited by competition. Tadros (1957) found that toxins released by the higher densities of certain soil microorganisms on nonserpentine soil prevent the herbaceous, serpentine endemic *Emmenanthe rosea* from establishing on normal soil. McMillan (1956) showed that some serpentine taxa grow as well on serpentine soil as on nonserpentine soil, even where interspecific competition is not a factor. Thus, there is more than one way in which plants have adapted to serpentine, but tolerance as a means to avoid competition may be the major strategy.

Halophytes

Halophytes (literally, salt plants) are capable of growing in soil with more than 0.2% salt concentration (Barbour (1970*b*) pointed out that this is a conservative limit, and some researchers put the limit at 0.25% or even 0.5%). **Glycophytes** (literally, sweet plants) are intolerant of salinity above that necessary to supply essential nutrients, approximately 0.1% salt. Certainly, above 0.2% salinity their growth is severely reduced.

All halophytes are not equally tolerant of salt and a great number of terms have been coined to describe them (Waisel 1972). We will follow Ingram (1957), and use only three terms: **intolerant, facultative,** and **obligate**.

Intolerant halophytes show maximum growth at low salinity and declining growth with increasing salinity (Figure 5–3). Facultative halophytes show maximum growth at moderate salinity and diminished growth at both low and high salinity. Obligate halophytes show

maximum growth at moderate or high salinity and are unable to survive at low salinity (perhaps below 0.1% salinity, although there is no
agreement about this lower limit). As shown by the growth curves in
Figure 5–3, all three halophytes may have similar optima and upper
tolerance limits; they differ mainly in performance at low salinities.

Most halophytes are intolerant, whether we look at their germination, growth, or reproduction, whether they are mangroves, coastal
salt marsh herbs, beach plants which receive salt spray, or salt desert
herbs and shrubs. A few halophytes are facultative. Perhaps there are
no obligate halophytes (Barbour 1970b). This conclusion can be reached
from both field observations and manipulative experiments in growth
chambers.

For example, one of the working hypotheses of some arid zone
ecologists is that certain species are dependable indicators of saline
soil. Field data reveal, however, that these "indicators" often grow well
on nonsaline soil. Ungar (1966) determined soil salinity within some
100 saline sinks in Kansas and Oklahoma and correlated salinity with
species presence. Species which are traditionally referred to as highly
salt tolerant (species of *Distichlis, Polygonum, Sesuvium, Sporobolus,
Suaeda, Tamarix*), actually occurred on soils which ranged from 4.5 to
0.01% salinity. He concluded, "The wide range of salt tolerance found

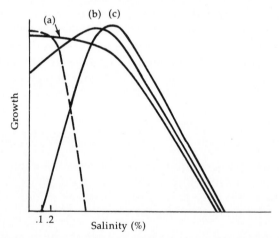

Figure 5–3. Response of glycophytes (dashed line) and of different types of
halophytes ((a), (b), and (c)) to increasing salinity. The growth of glycophytes
declines sharply above soil salinity of about 0.2%. (a) Intolerant halophytes
can exist on saline soils, but their growth declines as salinity rises. (b)
Facultative halophytes grow best at some intermediate salinity, yet can
survive at low salinity. (c) Obligate halophytes grow best at some intermediate salinity, and cannot exist at low salinities, perhaps below 0.1%
salinity. (From Ingram 1957 in *Seventh Symposium of the Society for General
Microbiology.* Reprinted by permission of Cambridge University Press.)

for the most salt tolerant species . . . indicates that none of these is an obligate halophyte. These species grow in highly saline areas because they do not compete as well in nonsaline wet or dry environments."

Observations of salt marsh and beach plants parallel those for salt desert plants. Taylor (1938, 1939) conducted a thorough study of salt marsh plants on Long Island and concluded that most species exhibited wide tolerance ranges, that is, they were not obligate halophytes. Cordgrass (*Spartina alterniflora*), for example, typically grew in ground water with 2.6% salt concentration, but he found it "growing far more luxuriantly" in several areas with only a 0.5% salt concentration. *Phragmites communis* grew equally well on dry land, standing fresh water, and nearly full strength sea water. Taylor concluded that it was dangerous to state that such plants prefer saline sites; rather, they merely tolerate them.

Taylor then grew nine species of salt marsh plants in sand at an outdoor site on Long Island and watered them from below with different concentrations of NaCl. He selected species which field observation had shown to have: (a) a wide tolerance range (*Distichlis spicata, Iva ovaria, Pluchea camphorata,* and *Phragmites communis*), or (b) a narrow, saline range (*Salicornia europaea, Spartina alterniflora,* and *Spartina patens*), or (c) a narrow, nonsaline range (*Baccharis halimifolia* and *Limonium carolinianum*). Taylor made general notes on growth and vigor during a period of 50 summer days and he compared his test plants with controls which were left in nature at approximately the same salinity. All species grew at 0% NaCl (tap water), but five species generally performed best at 0% NaCl: *Iva ovaria, Limonium carolinianum, Pluchea camphorata, Phragmites communis,* and *Spartina patens* (*Iva*, however, did not show diminished growth until beyond 1.5% NaCl). Three species showed a facultative response, with optimal growth at moderate salinities: *Baccharis halimifolia* and *Spartina alterniflora* at about 1% NaCl, *Salicornia europaea* at 3.1% NaCl. *Distichlis* anomalously exhibited two peaks of growth at 0% NaCl and 3.1% NaCl. Taylor's experiment was one of the few which closely related field observations to experimental results. If he had only used dilutions of sea water rather than NaCl, it would have been a near-perfect field experiment.

Chapman (1960) reviewed many other germination experiments and he concluded that most halophytes germinate best under freshwater conditions. He pointed out that a reduction in soil salinity sufficient to promote germination occurs in spring or winter. Germinating halophytes can be just as intolerant of salinity as glycophytes (Rozema 1975).

The annual beach plant sea rocket (*Cakile maritima*) is restricted to the open, harsh beach environment with high salt spray levels. Barbour et al. (1973) found that it could grow better in more protected coastal grassland sites if competition was removed. They sowed seeds of *Cakile* on the coastal strand (the normal habitat), in nearby undisturbed grassland soil, in disturbed grassland with the top 5 cm of soil replaced with beach sand, and in disturbed grassland with all grassland plants removed.

After 6 weeks (Figure 5-4), the best growth and survival rates were exhibited by seedlings in grassland soil weeded of all competing plants. Growth in the undisturbed grassland plot was poor; seedlings were spindly, weak, and pale, and leaves were not well developed—all symptoms of low light intensity. Two months later, none of these etiolated plants remained.

Competition and Niche

Since competition involves two organisms utilizing the same resource, it is obvious that competing organisms must have, to some extent, overlapping niches. Intraspecific competition is necessarily more intense than interspecific competition, because niche overlap is greater. Current evolutionary theory holds that selection pressure drives species within a community to partition the environment (utilize different parts of the environment), with the result that competition is minimized.

Height (cm)

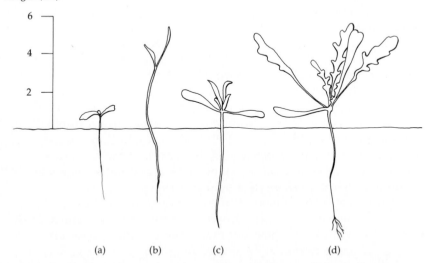

Figure 5-4. Appearance of sea rocket *(Cakile maritima)* seedlings of the same age grown in four different habitats. (a) The normal beach habitat, (b) unweeded grassland, (c) weeded grassland with sand topsoil, and (d) weeded grassland with grassland soil. The horizontal line represents ground level, and not all root matter is shown. (From Barbour et al. 1973. *Coastal Ecology: Bodega Head.* By permission of the University of California Press, Berkeley, CA.)

The concept of "one niche, one species" stemmed from laboratory experiments performed by the Russian microbiologist G. F. Gause (1934). He placed pairs of closely related protozoan or yeast species together in homogeneous microenvironments, and noted population growth rates *(dN/dt)*. Initially, growth rates of both species were depressed, compared to rates when grown in isolation, indicating that competition was occurring. Ultimately, however, there was a "winner" and a "loser," the winner coming to dominate the mixture and the loser reaching extinction. He concluded that species which coexist in nature must evolve ecological differences, for otherwise they would not be able to remain coexisting. The results of a competition experiment between two species of duckweed (Harper 1961) are shown in Figure 5–5.

The one niche, one species concept is often called **Gause's Competitive Exclusion Principle**, even though zoologists before him had published much the same conclusion (Krebs 1972). Laboratory work and field observations since his time, with animals of more complex life cycles than microbes and with environments more heterogeneous than culture bottles, have shown that minor niche differences exist even between closely related taxa, and these differences are sufficient to permit coexistence and release from some competition (see, for example, Ayala 1969 and MacArthur 1958).

Measurement of Competition

Recall, from Chapter 4, the logistic equation of population growth for a species in isolation:

$$dN/dt = rN \left(\frac{K - N}{K} \right)$$

where r is the natural, intrinsic rate of population growth, N is population size, and K is population size at saturation (the environment's carrying capacity). Multiplying out, we can see that the equation already contains an **intraspecific competition** component:

$$dN/dt = rN - \frac{rN^2}{K}$$

where rN is the unlimited rate of growth and rN^2/K is the decline caused by crowding (competition).

If we introduce a second species, we must introduce a second negative component, to account for **interspecific competition**. The equation for species 1 is:

$$dN_1/dt = r_1N_1 \frac{(K_1 - N_1 - \alpha N_2)}{K_1}$$

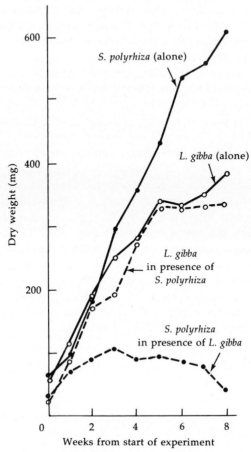

Figure 5–5. Results of a competition experiment performed between two species of duckweed *(Lemna gibba* and *Spirodela polyrhiza).* (Reprinted by permission from Harper 1961.)

where α is the inhibiting (competitive) effect on N_1 for every individual of N_2. Similarly, the equation for species N_2 is:

$$dN_2/dt = r_2 N_2 \frac{(K_2 - N_2 - \beta N_1)}{K_2}$$

where β is the inhibiting effect on N_2 for every individual of N_1.

  This pair of equations was developed independently by Lotka (1925) and Volterra (1926). They have a number of limitations, but still form a starting point for quantifying competition. Appropriate experiments have permitted some zoologists to calculate the components of the Lotka-Volterra equations, for example, α and β (Krebs 1972). If α and β are small, the interspecific components of the equations can be virtually ignored and the two populations can coexist. Coexistence will also result if $\alpha < K_1/K_2$ and $\beta < K_2/K_1$.

C. T. de Wit (1960, 1961) pioneered in the application of zoo-logical concepts of competition to plants. One of his techniques to mea-sure plant competition is to calculate input/output ratios for each species in a mixture, where:

$$\text{input ratio} = \frac{\text{propagules planted of species A}}{\text{propagules planted of species B}}$$

$$\text{output ratio} = \frac{\text{units produced of species A}}{\text{units produced of species B}}$$

Propagule can mean seeds, seedlings, potato eyes, bulbs, or lengths of rhizomes; a good synonym is diaspore. Units produced can mean seeds, number of shoots or tillers, length of new stolons, or plant weight. Whatever definition of propagule or unit is chosen, it must be the same for both species. Two species are planted in a replacement series, where the total number of initial propagules is a constant, but the ratio of the two species is changed from pure species A, to an even mix of A and B, and on to pure species B.

Figure 5–6 shows the four possible outcomes of any such experi-ment. First, species A can "win," in which case the input/output line is parallel to but above the equilibrium line. Second, species B can "win," in which case the ratio line is parallel to but below the equilibrium line (Figure 5–6(a)). Finally, species A and B can coexist, in which case the ratio line crosses the equilibrium line; that is, the slope is no longer 45° (Figure 5–6(b)). If the slope is less than 45°, the equilibrium point is stable because movement away from the equilibrium point tends to be

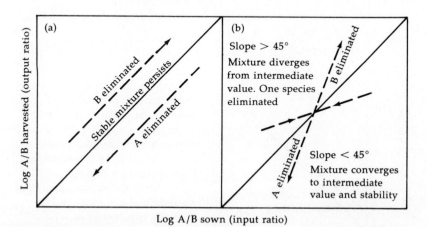

Log A/B sown (input ratio)

Figure 5–6. Input/output diagrams which illustrate the possible outcomes of competition between two species, A and B. (From *Environment and Plant Ecology.* Etherington. Copyright 1975 by John Wiley and Sons, Inc., LTD, London. Reprinted by permission of John Wiley and Sons, Inc., LTD, London.)

dampened by the populations' competitive qualities, and the populations drift back to equilibrium. If the slope is greater than 45°, the equilibrium is unstable because movement away from the equilibrium point tends to lead to further departure from stability and one species will decline to extinction.

Figure 5-7(a)–(c) shows actual data rather than hypothetical data. In Figure 5-7(a), the annual grasses barley and oats were sown in a replacement series and output measured as seed production. The results show that barley will outcompete oats at any mixture. Figure 5-7(b) shows competition between the perennials sweet vernal grass (*Anthoxanthum odoratum*) and timothy (*Phleum pratense*). The input/output (measured as tillers) line lies so close to the equilibrium line that we may predict the two species can coexist at any mixture. Figure 5-7(c) shows competition between two annual range plants from California grasslands: wild oat (*Avena fatua*) and slender wild oat (*A. barbata*), sown with a combined density of 1380 m^{-2}. The input/output

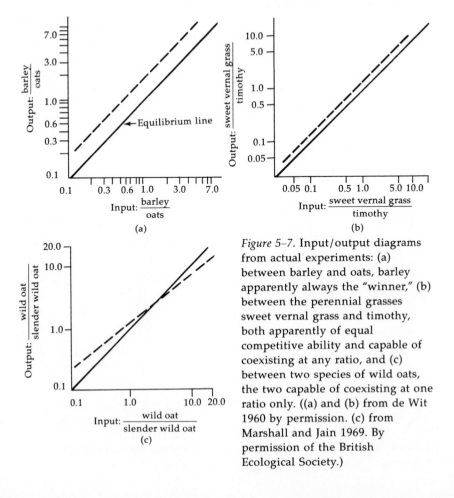

Figure 5-7. Input/output diagrams from actual experiments: (a) between barley and oats, barley apparently always the "winner," (b) between the perennial grasses sweet vernal grass and timothy, both apparently of equal competitive ability and capable of coexisting at any ratio, and (c) between two species of wild oats, the two capable of coexisting at one ratio only. ((a) and (b) from de Wit 1960 by permission. (c) from Marshall and Jain 1969. By permission of the British Ecological Society.)

line, with a slope of less than 45°, crosses the equilibrium line at a point corresponding to a mix of about 20% slender wild oat and 80% wild oat. One can conclude that these two species should be able to coexist in nature at this ratio.

Just as Gause's simple competition experiments did not reflect the true complexity of nature, so these replacement experiments may not bring us very close to the real world. Marshall and Jain (1969), for example, who conducted the wild oat study just described, attempted to validate their greenhouse results by examining natural wild oat populations. Did such natural populations at a density of about 1380 m^{-2} show the same sort of 20:80 equilibrium mix that their experiment predicted? No. About two-thirds of such stands in nature were 90–100% pure wild oats or slender wild oats; very few of the mixed stands were of the predicted 20:80 mix. Marshall and Jain believed the discrepancy was due to a far more complex situation in nature than they could imitate with their greenhouse study. For example, in nature the seeds would not be spread uniformly and at a constant depth as they were in the greenhouse study; also, other competing species from the field were missing in the greenhouse study; finally, the microenvironment and seed supply in nature are more heterogeneous than in the greenhouse study. However, de Wit's ideas and methods are an important, essential first step to the goal of quantifying and understanding competition.

A Final Word

It is one level of understanding to be able to document and quantify plant competition; it is another level to determine the resource being competed for—the very nature of the competitive interaction. The thoroughness of Harris' intermountain study, described earlier, which showed that competition for soil moisture was the ultimate factor, is unusual in the literature. Most ecologists make an educated guess as to the factors in short supply. If more than one resource is competed for, it may be difficult to demonstrate which is the most important one. As an example of this, consider a study by Hardy Shirley (1945) in north-central Minnesota.

The overstory trees in the Lake states are often the conifers white spruce (Picea glauca), white pine (Pinus strobus), red pine (P. resinosa), and/or jack pine (P. banksiana), yet these commercially valuable species do not reproduce in their own shade. Instead, hardwood seedlings and saplings grow up beneath the canopy. Why are conifer seedlings such poor competitors with the hardwoods? Shirley examined the effect of light on conifer seedlings by planting seedlings beneath screens that reduced full sun in steps from 2% to 89%, and found that survival and growth over a 4 year period increased as light intensity increased. Generally, growth was not satisfactory when light intensity was reduced more than 65% (Figure 5-8(a)). He also examined the combined effect of shade and root competition (Figure 5-8(b)). Results, after 4 years, were very complex. In dry areas, shade actually improved conifer seedling survival, though not seedling growth. In more mesic areas, removal of the overstory stimulated conifer seedling growth, weeding of ground vegetation further improved it, and trenching around the plot improved it even further. The conclusion is that competition for shade and soil moisture are intertwined and a single factor can not be pinpointed as the most important factor.

Furthermore, competition may be affected by other interactions in the real world, so the competitive potential of a species may not be expressed. Mack and Harper (1977) studied the intraspecific and interspecific abilities of four dune annuals in Wales and reached several important conclusions. One species, *Vulpia fasciculata*, showed strong competitive superiority in greenhouse experiments, yet in nature it did not exclude the other species and grow in pure stands. The reason for the anomaly may be herbivory by rabbits which preferentially graze on *Vulpia*. Another conclusion was that the typical response to competition by these annuals was a plastic reduction in size and seed production, rather than mortality (thinning).

AMENSALISM

Amensalism is an interaction which depresses one organism while the other remains stable. One example of amensalism is **allelochemic interaction**, the inhibition of one organism by another via the release of metabolic by-products into the environment. These by-products are selectively toxic, affecting some species but not others. Allelochemics are viewed by some biologists as simply the mechanism for an aggressive form of competition, but there is an important distinction to be made with competition: competition is an interaction brought about by the *removal* of a resource from the environment, and allelochemic interactions are brought about by the *addition* of a substance to the environment. A subset of allelochemic interactions involving plant species only is called **allelopathy**. In the 40 years since the word allelochemics was coined, a large body of literature has developed on the subject, seeming to demonstrate the pervasive chemical links between organisms within a community—decomposers, producers, and consumers (Sondheimer and Simeone 1970; Rice 1974; Whittaker and Feeny 1971).

However, the ecological significance of allelochemics is still not clear (Barbour 1973a). Before an interaction can be considered ecologi-

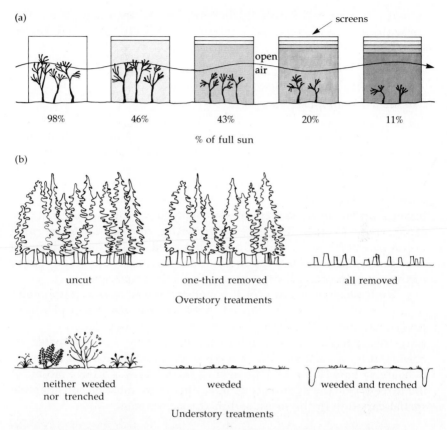

Figure 5-8. Diagrammatic summary of Shirley's experiments to determine the relative importance of competition for light and soil moisture to pine seedlings. (a) Pine seedlings were grown beneath different layers of screens to achieve different levels of sunlight. (b) Seedlings were grown beneath different overstories (closed canopy, one-third removed, all removed) and at the same time in three different understories (control, all other plants removed, all other plants removed and trenches dug around plot to sever the roots of adjacent plants).

cally significant, a series of steps should be followed. In the first step a correlation must be documented. For example, a negative association between two species of plants must be shown. The second step involves experimentation to try to determine cause and effect. If a chemical is involved, what is its identity and how does it affect the species involved? Much of this work can be done in the laboratory, but laboratory conditions should attempt to imitate field conditions. The third step returns to the field situation. Do the factors discovered in the laboratory operate in nature? Can the compounds be detected, and in what concentration? Can they remain viable in the soil system for long periods of time? The pitfalls in carrying a suspected interaction through all three steps are numerous and well illustrated in the literature.

Allelochemic Interactions at the Producer-Decomposer Level

To a large degree, decomposers in the soil and litter beneath a community are affected by the species of plants shedding the litter and penetrating the soil with roots. Soils beneath northern conifer forests are acidic because conifer litter is acidic and its decomposition influences soil pH. Fungi, as a result, dominate the soil microflora while bacteria dominate more neutral soil beneath deciduous forests (Eyre 1963). There are also differences even within one conifer forest: pine needles are much more acidic than spruce needles, and the soil beneath most pine species has less decomposer activity and is almost devoid of earthworms, in comparison with soil beneath spruce species.

Understory plants can further modify the soil microenvironment. Tappeiner and Alm (1975) showed that such features as soil pH, litter decay rate, and soil nutrient status were not only dependent on overstory tree species, but on whether the understory was dominated by hazel shrubs (Corylus cornuta), a mixture of other shrubs and herbs, or was largely bare (Table 5-3). Since the understory was often a mosaic of the three types, the authors concluded that there must be considerable spatial variation in the forest soil microenvironment.

Table 5-3. Effect of overstory and understory plants on soil pH, bulk density (g cm^{-3}), weights of nutrients in the top 3 cm (kg ha^{-1}), and turnover rates (in years) for all dry litter and for selected nutrients. (From Tappeiner and Alm 1975. Copyright 1975 by the Ecological Society of America.)

Overstory/ understory	pH	Bulk density (g cm^{-3})	Weight in soil					Turnover time (years)		
			Ca	N	K	Mg	P	Litter	Ca	K
Red pine/none	4.1	0.66	121	470	29	13	7	5.0	4.9	3.9
Red pine/hazel	4.1	0.54	111	535	29	14	7	3.2	2.5	3.1
Red pine/ herb-shrub	4.2	0.62	128	524	30	14	8	4.1	3.8	2.1
Birch/hazel	5.0	0.44	304	617	47	43	8	1.7	1.4	0.2
Birch/ herb-shrub	5.1	0.48	337	602	50	44	9	2.3	1.9	0.3

Table 5–4. Nutrient content in rainwater which falls directly to earth (rainfall) and in rainwater which falls through an oak *(Quercus petraea)* canopy (throughfall). Values are expressed in kg ha^{-1} yr^{-1}. (From Carlisle et al. 1966. By permission of the British Ecological Society.)

Nutrient	In rainfall (kg ha^{-1} yr^{-1})	In throughfall (kg ha^{-1} yr^{-1})
N	9.54	8.82
P	0.43	1.31
K	2.96	28.14
Ca	7.30	17.18
Mg	4.63	9.36
Na[a]	35.34	55.55
total	60.20	120.36

[a]an essential nutrient for some plants

Apart from pH and litter decay products, plants affect soil chemistry by passively contributing a variety of inorganic and organic compounds to the soil. Apparently, plants are very leaky systems. Carlisle et al. (1966) analyzed the nutrient content in rainwater falling directly on the ground and in rainwater falling through the leaf canopy of sessile oak *(Quercus petraea)*. Except for nitrogen, the rainwater falling through the oak leaf canopy contained a higher concentration of nutrients (Table 5–4). Tukey (1966) has shown that larger molecules can also be leached from leaves. He grew seedlings and cuttings of 150 species in nutrient culture with certain radioisotopes in it, then leached the plants by atomized mist or immersion in pure water for up to 24 hr. The leachate was channeled through an anion-cation exchange resin and analyzed. It contained 14 elements, including such essential nutrients as iron, calcium, phosphorus, potassium, nitrogen, and magnesium, seven sugars, some pectic substances, 23 amino acids, and 15 organic acids, including virtually all the acids in the Krebs cycle of respiration. In nature, these and larger molecules would enter the soil and they or their decomposition products would accumulate there.

Chemicals also leak out of roots. Smith (1976) collected root tips from mature trees in a deciduous forest of New Hampshire, leached them in distilled water much as Tukey treated shoots, and analyzed the leachate. Table 5–5 summarizes the impressive results (and Smith was careful to state that they are conservative estimates of root exudation because older roots were not examined, nor were many types of soluble organic materials searched for, and volatile substances were ignored).

Table 5-5. Root exudate materials (kg ha^{-1} yr^{-1}) released by young roots of the three principal tree species in a northern hardwood forest in New Hampshire. (From Smith 1976. Copyright 1976 by the Ecological Society of America.)

Category of material	Birch (*Betula alleghaniensis*)	Beech (*Fagus grandifolia*)	Maple (*Acer saccharum*)
sugars	1.66	1.27	0.25
amino acids/amides	0.22	0.29	0.05
organic acids	2.50	4.18	0.62
cations	23.56	18.52	4.90
anions	5.64	2.14	0.65
total	33.58	26.40	6.47

In nature, these substances would diffuse from the roots into the soil and thus the soil beneath each species would be distinctive.

Plant products may affect the activity of soil bacteria. Abandoned crop land in Oklahoma reverts to prairie through a sequence of successional communities that may require 30 or more years (Figure 5-9). Elroy Rice and his coworkers (Rice 1964, 1965; Blum and Rice 1969; Olmstead and Rice 1970; Wilson and Rice 1968) examined whether or not the course of succession could be directed by chemical interactions, in particular, interactions affecting nitrogen-fixing and nitrifying bacteria. They made aqueous extracts of flowers, leaves, roots, and stems of 14 pioneer weed species, and tested them for inhibitory effects on strains of *Azotobacter, Nitrobacter,* and *Rhizobium.* All of the extracts proved to be inhibitory to one or more of the bacterial strains. They identified the inhibitors in three of the weed species as chlorogenic acid and gallotannin, both polyphenols. Chlorogenic acid is a strong inhibitor of several enzyme systems (phosphorylase, peroxidase, and oxidase), and gallotannins are excellent protein precipitants. Soils beneath two weed species were found to contain large quantities of gallic and tannic acids (over 600 ppm tannic acid; in laboratory tests, less than 300 ppm tannic acid was effective in reducing symbiotic nitrogen fixation).

Rice claimed that the pioneer weed species had less of a demand for nitrogen than later successional species and that the weed stage prolonged itself by inhibiting certain bacteria, thus reducing nitrogen availability in the soil. However, the story is not ecologically complete. Chlorogenic acid and gallotannins are widespread in the plant kingdom (Robinson 1967) and might be expected in climax prairie grasses as well as in pioneer weeds. Indeed, Rice's data show that extracts of the climax prairie perennial grass *Andropogon scoparius* were just as toxic to the bacteria as many weed extracts. Further, if the inhibitors are stable in soil (and Rice presented some data to indicate this is so), then why is the weed stage not prolonged beyond 2–3 years? Possibly the weed stage is short because chlorogenic acid stimulates the growth of one of the plants in the second community, wire grass *(Aristida)*, even as it depresses the growth of the weedy pioneers. Not all plants are equally sensitive to chlorogenic acid.

The reverse of plants inhibiting bacteria can also occur: by-products of bacterial metabolism can inhibit higher plants. Brian (1957) listed 38 antibiotics which affect germination and plant growth in low concentration (1–10 ppm); many of these are thought to be formed and

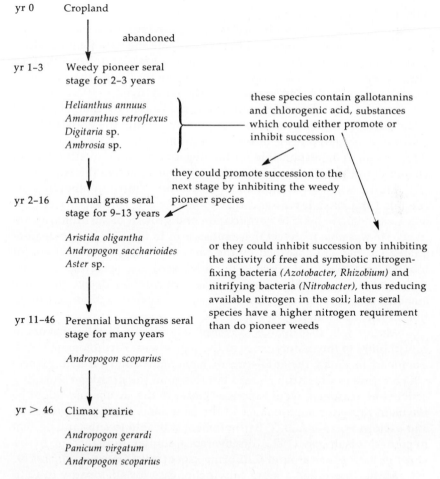

Figure 5–9. The sequence of plant succession in abandoned cropland of Kansas and Oklahoma, and the effect of allelopathic interactions detected by Rice. (Based on Booth 1941 and Perino and Risser 1972. By permission.)

released in the normal soil system. Although of relatively high molecular weight, they can be taken up by roots and translocated through a higher plant. Among his list are a number of metabolic inhibitors of great specificity and potency.

Allelopathy

A number of researchers have reported evidence for chemical control of plant distribution, spatial associations between species, and the course of community succession.

One of the most complete studies is that by Muller (1966) on the spatial relationship between coastal sage (*Salvia leucophylla*) and annual grassland in the Santa Ynez valley of southern California near Santa Barbara. A number of shrub species, including sage, dominate the foothills, while annual grasses and herbs dominate the valley floors. However, patches of sage shrubs may occur in the grassland. Beneath those shrubs, and for 1–2 m beyond the shrub canopy limits, the ground is nearly bare of herbs and grasses (Figure 5–10). Even 6–10 m from the canopy, annuals are stunted. Stunting is not caused by competition for water since shrub roots do not penetrate that far into the grassland, and stunting is observed even in the wettest times of the year. Soil factors do not seem to be responsible for the negative association either: major chemical and physical soil factors do not change across the bare zone.

Muller was able to show that *Salvia* shrubs emit a number of volatile oils from their leaves and that some of these (principally cineole and camphor) are toxic to germination and growth of surrounding annuals. He was able to detect these substances in the field and to demonstrate that they are adsorbed by the soil and can be retained there for months, and that they are able to enter seeds and seedlings through their waxy cuticles. However, he was not able to detect the same amounts of oils in natural soils that were necessary to produce inhibition in the laboratory.

Muller was also unable to completely eliminate other factors as contributors to the maintenance of the bare zones. Bartholomew (1970) examined in detail the influence of mammalian and bird herbivores which reside in the shrub clumps but forage in the grassland. Foraging activity, he reasoned, would increase closer to the shrubs and might be the main cause for maintenance of the bare zone. From seed predation and exclosure experiments, Bartholomew was able to substantiate that hypothesis. Halligan (1973) discovered a similar influence of herbivores in bare zones around California sagebrush (*Artemisia californica*).

Muller presented a more convincing case for allelopathy in California hard chaparral, a dense, scrubby vegetation dominated by such taxa as chamise (*Adenostoma fasciculatum*). Substances released from leaves of chamise clearly inhibited the germination and growth of herbs because all other environmental factors, such as light, moisture, and soil fertility, were successfully eliminated by experimental manipulation (Christensen and Muller 1975). The typically bare ground beneath hard chaparral is shown in Figure 5–11.

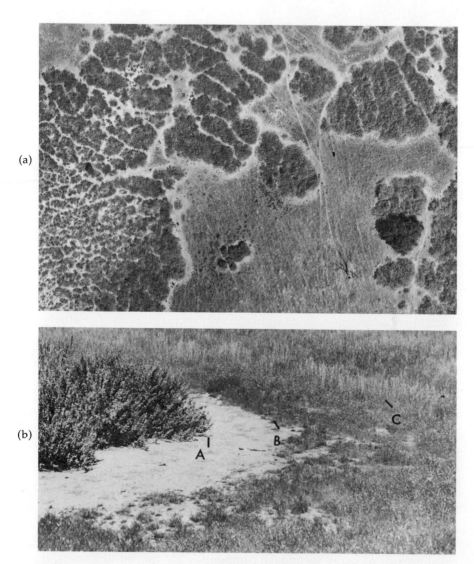

(a)

(b)

Figure 5–10. Nearly bare zones between soft chaparral and annual grassland near Santa Barbara, California. (a) Aerial photograph of *Salvia leucophylla* shrubs and adjoining grassland; the light bands beneath and next to the shrubs are devoid of all but a few species of small herbs. (b) Ground view of the same phenomenon. A indicates soft chaparral, B indicates the edge of the bare zone, and C indicates the grassland. (Courtesy of C. H. Muller.)

Figure 5-11. Photograph taken beneath the chamise shrub canopy of typical hard chaparral, showing the near absence of understory plants.

A few researchers have concerned themselves with the role of chemicals in directing community succession, and this aspect is discussed in the chapter on succession, Chapter 10.

Allelochemics at the Producer-Herbivore Level

For a given herbivore, all species of plants are not equally palatable. Many species are rejected totally, some are eaten preferentially, and others are eaten only when preferential species are absent. Grazing selectivity is easy to observe in large herbivores, but it is less obvious, though just as common, in small herbivores such as insects. The attainment of nonpalatability would confer major selective advantage to a plant species. It is quite likely that palatability depends upon the quality and quantity of certain metabolic by-products (Wallace and Mansell 1976).

In lengthy reviews, Brower and Brower (1964) and Ehrlich and Raven (1965) summarized the food preferences of butterfly groups throughout the world. Ehrlich and Raven concluded that many taxonomic groups feed exclusively on one to several families of flowering plants, and that ". . . secondary plant substances [alkaloids, quinones, essential oils, glycosides, flavonoids, and crystals] play the leading role in determining patterns of utilization. This seems true not only for butterflies but for all phytophagous groups and also for those parasitic on plants."

This may be too sweeping a statement. Although the physiological functions of many secondary compounds are indeed unknown, it may not be fair to take them out of the context of metabolic pathways. They may be intermediates in the synthesis of pigments, hormones, or other compounds of known function. There are conflicting reports on their rate of turnover (a high turnover rate for a substance indicates that it is an intermediate, not an end product) (see Seigler and Price 1976).

However, some plant products, whether they are intermediates or stable end products, do affect grazing intensity. At the same time, coevolution on the part of herbivores can include production of enzymes which break down the normally unpalatable or toxic plant products (Krieger et al. 1971).

Some temperate or tropic zone trees may lose their leaves as an adaptation preventing the build-up of herbivore populations and not for the avoidance of cold or drought (Janzen 1975). Even while leaves remain on the tree, their chemical composition can change, further limiting their suitability to herbivores. Feeny and Bostock (1968) found such a relationship between larvae of the winter moth *Operophtera brumata* and leaves of the deciduous oak *Quercus robur*. Larvae feed in spring on young, expanding leaves. They are intolerant of high tannin content in leaves, and the authors found that tannin content was minimal in spring, about 0.5% on a dry weight basis. By July, tannin reaches 2%, and by the end of September, just prior to leaf drop, it reaches 5%. Feeny and Bostock speculated that ". . . larvae of the winter moth will not mature satisfactorily on oak leaves 2 weeks older than those on which the larvae normally feed. If they hatch too early, they starve before the leaves appear; if they hatch too late, they are defeated by the tannins. . . ."

If inhibitory chemicals that act as herbivore defenses are in fact stable end products, rather than intermediates of metabolic pathways, then they represent an energetic cost to the plant. Cates (1975), among others, addressed the questions about how much this cost may reasonably be, and how herbivores respond. He studied wild ginger *(Asarum caudatum)*, an herbaceous perennial which grows in moist conifer forests from British Columbia, Canada, to the Santa Cruz Mountains of California. Within that range it is fed on by the slug *Ariolimax columbianus*, although slug density is higher in the north than the south. Field observation, coupled with feeding experiments, revealed two ecotypes of wild ginger. One was relatively palatable; in addition it had a faster growth rate, flowered earlier in the year, produced more seeds per fruit, and grew on moister sites than the other ecotype, which was relatively unpalatable. In laboratory trials, the slug removed more than 60% of the seeds or vegetative matter of the palatable ecotype, but less than 25% of the seeds or vegetative matter of the unpalatable ecotype. Cates hypothesized that the slow growth of the unpalatable ecotype was due to the cost of building chemical defenses (although its habitat preference of drier sites would further depress growth). He further predicted that if slug grazing pressure was not an environmental factor, the population density of the unpalatable ecotype would decline because the population density of the palatable ecotype would increase. Using the same slug as a good example of a "generalized" herbivore (one whose food plants cover a broad taxonomic spectrum), Cates and Orians (1975) tested the palatability of 100 plant species from the lowlands of western Washington; some were representative of pioneer or early successional communities (r-selected), while others were representative of later seral stages or of climax communities (K-selected). Pioneer species proved to be more palatable, supporting their hypothesis that r-selected taxa invest relatively little energy in herbivore defense, in contrast to K-selected taxa.

r- and K-selected taxa also differ in the type of allelochemic substances they contain. Those annuals which do invest energy in chemical herbivore defense appear to do so only with toxins, while woody perennials tend to produce tannins (Rhoades and Cates 1976). Toxins are small, relatively inexpensive molecules, of molecular weight 500 g mole^{-1} or less, and they are effective in small concentrations. Toxins usually account for less than 2% of leaf dry weight. They interfere with nerve and muscle activity, with hormone activity, or with liver or kidney functions. They are effective against generalist herbivores; specialist herbivores can develop means to detoxify the substances.

Woody perennials may contain toxins, but they invest considerably more energy into the manufacture of tannins. Tannins are large, complex molecules which reduce the digestibility of plant tissue by complexing with starch, proteins, or herbivore enzymes when the tissue is chewed. Tannin content may exceed 6% of leaf dry weight. Tannins are effective against both generalist and specialist herbivores. Only 17% of all annual plants tested have tannins, while more than 80% of all woody perennials have tannins.

Hormones as allelochemic agents

The most sophisticated and specific chemical interaction between plants and herbivores may involve hormones. Feeding preferences in some cases may be related to hormonal control rather than simple palatability. Ferns, for example, are generally less extensively eaten by insects than are flowering plants, and it may be no mere correlation that ferns evolved with some insect groups considerably earlier than the flowering plants. Soo Hoo and Fraenkel (1964) commented that larvae of the southern army worm reject ferns in general, and these authors performed some preliminary feeding experiments with water extracts of ground pieces of Boston fern (Nephrolepis exaltata). Shortly after, Kaplanis et al. (1967) extracted two major molting hormones from pinnae of bracken fern (Pteridium aquilinum): α-ecdysone and β-ecdysone (20-hydroxyecdysone). Ecdysones form a class of compounds similar in structure to cholesterol. They are ordinarily synthesized by the prothoracic gland of larvae, and they promote developmental reactions, such as pupation (Figure 5–12). However, Kaplanis cautioned, ". . . we do not have information on either the significance or function of these steroids in plants. Perhaps . . . these substances . . . interfere with the growth processes of insect predators."

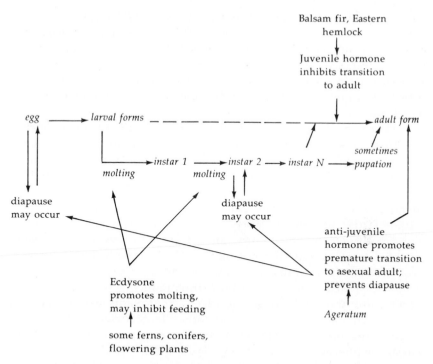

Figure 5–12. The possible role of insect hormones contained in plants, using an idealized insect life cycle.

Ecdysone has since been isolated from more than 10 species of conifers, 20 ferns, and 30 flowering plants, out of 1000 species surveyed (Williams 1970). A total of 28 different plant ecdysones are known, the most ubiquitous being β-ecdysone. The ecological significance of β-ecdysone in plants is unclear. It is not toxic when orally ingested (as feeding larvae would obtain it from a food plant), but there is some evidence that it could be a feeding deterrent in concentrations as low as 1 ppb. Perhaps it serves as a steroid base for other compounds once it is in an insect's metabolic system.

Another developmental hormone is the juvenile hormone, which predominates early in larval life (later, ecdysone predominates). It is a methyl ester of the epoxide of a fatty acid derivative, and there is some evidence that its structure differs in different groups of insects. By a series of coincidences, it was discovered that certain paper toweling prevented the European bug, *Pyrrhocoris apterus*, from developing into sexually mature adults. Instead, an extra one or two larval molts ensued, and all eventually died without being able to complete metamorphosis. The juvenility factor was traced to particular conifers used in American paper pulp, mainly *Abies balsamea*, *Tsuga canadensis*, *Taxus brevifolia*, and *Larix laricina*. The active principle was isolated and characterized. It has a structure similar to that of the juvenile hormone, and has been named juvabione. It proved to be effective on only one family of insects, the Pyrrhocoridae. Williams (1970), who has summarized the pyrrhocorid story, asked, "Have the plants in question undertaken these exorbitant syntheses just for fun? I think not. Present indications are that certain plants and more particularly the ferns and evergreen trees have gone in for an incredibly sophisticated self-defense against insect predation. . . ."

However, the pyrrhocorid story certainly fails to justify such a hypothesis. *Pyrrhocoris apterus* is a native of Europe, and the only plants which produce the very specific juvabione are natives of North America. The two simply do not occur together. Further, the bug and all members of its family feed by sucking the juices of weak herbs; they are not known to feed on any tree species. What ecological significance is there in the pyrrhocorid story?

Finally, and most recently, an anti-juvenile hormone has been extracted from ageratum, a common, annual garden plant *(Ageratum houstonianum)*. It has proven capable of inducing precocious molting of immature insects to sterile adults, thus reducing feeding time and reproduction at once (Bowers et al. 1976). Other plants may also possess such a hormone, and its effects on other insects could additionally include prevention of diapause (critical resting periods) and prevention of sex pheromone production (Figure 5–12).

SUMMARY

Species interactions may be negative or positive, and the spatial distribution of plants may give a first clue to the presence of an interaction. Burkholder recognized 10 types of interactions as being biologically and ecologically reasonable. Three of these appear to be poorly known and one (neutralism) may be rare. Two interactions, competition and amensalism, were discussed in this chapter.

The effect of competition on interacting organisms can be modeled mathematically with the Lotka-Volterra equations, or it can be expressed in evolutionary terms with Gause's Competitive Exclusion Principle, or it can be measured with de Wit's experimental designs which include replacement series plantings and calculation of input/output ratios. However, none of these approaches seems to get us very close to understanding competition in the real world. In addition, many competition studies do not reveal the resource being competed for. Competition may be an important factor in habitat restriction for some plants, such as serpentine endemics and halophytes.

Allelochemic interaction is one form of amensalism, and a review of the literature indicated that it may be an important interaction between plants and soil biota and between plants and herbivores. Allelochemics may play a role in plant distribution, plant succession, and plant palatability. Many substances are passively leached from leaves and roots by percolating rainwater; some of these may be allelochemic. Other metabolic products remain stored in plant tissue, including some which mimic insect hormones. The study of allelochemics as a strategy in herbivore defense, with costs and benefits, has barely begun. Most studies which purport to show allelochemic interactions seem to be incomplete either in a research sense or in making ecological sense; more research is necessary before we can assess the general importance of allelochemics.

CHAPTER 6

SPECIES INTERACTIONS:
Commensalism, Protocooperation, Mutualism, and Herbivory

In Chapter 5, two negative interactions resulting in negative spatial associations were examined. In this chapter, we will consider one other negative interaction and several interactions that result in positive spatial associations. These positive interactions have sometimes been lumped together under the term symbiosis. It will be seen that there is a fine line between a positive and a negative interaction, a line that is easily crossed in the course of time (evolution). Herbivory is an example of an interaction that is usually considered to have negative plant aspects, but biologists are coming to recognize the complex, subtle, and unexpected ways in which some plants can be stimulated by herbivores.

There is also a fine line between obligate interactions, where both partners fail in the absence of one another, and degrees of facultative interactions, where only one partner is benefited or where both are benefited slightly. This line, too, is easily crossed in the course of evolutionary time; the change is usually from a facultative interaction to an obligate interaction, not vice versa.

COMMENSALISM

Commensalism is an interaction which stimulates one organism but has no effect on the other. One example of commensalism is the growth of epiphytes on host trees (Figure 6–1). Epiphytes may be herbaceous perennials, such as orchids, ferns, bromeliads (such as Spanish "moss"), and cacti; or they may be lower plants, such as true mosses, algae, or lichens. In any event, epiphytes gain no nourishment from the host, using it only as a physical place of anchorage. Some have expanded leaf bases or unusual root surfaces which trap and retain rain water for later absorption. Others, like grandfather's beard (*Ramalina* *reticulata,* a lichen that festoons California oaks (Figure 6–1(c)) or Spanish "moss" *(Tillandsia usneoides,* a bromeliad common in the southeastern United States) (Figure 6–1(d)) absorb much of their water from humid air. *Ramalina*'s tissue moisture content fluctuates during the day,

(a)

(b)

Figure 6–1. Examples of epiphytes. (a) The bromeliad *Tillandsia usneoides* (Spanish "moss") in trees along the Gulf Coast. (b) A close-up of *Tillandsia.* (c) The lichen *Ramalina reticulata* (grandfather's beard) on a California oak tree.

(c)

closely reflecting the trend of the relative humidity of the air. Its net photosynthesis is positive only during the early morning hours (Rundel 1974). *Tillandsia*, on the other hand, cannot satisfy its entire water demand from the air even at 80% relative humidity, as Garth (1964) experimentally demonstrated; occasional rain is necessary. Garth found that the distribution of *Tillandsia* correlated very well with the average routes of storm paths which sweep east and north from Mexico, bringing frequent rain (Figure 6–2). The range limit of *Tillandsia* corresponds to areas with an average annual relative humidity of 64% or more.

It is likely that competition with other epiphytes and with the canopy of the host tree for space and light has induced each epiphytic species to specialize its habitat requirements, because censuses of tropical trees have shown that relatively distinct parts of the canopy can be associated with each epiphyte species (Figure 6–3; Longman and Jenik 1974; Janzen 1975). Epiphyte preferences for certain host species seem to reflect such physical attributes as openness of branching and roughness of bark. This is certainly true for Spanish "moss," which festoons cypress and some hardwood trees, but avoids pines. Pine avoidance is due to the sloughability of pine bark, and not to allelopathic or nutrient factors (Schlesinger and Marks 1977).

Epiphytism may easily grade into other types of interactions, such as mutualism and parasitism. Mutualism can result if the epiphyte produces nutrients which can be leached by rainwater running down the trunk and entering the soil around host roots. The upper canopy branches of trees in a Colombian rain forest (2700 m elevation, about 250 cm rain a year) are thickly covered with lichen epiphytes. Richard

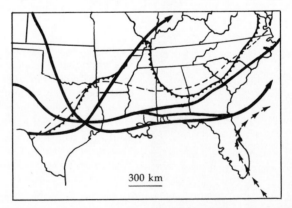

Figure 6–2. Major storm paths in the Coastal Plain of the southeastern United States. The distribution limits of Spanish "moss" (*Tillandsia usneoides*) closely correspond with the dashed line, which shows the isopleth of mean annual relative humidity = 64%. (From Garth 1964. Copyright 1964 by the Ecological Society of America.)

Figure 6–3. Microhabitats of epiphytes within an emergent tree of the tropical rain forest. Small epiphytes are common in zone 1, large epiphytes in zone 2, crustaceous lichen epiphytes in zone 3, and bryophytes in zones 4 and 5. (From *Tropical Rainforest and Its Environment* by K. A. Longman and J. Jenik. Copyright 1974 by Longman, London.)

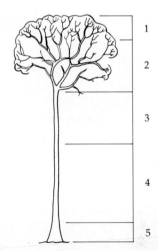

Forman (1975) carefully demonstrated that 86% of these lichens contain the blue-green alga *Nostoc* as the algal symbiont. *Nostoc* is capable of fixing atmospheric nitrogen into organic form, and Forman estimated that it does so in this forest at the rate of 1.5–8.0 kg N ha^{-1} yr^{-1}. This is equivalent to the amount of available nitrogen added to the forest from rainwater alone (N_2 is converted to NO_3^{--} during electrical storms and the nitrate is carried to earth in rainwater). The nitrogen fixed into lichen epiphytes is probably widely distributed as a result of lichen decomposition and perhaps of leaching from the living lichen.

Parasitism can result if the epiphyte's roots penetrate beneath the bark into phloem and xylem and develop absorbing organs called haustoria. There seem to be all possible intermediates between pure epiphytism and pure parasitism. For example, Hull and Leonard (1964; see also a review by Kuijt 1969) experimented with two genera of green mistletoe that parasitize a variety of conifers in the Sierra Nevada Mountains (Figure 6–4). They exposed host foliage to radioactive $^{14}CO_2$, then took autoradiographs of mistletoe foliage. They concluded that the genus *Phoradendron* derived very little if any carbohydrate from its host, while the dwarf mistletoe, *Arceuthobium*, drew heavily on the photosynthate of its host. (They didn't examine water utilization; undoubtedly both genera parasitized the host xylem.) Hull and Leonard found that the degree of parasitism closely correlated with chlorophyll content and photosynthetic rate of mistletoe tissue. *Phoradendron*, the hemiparasite, contained 0.93 mg chlorophyll per gram dry weight, about half the value of the host tissue, but *Arceuthobium* contained only 0.37 mg g^{-1}. In turn, the photosynthetic rate of *Arceuthobium* was very low, compared to the rate of the conifer host. It is not difficult to visualize a progression of increasing parasitism from Spanish "moss" (a typical epiphyte), to *Phoradendron*, then to *Arceuthobium*, and finally to a nearly nongreen parasite such as dodder (*Cuscuta*).

(a)   (b)

Figure 6–4. The mistletoes (a) *Arceuthobium* and (b) *Phoradendron*.

Epiphytes can also become parasites if their size or weight creates strain on the host. For example, the strangler fig (*Ficus aurea* in Florida, other species of *Ficus* elsewhere) germinates in the canopy of a host and begins life as a typical epiphyte. Then, aerial roots grow towards the soil, reach the ground, and thicken, engulfing the host trunk and preventing further growth. At the same time, the canopy of the strangler fig enlarges to overtop the host and deprive it of light. Eventually the host dies (Figure 6–5).

Another example of commensalism is the **nurse plant syndrome** shown by saguaro cactus *(Carnegiea gigantea)* (Figure 6–6). According to long-term studies by Niering et al. (1963) and Steenbergh and Lowe (1969), virtually all successful saguaro seedlings are found in close proximity to a shade-producing object. In a small percentage of cases the object is inanimate, but in most cases it is a perennial plant. Observations by these authors, coupled with experiments by Turner et al. (1966), show that the positive effects of the nurse plant are shading (reduces temperature and rate of soil drying), hiding the young cactus from rodent herbivores, and some protection from frost. In the Tucson, Arizona area, some 15 different species function as nurse plants. The frequency with which they appear as shade plants, however, is an approximate indication of their relative abundance and their percentage cover. In other words, saguaro has no preference for one species over another. Turner et al. showed that soils from beneath different nurse plant species do differ in color, salinity, and nutrient status, but their data indicated that such soil differences had little effect on cactus seedling mortality. So far, there is no evidence to show that the nurse plant is inhibited by growth of the cactus.

Went (1942), Muller (1953), and Muller and Muller (1956) showed that the nurse plant effect extended to some desert annuals. Species of *Malacothrix* and *Chaenactis*, for example, are positively associated with the canopies of burro bush *(Ambrosia dumosa)* and turpentine-broom *(Thamnosma montana)* shrubs. Apparently the reason for the association is that the growth form of these shrubs is a suitable trap for wind-blown organic debris. The debris collects beneath the canopies and this provides a better substrate for the annuals than soil in the open. It may be that seeds of the annuals are also trapped in abundance beneath the canopies. In any case, however, many desert annuals are restricted to association with a shrub. In the Great Basin desert of the western United States, bitterbrush *(Purshia tridentata)*, shadscale *(Atriplex confertifolia)*, and winterfat *(Eurotia lanata)* seedlings may also require nurse plants (West and Tueller 1971).

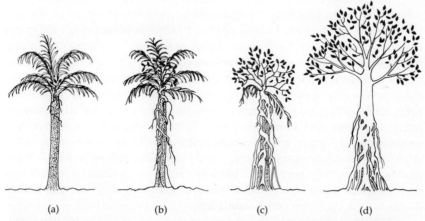

| (a) | (b) | (c) | (d) |

Figure 6–5. Four stages in the establishment of a strangler fig, *Ficus leprieuri,* on the palm *Elaeis quineensis.* (a) The young fig germinates high up in the palm and sends aerial roots down toward the soil. (b) The roots reach the soil and the fig shoot begins to expand. (c) The fig overtops the palm and the palm begins to senesce. (d) The palm has died but the fig tree remains. (From *Tropical Rainforest and Its Environment* by K. A. Longman and J. Jenik. Copyright 1974 by Longman, London.)

Figure 6-6. The nurse plant syndrome: saguaro cactus *(Carnegiea gigantea)* next to a palo verde *(Cercidium floridum).*

PROTOCOOPERATION

Protocooperation is an interaction which stimulates both partners, yet is not obligatory because stable growth continues in the absence of the interaction. One example of protocooperation is root grafts between members of the same or different species. As the roots of some trees grow through a soil and come in contact with each other, a natural graft or union may form. Apparently, this is a more common event than previously supposed: more than 160 tree species are known to form such grafts, and perhaps a fifth of them can form interspecific grafts (Graham and Bormann 1966). Some grafts can even be intergeneric: birch *(Betula alleghaniensis)* with elm *(Ulmus americana),* birch with maple *(Acer saccharinum),* and *Santalum album* with *Eugenia jambólana.* Grafting may occur between trees on relatively mesic sites or on sites as dry as chaparral slopes or woodland with rainfall below 50 cm a year (Saunier and Wagle 1965).

If both partners are equally successful and of the same life form, the relationship is protocooperation, with a mutual, balanced exchange of photosynthate (Figure 6-7(a)). There is evidence that hormones may also be transferred, resulting in more uniform physiognomy such as synchronous spring bud break. If one partner is smaller and suppressed, however, the relationship may be parasitic, with significantly more photosynthate going into the smaller tree from the larger than vice versa (Figure 6-7(b)).

Woods and Brock (1964) hypothesized that mycorrhizal hyphae (fungal mycelium filaments) may similarly link one tree with another beneath a forest soil surface. They found rapid translocation of labeled (radioactive) ^{45}Ca and ^{32}P from a red maple stump to the leaves of 19 other taxonomically diverse trees and shrubs in a North Carolina forest (Figure 6–8). They concluded, ". . . it would seem logical to regard the root mass of a forest . . . as a single functional unit. Inability of a new, invading species to enter into a 'mutual benefit society,' one in which minerals and other mobile materials are exchanged between roots, could have a negative survival value."

MUTUALISM

Mutualism is an obligate interaction; the absence of the interaction depresses both partners. Closely related organisms do not seem to form such an interaction, for common examples of mutualism are lichens (algae + fungi), mycorrhizae (fungi + higher plants), symbiotic nitrogen-fixation (bacteria or blue-green algae + higher plants), pollination (insects, birds, or mammals + flowering plants), zoochory (animal dispersal of plant propagules), and myrmecophytes (ants + woody plants).

Mycorrhizae

Mycorrhizae (singular, mycorrhiza) are fungal associations with the roots of higher plants. In some cases, the fungus covers the root exterior near the tip of the root with a thick mantle of hyphae; additional hyphae extend as far as 8 m out in all directions from the root

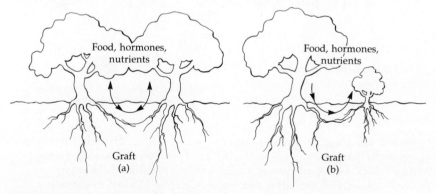

Figure 6–7. Two relationships between trees connected by a root graft, depending on the relative size of the trees. (a) The partners are both overstory trees and the flow of nutrients is equal in both directions, and so the interaction is protocooperation. (b) One partner is an understory tree and it parasitizes the larger tree.

into the soil, and other hyphae penetrate between cortical cells of the host root to form a nutrient absorbing network (Figure 6–9). These latter hyphae do not form haustoria that penetrate into the cells, but contact with root cells is nevertheless very close and metabolites are transferred in both directions (Scott 1969). Radioactive $^{14}CO_2$, for example, can be fixed in the leaves of the higher plant by photosynthesis and later be detected in the fungus. Roots passively exude such nutrients as amino acids, and these enter the fungus. Radioactive isotopes of phosphorus, calcium, and potassium can be shown to be taken up in greater amounts by plants with mycorrhizae than by plants without them, and so the growth of the higher plant is stimulated (Table 6–1). The relationship, then, is mutualistic. It is not, however, very species-specific. Even along one 10 cm segment of the root of one plant, as many as five different fungal species can form mycorrhizal unions. One tree may host as many as 100 different fungal species (Harley 1969; Hacskaylo 1971; Marks and Kozlowski 1973). Further, hyphae from one mycorrhiza can extend through the soil from one species to another.

The type of mycorrhizae just described are classed as **ectomycorrhizae**. The biology and ecology of ectomycorrhizae have most recently been reviewed by Harley (1969), Hacskaylo (1971), and

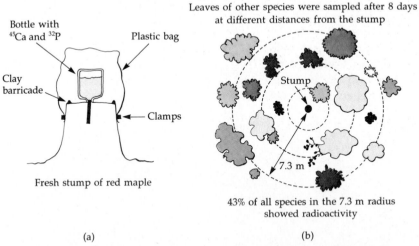

Bottle with ^{45}Ca and ^{32}P

Plastic bag

Clay barricade

Clamps

Fresh stump of red maple

(a)

Leaves of other species were sampled after 8 days at different distances from the stump

Stump

7.3 m

43% of all species in the 7.3 m radius showed radioactivity

(b)

Figure 6–8. The transfer of radioactive ^{45}Ca and ^{32}P from a freshly cut stump of red maple (see detail in (a)) to the leaves of surrounding woody plants. Within 8 days, 43% of all species within a radius of 7.3 m of the stump showed radioactivity, including red maple, Carolina hickory, mockernut hickory, fringe tree, persimmon, ash, holly, juniper, sweet gum, tulip tree, black gum, Virginia creeper, white oak, red oak, black oak, and blackjack oak. (From Woods and Brock 1964. Copyright 1964 by the Ecological Society of America.)

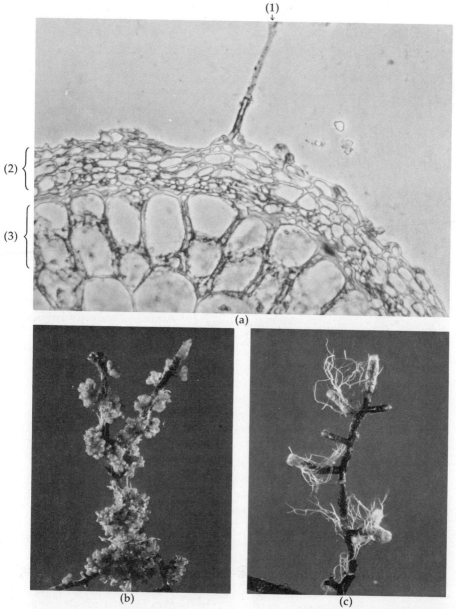

Figure 6-9. Ectomycorrhizae. (a) cross section of a root, showing (1) hyphae branching off into the soil, (2) the fungal mantle coating the outside of the root, and (3) hyphae forming a net between root cortex cells. (b) and (c) show the various shapes that mycorrhizae may take, depending on the species involved. The fungal mantle covers short, club-shaped lateral roots. ((a) courtesy of F. H. Meyer; (b) and (c) courtesy of B. Zak in Marks and Kozlowski 1973.)

Table 6-1. Growth and nutrient content of white pine seedlings *(Pinus strobus)* grown for one season with and without ectomycorrhizae. (Reprinted by permission from Gray 1959. Copyright 1959 by Holt-Dryden.)

Characteristic	With	Without
Seedling dry weight (mg)	405	302
Root/shoot ratio	0.78	1.04
N (% dry wt)	1.24	0.85
P (% dry wt)	0.20	0.07
K (% dry wt)	0.74	0.43
N uptake rate (mg N per mg dry wt root)	0.03	0.02
P uptake rate	0.005	0.001
K uptake rate	0.02	0.01

Marks and Kozlowski (1973). Ectomycorrhizae are commonly found on northern temperate zone trees in the oak, pine, willow, walnut, maple, basswood, and birch families. The fungi are typically Basidiomycetes.

Another large group of mycorrhizae are classed as **endomycorrhizae**. The fungal hyphae in this case do not form a sheathing mantle. Hyphae within the cortex do penetrate root cells, and other hyphae extend out into the soil, but not to the extent of ectomycorrhizae. The biology of endomycorrhizae has been recently reviewed by Gerdemann (1968) and Sanders et al. (1976). Endomycorrhizae are commonly found associated with a great diversity of higher plant life forms, ranging from grasses (all grasses appear to be endomycorrhizal) to orchids, heath and heather shrubs, and a variety of tropical and southern temperate zone trees. It is easier to list the families of higher plants which do not form endomycorrhizae than to list those which do: Cruciferae, Cyperaceae, Juncaceae, Pinaceae, and aquatic vascular plants do not form endomycorrhizae. The fungi may be Phycomycetes, Ascomycetes, or Basidiomycetes.

Ectomycorrhizae in particular appear to permit host plants to grow well in soils which otherwise would not permit good growth. The acidic, leached, nutrient-poor soil beneath the boreal coniferous forest is an example; all the dominant trees are ectomycorrhizal. Revegetation of mine spoil in this country occurs faster if planted conifers are innoculated with the hyphae of particular acid-tolerant mycorrhizae, as revealed by recent work of the U.S. Forestry Sciences Laboratory at Athens, Georgia. Indeed, experiments have shown that the mycorrhizal relationship does not maintain itself unless the environmental conditions are such that they limit the growth of fungi as independent organisms (Scott 1969).

In addition to improving host nutrition, mycorrhizae are essential to the normal development of some species. Seedlings of orchids and heaths fail to survive if the fungi are absent, and plantings of pines in areas where they have historically never grown (such as Australia), do poorly unless the mycorrhizal fungi are introduced to the soil. We noted in the last section that mycorrhizae may also serve to link the root systems of all members of a community, in this way evening out soil resources and making it difficult for other species to invade. The fungal floras of soils are known to change during succession, and to differ from vegetation type to vegetation type. There is some evidence that the narrow ecotone between grassland and forest in central North America is maintained, at least in part, by such fungal differences: tree mycorrhizal fungi are absent from grassland soils and tree seeds which are blown or carried into the grassland are incapable of producing seedlings that can compete with the grasses without their mycorrhizal association (White 1941; Goss 1960).

Symbiotic Nitrogen Fixation

Nitrogen fixation is the conversion of atmospheric nitrogen gas into organic ammonium. Only certain prokaryotic organisms are capable of this process. Some of these prokaryotes are free-living while others live in close association with eukaryotes, receiving sugars and other energy-rich molecules from the eukaryotic symbiont. Nitrogen fixation requires energy in the form of ATP and a local anaerobic environment:

$$4N_2 + 6\,H_2O \xrightarrow[\substack{\text{the enzyme Nitrogenase} \\ \text{no oxygen}}]{ATP \longrightarrow ADP} 4\,NH_3 + 3O_2$$

The association of *Rhizobium* bacteria with the root nodules of legumes is well known; however, other symbioses may have an equal or greater ecological importance. Species of the blue-green algae *Nostoc* and *Anabaena* can become associated with bryophyte gametophytes, root nodules of cycads, leaf tissue of the angiosperm salt marsh plant *Gunnera*, and leaf tissue of the aquatic fern *Azolla* (Peters 1978). The *Azolla-Anabaena* symbiosis has economic importance in addition to ecological significance. Agricultural trials in California have shown that three-fourths or more of all nitrogen requirements of rice can be met by cultivating *Azolla* in rice paddies.

Certain soil actinomycetes (filamentous bacteria that resemble fungi) are capable of invading the roots of some higher plants, causing elongate nodules to form. Within the nodules, nitrogen fixation occurs at a rate comparable to that of legume nodules. Some 160 species of woody plants, mainly of the temperate zones, have been shown to possess such nodules (Table 6–2). Many of the plants are pioneers in succession and occur in open habitats on acidic, saline, or sandy soils.

Pollination

Pollination is a very specialized form of mutualism which has developed in flowering plants; it may be the key to much of the variation and specialization in morphology of the angiosperms (Macior 1973). The transfer of pollen from the stamen to the stigma is essential to reproduction in outcrossing species of plants. Even plants that are self-fertile require the transfer of pollen because the male and female functions are not present in the same tissue. There is a physical space or gap between them which must be spanned for pollination to be effected.

A plant species is usually morphologically adapted to the specific behavioral and morphological characteristics of its pollinator, which is often an insect, bird, or bat. Adapted plants usually exhibit some or all

Table 6–2. Plants that form root nodules with nitrogen fixing actinomycetes. The third column shows the fraction of all species in the genus that exhibits nodules. (From Torrey 1978. Reprinted, with permission, from the September 1978 *BioScience* published by the American Institute of Biological Sciences.)

Genus	Family	Nodulated species/Total species
Alnus	Betulaceae	33/35
Casuarina	Casuarinaceae	24/25
Ceanothus	Rhamnaceae	31/55
Cercocarpus	Rosaceae	4/20
Colletia	Rhamnaceae	1/17
Comptonia	Myricaceae	1/1
Coriaria	Coriariaceae	13/15
Discaria	Rhamnaceae	2/10
Dryas	Rosaceae	3/4
Eleagnus	Eleagnaceae	16/45
Hippophae	Eleagnaceae	1/3
Myrica	Myricaceae	26/35
Purshia	Rosaceae	2/2
Rubus	Rosaceae	1/250
Shepherdia	Eleagnaceae	3/3

of the following characteristics: (a) petals, sepals or inflorescence attractive, either visually or olfactorily or both; (b) pollen grains often sculptured or sticky, sometimes massed together; (c) nectar or pollen of nutritive value to the pollinator; and (d) attractants available at blooming time (anthesis), which is correlated to activity patterns of the pollinator.

In temperate regions, insects are of primary importance as pollinators, especially insects of the order Hymenoptera (Moldenke 1975). Bees, both hive bees and the higher bumblebees (Apidae), have more blossom intelligence than any other insect group: they have a greater capacity for remembering blossom characteristics, discriminating among them, and remaining faithful pollinators (Faegri and van der Pijl 1971).

The significant characteristics of bee pollinated flowers are: (a) bilateral symmetry (zygomorphy); (b) landing platform, mechanically strong flower, often partly closed with sexual organs concealed; (c) bright colors, often blue or yellow, often with nectar guides; (d) moderate quantities of nectar, nectar sometimes partly concealed; and (e) many ovules per ovary and few stamens. The adaptations of the pollinator include: (a) the possession of good color reception in the blue, yellow, and ultraviolet ranges (Jones and Buchmann 1974); (b) a high degree of intelligence and a long memory; and (c) often a long proboscis capable of probing for nectar.

In the tropics, there is a dearth of insect pollen vectors, and birds assume a more important pollinator function. The sunbirds (Nectarinidae) of Africa and Asia, the honey-creepers (Drepanididae) of Hawaii, and the hummingbirds (Trochilidae) of both Americas, are all recognized as important pollinators. Many tropical plants have flowers that are adapted to bird pollination (ornithophily), for example, some species in the genera *Eucalyptus* and *Erythrina*. Old World flowers pollinated by birds are borne upon a stout stem or stout inflorescence branch to provide a landing place, because Old World birds which pollinate flowers do not hover as do the hummingbirds which can pollinate pendulous flowers. Table 6–3 presents a summary of the coevolved traits of birds and ornithophilous flowers which are important to their mutualistic interactions.

The obvious advantage conferred upon entomophilous (adapted for insect pollination), ornithophilous, or chiropterophilous (adapted for bat pollination) flowers is the possibility of pollen dispersal far from the host anther, allowing for outcrossing and therefore genetic variability. By providing a nutritional source for animals, flowers have influenced the evolution of pollinators and have evolved very specific traits to ensure pollination. The result is a series of mutualistic interactions, each with its own set of special adaptations.

Table 6–3. A comparison of flowers pollinated by birds and birds which pollinate flowers. (Modified from Faegri and van der Pijl, *The Principles of Pollination Ecology*, Pergamon Press, Oxford, 1971.)

Flowers pollinated by birds	Birds which pollinate flowers
Diurnal anthesis	Diurnal activity
Vivid, highly contrasting colors, often scarlet	Visual sensitivity for red, not for ultraviolet
Lip or margin absent or curved back, flower tubate and/or hanging	Too large to alight on the flower itself
Hard flower wall, filaments stiff or united, protected ovary, nectar stowed away but visible	Hard bill
Absence of odors	Scarcely any sense of smell
Nectar very abundant	Large, requiring a large amount of food
Capillary system bringing nectar up or preventing its flowing out	Long bill and tongue
Usually deep tube or spur, wider than in flowers pollinated by butterflies	Long bill and tongue
Distance between nectar source and sexual organs may be large	Large, long bill; large body
Nectar guide absent or plain	Intelligent in finding an entrance

HERBIVORY

Herbivory is the consumption of all, or part, of a plant by a consumer. If the consumer category is looked at very broadly, then it includes (a) parasitic microbes or plants, (b) saprophytic microbes which decompose dead tissue, (c) browsing and grazing animals which consume only a part of woody and herbaceous plants, respectively, and (d) animals which consume whole plants (or such a significant part of them that death ensues) or plant propagules (future, potential whole plants). Grazers and browsers are sometimes considered parasites, and whole-plant consumers can be called predators (Slobodkin et al. 1967). Wiegert and McGinnis (1975) additionally suggested that consumers of living tissue be called **biophages**, and consumers of dead tissue be called **saprophages** (Figure 6–10).

Herbivory may also be a part of other interactions already discussed in this chapter. For example, allelochemics may mediate or determine the amount and kind of plant material eaten, and herbivores such as pollinators, ants, or rodents which cache seeds, enter into mutualistic relationships with their food source.

The Impact of Herbivory: Consumption

A great number of studies have investigated the impact of herbivores on plant communities, and their results have been summarized in several recent publications (Odum 1971; Krebs 1972; Ricklefs 1973; Chew 1974; Larcher 1975; Whittaker 1975).

A few definitions are in order. **Gross primary productivity (GPP)** is the total amount of chemical energy fixed by photosynthesis for a given unit of land surface for a given unit of time. Typically, GPP is expressed as calories per square meter per year, but it may be expressed in biomass units also (recall from Chapter 4 that there are 4.2 kcal per gram dry weight of plant tissue, on the average). **Net primary productivity (NPP)** is gross primary productivity minus energy lost through plant respiration, which is equivalent to chemical energy stored or accumulated per unit area per unit time. For most terrestrial vegetation, NPP = 30–70% of GPP. Finally, **detritus** is a synonym for litter; it is dead plant material. If we are concerned with an annual plant, then virtually the entire plant body can be considered litter at the end of one year—only the seeds remain living. If we are concerned with a deciduous tree, then some bark, twigs, flowers, aborted fruit, and all the leaf matter produced in spring and summer (except the fraction already eaten by herbivores) will be returned to the soil surface as litter.

About 10% of NPP is consumed by biophage herbivores for typical terrestrial vegetation. The percentage varies with vegetation: 2–3% for desert scrub or arctic/alpine tundra, 4–7% for forest, 10–15% for temperate grasslands with minimal grazing, 30–60% for African grasslands or grasslands managed for domesticated animals. The percentage can also fluctuate from year to year: episodic outbreaks of herbivores such as tent caterpillars, locusts, budworms, rabbits, voles, lemmings,

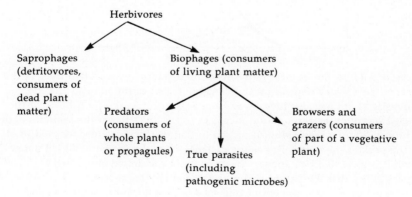

Figure 6–10. Categories of herbivores.

or pathogenic microbes may raise the consumption to 50–100% of NPP. Limited data show that vegetation may withstand considerable loss to herbivores without lasting damage. For example, tree wood growth rate is not reduced until more than 50% of the canopy's leaf surface is lost. Certainly, however, death can and does result from some herbivore outbreaks, particularly if they persist for more than one growing season.

If NPP is calculated for seed production only, then consumption by seed predators is typically well above 10% of NPP, and it often reaches 100%. It may be that herbivores exert their principal effect on vegetation in this way. Janzen (1970) concluded that intensive seed predation in the tropics may be the single most important factor which regulates tree populations. In the tropical rain forest, a high diversity of tree species coexist and neighboring trees are rarely of the same species. In addition, many seeds contain complex by-products which narrowly limit the kind of herbivores that can eat them. Janzen believes that the scattered tree pattern is a natural consequence of seed predation being heaviest close to the parent tree and lower further away. Figure 6–11 shows the kind of balance he visualizes between the dispersal ability of the seed and the intensity of predation. There is a balance point where the probability of seed life and germination is highest.

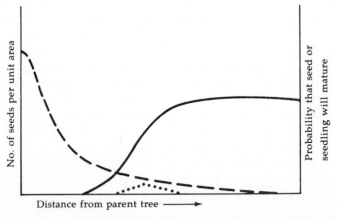

Figure 6–11. In the tropical rain forest, the distance between two trees of the same species may depend on the density of seed-eating animals. As the density of animals declines, away from a parent tree, the probability of the seed surviving to germination increases (solid line). But at the same time, the probability of a seed being dispersed to that distance declines (dashed line). The peak of the dotted line is the distance from the parent at which a new adult is most likely to appear; the area under the dotted curve is the probability that the parent will reproduce at all. (From Janzen, D. H. 1970. Herbivores and the number of tree species in tropical forests. *American Naturalist* 104:501–528. Copyright © 1970 by The University of Chicago.)

Finally, abscission of tissue (loss of plant energy ultimately to saprophage herbivores) can range from 20% of GPP in trees to over 90% in annual herbs. Except for the tropics, the annual rain of litter is not entirely consumed each year. Litter turnover time (time for complete decay) is two to three years in the temperate deciduous forest and five to fifteen years in the boreal coniferous forest. Consequently, the litter layer increases from year to year until some episodic disturbance such as fire brings the system back to equilibrium, releasing the store of trapped nutrients. As discussed in Chapter 15, ground fires result in a flush of growth of the remaining vegetation, indicating that some of the released nutrients are quickly cycled back into the biota. The rate of litter consumption depends on soil temperature, soil moisture, and on the chemical properties of the litter. Conifer foliage, for example, is acidic and is high in lignin, and both properties depress the activity of bacterial populations.

Figure 6-12 summarizes the average drain by herbivores on GPP for a typical plant or plant community. If GPP is given a value of 100 units per square meter per year, then respiration accounts for 50 units, biophage herbivores account for 15 units, litter (ultimately saprophage herbivores) accounts for 23 units, and stored plant energy or growth accounts for the remaining 12 units.

The Impact of Herbivory: Function

Many ecologists (see reviews by Noy-Meir (1975), Lee and Inman (1975), and Mattson and Addy (1975)) have concluded that the role of herbivores in temperate ecosystems is rather minor, that herbivores do not transfer much energy from one trophic level to another, and that their function is merely to dampen yearly oscillations in plant productivity, thus performing a homeostatic role. But such conclusions may be a bit too conservative, for there may be a more pervasive, positive influence of herbivores on plants. Robert Chew (1974) has written a stimulating review that shows what this influence might be (see also Chapter 11).

Herbivory is usually thought of as a negative interaction on the part of plants, a one-way flow of benefits to consumers. Chew discussed several benefits that may accrue to plants, and among them is the fact that an appropriate intensity of grazing can sometimes result in higher rates of plant photosynthesis and a resulting biomass that surpasses ungrazed, control plots. One reason for this response is that grazing reduces the canopy density to a point where all leaves receive sufficient light for efficient photosynthesis, while the ungrazed canopy has many heavily shaded and unproductive leaves. Also, shoot

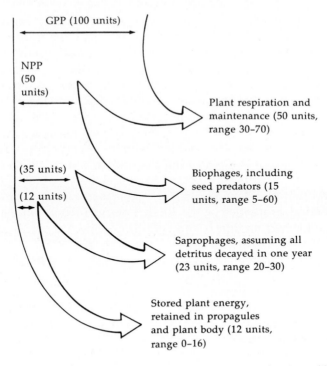

GPP (100 units)

NPP
(50
units)

Plant respiration and
maintenance (50 units,
range 30–70)

(35 units)

Biophages, including
seed predators (15
units, range 5–60)

(12 units)

Saprophages, assuming all
detritus decayed in one year
(23 units, range 20–30)

Stored plant energy,
retained in propagules
and plant body (12 units,
range 0–16)

Figure 6–12. The relative consumption of gross primary productivity (GPP)
and of net primary productivity (NPP) of a perennial plant by respiration
and herbivores. Typical values.

removal affects plant hormone balance and this in turn leads to initiation of new shoots and vigorous shoot growth. One can achieve the
same result with mechanical clipping, but there is some evidence that
animals provide an additional stimulus. Dyer and Bokhari (1976) found
that grasshoppers *(Melanoplus sanguinipes)* feeding on blue grama grass
(Bouteloua gracilis, a common central prairie plant) stimulated new shoot
growth beyond a response from mechanical clipping. They hypothesized that the additional response was due to thiamine in grasshopper
saliva being translocated down to the roots and there stimulating metabolic activity. Cattle are thought to have a similar effect (Reardon et al.
1972). Thiamine is a B vitamin normally produced in leaves and necessary for root growth (Bonner 1940).

Although Chew's ideas are exciting, they are not surprising.
Plants and their herbivores have coevolved over a period of thousands
and millions of years. Surely they have come to terms with each other
in ways as subtle and pervasive as the ways in which ecotypes have
come to terms with their microenvironments (Chapter 3). As an example of plant-herbivore coevolution, Table 6–4 summarizes a series of
plant defenses and herbivore adaptations which penetrate the defense.

The example involves several closely related species of tropical legumes and the associated weevils that feed on beans within their pods.

In the overall view, of course, herbivory has a negative impact on plant growth, reproduction, and survival. The evolution of herbivore defenses, whether chemical or mechanical or phenological, represents energetic costs to the plant. The few cases where herbivores have been

Table 6–4. Apparent plant defenses of some tropical legumes to seed weevil herbivores and corresponding adaptations by the seed weevil species to circumvent these plant defenses. (From Center and Johnson 1974. Copyright 1974 by the Ecological Society of America.)

Plant defense	Weevil adaptation
Gum production by seed pods following penetration of first larva from egg mass; this may push off remaining eggs or drown or otherwise obstruct young larvae.	A period of quiescence in embryonic development in the egg until seed maturation is completed. Resistance to these gummy fluids. Eggs laid singly instead of in clusters.
Dehiscence, fragmentation, or explosion of pods, scattering the seeds to escape from larvae coming through the pod walls.	Oviposition on seeds only after they have been scattered. Attachment of seeds to one another or to the pod valve.
Production of a pod free of surface cracks, which prevents eggs from adhering. Also, indehiscent pods, excluding weevils which oviposit only on exposed seeds.	Oviposition into the soft, fleshy exocarp. Chewing holes in the pod walls in which eggs are then laid. Gluing the eggs to the pod surface.
A layer of material on the seed surface that swells when the pod opens and detaches the attached eggs.	Attachment of the eggs by anchoring strands to allow for substrate expansion. Entry of the larva through the pod into the seed before the pod opens.
Production of allelochemic substances.	Mechanisms for avoidance and detoxification.
Flaking of the seed pod surface, which may remove eggs laid on that surface.	Oviposition beneath the flaking exocarp. Rapid embryonic development or feeding on immature seeds so that entry occurs before flaking begins.
Immature seeds remaining very small throughout the year and then abruptly growing to maturity just before being dispersed.	Entry and feeding upon immature seeds. Delay of embryonic maturation until seeds are mature.
Seeds so small or thin that weevils cannot mature in them.	Utilization of seed contents more completely. Feeding on several seeds.

shown to have a stimulatory effect on plant growth are exceptional. After all, if one plant is eaten back and another that is not eaten overtops it, then it doesn't matter if new growth is stimulated, for the grazed plant has already lost in the competition for light. A plant has a selective advantage over its neighbor only as long as the cost of avoiding herbivory is greater than the cost of being (partly) eaten.

SUMMARY

It is apparent that any given interaction between two organisms may well have attributes of other interactions. Herbivory, for example, can involve mutualism and amensalism. It is also clear that interactions fall on a continuum with many intermediate situations between Burkholder's ten types. For example, there are intermediate stages between commensalism (benefit for one organism, no effect on the other, interaction not obligatory) and parasitism (benefit for one organism, negative effect on the other, obligate interaction on the part of the parasite). We can also conclude that the environment is chemically very complex, with plant, animal, and microbial by-products and exudates playing an integral role in that environment and possibly mediating such important interactions as amensalism, mutualism, and herbivory.

Examples of commensalism are tropical epiphytes and the nurse plant requirement of several desert plants. Little ecological information on epiphytes exists. Lichen epiphytes which contain nitrogen-fixing algal symbionts may contribute significant amounts of nitrogen to their host, and thus some epiphytic relationships can be mutualistic. Others (for example, the strangler fig) can be parasitic.

Root grafts between trees result in protocooperation or parasitism, depending on the relative vigor of each tree. Root grafts between members of the same species are the usual case, but intergeneric grafts are known. Perhaps we should consider the root system beneath some plant communities to be a unit, a mutual benefit society which evens out the distribution of soil resources and photosynthate. Mycorrhizal connections between plant roots may contribute to the same end. It is possible that mycorrhizal relationships may be partly responsible for the grassland-forest ecotone in North America.

Herbivores can be subdivided into biophages, seed predators, and saprophages, among others, and the drain that each type places on plant productivity is quite different. Some researchers argue that because biophages remove only 10% of NPP (15% of GPP), they play a minor role in ecosystems, but other recent studies show that they may affect plant distribution and stimulate plant productivity. In addition, of course, many herbivores have mutualistic relationships with plants that have long been recognized (for example, pollination).

PART III

THE COMMUNITY AS AN ECOLOGICAL UNIT

In each type of habitat certain species group together as a community. Fossil records indicate that some of these groups (or very closely related precursors) have lived together for thousands and even millions of years. During that time, it is possible that an intricate balance has been fashioned. Community members share incoming solar radiation, soil water, and nutrients to produce a constant biomass, they recycle nutrients from the soil to living tissue and back again, and they alternate with each other in time and space. Synecologists attempt to determine what is involved in this balance between the species of a community and their environment.

First, how is the community to be measured and how are the measurements summarized for maximum, useful information? Second, why do some communities change more rapidly over time than others? How accurately can we predict future changes and reconstruct the past? Are there community traits which transcend traits of the individual species making up the community? A synecological view asks such questions, and the answers take us to a level of complexity beyond autecology.

CHAPTER 7

COMMUNITY CONCEPTS AND ATTRIBUTES

The term **community** is a very general one which can be applied to vegetation types of any size or durability. It can, for example, be applied to one stratum of plants in a very local area, such as the herbs, woody seedlings, and mosses on the floor of a streambank forest; or to a very widespread, regional vegetation type; or to a transitory plot of vegetation undergoing rapid change in the species which compose it; or to very stable vegetation which has exhibited no significant change for hundreds of years.

An **association** is a particular type of community, which has been described sufficiently and repeatedly in several locations such that we can conclude it has: (a) a relatively consistent floristic composition, (b) a uniform physiognomy, and (c) a distribution which is characteristic of a particular habitat. In terms of its basic importance to plant ecology and to the classification of vegetation, the association has been compared to the species of taxonomists. Just as a species is an abstract, somewhat artificial synthesis of many individual plants, so an association is a synthesis of many local examples of vegetation called **stands**.

It is not too difficult for an observant person to measure and describe vegetation in such a way that it can be subdivided neatly into associations (see Chapters 8 and 9). The question is: how natural are these units and how much can their existence be attributed to human bias and the need to classify? Some ecologists imply or state that associations are real, closed entities whose component species are interdependent. We think that this view—the organismic or discrete view of communities—needs more supporting evidence before it can be completely accepted.

IS THE ASSOCIATION AN INTEGRATED UNIT?

The Discrete View

One of the traits of an association, first accepted by the International Botanical Congress of 1910, is a relatively consistent floristic composition. Wherever a particular habitat repeats itself in a given region, the same cluster of associated taxa is found. This does not mean that every species repeats itself, nor even that the majority of taxa are repeated. Some species are extremely widespread, with broad tolerance ranges, and these can be found in many habitats and in many associations. Other species may have narrower range limits but nevertheless a few individuals can be found beyond the normal limits, and they are occasional members of many communities.

According to the procedures of one method of community classification, after accidental and ubiquitous taxa are removed from consideration, certain species will remain that show a large degree of association with each other and with a particular habitat. Associations are defined by the presence of such differential species, and a particular stand is placed in an association if it contains a significant fraction of all those differential species. This method—the table method of Braun-Blanquet and others—is described more fully in Chapters 8 and 9.

If clusters of species do repeatedly associate together, that is indirect evidence for strong interactions between them. Such evidence favors a view that communities are indeed integrated units—that the whole is somehow greater than the sum of its parts, much like an organism is greater than the sum of its cells, tissues, or organs. Some of the interactions described in Chapters 5 and 6, such as mycorrhizae, are also evidence that communities are units. Clements (1916, 1920) metaphorically equated associations with organisms. The pattern of species distribution and abundance predicted by this organismic view is diagrammed in Figure 7-1. The species in an association have similar distribution limits along the horizontal axis and many of them rise to maximum abundance at the same points (noda). The ecotones between adjacent associations are narrow, with very little overlap of species ranges, except for a few ubiquitous taxa found in many associations.

The Continuum View

In 1926, Henry Gleason published a carefully written paper entitled, "The individualistic concept of the plant association." In a letter written 27 years later, he recounted the furor that his proposal generated and the never-never world into which his ideas were cast for some time. Gradually his ideas did gain acceptance: the Ecological Society of America cited him as a "Distinguished Ecologist" in 1953 and as an "Eminent Ecologist" in 1959.

Gleason (1953) sampled forest vegetation along a north-south gradient in the Midwest and concluded that changes in species abundance and presence occurred so gradually it was not practical to divide the

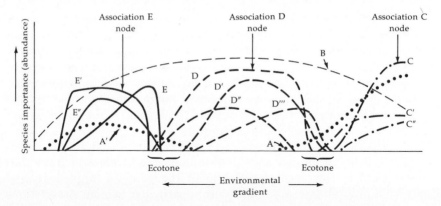

Figure 7–1. Patterns of species importance (abundance) along an environmental gradient as predicted by the discrete view of associations. Clusters of species (C's, D's, and E's) show similar distribution limits and abundance peaks; each cluster of associated species defines an association. Ecotones are narrow. A few species (A and A') have sufficiently broad ranges that they are found in adjacent associations, but in low numbers. A few other species (for example, B) have very broad ranges and are ubiquitous. Species A, A', and B are not differential species.

vegetation into associations. Even within a relatively homogeneous region there were subtle but important differences in the vegetation from one plot to another.

> The sole conclusion we can draw from all the foregoing considerations is that the vegetation of an area is merely the resultant of two factors, the fluctuating and fortuitous immigration of plants and an equally fluctuating and variable environment. As a result, there is no inherent reason why any two areas of the earth's surface should bear precisely the same vegetation, nor any reason for adhering to our old ideas of the definiteness and distinctness of plant associations. . . . Again, experience has shown that it is impossible for ecologists to agree on the scope of the plant association or on the method of classifying plant communities. Furthermore, it seems that the vegetation of a region is not capable of complete segregation into definite communities, but that there is a considerable development of vegetational mixtures. . . .

Gleason's conclusions have been supported by more extensive and sophisticated studies in such widely different vegetation types as desert scrub in Nevada (Billings 1949), upland forest in Wisconsin (Curtis and McIntosh 1951), montane forest in the Smokey Mountains in North Carolina and Tennessee (Whittaker 1956), and conifer forests in the Siskiyou Mountains of California (Whittaker 1960). These studies show that neither dominance of single taxa nor presence and abundance of groups of species changes abruptly along an environmental gradient; noda do not exist (Figure 7–2).

It is now apparent that the method used to sample vegetation will determine whether associations appear as distinct units or as arbitrary segments along a continuum. If sampling is subjective, such that stands are searched for which show the presence of differential species, then the discrete view will be supported, for only similar stands will be sampled, and intermediate stands will be ignored. It is possible that if only climax stands are sampled, many intermediate stands will also be ignored. If sampling is done less subjectively, ecotonal stands will be included and all the stands will be seen to lie along a continuum. Exceptions to the continuum view do exist, of course, where there are sudden discontinuities in the environment, such as a change in soil parent material, a change in elevation or slope aspect, the advance of fire, or the presence of landslide rubble.

The individualistic nature of species distribution is indirect evidence that communities are not larger than the sum of their parts. It suggests that the level of interactions and interdependence is relatively low, or at least nonspecific. For example, forest floor herbs may be well adapted to the phenology, shade cast, and leaf chemistry of an overstory canopy, but the exact species composition of that overstory is not critical. This was shown in a large scale, natural "experiment" 50 years ago. Between 1906 and 1930, virtually all of the chestnuts (*Castanea dentata*) in the eastern deciduous forest were killed by chestnut blight. Woods and Shanks (1959) showed that the canopy openings were largely filled by crown expansion or sapling growth of previously associated species (mainly oaks and red maple), and there was no additional major community upheaval. What had been an oak-chestnut community became an oak community, but the understory was little changed.

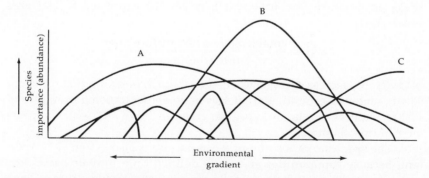

Figure 7–2. Patterns of species importance (abundance) along an environmental gradient as predicted by the continuum view of associations. Noda do not exist. If associations are recognized, based on peaks in abundance of dominant species such as A, B, or C, it can be seen that these associations are merely arbitrary segments along the continuum.

Associations can be recognized and in some way defined in the field, but we must appreciate their subjective, arbitrary boundaries, the nature of which depends on the bias of the investigator and the sampling methods. Whether associations are real or not, however, stands of vegetation certainly do exist. These stands exhibit attributes beyond those of the individual populations which make them up. Some of these attributes (Table 7–1) will be examined in the rest of this chapter and in the rest of this part.

SOME COMMUNITY ATTRIBUTES

Physiognomy, Species Composition, and Spatial Patterns

Physiognomy is a combination of the external appearance of vegetation, its vertical structure (its architecture or biomass structure), and the life forms of its dominant taxa. Chapter 8 discusses methods of sampling, measuring, and numerically describing the physiognomy of communities. Physiognomy is an emergent trait of communities. It could not, for example, be accurately estimated just from a list of all the taxa present within the community.

Life form (see Chapter 1) includes such plant features as size, life span, degree of woodiness, degree of independence, general morphology, leaf traits, the location of perennating buds, and phenology. Vertical structure refers to the height and canopy coverage of each layer within the community.

Canopy coverage can be expressed as the percent of ground covered by the canopy, when the edges of the canopy are mentally projected down to the surface. It may also be expressed as **leaf area index (LAI)**, where

$$LAI = \frac{\text{total leaf area, one surface only}}{\text{unit ground area}}$$

Many crops, such as corn, have a LAI of about 4, meaning that for every square meter of ground, 4 m^2 of leaves lie above it. Some examples of LAI for natural vegetation types are given in Table 7–3, which appears later in this chapter.

The species composition of a community is also extremely important, because communities are partly defined on a floristic basis. Several communities may belong to the same vegetation type (that is, have similar physiognomies), yet differ in the identity of dominants or other species. The abundance, importance, or dominance of each species can be expressed numerically, so that different communities can be compared on the basis of species similarities and differences (see Chapter 8).

Table 7–1. Some emergent characteristics of plant communities.

Physiognomy
   Architecture
   Life forms
   Cover, leaf area index (LAI)
   Phenology
Species composition
   Differential species
   Accidental and ubiquitous species
   Relative importance (cover, density, etc.)
Species patterns
   Spatial
   Niche breadth and overlap
Species diversity
   Richness
   Evenness
   Diversity (within stands and between stands)
Nutrient cycling
   Nutrient demand
   Storage capacity
   Rate of nutrient return to the soil
   Nutrient retention efficiency of the nutrient cycles
Change or development over time
   Succession
   Response to climatic change
   Evolution (?)
Productivity
   Biomass
   Annual net productivity
   Efficiency of net productivity
   Allocation of net production
Creation of and control over a microenvironment

The relative spatial arrangement of species within a community is another community trait. As described in Chapters 4 and 5, individuals within a species or individuals of different species may be distributed at random with respect to each other, clumped (positive interactions), or overdispersed (negative interactions).

The importance of species interactions and interdependence to a discrete view of the community has already been discussed in this chapter, and that assumption will be touched on again in Chapter 10. One theory of community stability discussed in Chapter 10 suggests that stable, long-lasting communities exhibit more species interactions and more species than transient, seral communities. If this is so, then the niches of species in stable communities must be narrower, with less overlap in niche boundaries, than those of less stable communities. Recently, some ecologists have tried to measure niche breadth and overlap to test such hypotheses (Huey and Pianka 1977; Parrish and Bazzaz 1976; Pickett and Bazzaz 1976), but the data are so few that generalities can't yet be made.

Part of a species' niche may include its unique metabolism. For example, flowering plants exhibit three different metabolic pathways of photosynthesis, C_3, C_4, and CAM, each of which is best fitted to different environmental conditions or to different species' phenologies (see Chapter 13). There have been some recent attempts to include such metabolic traits in desert and chaparral community descriptions (Mooney and Dunn 1970; Johnson 1976).

Community descriptions based on physiognomy, life form, niche overlap, and other functional traits (as opposed to taxonomic traits such as the identity of species in the communities) are useful because they permit comparison of widely disjunct stands which have little or no floristic similarity. These comparisons often show a convergence of vegetation types, given a similar macroenvironment. Chaparral stands in southern California and Chile, for example, have little floristic similarity even at the family level, but they share a similar Mediterranean-type climate and they exhibit similar total numbers of species, growth forms, leaf phenology and size, and percent canopy cover by succulent and spinescent species (Figure 7–3) (Parsons and Moldenke 1975).

Species Richness, Evenness, and Diversity

Species richness is simply the number of species in a community. Each species is not likely, however, to have the same number of individuals. One species may be represented by 1000 plants, another by 200, and a third by only a single plant. The distribution of individuals among the species is called **species evenness**, or species equitability. Evenness is maximum when all species have the same number of individuals. **Species diversity** is a product of richness and evenness; it is species richness weighted by species evenness, and formulae are available which permit the diversity of a community to be expressed in a single index number.

It is important to emphasize that richness and diversity are quite different. Although richness and diversity are often positively correlated, environmental gradients do exist along which a decrease in richness is accompanied by an increase in diversity (Hurlbert 1971). Community A, with five species but uneven numbers of individuals in each species, has a lower diversity than community B, with four species which have a very similar number of individuals in each. Community A has a higher species richness, however.

Many assumptions are made in every calculation of species diversity (Peet 1974). It is assumed that all individuals of one species are equal. This may not be true, especially in regard to animals of different sex or of different developmental stages, or to plants of different phenological stages (dormant, full-leaf, flowering, juvenile, senescent adult), different sizes (seedling, sapling, suppressed adult, overstory adult), or different ecotypes. For this reason, diversity is often calculated for each stratum in a community rather than for the entire community. It is also assumed that all species are equally different, whether the difference is in morphology or niche breadth. This assumption is generally made even when it is unlikely to be valid, simply because we have no way to quantify species differences. Finally, evenness is sometimes expressed as numbers of individuals, sometimes as biomass, sometimes as canopy cover, and sometimes in other ways. Obviously, plant diversity from community to community can be compared only if the same units of evenness are used. Some units are more appropriate to plants, and some to animals; possibly there is no one unit suitable for calculation of an ecosystem-wide diversity index (Hurlbert 1971).

A further assumption, and one that is made in any statistical procedure, is that the sample of the community was large enough to adequately represent the community. The data on numbers of species and

Figure 7–3. Chaparral vegetation in California. Even though chaparral vegetation stands in California and Chile differ in floristic composition, they share similar community characteristics.

the relative abundance of each are dependent on sample size. Given a large enough sample, there will be some species with few individuals, some species with many individuals, and many species with an intermediate number of individuals. The data will form a bell-shaped curve on a log-normal plot (Figure 7–4). A small sample, however, will probably not include the full spectrum of rare species. Notice, in Figure 7–5, how the relative abundance of lepidopteran insect species shifted and began to approach a bell-shaped curve as the sample collecting periods increased from less than 1 year to 4 years. The truncated appearance of small sample size curves is an artifact, then, and the truncated end is a "veil line" which hides additional categories of rarity (Preston 1948).

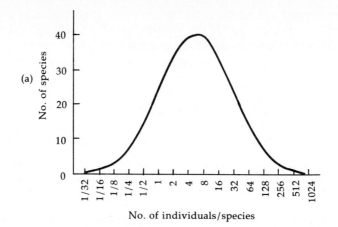

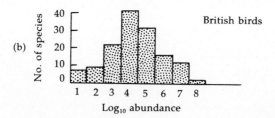

Figure 7–4. (a) The log-normal distribution of species, ranked according to the number of individuals in each species. Note that individuals along the horizontal axis are in categories of geometrically increasing numbers, plotted here on a log basis. Most species in this hypothetical example have a moderate number of individuals in the sample, about one to thirty individuals each, and very few species are rare or abundant. (b) A large sample of British birds, which approaches Preston's log-normal curve. ((a) from Preston 1948 (copyright 1948 by the Ecological Society of America) and Ricklefs 1973 (*Ecology,* Chiron Press, Inc., Newton, MA). (b) from Williams 1964 (*Patterns in the Balance of Nature,* by permission of Academic Press) and Krebs 1972 (*Ecology: The Experimental Analysis of Distribution and Abundance,* Harper and Row, Publishers, Inc., New York).)

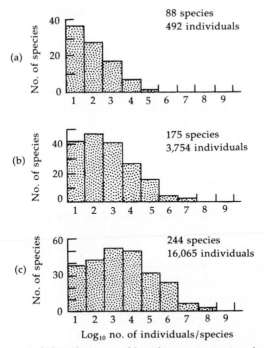

Figure 7–5. Log-normal distributions of lepidopteran species abundances as determined from sampling at Rothamsted Station, England, for periods of (a) 1/8 year, (b) 1 year, and (c) 4 years. As sample size increased, from 492 to 16,065 individuals trapped, the distribution approached Preston's log-normal curve. (From Williams 1964 (*Patterns in the Balance of Nature*, by permission of Academic Press) and Krebs 1972 (*Ecology: The Experimental Analysis of Distribution and Abundance*, Harper and Row, Publishers, Inc., New York).)

Diversity indices

Several indices have been proposed over the course of the last three decades (Peet 1974; DeJong 1975) but we will describe only two: **Simpson's index** (Simpson 1949) and **Shannon-Wiener's index** (Shannon and Weaver 1949; sometimes called the Shannon-Weaver index).

The most commonly written form of Simpson's index (and not the form in which it was originally proposed) is

$$D = 1 - \sum_{i=1}^{s} (p_i)^2$$

where D is the index number, s is the total number of species, Σ is a summation sign, and p_i is the proportion of all individuals in the sample which belong to species i.

If, for example, we have a community sample composed of 99 individuals of species A and 1 individual of species B,

$$D = 1 - [(0.99)^2 + (0.01)^2] = 0.02$$

The Simpson index of diversity gives very little weight to rare species (such as B, above) and is most sensitive to the numbers of abundant species. D can range from 0, for a community composed of one species only, to a maximum of $(1 - 1/s)$, that is, ≈ 1. Peet (1974) has recommended the reciprocal of this index $(1/D)$ as the "best" general index of diversity, because other indices, such as the Shannon-Weiner index, are more sensitive to rare species and this is where sampling error may be most pronounced.

The Shannon-Weiner index is

$$H' = -\sum_{i=1}^{s} (p_i)\,(\log_2 p_i)$$

where H' is the index number, s is the total number of species, Σ is a summation sign, p_i is the proportion of all individuals in the sample which belong to species i, and $\log_2 p_i$ is the log to the base 2 of that proportion. (Logs to the base 10 or e can also be used.)

If, for example, we have the same two-species community used to illustrate Simpson's index,

$$H' = -[(0.99)\,(\log_2 0.99) + (0.01)\,(\log_2 0.01)] = +0.08$$

It is sometimes easier to find tables of logarithms to the base 10 rather than the base 2; using that base, the formula is:

$$H' = (3.3219)\sum_{i=1}^{s} (p_i)\,(\log_{10} p_i)$$

where N is the total number of individuals of all species in the sample, and all other symbols are as previously described.

H' is thought to represent the "uncertainty" or "information" of a community. The more variable its composition, the more variable (the more uncertain or unpredictable) each sample of it would be. The index units have been called bits, bels, decits, or digits per individual, referring to the binomial units of information used in computers. H' varies from 0, for a community of one species only, to values of 7 or more in rich forests such as those of the Siskiyou Mountains of Oregon and California (DeJong 1975). The deciduous forest of the eastern United States is considerably less diverse, with H' values for tree species only ranging from >3.0 for the mixed mesophytic forest in the Cumberland and Allegheny Mountains, to values <2.0 at the western and northern edges (Figure 7–6).

Because the formulae are different, the absolute value of Simpson's and Shannon's index will differ for the same community. Table 7–2 compares the two indices for one forest stand. Note how Simpson's index almost ignores seven rare species.

What does diversity signify?

A lot of attention has been paid to species diversity in the ecological literature: first, in the development and comparison of several formulae; second, in searching for general trends in diversity along environmental gradients; and third, in trying to attribute functional explanations for diversity gradients and adaptive advantages that diversity might convey on communities. Despite all this attention, we think it best to treat an index of diversity as simply one descriptive attribute of a community, on a par with a list of all species found in the community, or an estimate of tons of biomass above the ground, or the leaf area index.

Generally, there is a gradient of increasing species diversity (and richness) from the poles to the equator, and from high elevations to low elevations. These gradients follow complex environmental gradients of increasing warmth, among other factors. It has been stated that diversity increases as any particular stress lessens (Krebs 1972), but this is not true for all stress gradients. An aridity stress gradient may show the opposite trend, with greater diversity in semiarid grassland and desert than in savanna, woodland, or forest (Hurlbert 1971). Diversity has been equated with productivity and stability, but some very diverse semi-arid grasslands and deserts are low in terms of productivity and stability, and there are many other exceptions (van Dobben and Lowe-McConnell 1975). Diversity has been taken as a reflection of the many interactions which supposedly characterize complex communities. Recently, however, a model of community structure which assumes no species interactions shows that species diversity ". . . may be maintained in spite of, rather than because of, such interactions" (Caswell 1976).

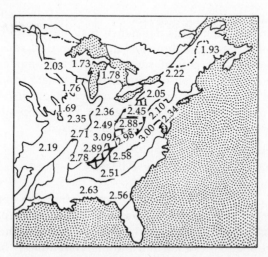

Figure 7-6. Tree species diversity (H') in different associations of the eastern United States deciduous forest formation. The cross-hatched region is mixed mesophytic forest. (From Monk 1966 (copyright 1966 by the Ecological Society of America) and Krebs 1972 (*Ecology: The Experimental Analysis of Distribution and Abundance*, Harper and Row, Publishers, Inc., New York).)

Table 7-2. Calculation of species diversity for a virgin hemlock-hardwood forest in northwestern Pennsylvania, using both the Simpson (*D*) and Shannon-Wiener (*H'*) indices. (Abundance data from Hough 1936. By permission.)

Tree species	Proportional abundance (p_i)	$(p_i)^2$	$-(p_i)(\log_2 p_i)$
Hemlock *(Tsuga canadensis)*	0.521	0.271	0.490
Beech *(Fagus americana)*	0.324	0.105	0.527
Yellow birch *(Betula allegheniensis)*	0.046	0.002	0.204
Sugar maple *(Acer saccharum)*	0.036	0.001	0.173
Black birch *(Betula lenta)*	0.026		0.137
Red maple *(Acer rubrum)*	0.025		0.133
Black cherry *(Prunus serotina)*	0.009		0.061
White ash *(Fraxinus americana)*	0.006	0.002	0.044
Basswood *(Tilia americana)*	0.004		0.032
Yellow poplar *(Liriodendron tulipifera)*	0.002		0.018
Magnolia *(Magnolia virginiana)*	0.001		0.010
totals	1.000	0.381	1.829

$$H' = 1.829$$
$$D = 1 - 0.381 = 0.619$$

It is fair to say that the significance of species diversity is not well understood at this time. This deficiency may be due to our lack of a good model of what a community is in a theoretical sense. Whittaker (1975) has shown that any of several theoretical models of species abundance and diversity can be supported with data from actual communities, and that a single model has not yet been developed which fits all communities. It is also possible that an appropriate formula to express species diversity has not yet been devised.

Nutrient Cycles and Allocation Patterns

Sixteen elements are known to be required for normal growth and development of all higher plants: carbon, hydrogen, oxygen, phosphorus, potassium, nitrogen, sulfur, calcium, magnesium, iron, boron, manganese, copper, zinc, chlorine, and molybdenum. A few other elements are required by certain plant groups, but not by most plants. Sodium, for example, is required in trace amounts by plants with the C_4 pathway of photosynthesis; silicon is required by horsetails *(Equisetum* sp.) (Epstein 1972). Some elements—heavy metals such as lead or gold, and sometimes sodium—are accumulated by certain plants to the point where their foliage is poisonous to livestock, but these elements are not required for normal plant metabolism.

Communities differ in their utilization of certain essential nutrients (that is, how much of each element is required for normal growth, or at least how much is absorbed from the soil solution and translocated to leaves and growing points); they also differ in the rate at which the nutrients are returned to the soil in litter fall, and the efficiency of the plant-soil-plant cycle. Early successional communities, for example, may require little soil nitrogen, accumulate very little of any nutrient in their tissues, and return nutrients rapidly to the soil, but erosion removes a large fraction of the returned nutrients because of their low cover or seasonal absence. Climax communities may require greater quantities of some nutrients, store great quantities of nutrients in wood, and return only a small fraction to the soil in leaf litter, but prevent erosive losses by shielding the soil with a permanent, closed canopy (see Chapter 10). Thus, climax communities have fewer leaks in nutrient cycles and more efficiently hold the nutrients in the plant-soil-plant cycle.

A few remarks about the carbon cycle will illustrate major community differences. If vegetation types are ranked according to the amount of **standing biomass** per hectare (B), they range from desert communities of only 100 kg ha^{-1} to tropical rain forests of 500,000 kg ha^{-1} (Figure 7-7). Generally, greater biomass indicates greater leaf area, which means that more radiant energy can be trapped each year, and a faster growth rate will result, with greater net productivity. **Net productivity** (P_n) ranges from 100 kg ha^{-1} yr^{-1} for desert communities to 40,000 kg ha^{-1} yr^{-1} for tidal zone, mangrove, marsh, and swamp communities (Figure 7-7). **Efficiency**, the fraction of radiant energy converted into kilocalories of tissue, ranges from 0.04% in deserts to 1.5% in tropical rain forests (Table 7-3).

The low efficiency and net productivity of desert communities does not mean that the component species are themselves inefficient or capable of only very low photosynthetic rates. Net productivity is a community trait, and to some extent a climatic trait, for it is affected by the leaf area index, temperatures during the growing season, distribution of rainfall, soil moisture storage, and the length of the growing season. If net productivity or efficiency is expressed on a 12-month basis, it is clear that the tropical rain forest, with a leaf area index of 10-11, and with warm temperatures and adequate moisture all year, should indeed have a greater net productivity and efficiency than a desert, with a leaf area index of 1 or less and a growing season (based on water supply) of only a few months. However, the photosynthetic rates of individual desert shrubs and shrubs of mesic forests are very similar (Barbour 1973b).

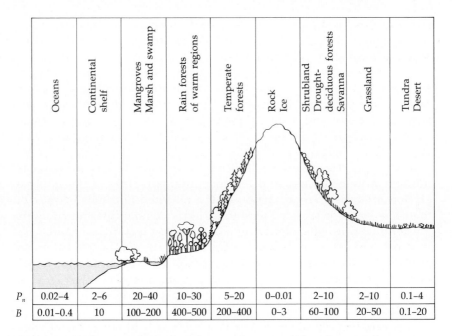

	Oceans	Continental shelf	Mangroves Marsh and swamp	Rain forests of warm regions	Temperate forests	Rock Ice	Shrubland Drought-deciduous forests Savanna	Grassland	Tundra Desert
P_n	0.02–4	2–6	20–40	10–30	5–20	0–0.01	2–10	2–10	0.1–4
B	0.01–0.4	10	100–200	400–500	200–400	0–3	60–100	20–50	0.1–20

Figure 7–7. Differences in net primary productivity (P_n) and in standing biomass (B) of various vegetation types. Data are in metric tons per hectare for B and metric tons per hectare per year for P_n . (From Larcher, W. 1975. *Physiological Plant Ecology.* By permission of Springer-Verlag.)

The **allocation** of net productivity to different organs also differs from community to community. Grassland communities channel much of their energy to belowground biomass, scrub communities less so, and forests even less so. If we compare the distribution of carbon in a subalpine conifer forest, a temperate broadleaf forest, and a tropical rain forest (Figure 7–8), there are significant differences. Most carbon in the subalpine forest is in the form of humus atop or in the surface soil. This is because the acidic, cold soil is not favorable to decomposers. Litter half-life is 10 or more years (Whittaker 1975). Most carbon in the tropical rain forest, in contrast, is locked up as inert wood. This means that leaf litter has a major function in the tropical rain forest, for it represents the only mobile, cycleable part of the nutrient bank. This is one reason why tropical soils cleared of forest vegetation soon become infertile: the annual litter rain has been stopped and what little soil reserves there are, are soon depleted by crops or leached from the soil. The litter decay rate is very rapid beneath the rain forest canopy; litter half-life is a fraction of a year. Nutrient and productivity relationships of plant communities are discussed further in Chapters 11 and 12.

Change Over Time

1–500 years: succession

All communities are dynamic, changing entities. Change, how-ever, is a relative term, and the time framework must be stated.

Plant communities which exhibit no cumulative, directional change for several centuries are considered to be in equilibrium with their environment, and are called **climax** communities (see Chapter 10). Other communities may exhibit significant changes in such a time period: some species decline in abundance and may disappear from the site; invasive species may increase in abundance; the vegetation type itself may change, for example, from a meadow to a forest, or from a pine forest to a hardwood forest. Such transient communities are called **successional**, or **seral**, communities. It may be possible to recognize and describe an entire sequence of successional communities which replace each other on one site, finally culminating in a climax commu-nity. This sequence of communities is called a **succession**, or **sere**.

Thousands of years: climatic change

Succession is thought to be driven by biological interactions, such as competition, which occur in the microenvironment created by the plants themselves; it is not driven by macroclimatic change. Climate

Table 7–3. The leaf area index (LAI, m^2 leaf surface per m^2 ground) and efficiency of net productivity for major vegetation types. Efficiency is calculated by converting grams of net production per unit area to kilocalories per unit area (there are approximately 4.2 kcal in each gram dry weight of plant tissue), then dividing by the total radiation received in a year. Only radiation between the wavelengths of 400 and 700 nm is entered in the calculation, because these are the only wavelengths usable for photosynthesis. (From Larcher, W. 1975. *Physiological Plant Ecology.* By permission of Springer-Verlag.) (See also pp. 247, 255, 450.)

Vegetation type	LAI	Efficiency (%)
Tropical rain forest	10–11	1.50
Deciduous forest	5–8	1.00
Boreal conifer forest	9–11 (7–38)	0.75
Grassland	5–8	0.50
Tundra	1–2	0.25
Semiarid desert	1	0.04
Agricultural	3–5	0.60

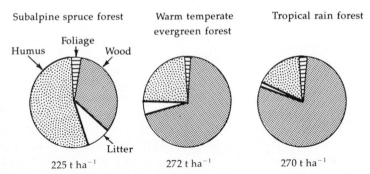

Figure 7-8. Content of organic carbon in the foliage, wood, litter, and humus for three forest types. (From Larcher, W. 1975. *Physiological Plant Ecology.* By permission of Springer-Verlag.)

is assumed to be constant when a succession is investigated and described.

Climate, however, has not been a constant throughout time (Figure 7–9) and it has always undergone significant fluctuations. Weather records, such as those taken by the United States Weather Bureau, only exist for the past 100 or so years. Nevertheless, even these records show statistically significant (though minor) changes. For example, rainfall has been declining and temperatures rising in the southwestern United States (Hastings and Turner 1965).

The climate before the nineteenth century can be inferred by several methods (see, for example, Flint 1957 and Strahler and Strahler 1974). One method, applicable to the southwestern United States, is the analysis of tree rings (**dendrochronology**). Climate of the past 8200 years has been estimated by examining sequences of growth ring widths in the wood of living and dead trunks of bristlecone pine *(Pinus longaeva = P. aristata*).* The data indicate cycles of warm, arid periods followed by cool, wet periods (Figure 7–10) (LaMarche 1974; Ferguson 1968). The period of time shown in Figure 7–10 begins with an exceptionally arid period called the **Xerothermic period**, which started approximately 8000 years ago and ended approximately 4000 years ago. This period is thought to be responsible for the present distribution limits of many southwestern vegetation types and communities (Axelrod 1977).

Relatively small changes in global temperatures have created striking climatic and vegetation changes. Most careful analyses suggest an average global surface temperature difference between full glacial eras and the present of only 4–6°C (Bryson 1974). This temperature difference, however, has been correlated with changes in cloudiness,

*The Latin name for bristlecone pine was recently changed. The new name is *Pinus longaeva*.

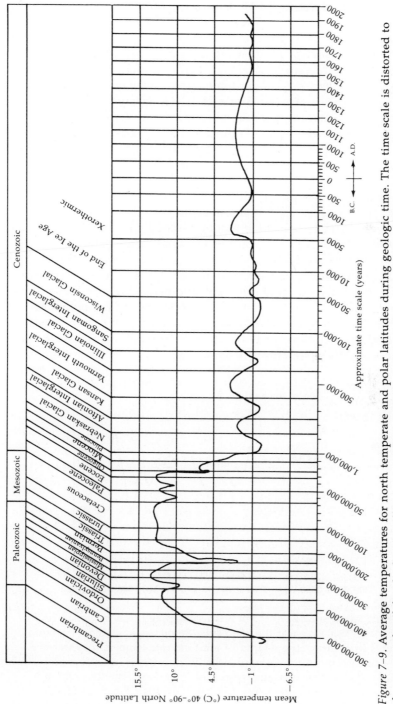

Figure 7–9. Average temperatures for north temperate and polar latitudes during geologic time. The time scale is distorted to show more detail for the last 1 million years. Notice the four glacial advances separated by warm interglacial retreats. The Xerothermic period (a very arid period) is shown here to peak at about 3000 B.C., which is equivalent to 5000 years before the present. (From Dorf 1960. Reprinted by permission of *American Scientist*, Journal of Sigma Xi, The Scientific Research Society.)

rainfall, length of the frost-free season, and other factors which have a large impact on vegetation (Figure 7–11).

The nature of vegetational change since the Ice Age can be documented in several other ways. In the arid southwestern United States, the dried remains of plants cached in underground middens by wood rats (*Neotoma* sp.) have in some cases remained intact and identifiable for thousands of years. Since the foraging activity of wood rats is restricted to a rather limited radius around the nest, the composition of the plant material gives some indication of the nearby vegetation at the time the midden was formed. The plant material can be carbon dated. This kind of evidence permitted Wells and Berger (1967), among others, to determine the changing elevation of vegetation zones following the retreat of the last glaciation in what is now the Mojave Desert.

A more widely used method of documentation utilizes pollen grains which accumulate at the bottom of slowly filling lakes or ponds. As pollen is shed and transported by wind, some falls onto a lake surface, sinks to the bottom, and becomes incorporated with silt and organic matter into sediment. Pollen of many species is resistant to decay in the anaerobic, cold sediment, and may remain intact for thousands of years. The family, genus, or even species of the pollen can be determined under the microscope. A core of sediment, then, reveals a chronological sequence of surrounding vegetation; the deeper the pollen occurs in the sediments, the older it is.

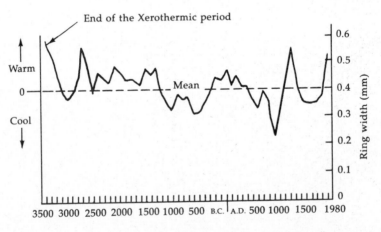

Figure 7–10. Average ring widths in bristlecone pine (*Pinus longaeva*) trunks dated back to 3500 B.C., near the end of the Xerothermic period. Positive departures from mean ring width (0.4 mm) indicate temperatures warmer than average during April–October; negative departures indicate cooler conditions. (From "Paleoclimatic Inferences from Long Tree Records," by LaMarche, V. C., Jr., in *Science*, Vol. 183, pp. 1043–1048, Fig. 5, 15 March 1974. Copyright 1974 by the American Association for the Advancement of Science.)

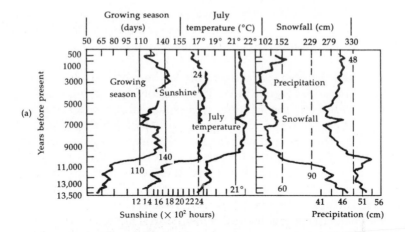

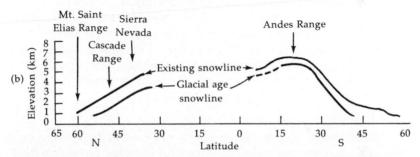

Figure 7-11. (a) A 13,500 year record of growing season length, hours of bright sunshine annually, mean July temperature, annual snowfall, and precipitation during the growing season at Kirchner Marsh, Minnesota. (b) The effect of climatic change over the past 13,000 years on the elevation of permanent snow (the upper limit of alpine vegetation) in western mountain ranges of North and South America. ((a) from "A Perspective on Climatic Change," by Bryson, R. A., *Science*, Vol. 184, pp. 753-759, Fig. 2, 17 May 1974. Copyright 1974 by the American Association for the Advancement of Science. (b) from Flint 1957 (*Glacial and Pleistocene Geology*, copyright 1957 John Wiley and Sons, Inc., reprinted by permission of John Wiley and Sons, Inc.) and Strahler and Strahler 1974 (*Introduction to Environmental Science*, Hamilton Publishing Co., Santa Barbara, CA).)

Within limits, ecologists assume that the abundance of pollen in the core is related to the abundance of species in the surrounding vegetation. This assumption, of course, is applied only to wind pollinated plants such as sedges, grasses, most trees, many shrubs, and some forbs (herbaceous plants other than grasses). The assumption is upheld by measurements of modern pollen "rain." Griffin (1975), for example, found that modern pollen rain in Minnesota plant communities correlated very well with the nearest community type. Pollen is not carried in significant quantities further than 50 km in forested regions (Livingstone 1968). If we consider a pond surrounded by vegetation in a radius of 50 km, then the pond can receive pollen from a total area of 7850 km². Most vegetation is not uniform over such an area, but it is likely that the pollen profile will give a good general picture of regional vegetation.

A **pollen profile** constructed from sediment beneath a Nova Scotia lake is shown in Figure 7–12. The present vegetation consists of a deciduous forest on the hilltops with sugar maple *(Acer saccharum)*, beech *(Fagus grandifolia)*, and yellow birch *(Betula lutea)*, and a coniferous forest in the valleys, with balsam fir *(Abies balsamea)* and white spruce *(Picea glauca)*. Figure 7–12 shows that this pattern of vegetation is relatively recent. Sediments at a depth of 5–6 m, corresponding to an age of about 9000 years before the present, reveal pollen mostly from herbs and shrubs characteristic of tundra vegetation near permanent ice. The nearest such vegetation today occurs 500 km to the north of the lake. Shallower depths, corresponding to an age of about 6000 years before the present, show peaks in spruce and fir, indicating the dominance of typical northern taiga forest, widespread today farther to the north. In yet shallower depths, the pollen of deciduous species (birch and maple) increases in abundance and the pollen of coniferous species declines, indicating a continual warming trend in climate.

Millions of years: evolutionary change

Microfossils, such as pollen grains, are used to document vegetational changes over the course of thousands of years, but **macrofossils**, such as leaf impressions, are used to document changes over millions of years. As with pollen, it is assumed that the abundance of fossils

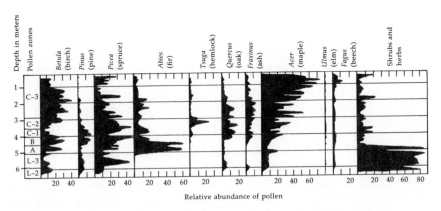

Figure 7–12. Pollen profile of sediment from a lake in Nova Scotia. The numbers along the horizontal axis refer to relative abundance of each type of pollen. Depth corresponds to age; layer L–2 has been carbon dated to 9000 years B.C. and the top of layer C–3 represents the present. (From Livingstone 1968. Copyright 1968 by the Ecological Society of America.)

represents the abundance of species in past vegetation. Care must be taken in interpreting site geology to determine whether plant material was deposited in place, or carried by water for some distance and then deposited.

By examining general leaf shape and the pattern of leaf venation, paleoecologists have been able to identify fossil species and to relate them to their nearest living species. The fossil species are usually extinct, but in many cases they are so close to living species that they can be written as, for example, *Pseudotsuga (menziesii)*, which means a fossil very closely related to the modern species of Douglas fir, *P. menziesii*. It is likely that fossil species were physiologically different from modern taxa, much as modern ecotypes of the same species differ from each other (Axelrod 1977). Nevertheless, the assumption is made that the present is the key to the past, and that modern relatives of fossil plants are in a climate similar to the climate which existed at the time and place the fossil material was deposited. In this way, past climates as well as past vegetation may be reconstructed.

One of the most complete records of fossil plants for western North America during the Cenozoic era (the past 65 million years) is in the John Day Basin of eastern Oregon (Chaney 1948). The record shows a cooling and drying trend. About 60 million years ago the prevailing community contained cinnamon, palms, figs, cycads, avocados, and tropical ferns, which are now found in cool mountain forests of Central America with an annual rainfall of 1500$^+$ mm and no frost. The leaves of these plants were large, with entire margins. About 40 million years ago there was a change to a mixed conifer-hardwood forest with birch, alder, oak, dawn redwood, elm, sycamore, beech, maple, chestnut, sweet gum, and others. The leaves of these plants were smaller, with dentate or convoluted margins, indicating a drier climate, and some trees were deciduous. This exact mixture does not appear anywhere today, but close approximations exist along the cool, wet California coast and on the Cumberland Plateau of Tennessee. Annual rainfall was still high, about 1250 mm, but the climate had grown cooler. About 25 million years ago there was a strong shift to winter-deciduous trees, such as oak, hickory, and maple. This indicates a climate like modern Indiana, with 1000 mm of rainfall annually and prolonged freezing temperatures in winter. Today, the John Day Basin is dominated by sagebrush. Trees are absent except along waterways, and precipitation is about 250 mm per year, including some snow in a cold winter period.

Fossil assemblages for other localities in North America have permitted paleoecologists to reconstruct past climate and vegetation zones on a continental basis (Figure 7–13). In general, vegetation zones have been shifted south and compressed over the past 40 million years; that is, environmental and vegetational gradients from pole to equator have become steeper.

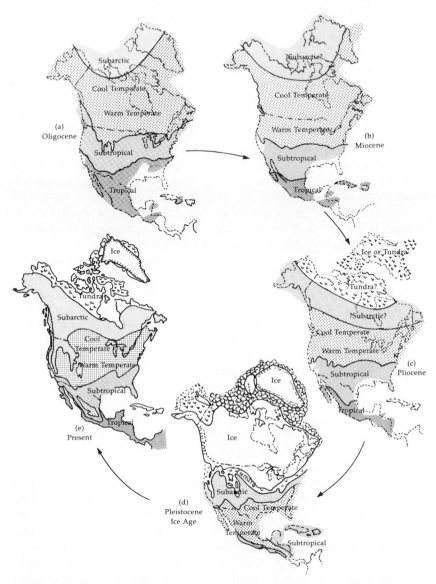

Figure 7-13. Climatic zones in North America during the past 40 million years. (a) Oligocene, about 35 million years ago; (b) Miocene, about 20 million years ago; (c) Pliocene, about 5 million years ago; (d) at the peak of the Pleistocene Ice Age, about 60,000 years ago; and (e) the present. (From Dorf 1960. Reprinted by permission of *American Scientist*, journal of Sigma Xi, The Scientific Research Society.)

Since the fossil record reveals that communities similar to those existing today have a history extending back millions of years, it is reasonable to ask if communities evolve. That question has not yet been resolved. Whittaker and Woodwell (1972) have suggested that evolution does occur at the community level. They develop their argument in this way: (1) all species evolve in communities rather than in isolation; thus, (2) the evolution of a community occurs as a process of coevolution of the associated species, making the whole community an interactive assemblage; therefore, (3) communities must change in structure and function from ancestral, simpler communities to modern, complex communities, as the component species become more interdependent; and finally, then, (4) communities have emergent characteristics corresponding to those of organisms, such as growth and maturity, structural differentiation, energy flow, material turnover, homeostasis (the tendency to remain stable or regain stability), adaptive optimization, and organization.

This view is supported by data on the convergence of widely separated communities that share a similar environment, such as chaparral vegetation in California, Chile, southern Australia, South Africa, and the Mediterranean region. The climate of the five regions is similar, and so is the physiognomy of the vegetation (see, for example, Figure 7–3, Mooney 1977b, and Parsons 1976). On a much smaller scale, Cody (1974) compared a field in Kansas with one in Chile, and he

Table 7–4. Comparison of vegetation and bird species in fields in Kansas and Chile. (From "Optimization in Ecology," by Cody, M. L., *Science*, Vol. 183, pp. 1156–1164, Table 1, 22 March 1974. Copyright 1974 by the American Association for the Advancement of Science.)

Attribute	Kansas	Chile
Vegetation structure:		
Mean height (m)	0.29	0.27
Vertical leaf density	5.10	6.23
Horizontal leaf density	8.68	9.26
Bird communities:		
Number of species	3	3
Mean between-species overlap in:		
Habitat	0.63	0.60
Feeding heights	0.78	0.89
Food and feeding behavior	0.18	0.21
Bird species comparisons:		
Eastern meadowlark and red-breasted meadowlark:		
Body length (mm)	236	264
Bill length (mm)	32.1	33.3
Grasshopper sparrow and yellow grass finch:		
Body length (mm)	118	125
Bill length (mm)	6.5	7.1
Horned lark and Chilean pipit:		
Body length (mm)	157	153
Bill length (mm)	11.2	13.0

found remarkable convergence in the vegetation and the bird populations (Table 7–4).

Examples of convergence in the morphology and/or behavior of widely separated and unrelated species, such as cacti in the southwestern United States and similar-looking euphorbs in Africa, have long served as examples of natural selection at work. Can we accept community convergence in the same light? Is there an optimal solution for community structure, given a certain climate, and does natural selection drive community development towards that solution? We are tempted to answer "yes," but Ricklefs (1973) and others answer "no." They reason that if communities were integrated units and a product of evolution, they would be closed systems like organisms, with sharp boundaries. In fact, however, as we have seen earlier in this chapter, groups of species do not parallel each other in their distribution curves, and when communities are recognized they are relatively arbitrary units, sharing broad ecotones with adjacent communities. Although pairs of species may have coevolved and are interdependent (pollinators and certain plants, parasitic plants and their hosts, mycorrhizal unions), there is no hard evidence that entire communities are integrated, interdependent units. Recall that the removal of chestnut trees from an oak-chestnut forest did not produce community-wide repercussions.

SUMMARY

The community concept is of general importance to synecology, just as the ecotype concept is central to autecology, and the species concept is central to taxonomy. The precise nature of the community is ambiguous because of the biases of individual ecologists and their sampling methods. This does not make the concept useless, however; we simply must appreciate its subjectivity. For the purposes of classification, stands can be grouped into associations which have a fixed floristic composition, physiognomy, and habitat range.

The discrete view of associations assumes that associations are closed systems, with interdependent species which synchronously peak in abundance, and with narrow ecotones; the discrete view assumes that the whole is greater than the sum of its parts. The individualistic view assumes that associations are open systems, with independent species that happen to associate together wherever their range limits and chance arrival of propagules overlap; consequently, associations are at most arbitrary units along a continuum.

Whether or not associations exist in the abstract sense, real stands (communities) do exist, and it is useful to consider the emergent attributes that communities exhibit which are beyond the attributes of the component species. These emergent attributes include physiognomy, species importance, spatial and niche patterns, species richness, species evenness, species diversity, the rate of nutrient cycling, the pattern of nutrient allocation to above and belowground parts, and community change over time.

The significance of species diversity to community stability, productivity, interdependence, and environmental stress is still not clear, possibly because: (a) we have not yet developed a reasonable method to measure diversity or (b) there is more than one model of community dynamics and response to stress. Simpson's index may provide the best general index of diversity, if for no other reason than the fact that it is relatively insensitive to sampling error.

Communities differ in their demand for essential nutrient elements, in the efficiency with which radiant energy is converted into net productivity, in the fraction of the nutrient pool which is stored, in the rate at which nutrients are returned to the soil in litter fall, and in the efficiency of the plant-soil-plant cycle.

Succession is community change in a period of up to 500 years, with climate and plant genomes assumed to be constant. In fact, however, climate has not been constant. There have been changes in temperature, rainfall, length of the growing season, and duration of sunshine that have accompanied glacial retreat in the temperate zone over the past 10,000 years. The resulting vegetation changes can be shown by pollen profiles in lake sediment. Vegetation changes over millions of years are documented in macrofossil deposits. These long-term changes, however, include genetic (evolutionary) change in the floras.

There is some argument and indirect evidence to suggest that communities evolve, directed by natural selection toward optimal solutions of environmental problems. The conservative opinion is that the most complex level at which natural selection has been shown to operate is with pairs of species, such as a parasite and a host.

For the most part, this chapter provides an introduction to Chapters 8–12, where all of these topics are discussed in more detail.

CHAPTER 8

METHODS OF SAMPLING THE PLANT COMMUNITY

Synecologists would like to understand the degree of species interdependence within communities, how the distribution of communities depends upon past and present environmental factors, and what the role of communities is in such ecosystem activities as energy transfer, nutrient cycling, and succession. However, communities must first be measured and summarized in some effective way before these questions can be addressed. The ongoing attempt to inventory the world's vegetation is based on a small sample of the total vegetation cover because of limitations in people, time, etc. These samples must be taken very carefully to ensure that the resulting estimates will be accurate and useful.

For the sampling to be done rationally and efficiently, the continuum of vegetation that covers the earth must be divided into discrete, describable community or vegetation types, just as the taxonomic continuum of individual plants has been divided into species. Most of the world's vegetation is known to science, but at a relatively simplistic level. It has been broadly classified (UNESCO 1973; Table 1-1), mapped for large areas (Riley and Young 1974; Figure 8-1), and photographed from airplanes or satellites (Figure 9-6). The dominant species and physiognomy of major vegetation types are generally known, but more detailed information about all the component species, the relative importance of each species, and relationships among the species is frequently lacking.

Even if one wishes to describe a particular plant community in a relatively circumscribed, accessible region, one will not usually make a complete census of the community, but instead will take measurements on perhaps only 1% of the total land on which the community exists. If the samples are chosen carefully, investigators feel confident in extrapolating from their sample data to estimate the true values of the **parameters** for the entire community. If the samples are not chosen carefully, the samples will not be representative of the true community parameters and they are said to be **biased**.

There are two approaches to locating representative samples. One approach is to locate the samples completely subjectively. A seasoned

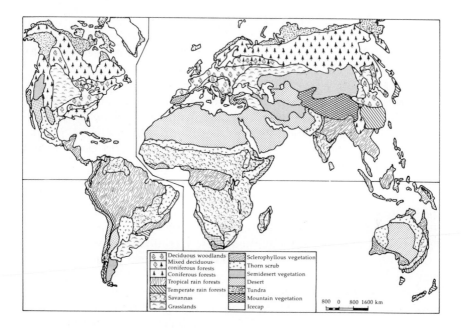

Legend:
- Deciduous woodlands
- Mixed deciduous-coniferous forests
- Coniferous forests
- Tropical rain forests
- Temperate rain forests
- Savannas
- Grasslands
- Sclerophyllous vegetation
- Thorn scrub
- Semidesert vegetation
- Desert
- Tundra
- Mountain vegetation
- Icecap

800   0   800 1600 km

Figure 8–1. Map of major vegetation types of the world. (From *World Vegetation,* by Riley and Young 1974. Reprinted by permission of Cambridge University Press.)

field worker who has traveled extensively in a region first formulates a concept of a particular community type. Representative stands of that type are found in the field and one or more sample quadrats are placed so that each quadrat encloses the essence of that stand. This is the **relevé** method. The other approach is to combine subjective selection of stands with an objective (random or regular) placement of sample quadrats within the stands. Theoretically, there is a third approach— random selection of stands and random placement of quadrats within them—but this is so time consuming that it is seldom done and so it need not concern us here. Completely randomized sampling will inevitably undersample rare, but interesting and ecologically informative kinds of vegetation.

In this chapter we will summarize several sampling methods, which is not a very complete review, but is enough to get a taste of the diversity among the methods. In Chapter 9 we will discuss six data analysis methods which can be used to convey the essence of a vegetation type to others who may be half a world away and have never seen the type, yet wish to compare it to types with which they are familiar.

THE RELEVÉ METHOD

The relevé method was largely codified, if not developed, by Josias Braun-Blanquet, an energetic Swiss ecologist who helped classify much of Europe's vegetation, wrote an impressive text on plant ecology in 1928,* founded and directed a center of synecology at Montpellier, France, called SIGMA (Station Internationale de Géobotanique Mediterranéene et Alpine), and was an active editor of the technical journal *Vegetatio* until he was 90. His methods of sampling and classification are sometimes called the **relevé**, **SIGMA**, **Braun-Blanquet**, or **Zurich-Montpellier (Z-M) school**.

Our description of the relevé method will be brief. Expanded discussions can be found in Mueller-Dombois and Ellenberg (1974), Shimwell (1971), Becking (1957), Küchler (1967), and Poore (1955*a,b*). Not many Americans have applied these techniques to North American vegetation; Henry Conard (1935), who helped translate Braun-Blanquet's book into English, is one of the few.

An investigator familiar with the vegetation of a region begins to develop concepts about the existence of certain community types that appear to repeat themselves in similar habitats. A number of stands that represent a given community are subjectively chosen. The investigator walks through as much of each stand as possible, compiling a list of all species encountered. Next, an area which best represents the community is located. It is then necessary to determine the **minimal area**—the smallest area within which the species of the community are adequately represented. The minimal area may be determined by a species-area curve. The resulting sample quadrat, based on the concept of minimal area, is called a **relevé**.

A **species-area curve** is compiled by placing larger and larger quadrats on the ground in such a way that each larger quadrat encompasses all the smaller ones, an arrangement called **nested quadrats** (Figure 8–2(a)). As each larger quadrat is located, a list is kept of additional species encountered. A point of diminishing return is eventually reached, beyond which increasing the quadrat area results in the addition of only a very few more species. The point on the curve where the slope most rapidly approaches the horizontal is called the minimal area (Figure 8–2(b)–(d)). Because this definition of minimal area is subjective, some define it instead as that area which contains some standard fraction of the total flora of the stand, for example 95%. Problems in defining minimal area have been discussed by Rice and Kelting (1955). Generally, the relevé used is somewhat larger than that which gives the graphical minimal area, for the sake of being conservative. Minimal

---

*His text was translated into English in 1932 by C. D. Fuller and H. S. Conard.

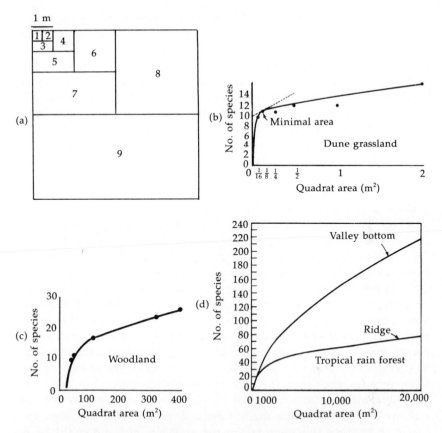

Figure 8–2. The species-area curve. (a) A system of nested plots for determining minimal area. (b) Minimal area for dune grassland in North Carolina is about 0.13 m². (c) Minimal area for an English woodland is about 100 m². (d) Minimal area for two stands of tropical rain forest in Brunei are 1000 m² (a ridge) and 20,000⁺ m² (a valley bottom). ((a) and (d) from *Aims and Methods of Vegetation Ecology.* Mueller-Dombois and Ellenberg. Copyright 1974 John Wiley and Sons, Inc. Reprinted by permission of John Wiley and Sons, Inc. (b) from Smith 1940. By permission of the Ecological Society of America. (c) from Hopkins 1957. By permission of the British Ecological Society.)

area is thought by some ecologists to be an important community trait which is just as characteristic of a community type as the species which make it up. Table 8–1 shows how minimal area correlates with vegetation type.

In the relevé method, each species is recorded and several parameters are estimated within the relevé: cover, sociability, vitality, periodicity, topographic characteristics, and environmental characteristics.

Cover is not measured precisely, but rather is placed in one of seven categories by a visual estimate (Table 8–2). Braun-Blanquet and others, such as Rexford Daubenmire (1968), recognize that plant cover is very heterogeneous from point to point and from time to time even within a small stand. They believe that an exact estimate at one place gives an aura of precision to community description that is not warranted. As another ecologist said, "The ecological world is a sloppy place" (Slobodkin 1974).

Another argument against overly precise cover estimates is that there is a differential bias from one individual to another and it is unlikely that any two estimates would agree closely. Schultz et al. (1961) dramatically demonstrated this bias by bringing an artificial quadrat (a 1 m² board with plants represented by discs of different size and color) to a national meeting of professional range management people and asking 100 of them to estimate the total cover of all the "plants" on the board. The resulting range of estimates was impressive: from 6 to 62%. Just as impressive was the fact that the average estimate (27% cover) was 33% in error of the true cover (20%). If each percentage estimate is converted to a class, such as those in Table 8–2, however, more than half of the percentage estimates will fall into the correct class. Seven classes do not provide as much precision as 100 percentage points, but using classes results in greater agreement among investigators. The range of percentage points within each class allows for each observer's deviance from the correct cover percentage.

Sociability is an estimate of the dispersion of members of a species, which does not necessarily have any relationship to cover. Two species may have the same cover, for example, but one could be re-

Table 8–1. Minimal areas for various vegetation or community types. (From *Aims and Methods of Vegetation Ecology.* Mueller-Dombois and Ellenberg. Copyright 1974 John Wiley and Sons, Inc. Reprinted by permission of John Wiley and Sons, Inc.)

Type	Minimal area (m²)
tropical rain forest	1000–50,000
temperate forest:	
overstory	200–500
undergrowth	50–200
dry temperate grassland	50–100
heath	10–25
wet meadow	5–10
moss and lichen communities	0.1–4

Table 8-2. Cover classes of Braun-Blanquet, Domin-Krajina, and Daubenmire. (From *Aims and Methods of Vegetation Ecology.* Mueller-Dombois and Ellenberg. Copyright 1974 John Wiley and Sons, Inc. Reprinted by permission of John Wiley and Sons, Inc.)

Braun-Blanquet			Domin-Krajina			Daubenmire		
Class	Range of cover (%)	Mean	Class	Range of cover (%)	Mean	Class	Range of cover (%)	Mean
5	75–100	87.5	10	100	100.0	6	95–100	97.5
4	50–75	62.5	9	75–99	87.0	5	75–95	85.0
3	25–50	37.5	8	50–75	62.5	4	50–75	62.5
2	5–25	15.0	7	33–50	41.5	3	25–50	37.5
1	1–5	2.5	6	25–33	29.0	2	5–25	15.0
†	<1	0.1	5	10–25	17.5	1	0–5	2.5
r	≪1	*	4	5–10	7.5			
			3	1–5	2.5			
			2	<1	0.5			
			1	≪1	*			
			†	≪≪1	*			

*individuals occurring seldom or only once; cover ignored and assumed to be insignificant.

stricted to a few, dense clumps of individuals while the other may be uniformly scattered throughout the quadrat or stand. Sociability is recorded on a scale of 1 to 5 (Table 8–3), and is written as a decimal addition to the cover value. Thus, species X on a data sheet may be represented by a number such as 2·1, which translates as 5–25% cover, with plants occurring singly.

The investigator also notes the **vitality** (vigor) and **periodicity** (seasonal importance) of species, as well as general topographic and environmental characteristics of the relevé. When all stands have been visited, a summary table of species X stands is prepared; this table will be discussed in Chapter 9.

The summary table reveals **synthetic traits**, which are traits of a community rather than of a single stand. Two synthetic traits are **presence** and **constance**. Presence is the percentage of all stands which contain a given species. If species X occurs in 8 of 10 stands, the species has 80% presence. Presence is calculated from the presence lists that were generated as the investigator walked through the stands. Constance, in contrast, is based on species encountered in relevés. One relevé, recall, is placed in each stand, and those relevés are all of equal area (though not necessarily of equal shape). Generally, presence is higher than constance. Species X may have been present in 8 stands, but in only 6 of the 10 relevés, thus having 60% constance (sometimes called constancy).

Table 8-3. Sociability scale of Braun-Blanquet.

Value	Meaning
5	growing in large, almost pure stands
4	growing in small colonies or carpets
3	forming small patches or cushions
2	forming small but dense clumps
1	growing singly

RANDOM QUADRAT METHODS

Most American ecologists select stands subjectively, but then sample within them by locating many random quadrats, rather than by subjectively locating a single, large quadrat as in the relevé method. The quadrats can be arranged in a completely random fashion or in a restricted (stratified) random fashion, as discussed in Chapter 4. The quadrats may also be arranged in a regular, nonrandom fashion, but then statistical conclusions cannot be reached.

Care must be taken in selecting the shape, size, and number of quadrats. A considerable body of literature developed on these subjects in the 1940s and 1950s. Some of that research was done on scale drawings of plots of real vegetation as seen from above, with miniature quadrats of various sizes and shapes placed randomly over the drawings. Some of these maps have been published, e.g., Curtis and Cottam 1962. Other research was done on map-like models of artificial vegetation, generally by placing discs of different size and color in random or other patterns, and then sampling these models with miniature quadrats. Schultz Developments* manufactures very rugged artificial vegetation "maps," suitable for class use.

In any case, whether the maps represent real or artificial vegetation, the point is that one knows the true number of plants and the true cover. Sample estimates of these parameters can then be compared for accuracy. The best sampling method will be both accurate and precise. **Accuracy** is close agreement of sample means with parameter means. In Figure 8-3, method A gives values which are very accurate, within 10-20% of the true mean. **Precision** is close agreement of sample means to each other, without reference to the true mean. In Figure 8-3, methods A and B are equally precise because their sample means cluster equally tightly. Method B, however, is much less accurate, sample

*2330 Walnut Blvd., Walnut Creek, CA 94596.

means being about 30% of the true mean. Method C is neither precise nor accurate.

Precision can be measured without knowing the true mean; it is equal to 1/(variance of sample means). Accuracy, however, can only be measured when the true values of the parameters are known, and in vegetation sampling this is never the case.

Quadrat Shape, Size, and Number

By sampling a map of forest vegetation in North Carolina, Bourdeau (1953) found that restricted random placement of quadrats yielded greater precision than completely random placement, but the two gave equally accurate estimates. From the same map, Bormann (1953) discovered that precision was best when quadrats were long, narrow rectangles which tended to cross contour lines. Square and round quadrats are often less precise because each one encompasses less heterogeneity within it than a long, narrow plot placed parallel to the major environmental gradient (see also Clapham 1932 and Lindsey et al. 1958).

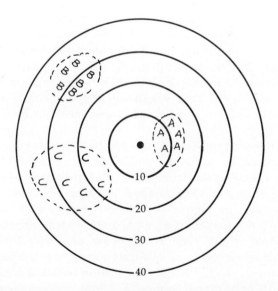

Figure 8-3. Accuracy and precision of sampling. The bull's eye represents the true cover for a stand of vegetation, and the radiating circles represent departures in accuracy, from 10 to 40% error. Means estimated by sampling methods A, B, and C are plotted according to their departure from the true cover. Samples taken by methods A and B are both precise (the points are in tight clusters), but A is more accurate. Samples taken by method C show poor precision and poor accuracy.

Accuracy, however, may decline as the plot lengthens because of the **edge effect**. The more perimeter there is to a quadrat, the more often an investigator will have to make subjective decisions as to whether a plant near the edge is "in" or "out," and these decisions are likely to be biased by the taxonomic knowledge of the investigator, how alert the investigator is that day, and how close it is to dinner time. In this respect, round quadrats are the most accurate because they have the smallest perimeter for a given area. They are also easier to define in the field with a tape measure and center stake, and so a large quadrat need not be carried around. Obviously, compromise choices on quadrat shape are often made.

The best quadrat size to use depends on the items to be measured. If cover alone is important, then size is not a factor. In fact, the quadrat may be shrunk to a line of one dimension or to a point of no dimension and cover can still be measured, as described later in this chapter. But if plant numbers per unit area or pattern of dispersal are to be measured, then quadrat size is critical, as discussed in Chapter 4. One rule of thumb is to use a quadrat at least twice as large as the average canopy spread of the largest species (Greig-Smith 1964); another is to use a quadrat size which permits only one or two species to occur in all quadrats (Daubenmire 1968); another is to use a quadrat size which permits a species to occur in no more than 80% of all quadrats (Blackman 1935). There are no fixed rules, however, and the choice is often made by combining intuition and convenience. One person working alone in a desert scrub might choose a quadrat 2 m long on each side, because the area can all be seen from one point, while a team of 2 to 3 people in the same vegetation might choose a quadrat twice that size. Table 4–2 lists some pragmatic suggestions by Cain and Castro (1959).

The number of quadrats to use can be determined empirically by plotting the data for any given parameter using different numbers of quadrats, and picking that number of quadrats which corresponds to a point where fluctuations are damped. For example, cover has been tallied in this way in Figure 8–4. Some species, such as bluebunch wheatgrass *(Agropyron spicatum)*, were under-represented by the first 10 quadrats, while others, such as sagebrush *(Artemisia tridentata)*, were over-represented. By 30 to 40 quadrats values had leveled out, but 50 quadrats did not give additional information with the exception of one species, needle grass *(Stipa comata)*. Therefore, one assumes that 35 to 40 quadrats will give a fairly accurate estimate for all additional sampling in this community type.

Alternatively, one can sample until the standard error of the quadrat data is within some previously decided, acceptable bounds. Some field workers suggest that the standard error be ±15–20% of the

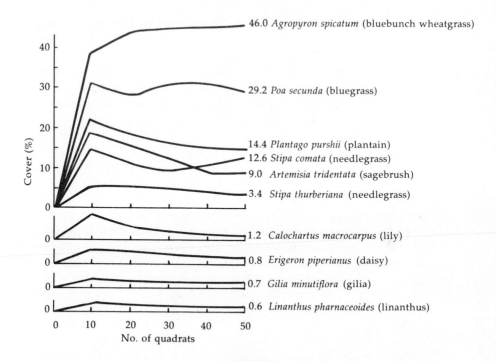

Figure 8–4. Fluctuations in cover are damped out as the number of quadrats increase. In this example from a sagebrush/grassland in the Idaho-Washington area, 35 to 40 quadrats might be sufficient to conduct the study, with diminishing reward for added effort beyond that point. (Reprinted by permission from Daubenmire 1959.)

mean (that is, two-thirds of all the quadrats supply data that fall within that range about the mean).

Most investigators manipulate quadrat size and number so that 1–20% of a stand is included in the sample. This area may be smaller than the area of a relevé so there is a greater possibility that some rare but ecologically significant species will be missed.

Cover, Density, Frequency, Dominance, and Importance

Cover (also called **coverage**) is the percentage of quadrat area beneath the canopy of a given species. The canopy of an overstory species creates a microenvironment which smaller, associated species must contend with. The overstory canopy, therefore, exerts a biotic control over the microclimate of the site. No doubt, the root system of the overstory species extends beneath the ground out to a perimeter corresponding with the canopy edge or even further, so the soil microenvironment is also under the biotic influence of the overstory species. It is assumed that a comparison of cover for each species in a given canopy layer will reveal the relative control or dominance that each species exerts on the community as a whole, such as the relative amount of nutrients or other resources each species commands.

For the practical measurement of cover, holes in the canopy may be viewed as nonexistent, and the canopy edge can be mentally "rounded out," the rationale being that such space is still under the root or shoot influence of the plant in question. The canopy of a plant rooted outside the quadrat is tallied to the extent that the canopy projects into the quadrat space. Thus, in Figure 8–5 and Table 8–4, shrub E does extend into the quadrat space when its canopy is rounded out (dashed lines), and it contributes 7.9% cover. Similarly, the radiating, basal leaves of B which project into the canopy space are tallied by rounding out the edges and estimating each plant separately, which

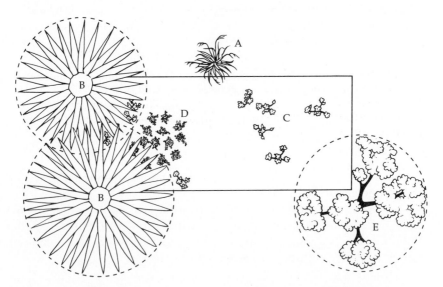

Figure 8–5. Estimation of cover. As seen from above, two members of species B contribute the most cover within the quadrat. Cover for both is estimated, even though only one was rooted in the quadrat. This species could be a plant such as *Yucca*, with radiating basal leaves; the leaf tips describe the perimeter of a circle and cover is estimated as though the circle was completely covered with leaves. Shrub E has a canopy which does not actually penetrate into the quadrat, yet if one fills in the canopy holes by connecting the radiating branches with an imaginary polygon or circle, then there is "cover." Cover values for this 0.1 m^2 quadrat are given in Table 8–4. (Reprinted by permission from Daubenmire 1959.)

Table 8–4. Absolute and relative cover and density, based on the 0.1 m^2 quadrat shown in Figure 8–5.

Species	Absolute cover (%)	Relative cover (%)	Absolute density per quadrat	per ha	Relative density (%)
A	0.2	0.4	0	0	0
B	33.2	68.8	1	100,000	4.2
C	0.7	1.5	9	900,000	37.5
D	6.2	12.9	14	1,400,000	58.3
E	7.9	16.4	0	0	0
total	48.2	100.0	24	2,400,000	100.0
overlap	5.3	—	—	—	—
bare ground	57.1	—	—	—	—

means that the overlap is counted twice. The D plants, some small herbs, are mentally grouped together and their cover is estimated separately from the overtopping B plants. In some vegetation, with many overlapping canopies, total cover could exceed 100% and there could still be bare ground. For this reason, bare ground cannot be estimated by subtracting total plant cover from 100%. Some ecologists, however, do not round out the canopies to the extent shown in Figure 8–5; they would award 0% cover for shrub E. Also, some ecologists do not count overlapping canopy areas in the same stratum twice.

Relative cover is the cover of a particular species as a percentage of total plant cover. Thus, relative cover will always total 100%, even when total absolute cover is quite low, as in the case of Figure 8–5 and Table 8–4.

Cover of tree canopies can be difficult to estimate, as well as painful after a few hours of neck bending. One solution is to use a "moosehorn" crown closure estimator (Garrison 1949). A periscope-like device is attached to a staff so that the eyepiece is easy to use for viewing while standing; the view of the canopy is seen superimposed on a template of dots. The percentage of dots "covered" by canopies is equivalent to percent cover. These readings could be taken at one location in each quadrat. A more elegant method is to take a picture of the canopy with a fish-eye lens from one location in each quadrat, then to analyze the photographs later for percent cover. Typically, however, canopy cover of trees is assumed to correlate with trunk cross-sectional area **(basal area, BA)** or with trunk **diameter at breast height (dbh)**. To obtain the basal area, the tree is usually measured with a special diameter tape that converts circumference to diameter units. Sometimes **relative dominance** is used as a synonym for relative basal area or relative cover.

Cover of shrubs and herbs is usually estimated to the nearest whole number or put into cover categories, but if greater detail is required the quadrats may be photographed from above or a scale drawing can be made with the help of a pantograph (Figure 8–6). Neither method works very well when the quadrat is larger than 1 m^2.

Density is the number of plants rooted within each quadrat. The average density per quadrat of each species can be extrapolated to any convenient unit area. For example, Figure 8–5 and Table 8–4 show that herb D had a density of 14 plants per 0.1 m^2 quadrat, which converts to 1.4 million plants per hectare. **Relative density** is the density of one species as a percent of total plant density. **Mean area** is plot area/density; it is the area per plant. Density is independent of cover. For example, many young, slender trees may have a higher density but a lower cover than a few older, branching trees. **Abundance** is a rather nebulous term, but often it is used as a synonym for density.

Frequency is the percentage of total quadrats which contains at least one rooted individual of a given species. It is partly a measure of the same thing the relevé investigators call sociability. Rarely, frequency is expressed on a cover basis: any plant, whether rooted in the quadrat or not, which contributes cover for species X is tallied as "present," and frequency becomes the percent of all quadrats in which the canopy of X was "present." **Relative frequency** is the frequency of one species as a percentage of total plant frequency.

As mentioned in previous chapters, frequency is an artifact of quadrat size, and in this respect it is a more artificial statistic than cover or density. If, however, the quadrat used is large enough and the plants are distributed randomly, then there is a relationship between frequency and density:

$$\% \text{ frequency} = 100(1 - e^{-m})$$

where e is the base of natural logarithms and m is the average density per quadrat (Curtis and McIntosh 1950). Obviously, here frequency must be *rooted* frequency. Usually, however, plants are not distributed randomly and then frequency and density are independent of each other. Thus, a clumped species may have a high density but a low frequency, while a much less abundant species distributed singly and regularly throughout a stand will have a low density but a high frequency.

It is unfortunate that such an important ecological term as **dominance** is still ambiguously defined by many ecologists. Generally, the dominant species of a community is that overstory species which contributes the most cover or basal area to the community, compared to other overstory species. This definition is based on physiognomy. If

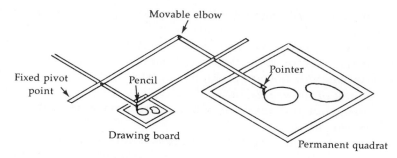

Figure 8–6. A pantograph, which is used to translate actual plant cover and plant location from a quadrat to a scale drawing.

oak has the highest relative cover in an eastern deciduous forest with oak, hickory, and elm in the overstory, then oak is said to be the dominant species. If all three species contribute about the same amount of cover, or if the balance shifts from one to the other depending on the stand, then the three species are codominants. In a savanna or a semidesert woodland, where tree canopies may contribute only 10–30% cover and understory plants such as grasses or shrubs contribute more, density enters the definition and some grass or shrub species will usually be considered the dominant species.

Another view of dominance is **sociologic dominance** (Kershaw 1973). Sociologic dominants control the reproduction and continued existence of a community, and they may be understory species. For example, the regeneration of ponderosa pine saplings in some ponderosa pine forests is inhibited by root competition for moisture by the understory grass *Festuca arizonica*, but not by another grass *Muhlenbergia montana* (Pearson 1942). The grasses are the sociologic dominants of the ponderosa pine community, even though ponderosa pine is the physiognomic dominant. For this reason, some methods of community description name communities by both overstory and understory species and separate the two species by a slash, for example, the *Pinus ponderosa/Muhlenbergia montana* community, or the *P. ponderosa/Purshia tridentata* community (Daubenmire 1952).

Foresters call any individual tree whose canopy is more than half exposed to full sun a dominant, even though it may not be a member of a species which is a physiognomic or sociologic dominant of that community. In this sense, dominant is a synonym for **emergent**, and the latter term should be used. In some other forest studies, dominance is equivalent to trunk basal area: the species with the most basal area per hectare is called the dominant. This use of the term dominant corresponds with our definition.

Finally, the term **aspect dominance** is applied to species which are very noticeable and at first glance appear to dominate a community by cover. Careful sampling would reveal, however, that other, less conspicuous species in the same canopy layer contribute more cover and are the actual dominants. Aspect dominance is most common in herbaceous communities, such as grasslands or meadows, where all members of one species will flower synchronously and in this way stand out from the rest of the vegetation.

Throughout this text, we will use the term dominant in the physiognomic sense.

Importance refers to the relative contribution of a species to the entire community. It can be used in a very nebulous, almost intuitive, informal sense, or it can be calculated in a precise way. At the investigator's pleasure, importance may be synonymous with any one measure—for example, density—but originally importance was defined as the sum of relative cover, relative density, and relative frequency (Curtis and McIntosh 1951). In the latter case, the **importance value (IV)** of any species in a community ranges between 0 and 300. Table 8–5 illustrates the calculation of IVs for all overstory trees in a Hawaiian rain forest. Notice that two species with similar IVs could have entirely different values for relative cover, density, and frequency; any differences are submerged in the addition process, and the one number that results is a synthetic index of importance. Other formulae for IV calculation have been developed, and they may sum only two relative values rather than three (Bray and Curtis 1957; Ayyad and Dix 1964), or sum more than three values (Lindsey 1956).

Table 8–5. Calculation of importance value (IV) for a rather open tropical rain forest at 450 m elevation near Honolulu, Hawaii. The four most abundant overstory trees are summarized below; common understory plants such as tree ferns *(Cibotium splendens)* are not included. In this study, cover is actually the basal area of all stems greater than 3 cm dbh. (From *Aims and Methods of Vegetation Ecology.* Mueller-Dombois and Ellenberg. Copyright 1974 John Wiley and Sons, Inc. Reprinted by permission of John Wiley and Sons, Inc.)

Species	Relative density	Relative cover	Relative frequency	IV	IV rank
koa tree *(Acacia koa)*	30.0	78.4	30.8	139.2	1
ohia lehua *(Metrosideros collina)*	20.0	13.9	23.1	57.0	3
ohia *(M. tremuloides)*	5.0	5.8	7.7	18.5	4
guava *(Psidium guajava)*	45.0	1.9	38.5	85.4	2

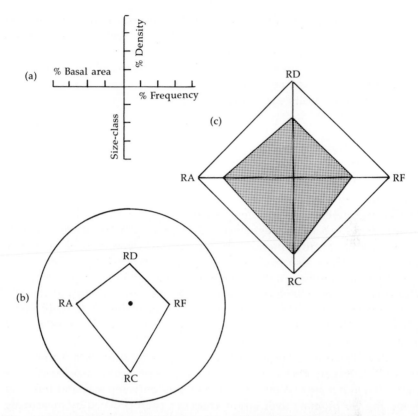

Figure 8–7. A phytograph. (a) Coordinates as originally designed by Lutz.
(b) A representation of sugar maple from an Ohio forest using relative
density (RD), relative frequency (RF), relative cover (RC), and relative basal
area of stems > 4 cm dbh (RA). (c) The same sugar maple data put on square
root axes, with diagonal reference lines added. ((a) reprinted from *The
Description and Classification of Vegetation* by D. W. Shimwell. Copyright 1971
by University of Washington Press, Seattle. (b) and (c) modified from
McCormick and Harcombe 1968. Copyright 1968 by the Ecological Society
of America.)

Importance may also be graphically presented in the form of a
four-sided phytograph (Figure 8–7). As originally proposed for tree
species by Lutz (1930) (Figure 8–7(a)), one axis shows relative density
of trees over 25 cm dbh, another shows frequency of trees over 25 cm
dbh, a third shows the number of individuals in predetermined size
(dbh) classes,* and a fourth shows relative basal area. The points on the
axis are then connected and species in a stand can be visually compared
by the shape and area of the resulting quadrangles. McCormick and

*Lutz's size classes are: 1 = up to 30 cm tall; 2 = 30 cm–4 m; 3 = saplings
2–8 cm dbh; 4 = poles 10–25 cm dbh; 5 = 25 cm dbh and over.

Harcombe (1968) proposed several modifications: (a) that the axes be on a square root scale, rather than an arithmetic scale, which in effect accentuates differences between 0 and 60 (Figure 8–7(c)); (b) that only relative values be used, rather than mixing relative and absolute scales; and (c) that the axes be relative density, basal area, cover, and frequency, or other traits such as biomass.

Biomass and Productivity

Biomass is the weight of vegetation per unit area; synonyms are standing crop and phytomass. The dominance or importance of any species can be expressed as the percentage of total biomass. For small quadrats in herbaceous vegetation, biomass may be measured by clipping all aboveground matter, drying it in an oven, and weighing it. Ideally, roots are also excavated, but they are often ignored; consequently, most biomass data represent only aboveground plant matter. Quadrat size and shape are just as important for obtaining accurate estimates of biomass as they are for other measures. Relatively large, circular quadrats may be the most efficient and give the most precise estimates (Van Dyne et al. 1963).

The clearing of large plots in woody vegetation is not practical. Instead, relatively few individuals of different age or size classes are harvested and a regression line is developed between size and biomass. Sampling for biomass over larger areas can then proceed by measuring plant size and the data can be converted to biomass.

Productivity is the rate of change in biomass per unit area over the course of a growing season or a year. Productivity and biomass may not be related. A mature forest has a large biomass but may exhibit a small productivity; a grassland has a smaller biomass but may exhibit a larger productivity. Productivity and biomass data may serve to characterize a particular vegetation type, as summarized by Rodin and Bazilevich (1967) and Whittaker (1975). Methods of biomass and productivity sampling have recently been thoroughly reviewed by Chapman (1976).

LINE INTERCEPT, STRIP TRANSECT, AND BISECT METHODS

H. L. Bauer (1943) developed the **line intercept** method for dense, shrub-dominated vegetation, which he found to be as accurate as traditional quadrat methods, but less time-consuming. If a quadrat is reduced to a single dimension it becomes a line. The line may be thought of as representing one edge of a vertical plane which is perpendicular to the ground; all plant canopies projecting through that plane, over the line, are tallied. The total fraction of the line covered by each species, multiplied by 100, is equal to its percent cover. Just as with quadrats, total cover can be more than 100%. Disadvantages of the method are the loss of density and frequency measures, because there is no area involved (although frequency can be expressed on a cover basis if the line is broken up into segments).

Often, a lengthy line intercept is combined with quadrats which run alongside it. Cover is measured along the line and density or frequency is noted in the quadrats. If the quadrats run continuously along the line, the method is called the **belt transect**, **strip transect**, or **line strip method**. These methods have been most often applied to forest vegetation (Lindsey 1955).

Bisects are scale drawings of the vegetation within line strips. The idea was originally applied to tropical forests (Davis and Richards 1933; Richards 1936; Beard 1946); Figure 8–8 is an example from the British West Indies. All plants in a strip approximately 60 m long and 8 m wide are shown, drawn as accurately as possible. For those who are not good artists, bisects can be drawn in highly diagrammatic fashion using symbols (Figure 8–9).

These three methods can record cover as a function of height above the ground, if the sampling is done carefully enough. When the data are summarized in bar graphs such as the one in Figure 8–10, striking differences between vegetation types become apparent. The eastern deciduous forest is seen to be composed of four canopy layers, with most cover being contributed by the overstory tree layer; in contrast, the boreal forest has three canopy layers, with the trees and ground (herbaceous) layers providing nearly continuous cover.

THE POINT METHOD

If a quadrat is reduced to no dimension, it becomes an infinitely small point. In practice, metal pins with sharp tips serve as the points, and cover is equal to the fraction of total pins which touch any plant part as the pin is lowered. Typically the pins are arranged in frames which rigidly limit the pin to a vertical path perpendicular to the ground (Figure 8–11). The frame may be located at several random places in a stand and 1–10 pins can be lowered at each place. For the best precision of estimated cover, lowering only one pin at each place is better than lowering several (Goodall 1957).

As the pin is lowered, the first plant it touches is recorded, then the pin is lowered more until it touches another leaf (of the same or a different plant than the first touch), and so on until bare ground is

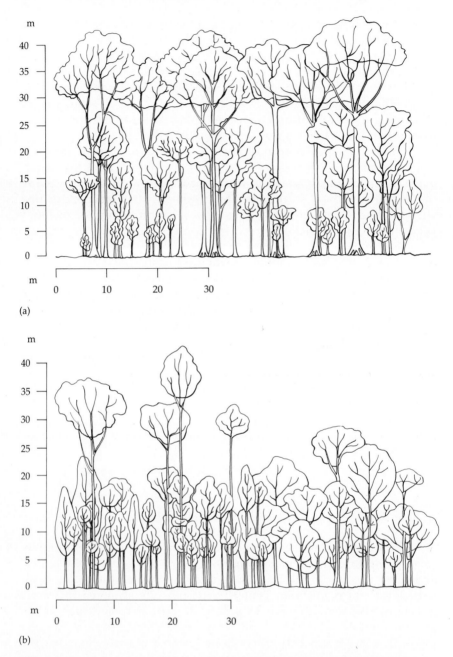

Figure 8–8. Bisects of tropical rain forest. (a) Trinidad, British West Indies. (b) Borneo. Both bisects represent all vegetation within a strip 61 m long and 7.6 m wide. ((a) from Beard 1946. By permission of the British Ecological Society; (b) from Richards 1936. By permission of the British Ecological Society.)

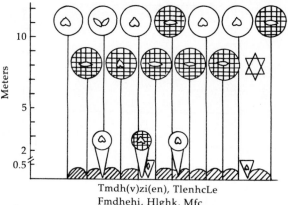

Tmdh(v)zi(en), TlenhcLe
Fmdhehi, Hlghk, Mfc

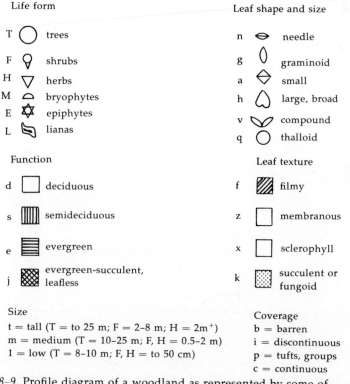

Life form

T	⬤	trees
F	♀	shrubs
H	▽	herbs
M	⌂	bryophytes
E	✡	epiphytes
L	🖙	lianas

Function

d	□	deciduous
s	⦀	semideciduous
e	▤	evergreen
j	▨	evergreen-succulent, leafless

Size

t = tall (T = to 25 m; F = 2–8 m; H = 2m$^+$)
m = medium (T = 10–25 m; F, H = 0.5–2 m)
1 = low (T = 8–10 m; F, H = to 50 cm)

Leaf shape and size

n	◉	needle
g	◊	graminoid
a	◇	small
h	⬠	large, broad
v	♈	compound
q	○	thalloid

Leaf texture

f	▨	filmy
z	□	membranous
x	□	sclerophyll
k	▦	succulent or fungoid

Coverage

b = barren
i = discontinuous
p = tufts, groups
c = continuous

Figure 8–9. Profile diagram of a woodland as represented by some of Dansereau's symbols. (From Dansereau 1951. By permission of the Ecological Society of America.)

reached. If no plant is hit, then the point is tallied as bare ground. These data permit two calculations. One is percent cover:

$$\% \text{ cover} = \frac{\text{no. of pins which hit species A as least once}}{\text{total no. of pins}} \times 100$$

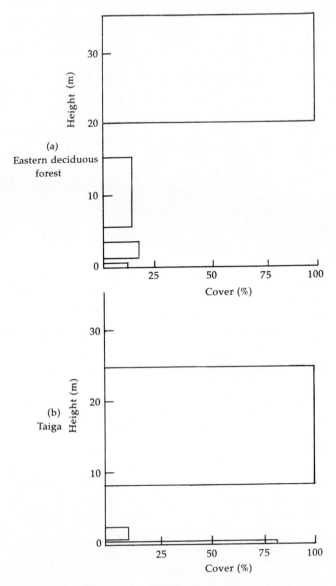

Figure 8-10. Canopy profiles. The height of the horizontal bars represents the average span of canopy height; the length of each bar represents the total cover by all species in that height range. (a) Typical eastern deciduous forest. (b) Typical conifer forest in the taiga across Canada.

The other calculation is percent of sward, which weights each species by its canopy thickness, or cover repetition, at each point:

$$\% \text{ sward} = \frac{\text{no. of contacts with species A}}{\text{total no. of contacts}} \times 100$$

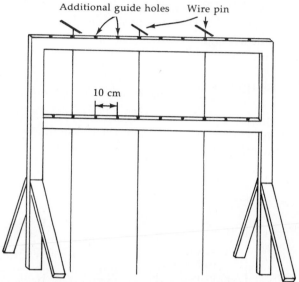

Figure 8-11. A point frame suitable for 1–10 pins. (From *Aims and Methods of Vegetation Ecology.* Mueller-Dombois and Ellenberg. Copyright 1974 John Wiley and Sons, Inc. Reprinted by permission of John Wiley and Sons, Inc.)

Disadvantages of the point method are that density and frequency cannot be measured (although cover frequency can), and it is limited to low vegetation, such as grassland, for obvious reasons. But for measuring cover of low vegetation, it may be the most trustworthy and objective method available (Goodall 1957).

DISTANCE (PLOTLESS) METHODS

Distance methods do not use quadrats, lines, or point frames. Only distances (from a random point to the nearest plant, or from plant to plant) are tallied. Average distance, multiplied by an empirically determined correction factor, becomes density. The basic distance methods were developed by Grant Cottam and John Curtis at the University of Wisconsin in the 1950s, and were tested and refined on maps of real and artificial forest vegetation. The five methods briefly described here have best been summarized and compared by Cottam and Curtis (1956), Lindsey et al. (1958), and Mueller-Dombois and Ellenberg (1974). Four methods are illustrated in Figure 8–12. These methods have been used with many different types of plants, but most often with trees.

Nearest Individual Method

Random points are located in a stand. At each point the distance to the nearest tree of any species is recorded, the species is identified, and its basal area is measured. Only one measurement is made from each random point. All distances for all species are summed and divided to yield one average distance. Density per hectare (10,000 m^2) for all trees is then:

$$\text{density} = \frac{10,000}{2 \, (\text{average distance, in meters})^2}$$

If distance is measured in feet, then the numerator is 43,560, the number of square feet in an acre. The 2 in the denominator is a constant correction factor. Relative density of each species is:

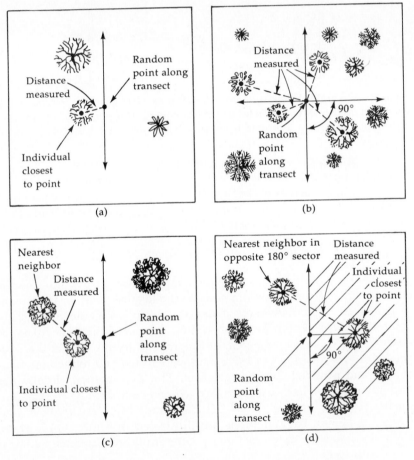

Figure 8–12. Distances measured by four distance methods. (a) Nearest individual method; (b) point-centered quarter method; (c) nearest neighbor method; and (d) random pairs method. The shaded area of (d) is an excluded area which contains the first plant, but cannot contain the second.

$$\text{relative density of species A} = \frac{\text{no. of trees of species A encountered}}{\text{no. of trees encountered}} \times \text{density for all trees}$$

Cover or dominance of each species can then be calculated as its relative density times its average basal area.

Point-Centered Quarter Method

Again, random points are located. The area around each point is divided into four 90° quarters of the compass, and the nearest tree in each quarter is sought. Each tree is identified, its basal area is measured, and its distance from the random point is measured. Again, average distance for all trees taken together is computed and this is converted to total density by the formula given for the nearest individual method, except that the correction factor vanishes. The correction factor is 1, not 2. Since more information is gained at each point than in the nearest individual method, the point-centered quarter method is more efficient and it requires only about ¼ the number of points to achieve the same level of accuracy and precision.

Nearest Neighbor Method

Random points are located in a stand and the nearest plant is located. The distance measured, in this case, is from that plant to its nearest neighbor (of any species). Total density is calculated as for the nearest individual method, except that the correction factor is 1.67. This nearest neighbor method has also been used to determine whether trees of the same species are distributed at random, are clumped, or are regular (Clark and Evans 1954; Pielou 1961).

Random Pairs Method

In the random pairs method, the nearest plant to a point is located. A line from point to plant is imagined. Perpendicular to it and passing through the point, is an exclusion line. In Figure 18–12(d) the exclusion line happens to correspond with the transect. A nearest neighbor is now searched for, but it cannot be on the same side of this exclusion line as the first tree. The conversion factor in the density formula is 0.8.

Bitterlich Variable Plot Method

A final distance method can be used to calculate basal area only, but basal area is important in calculating board feet of lumber and the method is extremely fast and has been widely adopted by foresters. It yields more reliable data for less field time than quadrat methods or other distance methods (Lindsey et al. 1958). The method is named after its German inventor,* who originally used a sighting stick 100 cm long with a crosspiece at one end, 1.4 cm across (Figure 8–13(a)). The stick was held horizontally, with the plain end at one eye, and the viewer would slowly turn in a complete circle. Every tree whose trunk was seen in the line of sight on the circuit was tallied and identified as to species if its trunk appeared to exceed the width of the crosspiece; all other trees were ignored. Using a stick of these dimensions, and based on geometric principles, the total basal area in $m^2\,ha^{-1}$ for any species is equivalent to the number of trees of that species tallied, divided by 2. If English units are preferred, then a stick 33 inches long with a crosspiece 1 inch across will give basal area in $ft^2\,acre^{-1}$ if the trees tallied are multiplied by 10.

As shown in Figure 8–13(b), an angle is being projected—an angle whose size depends on the relative lengths of the stick and the crosspiece. For example, the 33 inches $\times$ 1 inch arrangement produces an angle of 1°45′ and an English units basal area factor (BAF) of 10. BAF is the number which is multiplied by the number of tallies to obtain basal area. If the angle becomes smaller, more trees will be tallied and the BAF will become smaller; for example, an angle of 0°33′ has an English units BAF of 1. If the angle becomes larger, then fewer trees are tallied and the BAF will become larger; this is useful in a dense forest to avoid miscounting. In this country, angles which give BAFs of 5–20 are commonly used.

More recently, small, hand held prisms have replaced sighting sticks (Figure 8–13(c)). Looking both through and over the top of a prism, the lower trunk will appear to be offset more or less than the trunk diameter; if the lower trunk appears smaller than the trunk diameter, the tree is tallied.

SUMMARY

A major objective of synecologists is to complete an inventory of the earth's plant resources: not an inventory of individual species, but an inventory of communities and vegetation types. The results of such an inventory have application to applied and basic science, and to autecology as well, for the environment of a plant includes adjacent organisms as well as the physical factors of climate and soil. The inventory is far from complete, because of limitations in researchers, scientific interest, and accessibility of some areas. Conclusions will have to be based on samples representing 1% or less of the earth's surface vegeta-

*For an English article describing the method and the geometry behind it, see Grosenbaugh 1952.

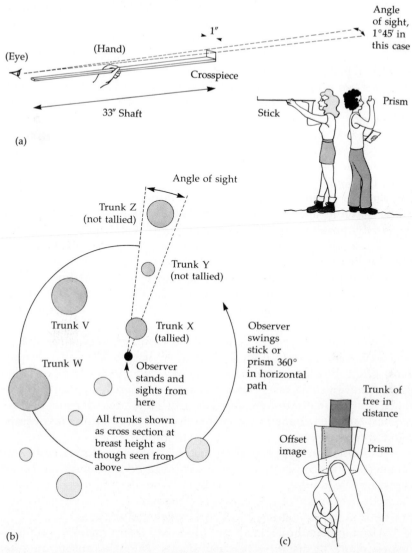

Figure 8-13. The Bitterlich variable plot method for calculation of basal area. (a) The original Bitterlich stick sighting device. This particular one will give total basal area in ft^2 acre^{-1} if the number of trees tallied is multiplied by 10. (b) A bird's eye view of trees which would be tallied as an observer at the center point turns 360°. Tree X is counted because its trunk diameter exceeds the angle projected by the sighting device, but trees Y and Z are not tallied. (c) The prism type of sighting device. In this case, the lower trunk appears larger than the upper trunk and the tree would not be counted.

tion. If the estimates are to be accurate and precise, the samples must be chosen carefully. This chapter described procedures for choosing and measuring those samples.

The relevé method uses a more subjective choice of sample locations than any other method, but the process of recording data is relatively rapid and nonmathematical, and its widespread use in the non-English speaking world makes it an attractive sampling method. Each stand is represented by one large quadrat whose size must meet minimal area requirements. Data recorded include cover, sociability, vitality, and periodicity. Important synthetic values are presence and constance.

Random quadrats methods involve fewer subjective decisions than the relevé method. Questions that have to do with the size, shape, number, and placement of quadrats are time consuming to answer, and there is no one best solution to them. A statistical level of confidence can only be applied to data from random quadrats. Such data can include absolute and relative cover, density, frequency, and biomass. Generally these four measurements are unrelated. If plants are distributed randomly, density and frequency are related, but plants are usually not distributed randomly. Cover may be the most meaningful of the four. It can be estimated as actual crown cover, but for trees it is assumed to correlate with trunk basal area. Cover and basal area are both aspects of size. Frequency is the most artificial of the four, being an artifact of quadrat size.

Two synthetic values from quadrat data are dominance and importance. A single definition of dominance has not yet been accepted. Most often, however, dominance is related to the cover of overstory species, but in some cases density is also considered. Typically, importance is a summation of relative cover or basal area, density, and frequency, but the summation may involve one more or one less component, and it can be expressed as a phytograph.

The line intercept sampling method reduces quadrats to a single dimension, and consequently only cover can be measured. Line intercepts may be combined with quadrats, as in strip transect and bisect methods. The point method reduces quadrats to no dimension; it is most useful for estimating cover in low vegetation types.

Distance methods measure distances from random points to nearest plants or distances between plants. The data can be converted to plant density, and in some cases the data can reveal whether plants are randomly or nonrandomly distributed. The most efficient distance method is the point-centered quarter method. The Bitterlich variable plot method was treated as a distance method, even though it only permits the calculation of basal area, because the geometric assumptions behind it relate to distance from observer to tree, and to tree diameter.

CHAPTER 9

METHODS OF DESCRIBING THE PLANT COMMUNITY

No matter how sampling is done—by quadrats, line intercept, points, or distances—the raw data are too voluminous or disorganized to mean much to the investigator, let alone to someone half a world away with whom the investigator would like to communicate. This chapter describes the commonly used methods of summarizing sampling data.

Sampling and data analysis, of course, are not independent of each other. Some methods of analysis require certain methods of sampling. For example, association analysis requires the use of random quadrats, and the Braun-Blanquet table method requires relevé information.

The biases and objectives of the investigator also influence the choice of a data analysis method. If the aim is to draw lines and describe discrete entities, then the table method and vegetation mapping are good choices. If the aim is to let the reader draw the lines between entities, then cluster analysis and association analysis are good choices. If the continuum of vegetation is to be emphasized, then gradient analysis and ordination are good choices.

A word of caution: the statistical tests and the mathematical rigor behind them are only superficially presented in Chapter 8 and in this chapter. The old saying, "A little knowledge can be a dangerous thing," fits the area of sampling and analysis very well. Although we will continue to cite the original literature in this chapter, these citations may not be helpful unless you possess some mathematical background, for their subject matter is technical and the reading is sometimes difficult. Our objective is to survey some commonly used procedures, so that you are reasonably familiar with the options available. Many professional ecologists find the mathematics and computer requirements behind certain methods formidable and do not use them. Other techniques, such as ordination, are still being amended and refined, since even mathematically oriented ecologists are not satisfied with them.

THE TABLE METHOD

Once sufficient stands thought to represent one or more communities have been sampled with relevés (see Chapter 8), the results are summarized in a **primary data table** (matrix) such as Table 9–1, which gives data for English pastures. In such a primary matrix, the stands and species are listed in the order in which they were sampled and encountered. The objective now becomes to generate a second, **differentiated table**, where similar stands lie near each other and species with similar distributions also lie near each other. As the rows (species) and columns are shifted about with scissors and tape, it soon becomes apparent that some species are of little use in differentiating groups of stands. Some species occur so rarely that only one or two stands show them; others occur so ubiquitously that nearly every stand shows them; still others do not form species groups. Such species will be eliminated from the differentiated table. In the case of Table 9–1, 24 of the total 39 taxa were in this unusable category, and they are not included in Table 9–2.

Table 9–1. The raw data table, listing species (down) chronologically or phylogenetically, and listing relevés (across) in numerical sequence. The units digit represents cover (see Table 8–2) and the tenths digit represents sociability (see Table 8–3). (Reprinted from *The Description and Classification of Vegetation* by D. W. Shimwell. Copyright 1971 by University of Washington Press, Seattle.)

Relevé number	1	2	3	4	5	6	7	8	9	10	11	12	13	14	15	16	17	18	19	20
Phanerogams[a]																				
Hippophae rhamnoides	5·1	5·1	5·1	5·1	5·1	3·2	2·2	4·3	2·2	5·1	3·2	5·1	3·3	4·3	3·2	5·1	5·1	5·1	5·1	5·1
Senecio jacobaea	1·1	†	1·1	†	†		†		†		†		†			1·1			†	1·1
Solanum dulcamara	2·1	2·1	†	†	†		†			1·1	†	†					†		1·1	1·1
Rubus fruticosus s.l	†	1·1		†	1·1				†										†3	2·3
Urtica dioica	3·3	1·3	1·3				†		†				†	†				1·1	†	
Rumex crispus	†	†	†						†							4·5	2·3	1·3		
Montia perfoliata				3·4	4·4						4·5		2·3				2·3	1·2		
Stellaria media				1·2	†						3·4									
Festuca rubra						1·1	†	1·1	†		2·3		3·3	1·3	1·3					
Agropyron repens							†	2·3	†	3·3			2·3	†2						
Ammophila arenaria						2·3	4·3	2·3	†		3·3		1·3	†3	4·3					
Sonchus arvensis						†	†	†	†		1·1			†					†	
Ononis repens						1·1														
Galium verum						†								†						
Calystegia soldanella							†	1·1	†	†		†		1·1	†		†			
Poa pratensis								†	†		1·1		†	†	†					
Agrostis stolonifera								†	†2					†	†2		†			†
Ranunculus bulbosus								†	†		†		†	†						
Plantago lanceolata									†				1·1	†	1·1					
Veronica chamaedrys								†	†											
Chamaenerion angustifolium									†			1·1		†		†			2·3	†
Cerastium vulgatum									†			†		†		†	1·1	1·1		
Sambucus nigra														†						†2
Cirsium vulgare														†			†			
Heracleum sphondylium														†						
Inula conyza														†			1·1		†	
Cardamine hirsuta														†			†			
Hypochaeris radicata																		†		
Arrhenatherum elatius																		1·1		
Sonchus asper																				†
Cryptogams[b]																				
Eurynchium praelongum	1·3	†3	1·3																1·3	1·3
Hypnum cupressiforme	†	†2	†2		†				†					†			†			
Brachythecium rutabulum		†	1·3	†	†				†3		†							†	†3	
Geastrum fornicatum				†	†															
Brachythecium albicans						†3			1·3		†3			†						
Bryum inclinatum						†3								†						
Tortula ruraliformis								†3						†3						
Cladonia rangiformis													†							
Bovista nigrescens														†						
Lophocolea heterophylla																				†
Number of Species	8	9	8	8	9	11	7	10	14	12	12	10	10	22	10	11	12	12	12	13

[a]Seed plants.

[b]Seedless plants.

Table 9–2. The differentiated table. Species and relevés shown in Table 9–1 have been rearranged so that similar groups (associations A, B, C, boxed) stand together. A large group of species which did not contribute to defining associations has been left out of the table. Also, relevé 14 has been omitted because it was a disturbed site. The units digit represents cover (see Table 8–2) and the tenths digit represents sociability (see Table 8–3). (Reprinted from *The Description and Classification of Vegetation* by D. W. Shimwell. Copyright 1971 by University of Washington Press, Seattle.)

Revised relevé order	1	2	3	19	20	4	5	10	12	16	17	18	7	6	8	9	11	13	15
Group A																			
Urtica dioica	3.3	1.3	3.3	†3	2.3														
Eurynchium praelongum	1.3	†3	1.3	1.3	1.3														
Group B																			
Montia perfoliata						3.4	4.4	4.5	2.3	4.5	2.3	1.3							
Stellaria media						1.2	†	†	3.4	†	2.3	1.2							
Geastrum fornicatum						†	†				†	†							
Cerastium vulgatum								†	†	†	1.1	1.1							
Cirsium vulgare											†	†							
Cardamine hirsuta										†	†								
Group C																			
Festuca rubra							†						†	1.1	1.1	†	2.3	3.3	1.3
Agropyron repens													2.3	†	†	3.3	†	†	†2
Ammophila arenaria													4.3	2.3	2.3	†	3.3	1.3	4.3
Poa pratensis															†	†	1.1	†	†
Plantago lanceolata																†	†	1.1	1.1
Brachythecium albicans														†3			1.3	†3	
Ranunculus bulbosus																†	†	†	

The remaining species are called differential or characteristic species. Their **fidelity** (faithfulness) to a given association can be expressed on the basis of how few stands outside the association contain them. The exact level of fidelity demanded varies from investigator to investigator, but a general rule is that a species which helps to define an association cannot occur in more than 20% of the stands outside that association.

Fidelity is not related to constancy. A species may be restricted to association X, but it may occur rarely even there and have a constancy of only 10%. Useful species must have both moderately high fidelity and moderately high constancy. Again, the required level of constancy varies, but in general it must be $50^+\%$.

As characteristic species are searched for, a reciprocal problem must be solved simultaneously: how many characteristic species must be shared by any two stands before they are considered part of the same association? A common answer is that $50^+\%$ of the total list of characteristic species for association X must be present in any of the stands belonging to it.

All of these criteria have been met in the differentiated table shown in Table 9–2. Three associations (Groups A, B, C) are outlined in boxes. Recently, a computer program has been developed to accomplish the preliminary task of rearranging columns and rows to obtain a differentiated table (Ceska and Roemer 1971).

Once there has been sufficient sampling of associations in a region, it may be possible to group similar associations together into a higher level of classification, analogous to the way many species can be grouped into a genus. This higher level is called an **alliance**. Alliances can then be grouped into **orders**, and orders into **classes**. In this way, a hierarchical classification of all vegetation in a region is possible, based largely on floristic similarities. The accuracy and usefulness of such a classification depends on trained, perceptive ecological imagination. Since Braun-Blanquet possessed this trait to an extraordinary degree, his system has been widely adopted.

ORDINATION

Ordination attempts to summarize sampling data in a simpler, less space-consuming fashion than the table method. Even a rather small differentiated table, such as shown in Table 9–2, contains 285 cells or bits of data (19 stands $\times$ 15 differential species). An ordination of the same data could be one small graph showing 19 points spread out in space. Each point represents a stand, and the distance between points represents their degree of similarity or difference. At a glance, one can see if there are any patterns of relatedness. Are some points (stands) clustered together; do others seem to form a continuous progression from one extreme to another? The objective of ordination is not to draw lines around similar stands and label them part of an association; rather, it is to show a pattern of continuous relationships. Obviously, much of the information contained in the original data is lost in the ordination diagram, but this loss is a consequence of any kind of data reduction, not just of ordination.

The first step in ordination is to express the similarity between any two stands in a single number, called the **community coefficient (CC)**. There are several ways to calculate CC, as Table 9–3 shows (see also Goodall 1973), but basically all the formulae indicate, in some way, the number of species shared by two quadrats. A CC of 100 represents identity, while a CC of 0 represents complete difference. Because of the variation in vegetation from place to place, even two relevés or plots in one stand would probably not have a CC of 100. Any two plots from the same association would, however, show a CC of 50^+. When evaluating data by the ordination method, a matrix of CC values is prepared for every pair of stands. For n stands, there will be $(n)(n - 1)/2$ different CC calculations. Figure 9–1(a) shows hypothetical data for seven stands, A–G; the CC values range from 90 (stands A and G) to 20 (stands A and B).

The second step is calculation of a matrix of dissimilarity rather than similarity. Each pair of stands has an **index of difference (ID)**, which is equal to $100 - CC$ (Figure 9–1(b)).

The third step is a transfer of the ID values to a graph. There are several ways to make the transfer and some require the use of comput-

Table 9–3. Four methods of calculating the community coefficient (CC) from presence and % cover data.

Species	Presence and % cover data	
	Stand (quadrat) A	Stand B
no. 1	10	20
no. 2	4	12
no. 3	—	7
no. 4	—	15
no. 5	32	15
no. 6	15	—
no. 7	2	—
no. 8	1	1
Total % cover	64	70

Calculation methods

Formula	Calculation	Community coefficient
Jaccard, presence only	$\dfrac{C}{A + B - C} \times 100 = \dfrac{4}{6 + 6 - 4} \times 100 =$	50
Jaccard, weighted by cover	$\dfrac{MC}{MA + MB} \times 100 = \dfrac{(10 + 4 + 15 + 1)}{64 + 70} \times 100 =$	22
Sorensen, presence only	$\dfrac{2C}{A + B} \times 100 = \dfrac{2(4)}{6 + 6} \times 100 =$	75
Sorensen, weighted by cover	$\dfrac{2MC}{MA + MB} \times 100 = \dfrac{2(10 + 4 + 15 + 1)}{64 + 70} \times 100 =$	45

where A = total number of species in stand A
    B = total number of species in stand B
    C = total number of species in both stand A and stand B
    MA = total % cover of species in stand A
    MB = total % cover of species in stand B
    MC = total % cover of species in both stand A and stand B,
        using the lower % cover figure for each species

ers. In the simplest method, called **polar ordination**, two highly dissimilar stands are selected as end points on a horizontal axis. Figure 9–1(c) shows a hypothetical example, where stands A and B have an ID of 80, which is the highest of all pairings for stands A–G. The ID of 80 becomes 80 graph units on axis 1. All other stands are now placed on the same axis by plotting their IDs with reference to A and B. For example, stand C has an ID of 67 with A, and an ID of 63 with B. A compass with a radius of 67 units is swung from A, and a compass with a radius of 63 units is swung from B, forming two arcs. The intersections of the two arcs define a line perpendicular to the axis. Where that

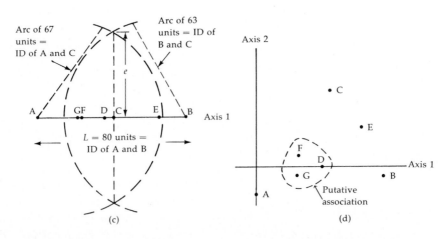

	A	B	C	D	E	F	G
A	—	20	33	65	30	86	90
B		—	37	60	70	52	50
C			—	55	68	49	45
D				—	70	70	75
E					—	45	47
F						—	40

CC values
(a)

	A	B	C	D	E	F	G
A		80	67	35	70	14	10
B			63	40	30	48	50
C				45	32	51	55
D					30	30	25
E						55	53
F							60

ID values
(b)

Arc of 67 units = ID of A and C

Arc of 63 units = ID of B and C

L = 80 units = ID of A and B

(c)

Axis 2

Axis 1

Putative association

(d)

Figure 9–1. The location of stands in an ordination figure. (a) Community similarity matrix for seven hypothetical stands, A–G, showing community coefficients (CC). (b) Community dissimilarity matrix for the seven stands, showing indices of difference (ID). (c) After the end point stands A and B have been located on axis 1, all other stands are located between them by swinging arcs that correspond to their ID values with stands A and B. Stand C, for example, has an ID of 63 with B and an ID of 67 with A. The crossing arcs define a line perpendicular to the axis; where that line crosses the axis is the location of stand C. The poorness of fit is the distance *e*, from the crossing arcs to the axis. (d) Creation of a second axis serves to pull the stands apart much better than did the first axis alone. Stands F, G, and D (within the dashed line) appear to cluster together and they represent an approximately homogeneous unit, perhaps an association.

line crosses the axis is the position of stand C on that axis. The length of the line from where the arcs cross to the axis (e) is a measure of the poorness of fit for stand C to this one-dimensional graph. The location of stands on the axis can also be calculated instead of plotted. The location of any stand, for example, C, from the left end of the axis (point A) is:

$$x = \frac{L^2 + dAC^2 - dBC^2}{2L}$$

where x is the distance along the axis from the left end, L is the ID of the end point stands (A and B), dAC is the ID between stands A and C, and dBC is the ID between stands B and C. The poorness of fit, e, for stand C is:

$$e = \sqrt{dAC^2 - x^2}$$

e varies for each point on the line, which means that the quality of fit for each point varies. Two points may be next to each other on the x axis because they have similar IDs with the end points. However, if one point has a high e value, it does not fit the line well, and therefore does not really belong next to the other point; the proximity of the two points is an artifact created by this method. A two-dimensional graph would provide a more accurate analysis of the data.

To spread the points in two dimensions, another axis must be generated with its own reference stands. One reference stand may be that stand with the largest e value, as calculated above. In the hypothetical example of Figure 9–1, stand C fills that criterion. The other reference stand is now that stand which has a large ID value with stand C. However, the objective of creating a second axis is to spread out stands that are close together on axis 1 (the x axis), so by agreement the search for the other reference stand is limited to $\pm 0.1L$ units away from C on axis 1. In this case, L is 80 units long, so stands ± 8 units from C are examined. D then becomes the other reference point of axis 2. Now all other stands, R, can be placed along axis 2 using a formula similar to that used for axis 1:

$$y = \frac{L'^2 + dCR^2 - dDR^2}{2L'}$$

where y is the number of units along the second axis from stand D in our example, dCR is the ID between the new stand, R, and the C stand at one end, dDR is the ID between the new stand, R, and the D stand at the other end and L' is the ID between the reference stands, C and D. The values of x and y then become the coordinates for locating the stands on a two-dimensional graph (Figure 9–1(d)).

A third axis may also be constructed, generally at right angles to the first two axes, but often two dimensions serve to spread the stands out well enough and to account for most inter-stand variability. Step-by-step procedures for this method are well illustrated in a laboratory manual by Cox (1976).

Some Limitations

This method of ordination is called polar ordination to emphasize that the reference points of each axis, the poles, are chosen by the investigator. It was the first ordination method to be widely applied, and it was developed by Bray and Curtis (1957). There are some important limitations to the method.

First, the ordination figure that results does not explain the "why" of vegetation any better than the table method. If the axes do lead to a good separation of stands, then the next level of questioning concerns the meaning of the axes. Do they correlate with certain environmental factors; do they represent environmental gradients of major importance to plant and community distribution? To answer that, one has to go back to the stands in nature and start looking at the microenvironment of each. Further, it will be difficult to relate environmental factors in any exact way to the axes, because species and stands will probably not respond to any factor in a linear fashion.

Second, since the objective of ordination is not the separation of stands into associations, the results do not lend themselves to the aims of community classification. If some stands appear to cluster together, as stands D, F, and G do in Figure 9–1(d), we may be tempted to put a dashed line around them and call them representatives of a relatively homogeneous association, but there are no guides to draw such a dashed line with any precision. (For example, should stand A also be included?)

Third, the original data of stands and species are lost to the reader. Most research papers which include ordinations do not also include the raw data, so the reader cannot get much more from the ordination graph than what the author concluded from it. There is no way for someone with a different point of view to reach an independent conclusion, and this is not good for science. After all, no one yet has the final vision of "truth;" all we do is improve our approximation of it. The truncated data of ordination may not permit that improvement process. In fairness, however, it must be added that this criticism applies to some extent to all the data analysis methods described in this chapter.

Fourth, the array of stands in an ordination may be changed by selecting a different CC formula or by choosing the reference stands differently. If some few stands are quite unlike any others, the polar ordination method just described will not yield much separation of stands. The end points will be far apart, but all other stands will cluster between them. Some subjectivity has to go into the selection of reference stands. Ordination works best when the extremes are not too far apart. A partial solution to this problem of subjectivity is the use of a different ordination method called **principal components ordination**. In this method, a computer is used to select the reference stands such that the spread of all stands is maximal (Orloci 1966; Whittaker 1973*b*). In this case, there will be only one ordination array possible, but computer facilities are necessary and some ecologists may lose an intuitive understanding of the process and the result. Other sophisticated, computer based data analysis techniques are available to ordinate both biotic and abiotic distance matrices.

CLUSTER ANALYSIS

As with ordination, the objective of **cluster analysis** is to simplify the data and to present them in graphical form. The resulting figure is not an array of points, but instead is a dendrogram of stands (Figure 9–2) similar to a dendrogram of species that a numerical taxonomist constructs (Figure 3–4).

CC values are computed for every pair of stands. The two stands with the highest CC (closest similarity) are plotted on a graph as vertical lines which are joined by a horizontal line at that CC value. In Figure 9–2, stands 1 and 4 have the highest CC, 41, and they are joined at that level. Now these two stands are lumped into a new, artificial, second level stand, and CC values are computed all over again. In this case, it turned out that the highest CC value in the second calculation was for stands 13 and 22. They are joined, CC values are recomputed, and so on until all stands are joined at some, generally low, CC value (a CC of 8 in Figure 9–2).

A cluster analysis dendrogram can be used for classification if the investigator selects some threshold value at which to define associations. As already mentioned, stands of one association are often expected to share a CC of 50[+]. Using that criterion, every one of the 25 stands in Figure 9–2 represents a different association. If a threshold value of 30 is chosen instead of 50, then 15 associations would result, as represented by the 15 vertical lines projecting up through the dashed Threshold I line in Figure 9–2. If a threshold of 20 is selected instead, then seven associations would result.

The advantage of cluster analysis over the table method and ordination is that classification is possible, even though it is arbitrary. Although it shows classification to be subjective, at least cluster analysis quantifies the classification process, because some threshold value is chosen as the lower limit to an association. Also, the relatedness of different associations can be quantified.

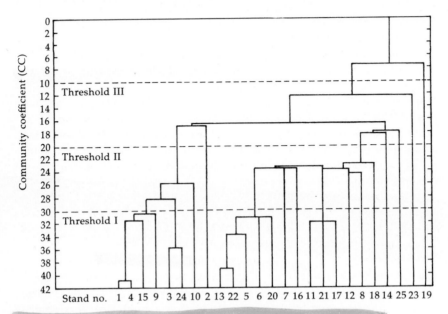

Figure 9–2. A dendrogram resulting from cluster analysis of 25 stands. If one decides that members of an association must exhibit a community coefficient (CC) of 30 or more, then this dendrogram indicates that 15 associations exist, because 15 lines project up through the dashed line called Threshold I. If a CC of 10 were instead selected, there would be only 2 associations (two lines project through the dashed line called Threshold III). (Simplified from *Aims and Methods of Vegetation Ecology*. Mueller-Dombois and Ellenberg. Copyright 1974 John Wiley and Sons, Inc. Reprinted by permission of John Wiley and Sons, Inc.)

ASSOCIATION ANALYSIS

Association analysis builds a dendrogram from the top down, rather than from the bottom up, as in clustering. Associations are divided on the basis of differential species just as in the table method, but the selection of differential species is based on probabilistic, statistical equations and not on fidelity and constancy.

Recall from Chapter 5 that a positive or negative association between two species can be revealed with a contingency table and a chi-square calculation, and the higher the chi-square value, the stronger the positive or negative association. Use of the chi-square formula requires the placement of random quadrats, so this method of analysis cannot be applied to relevé data. The first step is to compute chi-square values for every pair of species. The species having the highest sum of chi-square values with all other species is selected as the first differential species. In the salt marsh example shown in Figure 9–3, with 77 species, *Puccinellia maritima* (species no. 32) is the primary differential species, with a chi-square sum of 48. The 77 species which were encountered in the 70 quadrats are represented as a vertical line that splits at a chi-square sum of 48; at the left end of the horizontal line are those quadrats which contain *Puccinellia*, and at the right end are those quadrats without this species.

In those quadrats with *Puccinellia*, the chi-square procedure is repeated to find a second differential species, which is *Spergularia media* (species no. 38) in this example. A horizontal line at a chi-square value of 10 separates the quadrats once again: at the left end are quadrats with both *Puccinellia* and *Spergularia*, and at the right end are quadrats with *Puccinellia* but not *Spergularia*. Each of these groups is analyzed by the chi-square procedure, but in this example no species with a significantly high value was found (a 99% level of significance was used, which is represented by the dashed line in Figure 9–3). The chi-square procedure is repeated for those quadrats without *Puccinellia*, resulting in a total of eight groups of quadrats. Each group is relatively homoge-

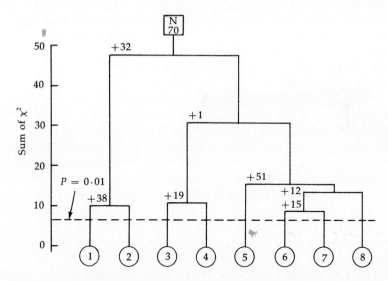

Figure 9–3. A dendrogram resulting from normal association analysis of 77 species in 70 quadrats in a salt marsh. The +32, in the upper left, represents a line of quadrats which all contain species no. 32, *Puccinellia maritima*. The +38 represents *Spergularia media*, and so on. Each of the seven species which are numbered (*Puccinellia, Spergularia*, etc.) serves to separate quadrat groups, and each species exhibited positive and/or negative associations with other species, as indicated by the high chi-squared values along the left axis. A chi-square value below 7 was considered nonsignificant in this example. The result was a division of the quadrats into 8 groups, each group representing an association. (From Ivimey-Cook and Proctor 1966. By permission of the British Ecological Society.)

neous and differs from other groups by the inclusion or exclusion of one of the seven differential species. Each group may be considered an association.

This method was first described in a slightly different form by Goodall (1953) in the days before computers. It was elaborated and given the name association analysis by Williams and Lambert (1959); their approach requires computer facilities. Still later (1961), they referred to this method as **normal association analysis** and they presented a complementary procedure called **inverse association analysis.** Inverse association analysis results in groupings of species rather than quadrats (see also Ivimey-Cook and Proctor 1966 and Goldsmith and Harrison 1976). Together, normal association analysis and inverse association analysis do the same thing that the table method does—that is, species and stands, or quadrats, are rearranged so that similar ones lie near each other in the summary picture—but these methods do at least give a numerical value to the degree of similarity and they invite the reader to decide what an association is. The result is just as subjective as a differentiated relevé table, but it is much more condensed and the relationships are reduced to numbers, rather than to lists of species.

GRADIENT ANALYSIS

Direct gradient analysis is a type of ordination, but the stands are not spread out along mathematically determined axes. At the very beginning of the sampling, the investigator determines that the axes shall represent some obvious environmental gradient, and the quadrats are located regularly along that gradient, rather than randomly. The environmental gradients are generally complex. For example, quadrats located along an elevational cline from a shaded, moist stream bank to an exposed ridge with dry, shallow soil lie on a gradient that involves changes in temperature, moisture, winter snow cover, light intensity, and probably soil nutrient status. Each quadrat along that gradient is given a relative, but numerical, gradient position value.

The importance value or other traits of major species are then plotted on a graph, in relation to quadrat position along the gradient. Figure 9-4 shows how the density of many tree species changes along moisture gradients in the Siskiyou Mountains of Oregon and the Santa Catalina Mountains of Arizona. This method of sampling and data analysis generally cannot be used for classification because few species appear to share the same peaks and end points in their curves on the graph.

Direct gradient analysis was first used by Paczowski at the turn of the century, but it was not described in English and applied to American vegetation until the 1950s (Curtis and McIntosh 1951). It has been

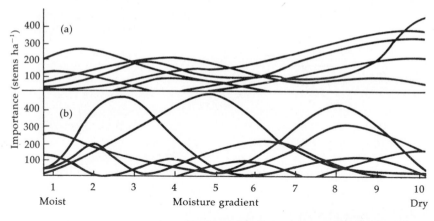

Figure 9–4. The direct gradient analysis method. The importance (measured as density, stems ha^{-1}) of several species is plotted as a function of stand position along a complex environmental gradient in the Siskiyou Mountains of Oregon (a) and the Santa Catalina Mountains of Arizona (b). Each line represents a different species. The gradient (horizontal axis) is labeled as a moisture gradient, but it actually relates to several other factors as well. (Reprinted with permission of Macmillan Publishing Co., Inc. from *Communities and Ecosystems* by R. H. Whittaker. Copyright 1975 by Robert H. Whittaker.)

most widely used by Robert Whittaker (1967, 1973, 1975, etc.), among many others. The results of direct gradient analysis have contributed to the individualistic concept of the community as promoted by Gleason (Chapter 7).

VEGETATION MAPPING

The most generally useful vegetation maps show both physiognomic and floristic information. In such maps, both the life form and the identity of the dominants of each type are indicated. Needleleaf forests are separated from broadleaf forests, and pine-dominated needleleaf forests are separated from fir-dominated needleleaf forests. If physiognomy alone is mapped, then considerable information is lost, as shown in Figure 9–5, which shows two vegetation maps of a region near Montreal, Canada. Based on dominant species, 30 vegetation types have been mapped (Figure 9–5(a)), but there are only 6 physiognomic types to map (Figure 9–5(b)).

The most sophisticated maps, such as those of Henri Gaussen, use symbols and colors to provide environmental information in addition to vegetational information. A temperate zone deciduous forest with moderately high humidity and moderately warm summer temperatures, for example, would be colored light blue (for moderately high

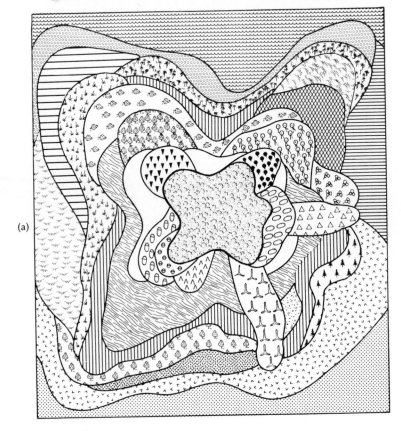

(a)

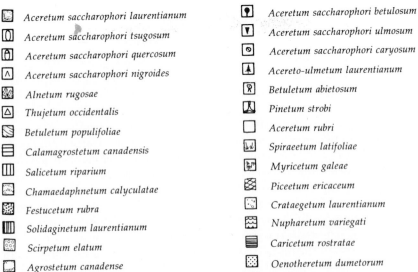

▦	*Aceretum saccharophori laurentianum*	⬤	*Aceretum saccharophori betulosum*
◫	*Aceretum saccharophori tsugosum*	▼	*Aceretum saccharophori ulmosum*
▣	*Aceretum saccharophori quercosum*	◉	*Aceretum saccharophori caryosum*
⋀	*Aceretum saccharophori nigroides*	⬆	*Acereto-ulmetum laurentianum*
▨	*Alnetum rugosae*	⏻	*Betuletum abietosum*
△	*Thujetum occidentalis*	⬛	*Pinetum strobi*
▧	*Betuletum populifoliae*	▢	*Aceretum rubri*
▤	*Calamagrostetum canadensis*	⬓	*Spiraeetum latifoliae*
▥	*Salicetum riparium*	▨	*Myricetum galeae*
▩	*Chamaedaphnetum calyculatae*	▨	*Piceetum ericaceum*
▦	*Festucetum rubra*	⬚	*Crataegetum laurentianum*
▥	*Solidaginetum laurentianum*	≋	*Nupharetum variegati*
▦	*Scirpetum elatum*	▤	*Caricetum rostratae*
▢	*Agrostetum canadense*	▦	*Oenotheretum dumetorum*
▨	*Danthonietum spicatae*	◉	*Trifolietum repentis*

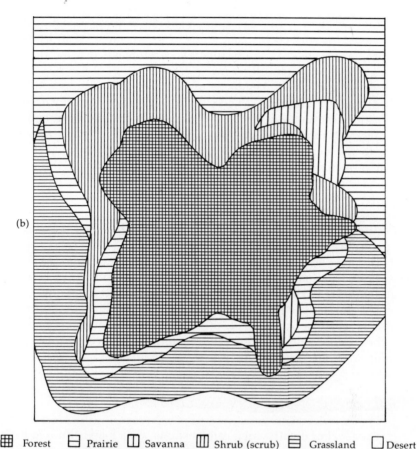

⊞ Forest   ⊟ Prairie   ⊞ Savanna   Ⅲ Shrub (scrub)   ⊟ Grassland   ☐ Desert

Figure 9–5. Vegetation mapping by species and by physiognomy. (a) (left) A map of 30 plant associations on the Montreal Plain, based on the dominant species of each association. (b) (above) A map of the same area, showing only 6 physiognomic vegetation types. ((a) from Dansereau 1961 by permission of the Centre National de la Recherche Scientifique, Paris. (b) from *Vegetation Mapping* by A. W. Küchler 1967. By permission of Ronald Press, New York.)

humidity) applied in a flat tint (for a forest), plus yellow (for moderate temperatures), giving an overall effect of light green. An overlay symbol of ⚲ indicates that the dominant species are broadleaf, deciduous trees. A warm desert scrub would be colored in red lines (red for very hot, lines for scrub), plus orange (very dry), with an overlay symbol of ⩊ to indicate xeromorphic shrubs. UNESCO adopted this method and published a booklet which describes the symbols in some detail (UNESCO 1973). Most vegetation maps, however, use color only to help the reader separate one vegetation type from another; Küchler's maps of the conterminous United States (1964), of Kansas (1974), and of California (1977) are good examples.

One problem with vegetation mapping is that vegetation types are generally described in paragraphs of prose, rather than by tables, numbers, or graphs, and the results reflect the biases of what the mapper regards as important differences.

Mapping generally begins by examining aerial photographs, such as that shown in Figure 9–6. These pictures are taken directly above an area with black and white, color, or infrared-sensitive film, and they are generally timed so that there is considerable overlap from one frame to the next. When prints from adjacent frames are placed together correctly and examined with a stereoscopic viewer, the vegetation and topography take on a three-dimensional quality that makes mapping easier. From the photographs, the investigator tries to distinguish as many different communities or vegetation types as possible, outlining the boundaries of each on acetate sheets placed on top of the prints.

The next step is to go back to the field and examine representatives of each tentatively mapped type in order to identify the dominants and check the reliability of the interpretation of the photographs. This phase is often called ground truth. A short description of each mapped type is prepared, the boundary lines may be amended, and then the acetate overlays are transferred to some political, topographic, or other kind of base map.

Limitations Imposed by Scale and Objectives

The level of detail which can be shown on a map is limited by its scale. The smallest area on any map which can be easily seen by a reader is a circle 1 mm in diameter. At a scale of 1:1,000,000, this corresponds to a real area of 247 acres (100 ha). Obviously, such a scale is inappropriate to show the location of associations which might have a mean area of 1 acre or less; for that level of detail, a larger scale would have to be used, perhaps 1:100,000.

Many of the boundary lines drawn on vegetation maps do not exist in nature. One community or vegetation type will typically change gradually into an adjacent one through a broad ecotone. The line on the map may represent the midpoint of that ecotone. If the ecotone is unusually large, or if several types coexist in some complex mosaic, then the ecotone or mosaic may be mapped as a separate unit in its own right.

A final word of caution: some vegetation maps represent actual vegetation, existing at the time the map is made; other maps represent virgin, prehuman vegetation, and others represent the potential vegetation which could return if human activities ceased in the area. Depending on the objective, the results may be quite different. Küchler's

Figure 9–6. Aerial photograph, showing several contiguous vegetation types or communities. A synecologist has marked the boundaries of each type with ink lines in preparation for publishing a vegetation map. (From *Vegetation Mapping* by A. W. Küchler 1967. By permission of Ronald Press, New York.)

1977 map of California vegetation is a blend of all three types of maps. For example, the central California valley is mapped as bunchgrass prairie, which is a primeval, virgin type. Today the area is either culti- vated land or an annual grassland dominated by aggressive, introduced species and it is highly unlikely that the original grassland could re- turn even if man and his domesticated animals were to vanish. Sierran montane brush fields which exist today on thousands of hectares are mapped instead as forest. Fire and logging have removed the original forest cover, but young conifers are slowly growing up through the brush and forest is the potential vegetation if given enough time. In contrast, many other montane areas are mapped according to existing vegetation. It is likely that most vegetation maps show a bit of all three approaches and are no more inconsistent than this California map. Küchler and McCormick (1965) have compiled a valuable reference to all published vegetation maps for the world, and Küchler (1967) has written a book dealing with the procedures of vegetation mapping.

SUMMARY

The investigator's biases and objectives affect the choice of method for data analysis. Further, some methods of analysis require certain sampling procedures. In the table method, associations are de- fined on the basis of differential or characteristic species which have high fidelity and constance values. Associations are presented in a large differentiated table which manages to preserve most of the origi- nal sampling data of species and stands.

In contrast, ordination reduces the sampling data to one or two graphs that show stands as points in space. The distance between stands on a graph represents their degree of similarity, and the graph axes may correspond to gradients of environmental factors. Some limi- tations to the simplest form of ordination are partially correctable but at a cost of increased calculation, and sometimes the result is difficult to interpret ecologically.

Direct gradient analysis is a form of ordination that requires the investigator to sample along complex environmental gradients, assign- ing each quadrat or stand a numerical position along that gradient. The direct gradient graph plots the importance of species as a function of each stand's gradient position. Generally, nonsynchronous curves for all the species result; hence, the graph is not useful for classification.

Cluster analysis uses community coefficient (CC) values of stand pairs to construct dendrograms which show the relatedness of stands. In contrast to the table method, the result invites readers to choose their own criteria for defining associations and it quantifies the amount of relatedness for stands of the same association and between associations. However, the table method and the cluster analysis method are equally subjective.

Association analysis also builds a dendrogram of stand-to-stand relationships, but its construction is based on differential species rather than on CC values. The differential species chosen are those which show the greatest degree of nonrandom association with other species; they are not chosen on the basis of fidelity and constancy, as in the table method. Normal and inverse association analyses, however, accomplish the same effect as the table method.

Sampling results can also be summarized in the form of vegetation maps. Some mapping schemes permit environmental data to be shown as well as vegetational data. The sources of mapping data are aerial photographs and ground truth. The level of detail is dependent upon map scale and the skill and biases of the cartographer. Some vegetation maps show only existing vegetation, but others may include potential vegetation and virgin (prehuman, climax) vegetation.

CHAPTER 10

SUCCESSION

Plant succession is a directional, cumulative change in the species which occupy a given area, through time. This definition must immediately be qualified by putting limits on the time involved. Many communities undergo significant, spectacular change through the seasons of one year. A grassland, for example, may be dominated by annual or perennial dicotyledonous herbs in the spring, when the grasses are just beginning their growth; then, in mid- to late-summer, the taller grasses dominate (Figure 10-1). In the hot deserts, annual herbs may form a relatively dense ground cover in spring or summer, providing sufficient rain has fallen some months earlier to trigger germination. The density of desert annuals is notoriously variable from year to year, depending on rainfall, and an additional source of variation is the taxonomic difference between annuals which flower in spring and those which flower in summer (Mulroy and Rundel 1977). Such seasonal changes are not usually included in a definition of succession.

If we look at the other extreme of time—thousands or millions of years—we introduce the factors of climatic change and evolutionary change. There is evidence, of course, that climatic change and genetic change (the formation of new ecotypes, for example) have occurred over shorter periods of time, such as hundreds of years, but these changes are minor compared to climatic changes in the temperate zones over the past 10,000 years or to floristic changes everywhere over the past 10,000,000 years. Such long-term changes are usually not included in a definition of succession.

Most of the thousands of words that have been written about succession have discussed changes which may occur in a time span of 1–500 years. If significant changes in species composition for a given area do not occur within such a period, the community is said to be a mature, or **climax community**. Climax communities are not static. Changes do occur, but they are not cumulative in their effect. Instead, the random, small changes in plant numbers or even in the flora merely result in fluctuations about some long-term mean. This is a state of **dynamic equilibrium**, similar to a chemical balance in a solution.

If a community does exhibit some directional, cumulative, nonrandom change in a period of 1–500 years, it is said to be a **successional** or **seral community**. It is often possible to estimate a communi-

(a)   (b)

Figure 10–1. Seasonal changes of dominance in a California grassland.
(a) Spring aspect, dominance by several annual forbs; (b) summer aspect,
dominance by grasses.

ty's future composition by extrapolating from changes measured in a
short time, by comparing other communities which have plants of different ages, or by noting differences between overstory plants and understory seedlings. Seral communities or species will replace one another until a climax community is achieved. The entire progression of
seral stages, from the first one which occupies bare ground (the pioneer community) to the climax community, is called a **succession** or a
sere (Figure 10–2).

The phenomenon of succession has been investigated and
documented for hundreds of years. Clements devoted an entire book to
the subject in 1916 when the term succession and its prevalence in
nature came to be widely recognized. Clements described succession in
such a dramatic, authoritarian manner that a fresh, unbiased examination of succession did not begin until recently, more than a half century after his book was published (see, for example, reviews in Golley
1977).

Clements equated formations (regional climax communities) with
organisms. He compared the seral stages which led to formations, and
the disturbances (fire, etc.) which destroyed them, to the developmental stages of an organism:

> . . . the unit or climax formation is an organic entity. . . . As an organ-
> ism the formation arises, grows, matures, and dies. . . . Furthermore, each
> climax formation is able to reproduce itself, repeating with essential fi-
> delity the stages of its development. The life-history of a formation is a
> complex but definite process, comparable in its chief features with the
> life-history of an individual plant.

The parallel between a community and an organism is not, however, as close as Clements proposed, and the metaphor has muddied some ecological theories about succession.

TYPES OF SUCCESSION

Primary versus Secondary

New land is continuously being exposed to colonization by plants, such as when volcanic explosions create new islands or cover a pre-existing landscape with a new surface of ash or lava, when a landslide buries a swath of vegetation beneath coarse rubble, or when an advancing sand dune smothers a coastal forest. New land is also exposed by mountain building, by the filling of ponds, by the changing of river courses, and by the expansion of tropical deltas under the influence of silt-trapping mangroves.

The establishment of plants on land not previously vegetated is called **primary succession**. If the pioneer community becomes established on a wet substrate, such as the edge of a gradually filling bog, then the invasion is a **hydrarch** primary succession. If the pioneer community becomes established on a dry substrate, such as exposed granite, then the invasion is a **xerarch** primary succession. In either case, the direction of succession is usually towards the most mesic site and the most mesophytic community possible, given the limitations of regional climate, topography, and soil parent material.

Secondary succession is the invasion of land that has been previously vegetated, the pre-existing vegetation having been destroyed by natural or human disturbances, such as wind-throw,* fire, logging, or cultivation. However, the barren surface is not as severe as the surface in primary succession because much of the soil remains (although it is sometimes depleted of nitrogen or other nutrients), and many plant propagule (seeds, bits of rhizomes, etc.) are already present in the soil. Consequently, secondary successions often proceed five to ten times faster than primary successions (Mueller-Dombois and Ellenberg 1974; Major 1974). Secondary succession on abandoned cropland is called **old-field succession.**

Autogenic versus Allogenic

One of the driving forces behind succession is the effect plants may have on their habitat. Plants cast shade, add to the litter, dampen temperature oscillations, increase the humidity, and their roots change

---

*Wind-throw is the uprooting or breaking of vegetation by high winds.

the soil structure and chemistry. Some of these modifications put the seedlings of overstory species that are less well-adapted to shade at a competitive disadvantage with the seedlings of other species better adapted to the moderated conditions. With time, then, a different association of species will come to dominate the site. Tansley (1935) called succession that is driven in this way **autogenic (biotic) succession**. Both the environment and the community change, and this metamorphosis is due to the activities of the organisms themselves.

Allogenic succession, on the other hand, is due to major environmental changes beyond the control of the indigenous organisms. A slight drying trend over the past 100 years in the southwestern United States may, for example, be *partly* responsible for succession from desert grassland to desert scrub. During the Dust Bowl, accelerated cycles of erosion and deposition of loess (wind blown material) changed the pattern of vegetation. River meanders which fluctuate in time may serve to move the succession process back to a meadow or tundra stage. Changes in sea level or topography may give some coastal sites additional protection from wind and salt spray, permitting the invasion of more inland species. The introduction of exotic, weedy species has changed western grasslands over the past two hundred years. These are all examples of allogenic factors, which initiate allogenic succession.

In this chapter we will emphasize examples of autogenic succession.

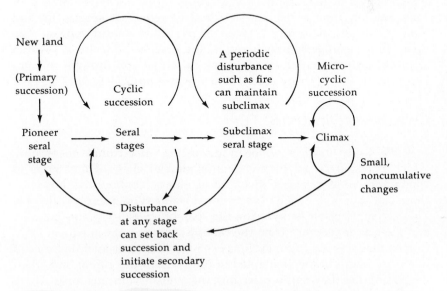

Figure 10-2. Diagrammatic pathway of different types of succession: primary, secondary, and cyclic. The climax stage is in a state of dynamic equilibrium.

Progressive versus Retrogressive

Succession often leads to communities with greater and greater complexity and biomass and to habitats which are progressively more and more mesic (moist). This type of succession is called **progressive succession**.

Retrogressive succession leads in the opposite direction, toward simpler, more depauperate communities (communities with fewer species) and toward either a more hydric (wet) or a more xeric (dry) habitat. Some retrogressive successions are allogenic. For example, the introduction of cattle, weedy annuals, and fire to Great Basin sagebrush steppe vegetation has resulted in degenerated rangeland (Young et al. 1975). Other retrogressive successions are autogenic. Drury (1956) noted that succession in Alaskan flood plains may at first be progressive, leading from a sedge meadow to a white spruce forest with low shrubs of cranberry and blueberry. The dense shade, however, encourages the growth of a dense moss carpet and the encroachment of a shallow permafrost (frozen soil and water) layer. As the soil moisture rises, sphagnum moss invades, white spruce is replaced by black spruce, and ultimately retrogression to a sedge meadow can result (Figure 10–3).

Another example of retrogressive succession comes from the Mendocino coast of California. There is a series of coastal terraces, elevated and exposed to plant invasion over a period of 500,000 years, which shows degeneration in vegetation and soil environment over time, leading from a rich, mesic forest of redwood, spruce, fir, and hemlock to an open, dwarf forest of pygmy pine and cypress. The dwarf forest is underlain by a very acidic, leached podzolic soil with a shallow hardpan that creates flooding in winter and drought in summer (Figure 10–4) (Jenny et al. 1969) (see also pp. 254–5, 414–7).

Cyclic versus Directional

We have, until now, been discussing **directional succession**, which is characterized by an accumulation of changes that leads to community-wide changes. Even in a climax community, however, there continue to be **cyclic** successional changes on a very local scale. These changes occur because the life span of overstory plants is finite, and their disappearance from the canopy may open the site to an invasion by new species. In some climax communities, the juvenile forms of overstory plants are well adapted to life beneath the parent and when the parent dies they will replace it in the overstory; in such cases, there will be no local, cyclic succession. In other communities, however, the overstory may inhibit the growth of juveniles beneath it—juveniles of

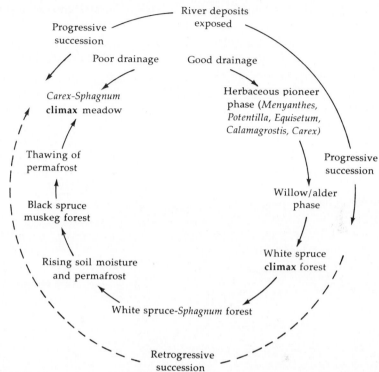

Figure 10-3. Progressive and retrogressive succession on Alaskan flood plains. (From Drury, W. H., Jr. 1956. Bog flats and physiographic processes in the upper Kuskowin River region, Alaska. *Contributions from the Gray Herbarium* 178:1-30. By permission of the Gray Herbarium.)

its own kind or juveniles of any species—and in such cases, local, cyclic succession will occur when the overstory plant dies.

Shrub-dominated communities often exhibit cyclic succession. Open areas within desert scrub in Texas, for example, appear to go through a short cycle of invasion by creosote bush *(Larrea tridentata)* followed by invasion by a cactus, Christmas tree cholla *(Opuntia leptocaulis),* followed by a reversion back to bare ground (Figure 10-5) (Yeaton 1978). Bare sites may be invaded by *Larrea* seedlings because the small seeds and fruits (Fig. 19-36) are abundant and may be widely wind-dispersed. Once a shrub is established, it may attract birds and rodents that scatter the seeds and fruits of *Opuntia.* As the cactus grows, its roots may compete for soil moisture with *Larrea,* leading to *Larrea* mortality. Now removed from the protective influence of a shrub canopy, the shallow root system of the cactus may be subject to erosive forces. Large *Opuntia* plants also attract burrowing rodents that further weaken the root system, and the plant dies. Now open space is available for *Larrea* seedlings to invade.

A northern hardwoods climax forest in New Hampshire offers another example of local, cyclic succession. Seedlings of the sugar maple *(Acer saccharum)*, American beech *(Fagus grandifolia)*, and yellow birch *(Betula alleghaniensis)* are not positively associated with overstory

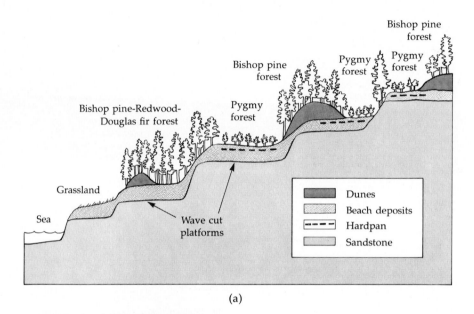

(a)

(b)

Figure 10-4. (a) The sequence of older terraces along part of the Mendocino coast of California. (b) On the oldest (highest) terraces, soil development has led to the formation of a shallow hardpan; above such soil is the climax community—an open, dwarf forest of pine and cypress, much more depauperate than earlier seral stages of spruce, fir, hemlock, and redwood. ((a) Reprinted by permission from Jenny et al. 1969; (b) Courtesy of H. Jenny.)

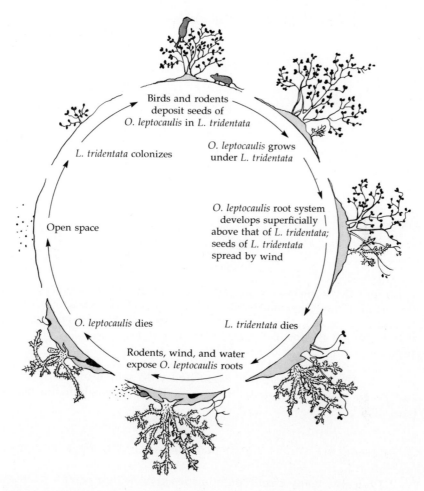

Figure 10–5. Cyclic succession in a desert scrub in Texas (*Opuntia leptocaulis* and *Larrea tridentata*). (From Yeaton 1978. By permission of the British Ecological Society.)

parents (Forcier 1975). Beech seedlings and saplings, for example, are positively associated with overstory sugar maple but are negatively associated with overstory beech. This means that when a beech tree dies, its space in the canopy will not immediately be filled by another beech. Forcier concluded that the following cyclic microsuccession would occur (Figure 10–6): yellow birch, whose seedlings were the most widespread of all three species, would most likely be the first species to fill the gap. It would grow rapidly and a sugar maple understory would develop beneath it (sugar maple seedlings are positively associated with birch overstory). When the relatively short-lived birch died, it would be supplanted by a sugar maple tree. However,

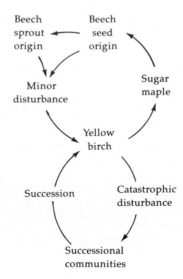

Figure 10-6. Cyclic microsuccession in a climax hardwood forest in New Hampshire. (From "Reproductive Strategies and the Co-occurrence of Climax Tree Species," by L. K. Forcier, *Science* Vol. 189, pp. 808–810, 5 September 1975. Copyright 1975 by the American Association for the Advancement of Science.)

beech seedlings are positively associated with maple overstory trees, and beech would ultimately succeed to the canopy when the maple later died.

Chronosequence versus Toposequence

Typically, many plant communities coexist in a complex mosaic pattern. That is, one climax community does not cover an entire region. Sometimes the mosiac reflects a periodic, local disturbance, such as fire, or it might reflect the progressive exposure of new land, as behind a retreating glacier. In the boreal forest of Canada, for example, a spruce-fir forest forms a matrix within which localized patches of meadow, aspen, and aspen with an understory of spruce-fir occur. These patches represent different stages of recovery (seral stages) from fire, wind-throw, or other disturbances to the matrix type. The mosaic expresses a successional relationship, and is called a **chronosequence**.

In other cases the mosaic reflects topographic differences, such as south-facing versus north-facing slopes, basins with poor drainage and fine textured soil versus upland slopes with good drainage and coarser soil, or different distances from a stress such as salt spray. In such cases, the communities within the mosaic do not bear a successional relation-ship to one another; they constitute a **toposequence**.

Each community in a toposequence may, in fact, be a climax community. A climax community on level mesic ground which reflects the regional climate is called a **climatic climax community**. Another community may be on atypical parent material or in a poorly drained depression so that the soil will not support the climatic climax; this is an **edaphic climax community**. Still other climax communities occur on north- or south-facing slopes where a unique microclimate will not support the regional (climatic) climax. Each region, then, is characterized by a mosaic of climax types; each region has a **polyclimax** landscape (see reviews by Whittaker 1953, 1973a).

Some ecologists have attempted to fit toposequences into successional schemes, but the results stretch the definition of succession too far and can be misleading. For example, Cooper (1919) and McBride and Stone (1976) summarized succession on dunes in the Monterey Bay region of California as shown in Figure 10-7. In actuality, however, the diagram summarizes a spatial zonation of communities inland from the beach. Open beach and dune pioneer communities characterize areas nearest to shore, with highest winds, sand blast, and salt spray. Dune scrub occurs further inland, and "climax" pine or oak forest occurs yet further inland. "Climax" forest will never invade the scrub or the beach, no matter how much time goes by, because the microenvironment near the beach is too severe. Succession would occur only if the sea level were to drop and the tide line were to regress some distance out to sea, giving more protection from the wind.

Indeed, sites with a particularly severe microenvironment, such as beaches, salt pans, outcrops of hard, nutritionally poor rock, or deserts (Shreve 1942), do not support succession. The few, scattered pioneer species able to colonize the habitat constitute the climax species as well, although cyclic succession on a very local scale may occur (Yeaton 1978).

METHODS OF
DOCUMENTING SUCCESSION

Repeated Measures on One Plot

The most direct, unambiguous way to document succession is to make repeated observations of the same area over time. Permanent quadrats or exclosures can be established, and measurements of plant cover, biomass, density, diversity, demography, or the like can be taken every year, every decade, or over longer periods of time. The establishment and initial sampling of such plots takes a large measure of unselfish foresight, for it is likely that only sampling done by some future investigator 50–100 years later will reveal a pattern of late succession in

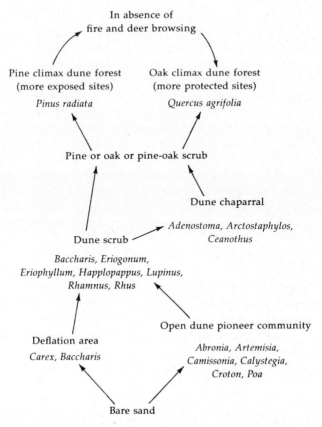

Figure 10-7. Presumed path of primary succession on dunes near Monterey Bay, California, according to Cooper (1919) and McBride and Stone (1976). It might be better to consider the communities as part of a toposequence, rather than successionally related in a chronosequence, for succession may not occur close to shore and oak or pine forest may only be possible some distance inland. (Reprinted from Cooper 1919, courtesy of the Carnegie Institution of Washington. McBride and Stone 1976 reprinted by permission of *American Midland Naturalist.*)

a long sere. A great deal of demographic information, of course, can be obtained in just a few years (see, for example, Werner 1976). If the area to be followed is relatively large and homogeneous, one may periodically sample a collection of random quadrats within it, a different collection of quadrats serving as the sample each time.

The notes and photographs of nineteenth century boundary and land surveyors have been used as general summaries of past vegetation, and then compared to modern vegetation for evidence of a succession (Hastings and Turner 1965; Potzger et al. 1956). Surveyors' notes, which list the identities of "witness" trees marked in the field at intervals, cannot be taken as random samples without bias, however. Certain species may have been favored because of their large size, long life, commercial importance, prominent position, or simply because the surveyor could identify those and not others (Lorimer 1977). Paired photographs, taken decades apart in time, avoid these biases (see, for example, Hastings and Turner 1965).

Succession may take place rapidly enough to occur within the memory span of residents still living. If so, their passive observations can be accumulated by an investigator as additional documentation of succession. Kerner (1863) used this method in his pioneering description of succession in central Europe.

In forested areas, a comparison of sapling density to overstory tree density by species may reveal successional patterns. Figure 10-8 shows that an oak-beech forest near Washington, D.C., is unlikely to remain stable, for there are no saplings of the most common overstory oak, and very few saplings of the other oaks; instead, beech and maples dominate the sapling layer. One could extrapolate from this information, collected at one moment in time, and predict that the oak-beech forest is seral to a beech-maple forest, but it is important to realize that two assumptions are being made.

One assumption is that the sapling population will not significantly change from year to year or decade to decade. However, it is not always safe to assume that the pattern of reproduction for all tree species through time is the same. It is possible that some years will

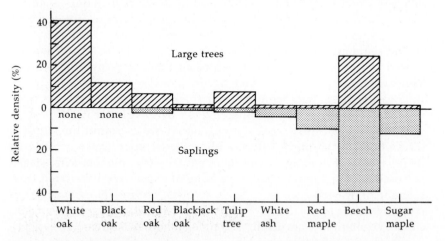

Figure 10-8. Relative density of overstory trees and saplings, by species, for an oak forest near Washington, D.C. The sapling stratum is dominated by beech and maple, possibly indicating that the oak forest (large trees) is seral to a beech-maple forest. (From Dix 1957. Copyright 1957 by the Ecological Society of America.)

yield a large seed crop for one species and many seedlings and saplings for several years thereafter, while another species has a complementary cycle.

Another assumption is that the mortality rate of saplings is uniform from species to species. This assumption is not always appropriate. For example, the spruce-fir forests of the Appalachians are dominated by spruce in the overstory but by fir in the understory. Nevertheless, the overstory balance does not shift to fir, because fir seedlings are more susceptible to damping-off disease and grow less well in dense shade than spruce (Oosting and Billings 1951).

Observations on Nearby Plots of Different Successional Ages

Most successional schemes have been intuitively determined from observations on nearby plots of different successional ages, rather than from observations on one plot over time. The objective is to find a series of plots which have been exposed to primary succession or disturbed and opened to secondary succession at different, known times. Obviously, the time since succession began varies for each plot, but one must assume or demonstrate that all other factors, such as slope, aspect, parent material, and macroclimate, are uniform.

Sometimes the plots of different ages are adjacent to each other, and zoned in obvious bands. Such is the case behind a retreating glacier, and Crocker and Major (1955) used these zones to deduce a path of primary xerarch succession leading from open tundra to spruce forest, as described in the next section. The assumption was made that the spatial sequence of communities back from the glacier's front was repeated in time—or would be repeated in time—for any one location. Thus, where a spruce forest now occurs as a climax community, sometime earlier there had been an alder thicket, and before that an open meadow, and before that ground covered by the glacier itself.

Similar concentric zones of vegetation occur around the edges of filling bogs in the northeastern United States: open water is followed by (a) a zone of floating or emergent aquatics (the pioneer seral stage), (b) a sedge or sphagnum mat, (c) a zone of shrubs with scattered trees tolerant of waterlogged soil, such as black spruce *(Picea mariana)* or tamarack *(Larix laricina)*, and finally (d) the matrix of more mesophytic, climax vegetation of spruce-fir or maple and other hardwoods, depending on the latitude (Figure 10–9).

In many other cases, however, the sites of different ages are randomly scattered throughout a region, and their age sequence must be determined from land use records. Oosting (1942) used this method in documenting secondary succession, as described later in this chapter.

Figure 10–9. (a) The edge of concentric circles of vegetation around a filling lake.
(b) Diagrammatic summary. Presumably, the communities from left to right represent seral stages in a primary hydrarch succession, the floating aquatics being the pioneer stage and the surrounding forest matrix being the climax. (Reprinted with modifications from *Plant Communities* by Rexford Daubenmire 1968*a*. Copyright 1968 by Harper and Row Publishing Co.)

(a)

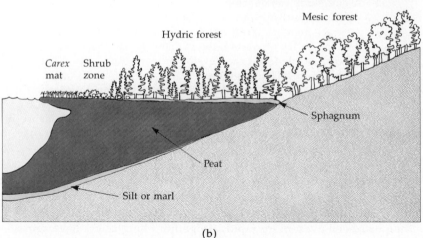

(b)

He located sites which had been abandoned as cropland 1–100⁺ years before. The assumption was made that the older plots of vegetation had all progressed through seral stages still present on younger plots. In many western vegetation types, fire, rather than cultivation, is the disturbance which must be dated. Large, recent fires can be dated with assurance from Forest or Park Service records, but older or more local fires must be dated indirectly by noting fire scars on nearby, surviving trees or by carbon dating beds of charcoal that remain in the soil. A series of charcoal and silt deposits beneath a coast redwood grove *(Sequoia sempervirens)* in northern California has permitted a history of fire and flood disturbance to be traced back 1000 years (Zinke 1977).

An example of primary xerarch succession

Crocker and Major (1955) conducted an elegant, thorough study of succession behind a retreating glacier in Glacier Bay, Alaska, describing soil and vegetation changes which take place in a period of 200 years, from pioneer mosses on bare rock glacial debris to a spruce-hemlock climax forest. Their work built on an earlier description of the area by Cooper (e.g., 1923 and 1931).

A period of 200 years is exceptionally short for primary succession; the short time no doubt reflects (a) the fact that glacial till decomposes easily, (b) the wet, moderate climate, and (c) the presence of nitrogen-fixing plants *(Alnus, Dryas)* in the sere. Reference to primary succession rates elsewhere helps to put 200 years into perspective:

1. Rainforest can develop on fresh lava in Hawaii in 400 years (Atkinson 1970).
2. A pine scrub can develop on bare granite outcrops in Georgia in 700 years (Burbank and Platt 1964).
3. A spruce-hemlock forest can develop on river terrace sediment in the Hoh Valley of Washington in 750 years (Fonda 1974).
4. A beech-maple forest can develop on dune sand around Lake Michigan in 1000 years (Olson 1958).
5. A sagebrush-bitterbrush desert scrub can develop on inland dunes in Idaho in 1000–4000 years (Chadwick and Dalke 1965).
6. A moss-birch-tussock grass tundra can develop on glacial debris in Alaska in 5000 years (Viereck 1966).

The retreat of several glaciers in Glacier Bay has been well documented since 1760 by human records according to Crocker and Major (1955). John Muir, for example, built a cabin near the ice front in the 1890s and that site is now 24 km from the ice front. Retreat prior to then had been even more rapid; altogether, the glaciers have retreated nearly 100 km in 200 years.

The zonation of communities back from the present position of the glacier suggests the sere and the accompanying soil changes which are summarized in Table 10–1. Shortly after the parent rock is exposed it is colonized by pioneer mosses and vascular plants, the most important of which is *Dryas drummondii* (Figure 10–10), which forms extensive mats and harbors nitrogen-fixing actinomycetes in root nodules. About 15 years after the substrate has been exposed, these pioneers are joined by patches of willow and alder shrubs and an occasional cottonwood tree. Gradually, the alder becomes dominant and the can-

Table 10–1. Primary xerarch succession at Glacier Bay, Alaska. The table includes dominant species, canopy cover, soil bulk density, soil pH, and biomass of litter on the soil surface. (From Crocker and Major 1955. By permission of the British Ecological Society.)

Year	Community name and dominant species	Cover (%)	Bulk density $(g\ cm^{-3})$	Soil pH	Litter $(kg\ m^{-2})$
0–15	Pioneer: moss (*Rhacomitrium*), fireweed (*Epilobium*), horsetail (*Equisetum*), *Dryas drummondii*	<50	1.5	8.0	0
15–35	Willow: willow shrubs (*Salix*, 3 species), alder (*Alnus*), *Dryas*, cottonwood (*Populus*)	85	1.3	7.5	1.0
35–80	Alder thicket: alder (*Alnus crispa* ssp. *sinuata*)	100	1.2	6.0–5.0	3.0
80–115	Transition to forest: alder (*Alnus*), Sitka spruce (*Picea sitchensis*)	100	0.9	5.0	6.0
115–200	Spruce forest: Sitka spruce (*Picea sitchensis*)	100	0.8	4.8	9.0
>200	Hemlock-spruce forest: hemlock (*Tsuga heterophylla*, *T. mertensiana*), spruce (*Picea*), *Vaccinium* shrubs	100	?	?	?

opy closes to give an alder thicket, about 35 years after the onset of succession. Soil changes occur most rapidly in this seral stage because alder root nodules also contain nitrogen-fixing actinomycetes, and a great amount of litter is added to the soil. Sitka spruce (*Picea sitchensis*) saplings become established and grow up through the alder canopy; as they form a closed overstory, the alders senesce. This long transition period from alder thicket to spruce forest takes place 40–115 years after the start of succession. About 200 years after the pioneer stage, the spruce forest is invaded by two species of hemlock which ultimately become dominant. An understory of mosses and low ericaceous shrubs* completes the physiognomy (Figure 10–11).

*Shrubs in or closely related to the Family Ericaceae.

(a)

Figure 10–10. (a) *Dryas drummondii* (× ½). (b) Scattered *Dryas drummondii* discs and small thickets of *Alnus crispa sinuata* on a 15-year-old surface near Glacier Bay, Alaska. *Dryas* is a dominant of the pioneer community and a plant which contains symbiotic nitrogen-fixing actinomycetes in root nodules. (Photo courtesy of D. B. Lawrence, Professor Emeritus, Dept. of Botany, University of Minnesota, St. Paul, MN.)

(b)

Soil factors which Crocker and Major (1955) examined along the sere include pH, nitrogen content, and bulk density. Bulk density declines as the porosity of the soil increases. In this sere, bulk density was cut in half, revealing an improvement in soil properties favorable for root growth. Litter biomass and soil nitrogen also increased, though nitrogen reached a peak prior to the climax (Figure 10–12). Soil pH shifted from a very basic reaction (the marble till has a high calcium content) at the start of succession to a very acidic reaction (caused in part by 100–200 years of soil leaching and the acidic reaction of decomposing conifer needles). Successional studies elsewhere by other researchers have shown that additional soil features, such as texture, depth, and water holding capacity, also change during succession. The soil microflora also undergoes succession (Webley et al. 1952).

An example of secondary succession

The Piedmont (literally, foothill) region to the east of the Appalachians is a gently rolling, still heavily forested area with a mosaic of hardwood forest, pine forest, recently abandoned fields, and many cultivated fields. Unlike the large farms of the Midwest and West, farms in the Piedmont region may extend for only a few hectares. Most farms are planted with corn and cash crops such as tobacco. Secondary succession, leading from freshly abandoned cropland to climax forest, has been examined in the Piedmont region by several ecologists (Bard 1952; Billings 1938; Keever 1950; Oosting 1942) and it is reasonably well documented.

Oosting (1942) sampled abandoned fields of varying age in North Carolina, trying to replicate each age with fields as similar as possible in soil, slope, aspect, and other physical features. His conclusions as to the vegetation on one-year-old fields, for example, were based on five fields. On older, forested plots, he compared sapling density with tree density as an additional source of indirect evidence for succession. The sere is summarized in Table 10–2.

One-year-old fields, sampled in the early summer one year after abandonment, show a total of 35 species, all annual or perennial herbs.

Figure 10–11. Sitka spruce forest on the terminal moraine from which ice receded about 200 years earlier. Younger tiers of trees are on an emerging shore that is rising 2.5 cm per year due to release from ice loading. Glacier Bay, Alaska. (Photo courtesy of D. B. Lawrence, Professor Emeritus, Dept. of Botany, University of Minnesota, St. Paul, MN.)

Not every field exhibits all 35 species, but two species have the highest consistent density and frequency on all fields: crabgrass *(Digitaria sanguinalis)* and horseweed *(Conyza canadensis)*. These are the pioneer dominants. Two-year-old fields still contain these two species, as well as virtually all the first year species, but the dominants are aster *(Aster ericoides,* not in first year fields) and ragweed *(Ambrosia artemisiifolia,* present but not dominant in first year fields). Despite the addition of 26 new species in second year fields, the Sorensen community co-efficient calculated by presence for first and second year fields is high, 0.63.

Three-year-old fields decline in species richness because of the almost complete dominance of large clumps of the perennial grass broomsedge *(Andropogon virginicus);* (Figure 10–13). *Andropogon* continues its dominance for several years, but pine saplings gradually overtop the grass and form a closed canopy 10 years after abandonment. The pine seedlings first appear 3–5 years after abandonment, scattered between the clumps of broomsedge, and in any given field they are generally all of one species: either loblolly pine *(Pinus taeda),* shortleaf pine *(P. echinata),* or Virginia pine *(P. virginiana).* Upland sites

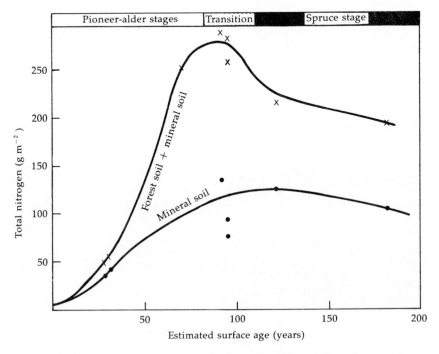

Figure 10–12. Total nitrogen in the top 45 cm of soil beneath various seral communities at Glacier Bay, Alaska. (From Crocker and Major 1955. By permission of the British Ecological Society.)

Table 10–2. Secondary (old-field) succession in the Piedmont region. Only the common names of the dominant species are listed; many other taxa are associated with each seral stage. (From Oosting 1942 (reprinted by permission of *American Midland Naturalist*), Bard 1952 (by permission of the Ecological Society of America), and Richardson 1977 (*Dimensions of Ecology*, by permission of Williams and Wilkins, Baltimore).)

Years after abandonment	North Carolina Piedmont
0	Cropland
	↓
1	Crabgrass, horseweed
	↓
2	White aster, ragweed
	↓
3	Broomsedge
	↓
5	Broomsedge, pine seedlings
	↓
10	Young pines, Broomsedge

Shortleaf pine ← / → Loblolly pine

	(drier sites)	(moister sites)
20		
30		
40		
60	Shortleaf pine hardwood understory	Loblolly pine, hardwood understory
100		
150	White oak, post oak, hickory, dogwood, etc.	White oak, many hickories, dogwood, sourwood, etc.

usually come to be dominated by pure stands of loblolly, with all the trees the same age. Pines reproduce poorly in their own shade, however (possibly because of an obligate requirement for relatively high light intensity or because they are poor competitors for soil moisture; see Shirley 1945), and so a 20 year-old stand will have a well developed

Figure 10–13. Old fields dominated by broomsedge *(Andropogon virginicus),* typical of secondary succession three to six years after abandonment and prior to dominance by pine. *(Introduction to Ecology.* Colinvaux. Copyright 1973 John Wiley and Sons, Inc. Reprinted by permission of John Wiley and Sons, Inc.)

understory of hardwoods. Many of the hardwoods such as dogwood *(Cornus florida)* and redbud *(Cercis canadensis),* are genetically limited to being understory trees throughout their life span, but others, such as hickory *(Carya* spp.) and oak *(Quercus* spp.) are future overstory trees, able to replace senile pines.

Pine stands 50–75 years old are over 25 m tall on good sites; beneath them are the oaks and hickories, perhaps 10 m tall, and beneath them are scattered understory hardwoods (Figure 10–14). As shown in Figure 10–15, stands 100 years of age show as many hardwoods in the overstory as pines, and the balance shifts more and more towards hardwoods with time. Climax hardwood stands 200+ years old still have scattered pines in the overstory and a few seedlings and saplings of pine below, but the great majority of overstory trees, understory trees, and saplings are hardwoods.

Secondary succession can stop at the pine seral stage if ground fires sweep the area every several years (and there is evidence that natural fires of this frequency did occur along the southeastern coastal plain centuries ago; see Chapter 15). Hardwood saplings are susceptible to relatively "cool" surface fires, but pines 10 years old or older have a thick enough bark and high enough canopy to withstand such fires without injury.

GENERAL TRENDS DURING SUCCESSION

For half a century, ecologists have been intent on documenting (or inferring) successional pathways in many types of vegetation. Only recently has there been an attempt to synthesize this information and to formulate general theories on the process of succession. Odum (1969) was among the first to attempt to make the transition from taxonomic specifics to functional generalities. He presented an impressive table of 24 ecosystem traits which he thought changed significantly during succession. He was careful to explain that documentation for these trends was very scanty, in some cases coming from laboratory studies of microsuccessions in aquaria. He presented his table more in the spirit of a hypothesis which invited testing, rather than as established fact.

Odum's table has been a great stimulus to further research and thought on succession (see, for example, Drury and Nisbet 1973). Some of his hypotheses have not been supported by this later work, others have been amended, some still appear to be valid, and a few have been added. We have chosen 13 community and ecosystem attributes to illustrate successional trends (Table 10–3). Some items are paraphrased

Figure 10-14. A loblolly pine stand with a hardwoods understory, typical of secondary succession in the Piedmont 50 years after abandonment.

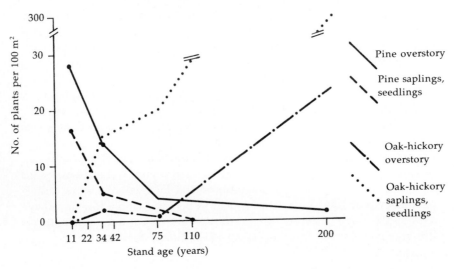

Figure 10–15. Density (per 100 m²) of overstory and understory pines and oaks and hickories throughout secondary succession in the Piedmont of North Carolina. (From Oosting 1942. Reprinted by permission of *American Midland Naturalist.*)

from Odum's table. Each trait will be discussed briefly in the text following. It is important to remember that the table is only concerned with progressive, directional succession, that early seral stages are being compared to later stages (not pioneer to climax, for some trends peak in mid-to-late succession rather than in the climax stage), that there are exceptions to every item, and that the rate of change in any factor is probably not uniform throughout succession (Major 1974).

Vegetation and Site Quality

Biomass increases during succession. Plant cover, the density of foliage above the ground (the leaf area index; see Chapter 7), and the height of the plants also increase. If succession leads from herbs or shrubs to trees, the percentage of total biomass that is below the ground may decline, but the total biomass still increases.

Physiognomy increases in complexity because the variety of growth forms increases as succession proceeds. If successional stages involve trees, the structure and leaf orientation of those trees may change (Horn 1971 and 1975). Tree species characteristic of early successional stages, such as aspens or pines, tend to be tall, thin, conical, and capable of rapid growth; their leaves are small, numerous, randomly oriented, and are borne in such a way that many leaves are shaded by others above them (multilayered) (Figure 10–16(a)). Tree species characteristic of the climax community may have a similar profile but grow more slowly, have fewer and larger leaves, and bear those leaves in a planar fashion such that self-shading is minimized (monolayered) (Figure 10–16(b)).

The major site of nutrient storage in the ecosystem shifts from soil to plant biomass (perennial roots, storage organs, and tree trunks). The role of detritus (leaves, twigs, and other litter) in nutrient cycles thus becomes more important during succession, because soil nutrients have been depleted and are being stored for long periods of time in plant biomass. Detritus carries at least a fraction of the nutrient pool back to the soil where it can be tapped to satisfy new plant growth demands the next season.

The speed with which nutrients cycle from soil to plant and back again slows down during succession because many nutrients are stored in long-lived but inert parts of plants. Such storage does, however, prevent erosional losses of nutrients out of the ecosystem, so we can

Table 10–3. Some vegetation and ecosystem traits which often change during progressive succession. The status of each trait is shown for early and late stages of succession (*not* for pioneer and climax stages necessarily, for some trends peak at intermediate seral stages). Each trend is briefly discussed in the text.

Trait	Early stages	Late stages
Biomass	small	large
Physiognomy	simple	complex
Leaf orientation	multilayered	monolayered
Major site of nutrient storage	soil	biomass
Role of detritus	minor	important
Mineral cycles	open (leaky), rapid transfer	closed (tight), slow transfer
Net primary production	high	low
Site quality	extreme	mesic
Importance of the macroenvironment	great	moderated and dampened; less
Stability (absence or slowness of change)	low	high
Plant species diversity	low	high
Species life history character	r	K
Propagule dispersal vector	wind	animals
Propagule longevity	long	short

hypothesize that the cycling of nutrients is more efficient later in succession, even though it occurs very slowly.

Net primary production (photosynthesis minus respiration) may decline with succession for a variety of reasons: (a) there is a great deal of supporting but nonphotosynthetic tissue whose maintenance (respi-

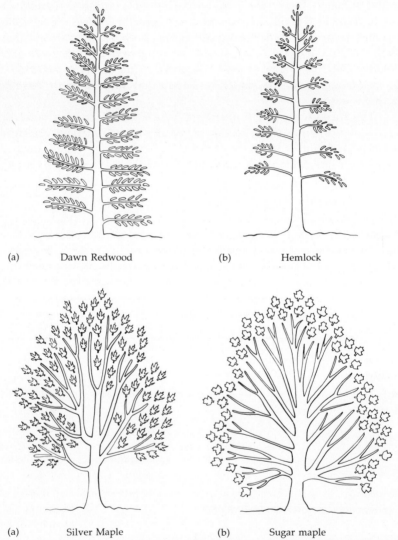

(a)   Dawn Redwood   (b)   Hemlock

(a)   Silver Maple   (b)   Sugar maple

Figure 10–16. Distribution of leaves in multilayered trees (inner leaves shaded) and monolayered trees (self shading minimized). (a) Dawn redwood and silver maple, which are multilayered; and (b) hemlock and sugar maple, which are monolayered. (Reprinted by permission from Forest succession by H. S. Horn. Copyright © 1975 by Scientific American, Inc. All rights reserved.)

ration) reduces net production; (b) nutrients may be limiting because of their storage in inert tissue; (c) many plants in the overstory may be senescent, with lower photosynthetic rates than young plants; and (d) the leaf orientation of climax trees may be inappropriate for high photosynthetic rates (Horn 1974).

The environment becomes more mesic during succession, and the effect of the macroclimate is dampened. As the canopy closes, diurnal fluctuations in temperature and humidity are moderated. Increased humus, increased soil depth, and a finer soil texture in later successional stages leads to greater soil moisture retention and a buffering of seasonal changes in precipitation.

Stability and Diversity

Stability has many definitions. Depending on the definition, stability may increase or decrease with succession (Colinvaux 1973; Horn 1974; Ricklefs 1973; Woodwell and Smith 1969). If we use the most straightforward definition, that stability is equivalent to lack of change, then stability increases with succession: the climax species are longer-lived, and any changes occurring in the undisturbed climax community are in the form of random, minor fluctuations around some long-term mean. Table 10-3 uses this definition. If stability is defined as resistance to minor changes in the macroenvironment, then stability also increases with succession because the plant cover dampens fluctuations and extremes in the macroenvironment. If stability is defined as the ability to return rapidly to a balance point **(homeostasis)** following a major recurring disturbance such as a fire or windstorm, then preclimax communities are the most stable while climax communities are the least stable, requiring hundreds of years to return. If we adopt such a definition, then we have to agree with Henry Horn (1974) that climax communities are fragile:

> Conservationists . . . have often cited the conventional generalization that diversity conveys stability, arguing that diverse [climax] natural communities should be conserved for their stabilizing influence. . . . One could equally argue that if complex [climax] systems were inherently stable, they should need no protection. The opposite view, that . . . climax communities are inherently fragile, is a much more powerful reason for their requiring protection. . . .

Plant species diversity (see Chapter 7) increases throughout early succession, but it decreases in the temperate zone in late succession as the canopy closes and a few species become major dominants. Thus, in the temperate zone, a periodic local disturbance which sets succession back to earlier seral stages is required to maintain maximum diversity (Loucks 1970). There are exceptions to this trend of increasing diversity, just as there are exceptions to all the trends listed in Table 10–3. Old-field succession in Oklahoma, for example, shows a steady decline in species diversity from pioneer weeds through climax prairie (Perino and Risser 1972). Diversity is nearly halved, even though species richness (the total number of species) rises throughout the sere.

Autecology

Pioneer species are typically r-selected, with rapid growth rates, short life spans, relatively large energy allocations to sexual reproduction, and high photosynthetic rates. Climax species are typically K-selected, with slow growth rates, long life spans, small energy allocation to reproduction in any one year, lower photosynthetic rates, and heavy, animal dispersed seeds (see Chapter 4). However, r-selected species can still exist in climax communities: they may occupy sites beneath openings in the overstory and their seeds can lie dormant in the soil for long periods of time until such openings occur. Or, they may have light, abundant seeds which are imprecisely dispersed by wind.

It has been hypothesized that, in general, more plant-animal, plant-plant, and plant-microbe interactions (what Margalef (1968) calls information) occur in later successional stages than in early ones. There is no conclusive evidence to show that this hypothesis is correct. In fact, some pioneer communities composed of lichens or nitrogen-fixing plants may show a more complete dependence on interactions than climax communities. Certainly, many of the examples of allelochemics and other types of interactions discussed in Chapters 5 and 6 come from seral rather than climax communities. This could be just coincidental, however, reflecting greater research interest in seral communities.

DRIVING FORCES OF SUCCESSION

Clements and Relay Floristics

Clements wrote of succession as a six step process:

1. **Nudation,** the exposure of a new surface in primary succession or the clearing away of previous vegetation in secondary succession.

2. **Migration** of seeds, spores, or vegetative propagules from adjacent areas, though in secondary succession many of these are already present in the soil.

3. **Ecesis,** the germination, early growth, and establishment of plants.

4. **Competition** between the established plants.

5. **Reaction,** the autogenic effects of plants on the habitat.
6. **Stabilization**, the climax.

Clements visualized succession as composed of several discrete seral communities. The associated species of each community undergo steps 2–5 together, changing the habitat in such a way that new species (associates of the next seral community) are at a competitive advantage and thus supplant the previous community. The process is repeated until a climax community develops which can propagate itself continuously and prevent the establishment of additional taxa.

This view of succession is called **relay floristics** (Egler 1954; McCormick 1968) and it is diagrammed in Figure 10–17. Each seral community relays the site to the next community. The driving force behind succession, then, is the reaction of the site to the plants living on it.

An example of relay floristics in primary succession comes from a study of rock outcrops in the foothills of the southern Appalachians by Sharitz and McCormick (1973). Small islands of vegetation, 3–5 m in diameter, are scattered over the rock surface (Figure 10–18), and each is made up of concentric zones of seral communities. The outermost zone, on bare rock, consists of mosses and lichens; the next zones on 1–4 cm

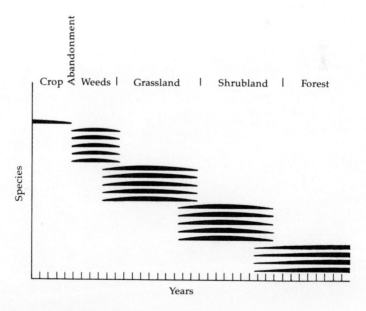

Figure 10–17. Diagrammatic summary of the relay floristics model of succession. The thicker the line, the more important the species at a given time. (From Vegetational Science Concepts. I. Initial floristic composition, a factor in old-field vegetation development. By F. E. Egler. *Vegetatio* 4:412–417. Reproduced by permission of Dr. W. Junk BV.)

of soil are dominated by annuals; more interior zones on deeper soils up to 15 cm thick are dominated by herbaceous and woody perennials. The spatial distribution of some species can be quite distinct. The pioneer species modify the habitat by contributing to the formation of soil, and thereby aid in the establishment of plants in the following seral stage. Sharitz and McCormick showed that competition for soil moisture is responsible for the abruptness of early seral stages. The stonecrop *Sedum smallii*, for example, which dominates the community on the shallowest soil, has a high tolerance for low soil moisture and soils less than 4 cm deep. Stonecrop is competitively inferior to *Minuartia uniflora*, however, which dominates the adjacent interior community on soils 4–10 cm deep.

Chance and Initial Floristic Composition (IFC)

If the relay floristics model of succession is entirely correct, then we should expect to find groups of species appearing and disappearing together during the course of succession. We would also have to agree with Clements that the course of succession is predictable, for each seral stage relies upon the previous one. Some recent work on primary and secondary succession shows that these assumptions are not always correct.

Figure 10–18. The edge of an island of vegetation on a rock outcrop, showing several seral stages. The outermost community dominated by lichens and mosses (0) lies beyond one dominated by *Sedum smallii* (1) which is followed by a community dominated by *Minuartia uniflora* (2). More interior communities (3) are dominated by perennial plants. (From Sharitz and McCormick 1973. Copyright 1973 by the Ecological Society of America.)

In England, Walker (1970) took corings of sediment from the edges of 66 lakes which had been undergoing hydrarch succession for hundreds of years. Sediment had been laid down in the order of the succession that had actually occurred, with the deepest sediments harboring the oldest plants. Because of the slow rate of decomposition, plant remains could be identified throughout the cores. Walker examined the sequence of sediments in each core to see if the sediments corresponded to the traditional sequence of succession. His results showed that there was considerable variability in the progression of seral communities. In fact, the "traditional" sere was never followed.

Frank Egler (1954) initiated plots of secondary succession on his forested estate in Connecticut, and similarly concluded that the progression of seral communities was neither fixed nor predictable. He concluded that the path of succession (at least in the early stages) is driven by chance and the differential longevity of plants. Chance determines which propagules are in the soil or soon reach the site at the time succession begins; he referred to this collection of starting propagules as the **initial floristic composition (IFC)** of the site. All of the pioneer species, many of the seral species, and some of the climax species are present in this initial floristic composition. Some of them germinate and become established quickly, others germinate quickly but grow more slowly and for a longer period of time, and others may become established even later. The larger, longer-lived, slower-growing species ultimately outcompete the smaller pioneer species, and dominance shifts. The site is not relayed from one collection of species to another in this view of secondary succession; rather, there is a sorting out of species one at a time, based mainly on their longevity (Figure 10–19).

Egler did not present any data to document his IFC theory of succession, but others subsequently did. McCormick (1968) manipulated 36 plots of abandoned cropland at an experiment station in Pennsylvania and probably accumulated more data on old-field succession than exist anywhere else. Unfortunately, many of his findings have not yet been published. He found a great deal of variability in the composition of early seral communities. Some first year plots were dominated by young woody vines such as honeysuckle (*Lonicera japonica*) and blackberry (*Rubus* species), rather than by annual herbs. Fourth year plots "should" have been dominated by *Solidago* and *Aster* species, according to traditional views (see Table 10–2), but very few plots were so dominated. To test the theory that early stages prepare a site for later stages, McCormick weeded out annual herbs from some plots as soon as the seedlings could be identified. Then he compared the vigor of perennial herbs that dominated subsequent seral communities in both weeded and nonweeded plots. He found that the perennials grew more vigorously in the weeded plots, not less vigorously, as the relay floristics model would predict.

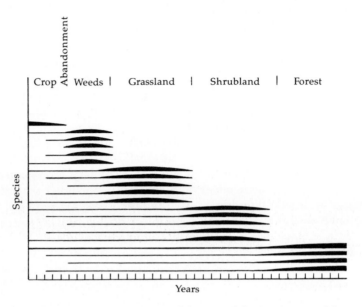

Figure 10-19. Diagrammatic summary of the initial floristic composition model of succession. The thicker the line, the more important the species at a given time. (From Vegetational Science Concepts. I. Initial floristic composition, a factor in old-field vegetation development. By F. E. Egler. *Vegetatio* 4:412–417. Reproduced by permission of Dr. W. Junk BV.)

A close examination of species lists for old fields that Oosting (1942) compiled shows some agreement with the IFC model. Dominants of later stages were sometimes present in earlier stages, and many species were shared among the first several seral communities, giving a high community coefficient. However, pine seedlings did not appear until broomsedge became dominant, and hardwoods did not appear until a pine overstory became established.

Another example of an overlapping species replacement pattern, rather than an abrupt relay pattern, comes from a primary sere on glacial outwash in Alaska (Figure 10–20). Most of the dominants for each of the five seral stages were present in more than one stage. The moss *Hylocomium,* as an extreme example, was rare but present in the pioneer stage, was very abundant in the next three seral stages, and still provided 5–25% cover in the climax tundra. Only *Dryas* and *Eriophorum* were restricted to one stage.

It is likely that there is some value in both the relay and the initial floristic composition models of succession, but more research is needed to formulate a realistic, comprehensive model of succession. Two other driving forces that have been proposed are species interactions (especially allelopathy) and natural selection for maximum efficiency in the cycling of energy and nutrients.

Is Allelopathy a Driving Force?

There is evidence that allelopathy plays a role in some seres (Chapter 5), but that evidence is so scattered and sometimes contradictory that it is not possible to reach a general conclusion.

Catherine Keever (1950), for example, initiated a doctoral dissertation to determine whether allelopathy might be of major importance to old-field succession in the Piedmont. Of the many species and seral

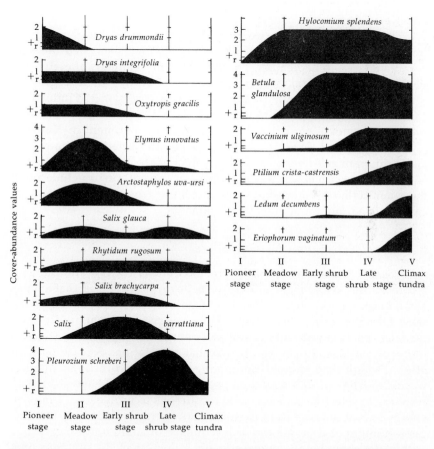

Figure 10–20. The importance (according to the cover-abundance scale of the Braun-Blanquet system) of representative species in a primary sere from glacial outwash to tundra in Alaska. r = rare, + = less than 5% cover and found only occasionally, 1 = less than 5% cover, but commonly found, 2 = 5–25% cover, 3 = 25–50% cover, 4 = 50–75% cover, and 5 = 75–100% cover. (From Viereck 1966. Copyright 1966 by the Ecological Society of America.)

stages examined, she found allelopathy to be a factor in only one case: the decaying roots of horseweed *(Conyza canadensis)* inhibited the growth of its own seedlings. Consequently, the horseweed seral stage is self-limiting. She found all other species replacements in succession to be due to competition for light or soil moisture, or to differences in longevity, time of germination, and pattern of seed dispersal. *Aster*, for example, is a poor competitor for soil moisture with *Andropogon;* biennials which are insignificant the first year as rosettes, outcompete and overtop annuals a second year; pine seeds are not as widely distributed as seeds of ragweed or rhizome fragments of other herbs.

Other manipulative experiments in the Piedmont region corroborate Keever's general conclusion of the limited importance of allelopathy. Pinder (1975), for example, removed all *Andropogon* clumps from old-field plots in Georgia early in July, then harvested the subdominant forbs in September. These forbs were remnants of early seral stages. Removal of *Andropogon* resulted in a tripling of forb growth, giving a yield comparable to that for earlier seral stages. Pinder concluded that forb suppression during the *Andropogon* stage was due to competition, not allelopathy. A more complex experiment in New Jersey by Allen and Forman (1976) involved the selective removal of many different old-field species. The effect of the removals on growth of remaining species was not uniform, but the authors concluded that the magnitude of the effect depended on the growth form of the species removed, not on allelopathic interactions.

Is Natural Selection a Driving Force?

Clearly, succession occurs much too quickly for evolution to play a direct role. What some ecologists believe, however, is that climax communities have been subjected to natural selection in much the same way that populations have. They believe that community-level processes such as nutrient cycling and energy flow may act as selective forces at the community level, and that populations replace one another through time, thereby optimizing community functions. In other words, during evolutionary time, species and niches have shifted in response to selective pressure so that solutions to problems of time and energy allocation have been maximized or optimized (Cody 1974). Succession might also be viewed as a sequence of communities, each one within the chain being a more efficient system than the previous one.

It is ironic that this recent view of succession and the nature of climax communities—a view which treats communities as integrated entities which can be affected by natural selection, much like populations or organisms—brings us full circle back to the organismic view of Clements. As discussed in Chapter 7, there is no direct evidence that natural selection operates at the community or ecosystem level.

The simplest concept of succession is that succession is a population phenomenon, involving the gradual, inevitable replacement of opportunist species (r-selected species) with equilibrium species (K-selected species). The rate at which the transformation occurs, and the exact pathway it takes, simply reflects the rate at which propagules arrive and the nature of the macroenvironment. In the absence of any disturbance, equilibrium species are always at a competitive advantage, as dominants, over opportunists. This advantage is due to, or enhanced by, such features as longevity, large biomass, moderated microenvironment, and symbiotic or antagonistic interactions with other organisms (Drury and Nisbet 1973). The temporal and spatial frequency of disturbances, however, has always been great enough to maintain opportunist species as well as equilibrium species.

SUMMARY

Succession is a directional, cumulative change in the species composition of vegetation at one location over the course of about 1–500 years. Seral (successional) communities exhibit significant, cumulative change during such a period, while climax communities do not.

Primary succession begins with plant colonization of new land; secondary succession occurs on land which still supports some residual soil and plant propagules. A secondary sere may be completed in 50–300 years, but a primary sere requires 200–1000+ years because of the greater severity of the pioneer conditions. Progressive succession leads to more mesic sites and communities of greater complexity and biomass; however, there are many well documented cases of retrogressive succession (both allogenic and autogenic), which results in a more xeric habitat and simpler communities. There is also a type of local, nondirectional succession called cyclic succession, and in some severe habitats there is no succession, the pioneer community being the climax community as well. Some seres do not reach a climax because the periodicity of some disturbance, such as fire, is shorter than the time required for the progression of a complete sere. Most regions are characterized by a mosaic of climax types and seral stages leading to them.

Succession may be documented by repeated measures over time on a single plot, or by reference to historical records for that plot, but most seres have been inferred from indirect evidence. One indirect method is to sample vegetation on many separated plots which differ in age. Also, the species composition in seedling and sapling strata can be compared to the overstory stratum.

Many changes in soil properties, microclimate, and vegetation occur during progressive succession. Some trends peak in the climax and others peak in mid-to-late succession; most do not have a constant rate of change. The following attributes increase during succession: biomass, complexity of the physiognomy, nutrient storage in biomass, the importance of detritus in mineral cycles, the mesic and equable nature of the site, stability (in the sense of longer species and organism turnover times), species diversity, and the proportion of species which are K-selected. Net productivity may decline.

Several factors have been hypothesized to be the driving force behind succession. Clements theorized that successional species modify the site in such a way that they are at a competitive disadvantage with invading species of the next seral stage. The Clementsian model of succession is a series of discrete communities which relay the site to the climax community. Others view chance and differential longevity as the driving forces of succession. Species, rather than discrete communities, replace each other during succession, and seral communities are not discrete. Biotic interactions may also direct some successions, but there is not yet sufficient evidence to reach a general conclusion about the importance of these interactions.

Finally, some ecologists hypothesize that climax communities have been subjected to natural selection for optimum solutions to problems of time, space, or energy allocation, and that successional trends, as well, may be determined by selective forces operating above the population level. It is not known, however, whether natural selection can operate at the community or ecosystem level, nor what forces might drive succession towards levels of greater efficiency.

Possibly the best model of succession is a very simple one, which is that succession involves the gradual, inevitable replacement of opportunist (r-selected) species with equilibrium (K-selected) species. The rate and exact pathway of the replacement depends on the rate at which propagules arrive and the nature of the macroenvironment, indicating that succession is probably a population phenomenon, regulated by the invasive capacity of new species.

CHAPTER 11
PRODUCTIVITY

In this chapter we will consider the functioning of plant communities in the context of energy flow and the carbon cycle, contrasting the abilities of different communities to convert solar radiation into plant biomass. It is this biomass which represents the energy store for the entire biotic community. Therefore, the measurement and understanding of the factors controlling rates of biomass accumulation are basic not only to our interpretation of plant community dynamics but also to our understanding of factors which control the energy store of the biosphere.

CARBON CYCLE

The most active reservoir for carbon is the atmosphere (Figure 11-1), in which carbon dioxide occurs at concentrations of 300–350 ppm. Carbon dioxide is removed from the atmosphere by photosynthesis and released by respiration and combustion of fossil fuels. Photosynthetic uptake and respiratory release are essentially in equilibrium in prevailing climates. Because of the widespread use of fossil fuels, the concentration of carbon dioxide is increasing at an annual rate now exceeding 1 ppm (Figure 11-2), even though, to a large extent, the oceans buffer changes in ambient concentrations. Local concentrations may exceed 400 ppm on a daily basis in areas with topographic conditions which reduce atmospheric mixing and a high level of fossil fuel combustion. Such increases will continue as widespread use of fossil fuel continues and will decrease dramatically when supplies are depleted.

Increasing CO_2 levels support higher photosynthetic fixation rates by C_3 plants (most green plants). It is possible that we may observe changes in competitive interactions of some species due to CO_2 induced changes in their photosynthetic rates, or to temperature related responses caused by fluctuations in atmospheric CO_2 (CO_2 traps solar energy much like a greenhouse does). We might expect subtle changes in physiological response, vegetational structure, and vegetational composition as the carbon cycle fluctuates and finally reaches equilibrium.

ENERGY FLOW MODEL

Before we begin our discussion of terrestrial plant productivity, it is necessary to define some terms applicable to community energetics. Energy enters the ecosystem in the form of light and is fixed into chemical energy by photosynthesis. The total amount of energy fixed by photosynthesis per unit time is referred to as **gross primary productivity** (GPP; see Figure 11–3 and p. 123). Not all energy fixed by photosynthesis is converted to biomass; a significant part is released by respiration to supply energy for plant metabolic activities. Gross photosynthesis minus respiration is equal to **net primary productivity** (NPP), which is the energy that is stored in plant tissues. The dry weight of plant material present at any point in time is referred to as **standing crop biomass** (phytomass). Existing biomass is not a measure of community productivity, since biomass turnover rate is frequently

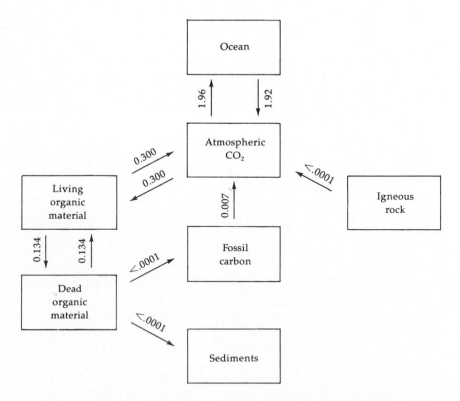

Figure 11–1. Major pathways of the carbon cycle. Numbers are carbon transfer rates in kg m^{-2} yr^{-1}. (Reprinted with permission of Macmillan Publishing Co., Inc. from *Communities and Ecosystems* by R. H. Whittaker (modified). Copyright 1975 by Robert H. Whittaker.)

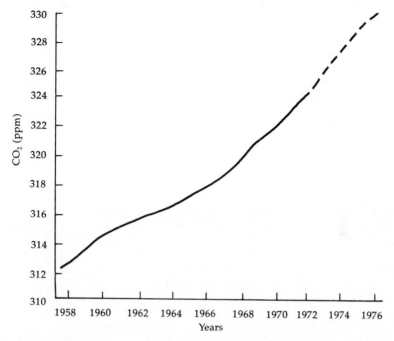

Figure 11–2. Mean yearly atmospheric carbon dioxide concentrations at Mauna Loa, Hawaii, 1958–1976. (From Machta 1972. Reprinted by permission of the American Meteorological Society.)

not related to the total standing crop. For example, productivity per kilogram of biomass in a temperate grassland may be 375 g yr^{-1}, but only 40 g yr^{-1} in a temperate deciduous forest; the larger amount of nonphotosynthetic biomass in a deciduous forest is responsible for most of the difference. Net primary productivity is a measure of the rate at which energy is stored or incorporated into living tissues. We can measure NPP as change in biomass through time, or as net community photosynthetic rate.

Part of NPP is the food source of herbivores and decomposers; the remainder accumulates within the community as standing crop biomass or is consumed by fire. The distribution and rate of turnover of biomass is closely related to the community physiognomy, the type of herbivores present, and the relative importance of decomposers in the system. These phenomena are discussed in more detail when we consider individual communities later in this chapter.

It is common practice to compare the efficiency by which a species or a stand transmits energy from one form or physical state to another. Efficiency values are the ratio of relative energy flow between various pathways of transfer within a community (Odum 1971). **Exploitation efficiency** (Table 11–1) is related to the ability of plants to

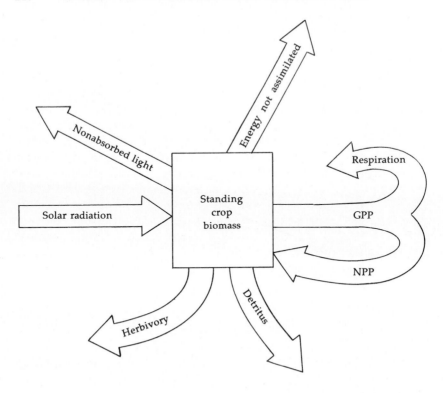

Figure 11–3. Potential pathways of energy partitioning at the primary trophic level.

intercept light. Characteristics important in this regard include latitude, topographic location, leaf area index (LAI) (see Chapter 7), and leaf orientation. **Assimilation efficiency** refers to the ability of plants to convert absorbed radiation into photosynthate. Factors modifying assimilation efficiency are those governing the photosynthetic process (see Chapter 13 for details), such as resistance to CO_2 assimilation, water and light availability, evaporative demands of the atmosphere, and temperature. **Net production efficiency** is a measure of the capacity to convert photosynthate into growth and reproductive biomass rather than utilizing it for maintenance respiration. The amount of energy used for maintenance depends on such factors as temperature (because of the direct effect on rates of chemical reactions) and on the amount of nonphotosynthetic biomass that must be supported. For a more complete evaluation of the concepts and terminology of efficiencies see Kozlovsky (1968). The general trends and variations to be expected for ecological efficiency values are discussed later in this chapter.

METHODS OF MEASURING
PRODUCTIVITY

The most accurate means of measuring net primary productivity is to first measure the net photosynthetic rates of photosynthetic tissues, then subtract the respiration rates of nonphotosynthetic tissues, and finally, extrapolate to the community level, using the net production per gram of biomass of each species in the community. This assessment of NPP is not yet possible on a large scale because we do not have photosynthesis and respiration measurements for all the species in any community, nor are we prepared to predict plant response to the wide range of conditions experienced by plants in the natural community. Consequently, we usually use methods which depend on the accumulation and disappearance of biomass through time.

Net primary productivity (NPP) is frequently measured by calculating the change in biomass through time as equal to:

$$NPP = (W_{t+1} - W_t) + D + H$$

where $W_{t+1} - W_t$ is the difference in standing crop biomass between two harvest times, D is the biomass lost to decomposition, and H is the biomass consumed by herbivores during the period between harvests. Productivity may be expressed as g m^{-2} yr^{-1} or, if the caloric content of the material is known, as cal m^{-2} yr^{-1}. The latter units are more meaningful when efficiency of light conversion or respiration is important, since these are measured in calories rather than grams.

Even though belowground productivity has been estimated to account for 10–75% of total productivity, we cannot yet accurately measure root production because of technological limitations. Attempts have been made to approximate root production by monitoring changes in biomass through time using root material extracted from soil cores (e.g., Kelly 1975) or by monitoring root growth in glass sided

Table 11–1. Formulae defining common efficiency values as applied to primary producers. (Modified from Ricklefs 1973, *Ecology*, by permission of Chiron Press, Inc., Newton, MA.)

Exploitation efficiency (%) =	$\dfrac{\text{Gross primary productivity}}{\text{Solar radiation}} \times 100$
Assimilation efficiency (%) =	$\dfrac{\text{Gross primary productivity}}{\text{Absorbed radiation}} \times 100$
Net production efficiency (%) =	$\dfrac{\text{Net primary productivity}}{\text{Gross primary productivity}} \times 100$

root pits (Lieth 1968; Newbould 1968). These methods are not adequate to accurately measure production because they do not take into account root turnover, which is often substantial in terrestrial systems. For example, estimates taken at consecutive harvests may show no net increase in root biomass when, in reality, a large portion of the roots may have been shed and replaced during the period. Most production studies either do not account for root production or make gross estimates from biomass samples or from data on the ratio of root and shoot growth as determined in controlled environments.

Aboveground biomass can be measured with little error in herbaceous vegetation by replicate samples harvested randomly from a grid. This technique is most effective with annual vegetation where little biomass is lost to decomposition during the growing season. If herbivore activity is significant, comparisons between replicate samples taken inside and outside herbivore exclosures are often employed. If the herbivores are primarily insects which cannot be excluded, the area of tissue or the number of reproductive parts consumed must be monitored by direct observation. See Singh et al. (1975) for a review of harvest techniques applied to grasslands.

Production studies must also account for loss of leaves, branches, etc., during the period between harvests. Several types of **litter traps** have been devised to quantify the litter production of shrubs and trees (Figure 11–4). The most effective trap design will depend upon local conditions and the nature of the material being sampled.

Dimension analysis is an alternate way of estimating productivity in those situations where the volume of individual plants is very large or regrowth is so slow that extensive damage may be caused by complete harvest of sample plots. The technique is based upon the assumption that some easily measured parameter such as plant height, diameter at breast height (dbh), or plant volume can be correlated with standing crop. Whittaker and Woodwell (1968) devised a sophisticated dimension analysis scheme in which branch or plant age, as determined by growth ring or bud scale scar analysis, is related to biomass, production, and surface relations of woody plants. A few individuals must be harvested to determine the slope of a regression line which may then be used to predict plant biomass from the easily measured parameter. For example, satisfactory predictability of standing crop has been obtained from basal diameter, dbh, or diameter^2 × height (Madgwick and Satoo 1975; Whittaker and Woodwell 1968) using the regression model proposed by Kittredge (1944):

$$\log_e Y = a + b \log_e X$$

where Y is the estimate of standing crop, X is the measured parameter, a is the point at which the regression line crosses the Y axis, and b is the slope of the regression line. The values of the constants are determined

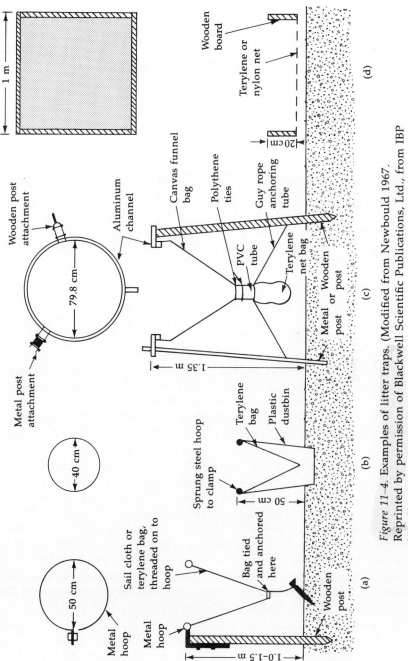

Figure 11–4. Examples of litter traps. (Modified from Newbould 1967. Reprinted by permission of Blackwell Scientific Publications, Ltd., from IBP Handbook No. 2.)

in a preliminary harvest. Errors in making estimates of biomass from log-log regression lines may be avoided by using the methods outlined by Beaucamp and Olson (1973).

WORLD DISTRIBUTION OF PRODUCTIVITY AND BIOMASS

Terrestrial vegetation occupies approximately 30% of the globe's surface, and provides 62% of the total world primary productivity; also, most of the world's biomass consists of terrestrial vegetation (Lieth 1973). Whittaker and Likens (1975) estimated world net primary productivity to be about 170×10^9 t yr^{-1}, and of this they estimated that $90-120 \times 10^9$ t yr^{-1}, or 53–71%, is produced in terrestrial systems. Russian estimates are somewhat higher, but the estimates quoted here are based on average values from actual productivity measurements and they are conservative estimates. Table 11–2 reveals the dramatic differences in productivity and biomass values reported for the major terrestrial ecosystems. Productivity estimates are lowest in deserts and highest in tropical rainforests, spanning the entire range between 0 and approximately 3000 g m^{-2} yr^{-1}. Whittaker suggests that 3000 to 3500 g m^{-2} yr^{-1} is a maximum productivity value for terrestrial systems. Biomass estimates range from 1.0 kg m^{-2} in desert and tundra ecosystems to 200 kg m^{-2} in some temperate rainforests. The usual range, however, is between 1.0 and 60 kg m^{-2}. A careful evaluation and comparison of these values later in this section will further refine our understanding of the distribution of productivity and biomass.

The relationship of biomass to productivity is often expressed as the **biomass accumulation ratio** (BAR) (Whittaker 1975). BAR is the ratio of dry weight biomass to annual net primary productivity. BAR values are a measure of the accumulation of primarily woody material, a characteristic related to environmental harshness and the potential age of dominant species. There is a broad overlap in BAR values between ecosystems. Representative values are 2–10 for deserts, 1.3–5 for grasslands, 3–12 for shrublands, 10–30 in woodlands, and 20–50 for mature forests. Whittaker and Niering (1975) calculated BAR values for an elevational gradient from desert to subalpine forest in the Santa Catalina Mountains of Arizona and found that in forest and woodland zones BAR decreased as a function of biomass; however, once the low elevation desert shrublands were encountered, BAR values and biomass did not change significantly with further drops in elevation.

Jordan (1971) also recognized the relationship between major environmental gradients and the biomass allocation patterns of the dominant plants. He calculated the ratio of wood production to litter production and found high correlations between the ratio and total

Table 11-2. Net primary productivity and related characteristics of terrestrial biomes. (From R. H. Whittaker and G. E. Likens, 1975. The Biosphere and Man. In *Primary Productivity of the Biosphere* edited by Lieth and Whittaker. By permission of Springer-Verlag, New York.)

Ecosystem type	Area (10^6 km^2)	Net primary productivity (dry matter)			Biomass (dry matter)			Leaf surface area	
		Normal range (g m^{-2} yr^{-1})	Mean (g m^{-2} yr^{-1})	Total (10^9 t yr^{-1})	Normal range (kg m^{-2})	Mean (kg m^{-2})	Total (10^9 t)	Mean (m^2 m^{-2})	Total (10^6 km^2)
Tropical rain forest	17.0	1000–3500	2200	37.4	6–80	45	765	8	136
Tropical seasonal forest	7.5	1000–2500	1600	12.0	6–60	35	260	5	38
Temperate forest:									
evergreen	5.0	600–2500	1300	6.5	6–200	35	175	12	60
deciduous	7.0	600–2500	1200	8.4	6–60	30	210	5	35
Boreal forest	12.0	400–2000	800	9.6	6–40	20	240	12	144
Woodland and shrubland	8.5	250–1200	700	6.0	2–20	6	50	4	34
Savanna	15.0	200–2000	900	13.5	0.2–15	4	60	4	60
Temperate grassland	9.0	200–1500	600	5.4	0.2–5	1.6	14	3.6	32
Tundra and alpine	8.0	10–400	140	1.1	0.1–3	0.6	5	2	16
Desert and semidesert scrub	18.0	10–250	90	1.6	0.1–4	0.7	13	1	18
Extreme desert: rock, sand, ice	24.0	0–10	3	0.07	0–0.2	0.02	0.5	0.05	1.2
Cultivated land	14.0	100–4000	650	9.1	0.4–12	1	14	4	56
Swamp and marsh	2.0	800–6000	3000	6.0	3–50	15	30	7	14
Lake and stream	2.0	100–1500	400	0.8	0–0.1	0.02	0.05	—	—
Total	149		782	117.5		12.2	1837	4.3	644

light energy available during the growing season (Figure 11–5), and between the ratio and annual precipitation.

In general, trends in world productivity are loosely related to biomass, in that *K*-selected plants in less harsh or more predictable environments are able to accumulate extensive root and branch systems. These form the bases for more efficient light interception, nutrient uptake, and in turn, productivity. It should be remembered, however, that much of the accumulation of biomass serves only a support function, which explains the different rates of change in productivity and biomass values observed from the poles toward the tropics.

When we consider the information on forested ecosystems in Table 11–2, several generalizations are apparent. Even though the productivity of tropical rainforests has been estimated to exceed 3000 g $m^{-2} yr^{-1}$, the range of productivity values overlaps all other forest and woodland types and may even include values for some temperate grassland systems. There is some evidence suggesting that temperate evergreen forests are more productive than their deciduous counterparts (e.g., Post 1970; Reiners 1972). At first thought, this seems improbable because evergreen tissues have 30–60% lower maximum photosynthetic rates per unit surface area than deciduous leaves

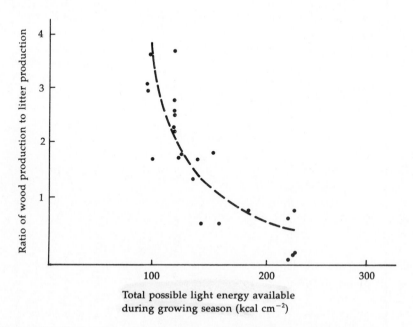

Figure 11–5. Ratio of wood production to litter production in forest communities as a function of amount of light energy available during the growing season. (From Jordan 1971. Reprinted by permission of *American Scientist*, journal of Sigma Xi, The Scientific Research Society.)

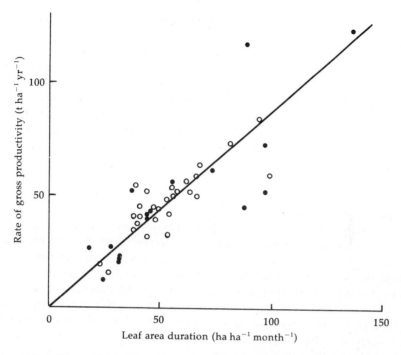

Figure 11–6. The relationship between leaf area index, evergreenness and gross productivity. Leaf area duration values are calculated as LAI times length of growing season in months. Closed circles are for broadleaved forests and open circles for needleleaved forests. (From Kira 1973 in *Photosynthesis and Productivity in Different Environments* IBP 3 edited by J. D. Cooper. Reprinted by permission of Cambridge University Press.)

(Mooney 1972); however, this is offset by high leaf area index (LAI) for evergreen forests. Kira (1975) explained that deciduous plants have very low LAI during periods of leaf development and may miss periods of favorable moisture and temperature following photoperiodically induced leaf drop. Kira also suggests that the high net productivity of tropical evergreen forests is related to the increased height and greater solar radiation which allow increased LAI, and therefore higher exploitation efficiency. Needleleafed evergreens frequently have nearly twice the LAI of deciduous trees of similar biomass and height, thus allowing plants with a lower photosynthetic capacity per unit leaf area to have a higher potential stand production value. This very close relationship between LAI, evergreenness, and gross productivity is plotted in Figure 11–6. A high LAI is made possible by large size, necessitating larger amounts of nonproductive structural tissue, thereby reducing the net production efficiency of forest trees. See also pp. 134, 145.

Grassland productivity is typically lower than in communities with trees because of the LAI relationships considered above. Ovington et al. (1963) compared prairie productivity with the productivity of adjacent savannah and oakwood communities and found grassland productivity to be 10–20% of the productivity of communities with woody plants. The absence of woody tissues results in low BAR values (1.3–5) and short-lived aboveground parts in grasslands. Kucera et al. (1967), Penfound (1964), and Hadley (1970), have observed dramatic increases in productivity when aboveground parts are removed mechanically or by fire. Rice and Parenti (1978) suggested that such increases may be in response to increased soil temperatures following biomass removal.

Arctic and alpine communities include life forms which range from shrubs to cryptogams and have LAIs comparable to other communities in severe climates (see pp. 134, 145). Net primary productivity is typically 100–150 g m^{-2} yr^{-1} with biomass as low or lower than desert communities. Miller and Tieszen (1972) modeled production processes and noted the same increases in production in response to removal of dead material in arctic and alpine communities as in grasslands. Lemmings remove large quantities of biomass in the natural system (Dennis and Johnson 1970), and may cause increases in production similar to the increases caused by fire in grasslands. The main restrictions on production in the tundra are low LAI, low temperatures, and low angle of incident radiation.

Primary production in arid lands varies from 10 to over 200 g m^{-2} yr^{-1} depending on the amount and pattern of rainfall. LAI is severely limited by stomatal closure during dry periods and variations occur seasonally in response to water availability (Chew and Chew 1965; Burk and Dick-Peddie 1973).

LITTER PRODUCTION
AND DECOMPOSITION

Biomass may remain living, serve a support function, be consumed directly by herbivores, or become detritus. The latter possibility is most likely; more than half of annual net productivity is deposited as litter. Litter is the food source for decomposers and detritivores and it is the means by which nutrients are returned to the cycling pool. It is important then, to consider the rate of accumulation and decomposition of dead plant parts. Olson (1963) calculated the ratio of litter production to litter accumulation as an expression of the rate of decomposition. The ratio is high in tropical environments, reflecting the high rate of leaf production and rapid decomposition, and decreases on a gradient toward the poles. Litter fall is generally related to LAI causing the ratio to decrease with latitude in a fashion similar to that of NPP. Within forests, litter fall decreases with latitude and/or altitude. However, Jordan (1971) pointed out that litter fall does not show the same relationship along a moisture gradient. No differences in litter production are found when grassland, old-field and tundra values are compared with forests at similar latitudes (Table 11–3).

Plant decomposition rates have been compiled by Singh and Gupta (1977) which show that decomposition varies with vegetation type and environment (Table 11–4). Broadleaf temperate forests require about 1 year for complete litter decomposition whereas coniferous forests typically require 3–5 years. The extremes reported by Singh and Gupta are represented by high mountain pine forests of California

Table 11–3. Rate of litter production in various forests, perennial herb, and grass ecosystems. (From data compiled by Jordan 1971. Reprinted by permission of *American Scientist*, journal of Sigma Xi, The Scientific Research Society.)

Community	Location	Litter fall $(g\ m^{-2}\ yr^{-1})$	Source
Tropical rainforest	Thailand	2322	Kira et al. 1967
Tropical rainforest	Average of several	1600	Rodin and Basilevic 1968
Subtropical forests	Average of several	1200	Rodin and Basilevic 1968
Dry savanna	Russia	290	Rodin and Basilevic 1968
Oak forest	Russia	350	Mina, cited in Rodin and Basilevic 1968
Fir-taiga	Russia	250–300	Rodin and Basilevic 1968
Oak-pine forest	New York	406	Whittaker and Woodwell 1969
Pine forest	Virginia	490	Madgwick 1968
Tropical seasonal forest	Ivory Coast	440	Muller and Nielsen, cited in Kira et al. 1967
10 Angiosperm forests	Europe	280	Bray and Gorhman 1964
10 Angiosperm forests	Tennessee	320	Whittaker 1966
13 Gymnosperm forests	Tennessee	267	Whittaker 1966
Old-field upland	Michigan	312	Weigert and Evans 1964
Old-field swale	Michigan	1003	Weigert and Evans 1964
Perennial herbs	Japan	1484	Iwaki et al. 1966
Tallgrass prairie	Missouri	520	Kucera et al. 1967; Dahlman and Kucera 1965
Mesic alpine tundra	Wyoming	162	Scott and Billings 1964

Table 11–4. Representative rates of plant litter decomposition. (From data compiled by Singh, J. S. and S. R. Gupta. 1977. Plant decomposition and soil respiration in terrestrial ecosystems. *Botanical Review* 43:449–528.)

Climate	Biome and location	Decomposition rate (% per day)	Source
Tropical	Rainforest: Trinidad	0.45	Cornforth 1970
	Grassland: India	0.30	Gupta and Singh 1977
	Other	0.17–1.5	
Temperate	Oak forest: Minnesota	0.018	Reiners and Reiners 1970
	Missouri	0.095	Rochow 1974
	New Jersey	0.018	Lang 1974
	England	0.30	Edwards and Heath 1963
	Pine forest: California	0.0027–0.0082	Jenney et al. 1949
	Missouri	0.036	Crosby 1961
	Southeastern United States	0.07	Olson 1963
	Other	0.0027–0.12	
	Deciduous forest: Eastern United States	0.057	Shanks and Olson 1961
	England	0.043–0.06	Anderson 1973
	Australia	0.04–0.15	Ashton 1975
	Grassland: North Dakota	0.082–0.11	Redmann 1975
	Missouri	0.14	Koelling and Kucera 1965
	Utah	0.082–0.14	Bleak 1970

which need over 30 years for complete degradation and tropical forests where complete breakdown may occur in less than 2 months. These turnover rates depend primarily on the chemical makeup of the litter, the temperature, and moisture conditions of the habitat.

Species specific differences in decomposition rates may be related to the relative amount of leachable material contained in the litter (Bocock et al. 1960; Gosz et al. 1973). Species with high leachability show rapid loss of weight in the first few weeks, followed by slower loss after leachable material has been removed (de la Cruz and Gabriel 1974). Species specific differences related to leachability were described by Gosz et al. (1973) for leaves of yellow birch *(Betula alleghe-niensis)*, sugar maple *(Acer saccharum)*, and beech *(Fagus grandifolia)* from the Hubbard Brook Experimental Forest in New Hampshire.

Beech had lower initial weight loss because of a higher level of non-leachable organic compounds than yellow birch or sugar maple (Melin 1930). Organic compounds leached from newly deposited litter provide a readily usable source of energy for decomposers. The increased activity of decomposers would also stimulate initial degradation.

Nitrogen is an essential nutrient for soil organisms and its availability is therefore related to rates of decomposition. Voight (1965) found that hardwood litter had a higher initial N content than conifer litter. Nye (1961) found that tropical forest litter had a higher N content than temperate forest litter. J. M. Anderson (1973) reported that chestnut *(Castanea sativa)* leaves had more N than beech *(Fagus sylvatica)* leaves. Although Anderson (1973) has warned that no cause and effect relationship should be drawn, leaf nitrogen content and availability are useful predictors of potential decomposition rates.

Moisture and temperature together exert a significant influence on rates of decomposition (e.g. Rochow 1974; Wiegert and Evans 1964). It is not possible to separate the effect of temperature and moisture because they are not environmentally independent, and extreme levels of either will exceed the tolerance of decomposers. Douglas and Tegrow (1959) found decomposition rates of only about $1.0\,t\,ha^{-1}\,yr^{-1}$ in arctic tundra communities. Such low rates are likely a result of the very short growing season. Contrast the eight times greater rate in tropical rainforests with continuous growing seasons (Wanner 1970). Bleak (1970) measured the disappearance of litter under winter snow in central Utah and found losses of up to 50% despite temperatures of $+1.2$ to $-2.5°C$. Fungi and bacteria were active at these temperatures and were the primary agents for breakdown. Winn (1977) examined decomposition rates in chaparral stands in California where temperatures rarely limit decomposer activity. Microbes responded quickly to small fluctuations in moisture condition and decomposition is apparently controlled by the level of hydration of litter.

Numerous other factors, such as secondary metabolites leached into the litter, soil characteristics, and the kinds of organisms in the detritus food chain, influence decomposition on a local basis. Many of these factors can be related to water, temperature, and the chemical nature of the litter.

ENVIRONMENTAL FACTORS AND PRODUCTIVITY

Carbon Dioxide

Earlier in this chapter we mentioned that the level of CO_2 in the atmosphere is gradually increasing but we did not consider the consequences for plant productivity. Carbon dioxide concentration is known to be a limiting factor in photosynthesis of C_3 plants (Chapter 13). The predominance of C_3 photosynthesis in the world flora would lead one to speculate that dramatic increases in production should occur as atmospheric levels of CO_2 increase. However, the problem is not that well-defined because CO_2 is not evenly distributed within the plant canopy (Saeki 1973) and its concentration depends on other factors such as light and wind. CO_2 concentrations (Figure 11–7) in windless daytime conditions are minimum near the top or middle of the canopy in the vicinity of the greatest leaf area. It is during windless, high light periods that CO_2 concentrations are most limiting (Lemon 1960). Maximum levels occur at the soil surface because of the respiratory activity of roots, soil inhabiting heterotrophs, and the low diffusivity of CO_2 near the surface. A gradual gradient from soil surface to ambient concentrations is rapidly reestablished even by low rates of air flow. M. C. Anderson (1973) reviewed the problems of understanding the impact of changes in CO_2 concentrations on productivity.

Experiments have been conducted concerning the photosynthetic capacity of some montane plants of southern California now exposed to periodic high concentrations of CO_2 (Wright 1974). As expected, all C_3 plants showed a dramatic increase in photosynthetic rate when concentrations of CO_2 were increased; however, the response was more pronounced in the angiosperms tested than in the conifers. If these photosynthetic responses are translated into higher productivity, and

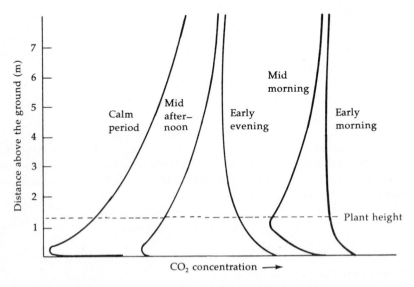

Figure 11-7. Idealized profiles of CO_2 in the air surrounding a photosynthesizing corn field as a function of height at different times during the day. (Lemon 1960. Reproduced from *Agronomy Journal*, Volume 52, pages 697–703, 1960 by permission of the American Society of Agronomy.)

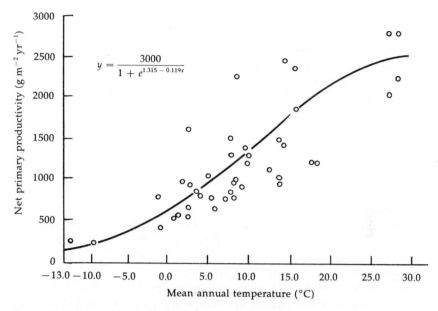

$$y = \frac{3000}{1 + e^{1.315 - 0.119x}}$$

Figure 11–8. Net primary productivity, including both above and belowground productivity, in relation to mean annual temperature. (From Lieth, Primary production: terrestrial ecosystems, *Human Ecology* 1:303–332, 1973.)

therefore greater competitive ability, we might expect both greater production as a result of increases in CO_2 and a change in species composition of our mixed forest types. The latter effect is compounded by the greater susceptibility of conifers to air pollutants such as ozone (Miller 1969), which are often associated with locally high CO_2 concentrations.

Light and Temperature

Global radiation varies with atmospheric conditions, latitude, and altitude. However, the effect of radiation on community productivity is indirect and is seen primarily through differences in growing season (Whittaker 1975) and temperature (Lieth 1973) (Figure 11–8). Productivity increases along the mean annual temperature gradient from the poles toward the equator. Optimum temperatures for productivity coincide with the 15–25°C optimum range of photosynthesis. This close correlation between the photosynthetic temperature optimum and productivity may be coincidental since length of growing season is perhaps a more important determinant than temperature itself. The wide spread of points in Figure 11–8 is indicative of the many environmental factors other than temperature which affect productivity.

Edaphic Factors

Soil characteristics such as texture, nutrient status, depth, etc., are well documented as important factors determining the competitive relationships and growth rates of plants in a wide variety of environments (Bradshaw 1969). Nutrient levels are frequently limiting to plant growth; consequently, when nutrients are added, both species composition and productivity change. Willis (1963) studied the nitrogen- and phosphorus-limited calcareous dune grasslands of England where fertilizer applications increased density, productivity, and the importance of grasses at the expense of broadleaved species.

A dramatic productivity response to edaphic factors is documented for the pygmy forests on the coastal terraces of Mendocino County, California. Pygmy cypress *(Cupressus pygmaea)* and Bolander pine *(Pinus contorta* var. *bolanderi)* form an open, stunted forest frequently less than 2.0 m tall on highly acidic (to pH 3.7), shallow (to 2 dm), podzolic soils (Chapter 16) (see Figure 10–4) (Vogl et al. 1977). Pygmy individuals 50–125 years old grow 10–30 m from redwoods *(Sequoia sempervirens)* of similar age which are up to 30 times their size. The Bolander pine is endemic to the pygmy barrens; however, where pygmy cypress grows in less limiting habitats it may reach 50 m in height. The restrictions on growth are clearly soil related and probably due to the combined effect of low pH, potential aluminum toxicity, poor drainage, and shallow soils (Westman 1975). Westman and Whittaker (1975) compared production samples from a pygmy forest, a bishop pine *(Pinus muricata)* forest, and a coastal redwood forest (Table 11–5). Productivity and biomass differences are impressive when one considers that these vegetation types occur in the same humid, maritime climate, on the same parent material, and that such great differences are due to edaphic factors. It is interesting to compare productivity per unit leaf area and to note that the pygmy forest has a high production efficiency for an evergreen forest (most fall between 60 and 120 g m^{-2}). Not all pygmy forests are the result of edaphic conditions. For example, stunted forests of the New Jersey Piedmont were once thought to be a result of soil nutrient influences, but recent studies by Good and Good (1975) show that physiological or genetic influences reduce the growth rates of individuals and that the overall growth form is a complex response to frequent fire (see also McCormick and Buell 1968).

Edaphic factors, then, are most important in determining the species composition of certain communities and only in isolated situations is there clearly an edaphic influence on community productivity. One must be aware, however, that in most habitats it is difficult to isolate the influence of soil factors. Calcareous dunes of the British Isles and

Table 11-5. Comparisons of productivity and biomass of a *Sequoia sempervirens* forest, a *Pinus muricata* dominated forest, and a pygmy cypress *(Cupressus pygmaea)* forest, all from Mendocino County, California. (From Westman and Whittaker 1975. By permission of the British Ecological Society.)

	Sequoia sempervirens forest	*Pinus muricata* forest	Pygmy cypress forest
Aboveground biomass (t ha^{-1})	3200	415	27
Aboveground net productivity (g m^{-2} yr^{-1})	1401	1089	307
Leaf area index (m^2 m^{-2})	20	16	2.1
Productivity/leaf area (g m^{-2})	72	115	147

the pygmy forests of California are examples of extreme edaphic circumstances and more subtle soil related controls of productivity remain to be identified.

Water

The environmental factor most directly correlated with productivity is water. Lieth (1973) (Figure 11-9(a)) showed the correlation between productivity and mean annual precipitation, and Rosenzweig (1968) (Figure 11-9(b)) has compared productivity with actual evapotranspiration. **Evapotranspiration** is a measure of the total amount of water lost by transpiration and evaporation. Precipitation and evapotranspiration approach equality in arid environments so that the relationship between productivity and annual rainfall is nearly linear at values below 500 mm. At higher levels of precipitation, more water is lost as runoff, or drains below the root zone where it no longer influences production. A fourfold increase in precipitation and associated cooler temperatures resulted in a species specific 100–600% increase in aboveground productivity for Mojave Desert shrubs (Bamberg et al. 1976). Cable (1975) studied the response of perennial grasses near Tucson, Arizona and developed a model in which the product of current precipitation and the past summer's precipitation was a 203 times more accurate predictor of growth than the current precipitation alone. The reason we see such close correlations between annual means and overall productivity (annual means are usually poor predictors of plant response) may be this delayed response which tends to reduce the impact of unusually wet or dry years.

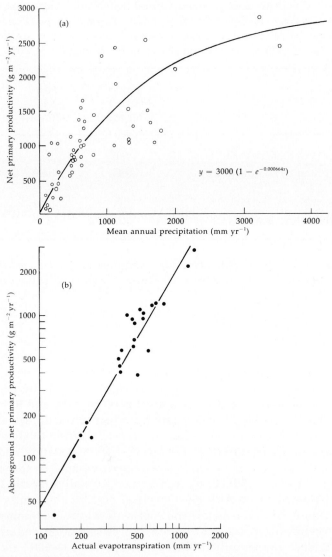

Figure 11–9. Patterns of terrestrial net primary productivity as predicted (a) by Lieth (1973) using mean annual precipitation and (b) by Rosenzweig (1968) using actual evapotranspiration. ((a) from Lieth, Primary production: terrestrial ecosystems, *Human Ecology* 1:303–332, 1973. (b) from Rosenzweig, M. L. 1968. Net primary productivity of terrestrial communities: prediction from climatological data. *American Naturalist* 102:67–74. Copyright © 1968 by The University of Chicago.)

   Direct evidence for productivity trends can be obtained by studying adjacent communities along a gradient of temperature, moisture, and evaporation. Whittaker and Niering (1975) conducted such a study in

the Santa Catalina Mountains near Tucson, Arizona, along an uninter-
rupted vegetational gradient from subalpine forest through woodlands
and grasslands, to desert. Figure 11–10 compares the Santa Catalina
Mountain results with those predicted by Rosenzweig (1968) and Lieth
(1973) for the relationship between actual evapotranspiration or mean
annual precipitation and net primary productivity. The actual measurements show a more complex relationship than predicted. There is an
abrupt change of slope (Figure 11–10(c)) at 400–500 g m^{-2} yr^{-1}. If one
were to guess at what point such a change might occur, the margin of
the desert would seem a reasonable place because of the dramatic
change in life form and the nature of precipitation extremes in that
ecotone. Interestingly, however, the transition does not occur at the
desert margin, but at the transition from open to dense woodland
where trees become the dominant producing life form. Net productivity of about 1300 g m^{-2} yr^{-1} represents the upper level measured and

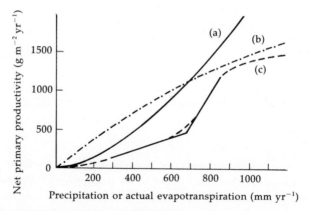

Figure 11–10. Three interpretations of the relation of net primary productivity
(dry g m^{-2} yr^{-1}) to precipitation and actual evapotranspiration. (a) The curve
fitted by Rosenzweig (1968) for aboveground net primary productivity of
forests and shrublands in relation to actual evapotranspiration (mm yr^{-1})
using the formula NPP $= 1.66 \log_{10}$AE $- 1.66$. (b) The curve fitted by Lieth
(1973) for total net primary productivity in relation to mean annual
precipitation (mm yr^{-1}) using the formula NPP $= 3000/(1 - e^{1.315 - 0.119 \text{ MAP}})$.
(c) A curve relating the two slopes of the Santa Catalina total net primary
productivity estimates to probable mean annual precipitation, and adding to
these a third, upper slope for limitation of climax temperate forest
productivity at around 1500 g m^{-2} yr^{-1}. The dashed portion of the curve is
hypothetical. ((a) from Rosenzweig, M. L. 1968. Net primary productivity
of terrestrial communities: prediction from climatological data. *American
Naturalist* 102:67–74. Copyright © 1968 by The University of Chicago.
(b) from Lieth, Primary production: terrestrial ecosystems, *Human Ecology*
1:303–332, 1973. (c) modified from Whittaker and Niering 1975. Copyright
1975 by the Ecological Society of America.)

must denote another abrupt change in slope since estimates of maximum productivity for climax temperate forests are approximately 1500 g m^{-2}yr^{-1} (note that the dashed portion of (c) is hypothetical). Assuming this sort of nonlinear relationship is typical, Whittaker and Niering offer some hypotheses. They suggest that the concave lower part of curve (c) resembles the theoretical curve (a) of Rosenzweig because, in these arid environments, evapotranspiration is essentially the same as precipitation; productivity falls lower because a leaf area index small enough to minimize water loss necessarily limits productivity. The steep central portion of curve (c) represents evergreen forests of environments humid enough to support a less limiting leaf area index, thereby allowing a more rapid increase in productivity per unit available moisture than in more arid habitats. Further increases in available moisture (above 800–900 mm) are associated with distinctly less rapid increases in net productivity. Here, factors such as nutrient turnover, balance of supporting and photosynthetic tissue, and light absorption may severely limit increases in productivity in these temperate climax forests. In these situations, precipitation alone is a better predictor of productivity than evapotranspiration.

Productivity responses along environmental gradients are complex and nonlinear. It will be necessary to accumulate more data on these relationships before the interactions of climate, life form, and productivity are understood and are able to be predicted.

Herbivores

The role of herbivores in ecosystems is extremely important, largely because of the consumption of photosynthetic tissues by the organisms (Petrusewicz and Grodzinski 1975). Activities such as pollination, seed dispersal, and soil aeration are no less important, but have less direct impact on productivity. Consumption of photosynthetic tissues does not always result in a predictable impact on primary productivity. For example, a reduction in tuber yield of only 1 to 2% was noted when 20% of potato plant leaves were consumed by the Colorado potato beetle. In contrast, Varley (1967) reported that the loss of wood production far surpassed the relative consumption of oak leaves by caterpillars. The impact of herbivores also depends on the kind of plant community in question. In grasslands, a larger proportion of the biomass is readily available for consumption than in forest communities; however, only 12–20% of grassland biomass passes through the grazing food chain. Herbivory is even less important in forest ecosystems: less than 10% of the biomass is consumed (Weigert and Owen 1971). Therefore, the kind of plants present and the quality of standing crop biomass are important determinants of the importance of herbivores in regulating primary productivity.

Herbivore activity influences primary productivity directly by reducing photosynthetic area and indirectly by modifying other environmental factors. For example, nutrients may leach more readily from the damaged foliage, and litter may have higher nutrient content when removed by herbivore activity. This is because natural abscission is preceded by remobilization and reabsorption of certain nutrients in the expendable part. Herbivore feces and urine make nutrients more readily available for decomposition, thus increasing the turnover rate. Availability of partially broken down plant material may stimulate microbial activity, thereby increasing the turnover of nutrients. Removal of living biomass may change species composition, increase light penetration into the canopy, or reduce competition for light, water, and nutrients, any of which will affect primary production.

Mattson and Addy (1975) considered the importance of plant-eating insects in regulating primary productivity of forest communities. They concluded that insects help maintain consistent, high primary productivity in natural systems by consuming biomass from less vigorous plants, thus opening the canopy for younger, more vigorous plants to establish themselves. Such insect activity may occur on a large scale as in the case of the Douglas fir tussock moth, the gypsy moth, the spruce budworm, and the southern pine beetle, which periodically cause widespread defoliation and destruction of trees (Figure 11–11). Insect outbreaks seem to be related to increased nutritional value of

Figure 11–11. Spruce forest in southern Colorado killed by insect damage. (Photo compliments of Harold Bradford.)

the insect food and lower host resistance brought about by tree age, marginally adequate habitat conditions such as drought, or low levels of soil nutrients. The result of defoliation is an increase in the availability of soil nutrients, and a reduction in competition for light and water. Similar changes in grasslands and chaparral are stimulated by periodic fire, and in forests by fire, wind-throw, or cutting. Loucks (1970) suggested that such drastic perturbations are essential to stimulate new waves of high productivity and species diversity existing in pre-climax forests.

Even a modest level of defoliation may stimulate vegetative and reproductive growth of plants. Reichle et al. (1973) analyzed insect consumption in a tulip poplar *(Liriodendron tulipifera)* forest and found that, on the average, 2.6% of the net primary production of foliage was consumed yearly. This nutrient return may be important in maintaining vigorous growth of the forest. Churchill et al. (1964) support this conclusion, showing that net production was greater following partial defoliation in aspen *(Populus tremuloides)*.

Data are accumulating which indicate that the revival of productivity by herbivores may be due not only to the passive changes in abiotic factors stimulated by foliage removal or increased nutrients but may also be due to direct chemical stimulation from the animals (Reardon et al. 1972, 1974; Dyer 1975; Harris 1974; Dyer and Bokhari 1976) (see Chapter 6).

There is abundant evidence (e.g., Rockwood 1974, 1976; Janzen 1973; Whittaker and Feeny 1971; Ehrlich and Raven 1964) that the activity of herbivores is very sensitive to the chemical makeup of the host. Herbivores have a direct influence on the active and potential productivity of plants by regulating the amount of photosynthetic tissue; also, the chemicals which control the quality and palatability of the food source are modified as a result of herbivore-plant interactions. These relationships and the relative stability of most natural plant-herbivore relationships are the result of a long coevolutionary history between plants and herbivores. Is it possible, as Mattson and Addy (1975) suggest, that foliage-eating insect-plant relations, are, in the long term, mutualistic? The data are, in an evolutionary sense, inconclusive because the presence of a broad range of herbivore defense mechanisms implies that the benefits which the relationship may confer to the plant do not offset the energetic costs of herbivory.

Succession

Productivity follows a pattern of change during succession similar to the pattern of change noted in Chapter 10 for species diversity. Figure 11–12 depicts this trend of gradually increasing productivity during the pioneer and early tree stages followed by decreasing pro-

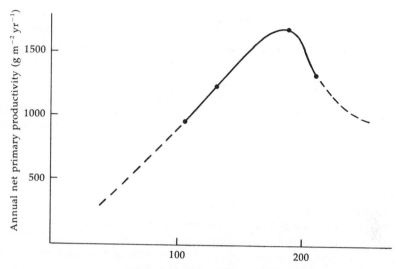

Figure 11–12. Annual net primary productivity, as estimated by tree stem measurements, plotted by age class of the oldest trees in each stand. (Recalculated from Loucks 1970, Evolution of diversity, efficiency, and community stability. *American Zoologist* 10:17–25.)

ductivity as the self-perpetuating climax trees reach maturity (Loucks 1970). Botkin et al. (1972) constructed a model of forest growth which predicts that standing crop reaches a peak within about 200 years, then drops 30–40% during the following 200 years. This biomass reduction could, in itself, account for some declining productivity. Other trends, such as a decline in photosynthetic efficiency of overmature individuals (Hellmers 1964), the allocation of a greater proportion of net productivity to nonphotosynthetic structural biomass, limitations imposed upon LAI by canopy form and leaf orientation (Horn 1974), and the binding of nutrients into structural biomass (Connell and Slatyer 1977), also lead to reduced productivity in mature communities. These changes in biomass allocation and physiological response lead to greater susceptibility of many communities to fire, insect attack, and wind-throw. Loucks (1970) suggested that these characteristics are adaptations to repeating patterns of changing environments which stimulate periodic* returns to states of high net primary productivity.

SUMMARY

The total energy fixed by photosynthesis per unit time (gross primary productivity) is used, in part, as maintenance energy, while the

---

*300–400 years in northern forests, 30–50 years in chaparral, and annually in some grasslands.

remainder (net primary productivity) is used for new plant biomass or reproduction, and is the food source of herbivores. The amount, distribution, and turnover rate of biomass determines community physiognomy, types of herbivores present, and the relative importance of decomposers in a community. The efficiency with which plants absorb light (exploitation efficiency), incorporate absorbed energy into photosynthate (assimilation efficiency), and convert photosynthate into structural components (net production efficiency) are important determinants of population and community characteristics.

Productivity is measured by determining photosynthetic and respiratory rates, by determining changes in biomass through time, or by dimension analysis, which estimates standing crop by determining the relationship between an easily measured parameter and biomass. Dimension analysis and harvest techniques are the most commonly and easily applied methods of studying productivity.

Terrestrial vegetation supports about 62% of the total world primary productivity, totaling about 100×10^9 t yr^{-1}. Productivity varies from a theoretical maximum of 3000–3500 g m^2 yr^{-1} in tropical rainforests to near 0 in some deserts. These trends in productivity are related to severity of habitat and are paralleled by changes in biomass. High levels of productivity occur in communities with maximum leaf area indices (LAI). LAI is maximized in trees but the large amounts of energy necessary to maintain nonproductive support biomass reduces the potential advantage of the tree growth form.

Litter production and decomposition rates have an important influence on mineral cycling and the composition and abundance of detritivores. Rates of decomposition vary with species, the chemical composition of litter, the availability of nitrogen and water, and temperature. Litter production and decomposition rates are highest in the tropics and generally decrease toward the poles.

Carbon dioxide, soil, light, temperature, water, consumers, and level of community development are important factors influencing productivity. Habitats with environmental factors in the optimal range for photosynthesis for the longest periods of time are most productive. Local imbalances of soil nutrients may severely restrain productivity, as in the pygmy forests of California. The successional stage of the community also affects productivity: productivity increases during the pioneer and early tree stages and decreases as the community reaches maturity.

CHAPTER 12

MINERAL CYCLES

Nutrients do not move through living systems in smooth, even-flowing transition, but in pulses, jerks, and floods. The cycling of matter is inherent in the functioning of ecosystems, and is integral to their structure. Both essential and nonessential materials move in cyclic fashion. In contrast, energy flows one way and is noncyclic: it is continually replaced by the sun and is lost from the system as heat, or exported as energy-rich plant and animal parts and wholes (Figure 12–1).

The goal of nutrient cycling research is to quantify: (a) the sum total of nutrients and nonessential elements present within a system; (b) the cycling times, turnover rates and residence times of nutrients within the system; and (c) the biotic and abiotic factors which drive nutrient cycles. Our approach in this chapter will be to compare and contrast several different ecosystems with respect to nutrient cycling, the tightness of the system (i.e. the efficiency with which nutrients are recycled), and the stability of the system.

INTRODUCTION

Biogeochemical Cycles

We refer to the movement of chemical elements within the environment as **biogeochemical cycles**. "Bio" refers to living systems, and "geo" to the rocks, water and air of the earth. Geochemistry deals with the exchange of elements among the physical components of the earth. Thus, biogeochemistry refers to the transfer or flux of materials back and forth between the living and nonliving components of the biosphere.

Nutrient cycling is the movement of those materials that are essential to organisms. Nutrient cycling is a homeostatic mechanism whereby plants build up an available (or exchangeable) pool of nutrients, incorporated as living or dead biomass. The accumulation of this pool through time accompanies succession (see Chapter 10 for a complete discussion). Conversely, a drastic reduction in the nutrient pool accompanies the degradation of an ecosystem.

Gaseous and Sedimentary Cycles

The movement of materials is largely by one of two basic avenues: (1) those elements which have a major gaseous phase are involved in gaseous cycles; (2) those elements which lack a major gaseous phase move in sedimentary cycles. The gaseous nutrients are part of what Odum (1971) terms perfect cycles, in that local perturbation is quickly compensated for, and equilibrium is rapidly reestablished.

An example of a gaseous cycle, the carbon dioxide cycle, is diagrammed in Figure 12-2. It is estimated that 7.3 billion metric tons of CO_2 are added to the atmospheric pool each year as a direct result of the burning of fossil fuels and of agricultural and silvicultural (forestry) practices. Woodwell et al. (1975) suggest that a major portion of the increased concentration of carbon dioxide in the atmosphere results from the destruction of forests. The bulk of this carbon dioxide is absorbed by the sea and stored as carbonates. The carbon dioxide cycle essentially has two large reservoirs—the atmosphere and the sea.

Those elements which do not have a prominent gaseous phase, such as calcium, phosphorus and iron, share the earth's crust as their reservoir. All of these elements ultimately tend to be deposited in the sea, in some pond, or in a lake, and only returned to the nutrient pool

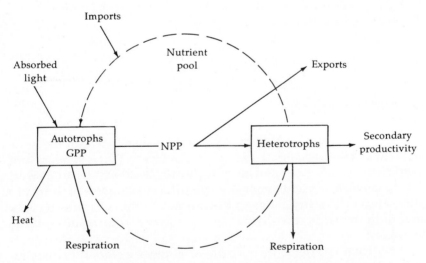

Figure 12-1. A biogeochemical cycle (dashed circle) superimposed on a simplified energy flow diagram contrasting the cycling of material and the one-way flow of energy. GPP = gross primary productivity and NPP = net primary productivity, which may be consumed within the system by heterotrophs or exported from the system. (After E. P. Odum 1963. Limits of remote ecosystems containing man. American Biology Teacher 25:429–443.)

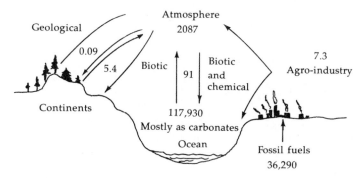

Figure 12–2. The carbon dioxide cycle. Figures are 10^9 metric tons CO_2. (From Odum: *Fundamentals of Ecology,* copyright 1971 by the W. B. Saunders Co., Philadelphia, PA.)

by geological activity—orogeny (mountain building) or some other tectonic phenomenon. Exceptions to this limitation include the short-term recycling which occurs when predatory birds return to land after feeding on ocean fishes. Their rookeries become covered with gleaming white deposits of *guano,* which is rich in phosphorus and other materials. Anadromous fishes (which have matured in the sea) return to spawn and die inland. This migration also recycles nutrients quickly, but the contribution of such transfer is minimal.

Those nutrients which are moved as gases (e.g., CO_2 and O_2) participate in regional and/or global cycles. Their residence time and flux rates must be largely inferred from simple models, and we shall not emphasize these cycles here. It is those nutrients that cycle primarily within one ecosystem—Ca, P, Fe, S, K, etc.—that we will concentrate on in this chapter. Some nutrients, such as S, may have both a gaseous and a sedimentary phase; however, they may be placed in a specific category depending on what phase is incorporated into living tissues.

Flux and Storage

Figure 12–3 gives a diagrammatical view of nutrient cycling. The nutrients are viewed here as compartmentalized, occurring as primary (soil) or secondary (rock) minerals (see Chapter 16), as available nutrients in the atmosphere, or in living or dead organic matter (Bormann et al. 1974). Odum (1971) points out that the flux of nutrients is more significant to the functioning of ecosystems than is the absolute amount present in a given compartment at a specified time. Transfer between compartments may occur through plant uptake of nutrients, leaching of nutrients from plants, biological decomposition, and exchange between the available nutrient compartment and the soil and rock compartment.

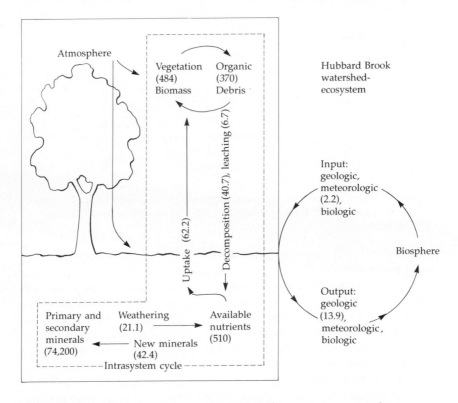

Figure 12–3. A nutrient cycling model. Within the ecosystem a nutrient element, such as calcium, may be found in the organic compartment, where it is a part of living and dead biomass; in the available nutrients compartment, where it occurs as a cation on exchange surfaces or is dissolved in the soil solution; or in the soil and rock minerals compartment, where it occurs as a component of primary or secondary minerals (see Chapter 16). Calcium does not reside in the atmospheric compartment. Nutrients flow between compartments along pathways such as nutrient uptake, decomposition of organic matter, weathering of minerals, or formation of new minerals. The ecosystem is connected to the larger biogeochemical cycles by meteorologic, geologic, and biologic vectors that move nutrients across ecosystem boundaries. Data are from the Hubbard Brook watershed ecosystem, where input of nutrients with sedimentary cycles is almost wholly meteorologic, while output is almost wholly geologic. All data are in kg ha^{-1} and kg ha^{-1} yr^{-1}. (Modified from Likens et al. 1977. *Biogeochemistry of a Forested Ecosystem.* By permission of Springer-Verlag.)

The vehicles that transport nutrients to and from the ecosystem include: (a) meteorologic factors, for example, rain bringing dissolved sulfates, carbonates and particulates, and wind bringing aerosols;*

---

*An aerosol is a suspension of fine particles, solid or liquid, in a gas.

(b) geologic forces, such as surface or subsurface water, which moves nutrients like phosphates, calcium, etc., into or out of a system; (c) biologic flux, which is generally caused by animal transport.

The time frame for nutrient cycling depends on the compartment. Metabolic processes may turn over elements in minutes. For example, in photorespiration carbon is moved from the atmosphere, into the plant, and back in a matter of minutes.* Nutrients in the leaf litter in a forest may turn over within an ecosystem annually, or require several years or decades to turn around. Some biogeochemical processess (e.g., deposition, tectonic movement which exposes sediments, and subsequent retrieval by some root system) may require hundreds of millions of years.

PLANT NUTRIENTS

Essential elements and inorganic compounds are categorized by plant physiologists according to relative quantities required by plants for adequate nutrition. Major nutrients (macronutrients) are those required in rather large amounts. Those required in only small or trace amounts are minor nutrients (micronutrients). An example of differences in amounts of macro- and micronutrients for corn plants is shown in Table 12-1.

Whether we speak of major or minor nutrients, the nutritional state of a plant in regard to elements present in the growth medium may be viewed as: (a) deficient, (b) adequate, or (c) injuriously excessive. Deficiencies may result in stunted growth or premature sexual maturity and senescence. Specific deficiencies, caused by lack of only one or a few of the vital elements, are often revealed by very characteristic symptoms. If nutrients exceed the limits of tolerance, even those nutrients essential to proper nutrition may be toxic. Examples of excess include high salinity in estuarine habitats, which limits the distribution of glycophytes and, at extremes, even halophytes such as cord grass *(Spartina alterniflora)* (see Chapter 5 for more details).

To illustrate the three nutritional states, Table 12-2 gives the standards used by citrus growers for determining the nutritional state of orange trees *(Citrus sinensis)* (Bidwell 1974) and gives the mean level of nutrients in land plants and in the soil. How greatly do these data differ? The standards used by citrus growers apply only to orange trees; the other information is derived from many sources and is an average for many land plants. Yet, if we compare the optimum for orange trees

*Photorespiration is a phenomenon involving the light-dependent production of glycolic acid (in chloroplasts) and its subsequent oxidation in peroxisomes (see Chapter 13).

and the mean concentrations in land plants, we find that the values are very similar, and so these data illustrate the similarities in nutrient needs of plants.

Nitrogen is a constituent of proteins, nucleic acids, and is structurally involved in most catalytic molecules. It accumulates in young tissues, seeds and storage organs. Phosphorus is involved in the structure of many vital molecules, such as nucleic acids and phospholipids. The energy that is used in the cell is released largely by hydrolysis of various phosphate bonds. Sulfur forms part of some amino acids, and is crucial to the stability of the tertiary structure of enzymes and other proteins. Potassium is very soluble and is easily leached from light or sandy soils. Although it has no apparent structural role, potassium plays a catalytic role: it is used primarily in meristematic zones in young tissues and is important in regulating guard cell turgor. Magnesium is a structural part of the chlorophyll molecule and serves to activate many of the enzymatic reactions which transfer phosphates. It tends to accumulate in young leaves, and will be translocated out before leaf abscission. Calcium is essential for the formation and metabolism of the mitochondria and nucleus. Cell membranes lose their integrity when there is a calcium deficiency. Apparently Ca^{++} also reduces the permeability of cell membranes to other minerals. Calcium is virtually immobile in the plant, due to its structural role, and is therefore usually released only by decomposition.

Table 12-1. Analysis of a corn plant. (Reprinted with permission of Macmillan Publishing Co., Inc. from *Plant Physiology* by R. G. S. Bidwell. Copyright 1974 by R. G. S. Bidwell. (Adapted from Miller 1938.))

Element	% of whole plant	Element	% of ash
O	44.4	N	25.9
C	43.6	P	3.6
H	6.2	K	16.4
Ash	5.8	Ca	4.0
		Mg	3.2
		S	3.0
		Fe	1.5
		Si	20.8
		Al	1.9
		Cl	2.5
		Mn	0.6
		Undetermined	16.6

Table 12–2. Standards for classification of the nutrient status of orange trees (*Citrus sinensis*) based on concentrations of mineral elements in 4- to 7-month old, spring-cycle leaves from nonfruiting terminals, compared to average content of mineral elements in plants and in the soil. (These data are on the basis of dry matter, and are given in percent or parts per million as shown.) (Compiled from Smith 1966 (in *Fruit Nutrition*, N. F. Childers, ed., Horticultural Publications, New Brunswick, NJ), Kalle 1958 (in *Handbuch der Pflanzenphysiologie*, Bd. IV, W. Ruhland, ed.; by permission of Springer-Verlag), Finck 1969 (*Pflanzenernahrung in Stichworten*, F. Hirt, Kiel), and Fortescue and Marten 1970 (in *Ecological Studies*, Vol. 1, pp. 173–198, D. E. Reichle, ed.; by permission of Springer-Verlag).)

	Orange trees			Land plants		Stored in soil
Element	Deficient, less than	Optimum range	Excess, more than	Range	Mean	Mean
Nitrogen (%)	2.2	2.5–2.7	3.0	1–5	2	1
Phosphorus (%)	0.09	0.12–0.16	0.30	0.1–0.8	0.2	0.7
Potassium (%)	0.7	1.2–1.7	2.4	0.5–5	1	14
Calcium (%)	1.5	3.0–4.5	7.0	0.5–5	1	14
Magnesium (%)	0.20	0.30–0.49	0.80	0.1–1	0.2	5
Sulfur (%)	0.14	0.20–0.39	0.60	0.05–0.8	0.1	0.7
Boron (ppm)	20	36–100	260	5–100	20	0.01
Iron (ppm)	35	50–120	250?	50–1000	100	38
Manganese (ppm)	18	25–49	1000	20–300	50	0.9
Zinc (ppm)	18	25–49	200	10–100	20	0.05
Copper (ppm)	3.6	5–12	20	2–20	6	0.02
Molybdenum (ppm)	0.05	0.10–1	100?	0.1–1	0.2	0.002
Chlorine (%)	?	<0.2	0.7	0.02–0.1	—	0.1

Iron is considered by some scientists to be a micronutrient, and by others to be a macronutrient or in a separate category (Bidwell 1974; Thompson and Troeh 1973). It is essential for the synthesis of chlorophyll, though it is not part of the structure of the molecule. It is the center of the porphyrin ring of the cytochromes, and so is involved both in the transformation of radiant energy and in the utilization of energy within the cell.

Deficiencies of copper and chlorine in the soil are rare, but deficiencies of boron, manganese and molybdenum occur often. Molybdenum is a minor nutrient, but is essential for nitrogen fixation. Manganese and copper are enzyme catalysts, and boron may exert influence on the activity of various enzymes. Zinc is essential to the synthesis of the important plant hormone, indole acetic acid (IAA), and may be involved in protein synthesis. Chloride, which probably is absorbed in ionic form and remains so, plays a vital role in photosynthesis.

Certain elements behave as analogs to nutrients, and will replace them to some extent. The uptake of strontium, which mimics calcium and will relieve deficiency symptoms for a time, has been of great interest to nuclear scientists and nutritionists in recent years. Radioactive strontium was released by atmospheric testing of nuclear devices and tended to be concentrated in the Northern Hemisphere. Lichens in the tundra took up the strontium as if it were calcium. Reindeer fed upon the lichens, and Laplanders, who were nourished by caribou (reindeer), showed higher concentrations of the harmful isotope than peoples of other lands.

FACTORS IN NUTRIENT CYCLING

Every physical force and organism impedes, hastens, or otherwise affects nutrient cycling. Our knowledge of general pathways of nutrient movement has preceeded our understanding of how much is moved, and in what time frame the transfer occurs. However, quantitative studies have appeared with increasing frequency (e.g., Woodwell et al. 1975). A brief look at some of the factors involved in nutrient flux will be of value.

The Hydrologic Cycle

The following mnemonic device

$$P = E + T + R + I$$

reminds us of the balance that must exist in the hydrologic cycle, such that P (precipitation) is equal to E (evaporation) plus T (transpiration) plus R (runoff) plus I (infiltration, the downward entry of water into the soil). Nutrient budgets are strongly correlated with the hydrologic cycle: precipitation carries nutrients in solution, runoff and infiltration remove nutrients from a system or move them down the soil column, and evapotranspiration of water concentrates and conserves nutrients. In Europe, the contribution of nutrients from the atmosphere (primarily Cl, Na, Ca, S, K, Mg and N) may range as high as 25–75 kg ha^{-1} yr^{-1}, compared to 19–40 kg ha^{-1} yr^{-1} for a deciduous forest in the eastern United States (Larcher 1975; Likens and Bormann 1972). Groundwater and water rising by capillary action transport nutrients into the root zone. Although transfer by runoff is often only to a region of lower elevation, on a local scale streamflow with its load of nutrients must be viewed as loss. This section will give several illustrations of the role of water in nutrient cycling.

Rainfall

The contribution of rainfall to nutrient input can be quantified easily by the use of precipitation gauges and subsequent chemical analysis (Likens et al. 1970; Cole et al. 1967). Precipitation brings sulfate and hydrogen ions, nitrate, and significant amounts of ammonium, chloride, sodium and calcium to the eastern deciduous forest (Table 12–3). Rainfall is often acidic, frequently with a pH of less than 4.0 (Likens and Bormann 1972). This is attributable, at least in part, to human activities. Increasing amounts of sulfur compounds are being put into the air, and these ultimately form sulfuric acid. When sulfuric acid is carried into the soil, cations such as K and Ca may be displaced by the hydrogen ions of the acid, and will be leached away by seepage and surface runoff. Under these circumstances, there is an apparent increase in export of nutrients in the eastern deciduous forest (Table 12–3). Is the nutrient flux the result of increased weathering because of the acid, or does it represent an actual loss of available nutrient from the system? We will consider this again when we discuss runoff.

Table 12–3. Weighted average concentrations of various dissolved substances in bulk precipitation and stream water for undisturbed watersheds 1–6, Hubbard Brook Experimental Forest, 1963–1969. (After Likens, G. and F. H. Bormann, "Nutrient Cycling in Ecosystems," in *Ecosystem Structure and Function*, copyright 1972 Oregon State University Press.)

	Precipitation	Stream water
	$(mg\ l^{-1})$	$(mg\ l^{-1})$
Calcium	0.21	1.58
Magnesium	0.06	0.39
Potassium	0.09	0.23
Sodium	0.12	0.92
Aluminum	*	0.24
Ammonium	0.22	0.05
Sulfate	3.10	6.40
Nitrate	1.31	1.14
Chloride	0.42	0.64
Bicarbonate	*	1.9[a]
Dissolved silica	*	4.61

*Not determined but very low.

[a]Watershed 4 only.

Leaching

Leaching refers to the removal of soluble constituents from the soil or litter by percolating water. The primary cause of nutrient loss from a system is leaching, and the most easily leached nutrients are K and Ca (Tappeiner and Alm 1975). Hunt (1977) has calculated that 45–65% of the labile component (the component which readily undergoes chemical change) may be leached from grassland soils and litter in as short a time as 4 hours. This process is positively correlated with increasing temperature. For example, up to 36% of the labile component will be transferred by leaching at 25°C.

Leaching from attached leaves, on the other hand, adds nutrients to the soil (see Chapter 5). As rain runs over the foliage, it passively absorbs nutrients. The nutrients are then carried to the ground in **throughfall*** and **stemflow.**† For example, leaching of foliage provides 12.33 kg Ca ha^{-1} yr^{-1} in the tulip tree *(Liriodendron)* forests in eastern Tennessee (Shugart et al. 1976). Mosses gain nutrients primarily from throughfall and stemflow. In some vegetation types with considerable moss cover on the ground, nutrients derived from throughfall must pass through mosses before they are available to trees (Foster and Morrison 1976).

Evapotranspiration, runoff, and infiltration

In a region with an impermeable geologic substrate, water loss by evapotranspiration may be inferred from the difference between precipitation and streamflow. Chemical concentrations in precipitation are a measure of nutrient input, and concentrations in stream water are a direct measure of nutrient output in such a region. At the Hubbard Brook Experimental Forest, stream-gauging stations were anchored to bedrock to meter nutrients in stream water flowing from a watershed underlain by an impermeable substrate. The losses shown in Table 12–3 appear enormous, but water lost by evapotranspiration accounts for much of the increased nutrient concentration (Likens and Bormann 1972). Nutrient and evapotranspirational loss is relatively stable from year to year in an undisturbed forest. The biotic contribution to this stability will be referred to again in a later section of this chapter. Rather than promoting nutrient loss, transpiration actually conserves mineral elements because excess water, which would promote loss through leaching, is removed without harm to the system, leaving the nutrients behind.

---

*Throughfall is rainwater which falls through a canopy.
†Stemflow is the fraction of precipitation that flows down the stem or trunk of a plant and then enters the soil.

In a region where the substrate is permeable, evapotranspirational losses and water movement through the soil may be estimated by the use of a **lysimeter** (Figure 12–4). In a tropical rain forest, which has a permeable substrate, lysimeters are employed to determine the volume of water running through the soil. Water thus collected can be used to determine the concentrations of nutrients leached. The product

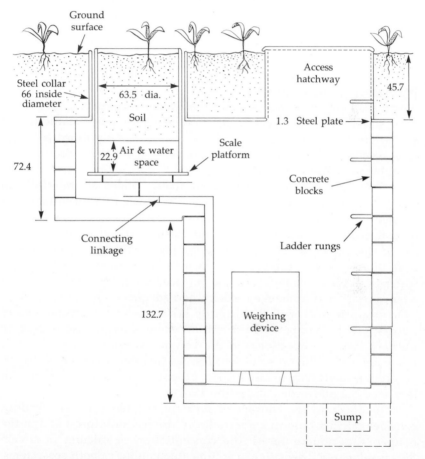

Figure 12–4. Diagram showing the principle of a weighing lysimeter (measurements are in cm). A lysimeter consists of a container holding a mass of soil mounted flush with the soil surface and arranged so it can move up and down on a weighing device (Modified from England and Lesesne 1962). The lysimeter pictured here uses a scale to follow changes in weight, but electronic weighing devices have advantages. For the results to be applicable to specific crops or stands of vegetation, the lysimeter must be surrounded by similar vegetation. Some problems of lysimetry are discussed by Hagan et al. (1967). (Reprinted by permission of McGraw-Hill Book Co. from *Plant and Soil Water Relationships: A Modern Synthesis* by Kramer, 1969.)

of the volume and the concentration of each nutrient is the loss rate for
that nutrient. Evapotranspirational losses are inferred from the difference between precipitation input and percolation output. One draw-
back of lysimetry is that the hydrologic regime is disturbed, sometimes
greatly so, by manipulation of the soil to put a lysimeter in place.

Nutrient budgets

In general, the successional status of community can be deduced
from nutritional budgets derived from studies such as those just described, which measure nutrient input and output. If elements are being stored, as indicated by a positive input/output balance, new net
biomass is being accumulated, and the system is aggrading (i.e., it is a
seral stage). A small negative balance may represent weathering of
parent material, and suggests that the system is at equilibrium (a mature, or climax community). The watersheds at Hubbard Brook are at
equilibrium (Likens et al. 1970).

The chemical concentrations in stream water at the Hubbard
Brook Experimental Forest are amazingly constant (concentration of
nutrients being independent of volume of water flow), especially those
of magnesium, sulfate, chloride and calcium (Likens and Bormann
1972). This constancy may be due to the high exchange capacity of the
soils there (see also Chapter 16). As ions are removed from the soil
water by roots or runoff, they are replaced from the soil. There is a
near-saturation of exchange capacity, and adequate nutrition for whatever grows in that soil. Rich, fertile systems show a greater loss of
nutrients than do impoverished soils as precipitation increases, because
nutrient loss is proportional to the volume of runoff. The acidity of
rainwater also influences nutrient loss, as was stated earlier. Nutrient-
unsaturated soils tend to lose a lower proportion of nutrients even with
increased runoff (Jordan and Kline 1972).

To illustrate the concept of nutrient constancy in stream flow,
consider calcium both in a Puerto Rican tropical rain forest and in the
temperate deciduous forest: the concentration of calcium in stream
flow is relatively insensitive to volume fluctuations in both ecosystems.
In the temperate deciduous forest, stream flow varies by more than
four orders of magnitude, and yet the calcium concentration in the
water remains unchanged (Likens and Bormann 1972). In the rain forest, increases of 100 times or more in the volume of soil water reduced
calcium concentration by less than 25% (Jordan and Kline 1972).

In contrast, the concentration of potassium is a very sensitive index of biological activity, for it may be reduced dramatically in stream
flow during periods of plant growth and then rebound during dormant times.

Salt spray

The nutrients that are received in a dune community from salt spray are essential to the continued functioning of that system. Sea water contains all the necessary mineral ions, except nitrates and phosphates (Boyce 1954; Salisbury 1952). Young dune soils may be as high as 99% sand, with very poor capacity for cation storage because they lack clay and organic matter. Dune soils are known to have a cation exchange capacity (CEC) of only about 10–15 meq 100 g^{-1} of soil. (A millequivalent (meq) is the amount of material that will combine with or replace one milligram of hydrogen.) For comparison, one may expect a CEC of 40–50 meq 100 g^{-1} of clay, and a CEC of 160–250 meq 100 g^{-1} of organic matter (Ranwell 1972).

The dune system at Cape Hatteras, North Carolina, receives the following weights (kg $ha^{-1} yr^{-1}$) from salt spray: Na, 250–1300; Mg, 37–120; Ca, 19–120; and K, 13–77. By contrast, rainfall contributes only 1.0–1.3, 21–26, 2.3–3.4, and 3.0–3.9 kg $ha^{-1} yr^{-1}$ of each nutrient, respectively (van der Valk 1974). In spite of the very heavy input, the habitat is not saline, due to the extreme leachability of these porous soils.

Other Abiotic Factors in Nutrient Cycling

Fire

Wildfire plays an important role in the maintenance and functioning of many communities. The reproduction of giant sequoia (*Sequoiadendron giganteum*), jack pine (*Pinus banksiana*), Bishop pine (*P. muricata*) and various species of California lilac (*Ceanothus*) is enhanced by fire of a certain intensity (see Chapter 15). Fire is also important to cycling of nutrients in certain regions. In the far north, low temperatures prevent decomposition, and the result is acid soil covered by raw humus. In Norway, for example, burning actually improves forest site quality by raising the pH, thus favoring the growth of populations of nitrogen-fixing bacteria (*Azotobacter, Rhizobium*). In chaparral regions, fire may serve to: (a) break down sclerophyllous litter, which is resistant to biotic decomposition; (b) remove inhibitors of microbial decomposition; and (c) alter wettability of the soil (see Chapter 15).

Nutrients are released in the form of soluble mineral ash by slash and burn agriculture. Thus, in a tropical rain forest, slash and burn temporarily enhances the nutrient regime. Crop yields are good at first, but diminish quickly. Reasons for this vary, but at least three should be mentioned here:

1. The fire destroys such recycling mechanisms as mycorrhizae within the soil.

2. The nutrients are released in a single large pulse, and their availability probably exceeds the exchange capacity of the soil.
3. Thus, nutrients are quickly leached out of the root zone (Jordan and Kline 1972).

Earth movement

Mountain building (orogeny), movement along a lateral fault, volcanism, and other types of tectonic activity exert an enormous influence on mineral cycling. The impact may be sudden or require eons of time. For example, the volcano Paracutin in Michoacan, Mexico, poured an estimated one billion tons of ash, cinders, and bombs* on the land in its first year. Often such material is easily dissolved, and in tropical regions it will enhance the nutrient regime within a short time (Shelton 1966; Walter 1973). From a different perspective, an atom of phosphorus may be lifted from the sea by tectonic movement over eons of time, then exposed to weathering, ultimately to enter the biosphere and resume cycling, at least for a time. The removal of soil may occur within a very short time. On 23 June 1925, some 38 million cubic meters of sandstone, a portion of a forested mountain slope, slid into the valley of the Gros Ventre River in Wyoming. The loss occurred within mere minutes, not just denuding the slope, but also completely removing the mineral substrate (Shelton 1966).

Such catastrophes are considered rare in our time frame, but they are common in geological time, and even in ecological time. In addition, the magnitude of the event counterbalances its infrequent occurrence. Geophysical phenomena rank as primary physical agents of biogeochemical cycling.

Biotic Factors

There are many biotic factors in mineral cycling. For example, in the calcium cycle of the *Liriodendron* forest mentioned previously, insect grazing accounts for 1.07 kg $ha^{-1} yr^{-1}$ of Ca flux (Reichle et al. 1973). The transfer of biomass, with its nutrient content, from producer to herbivore to carnivore to decomposer, etc., is well known. Most of us recognize that respiration, excretion, defecation (in animals), leaf fall (in plants), and death are integral components of nutrient cycling.

Litter

The role of litter in nutrient cycling must be viewed in an ecosystem framework (see also Chapter 11). Two factors determine the

*Volcanic bombs are molten material thrown from volcanoes which solidifies and falls as an igneous rock with a bomb-like shape.

amount of litter in an ecosystem: the total litter produced in a unit of time, and the rate at which it is decomposed. In boreal forests, such as the jack pine *(Pinus banksiana)* system in northern Ontario, Canada, the annual litter production is about 4000 kg ha^{-1} yr^{-1}. This organic material decomposes very slowly, requiring as much as 16 years for some elements to recycle (Foster and Morrison 1976). The acid soils that are produced contain a well-defined surface layer of organic material in which nutrients are taken up and immobilized in the biomass of microbes.

In contrast to the slowly turning mineral cycle of the boreal forest, litter in the tropical rain forest decomposes quickly, and there is nearly complete turnover each year. Litter production in a tropical rain forest can be staggeringly high; leaf fall alone is 45–126 kg ha^{-1} every day (Walter 1973). As Figure 7–8 shows, the bulk of the carbon in a subalpine conifer forest is in organic litter and humus on or in the soil, but it mainly resides in the wood of the tropical rain forest.

The deciduous forest of the temperate zone stands midway between these two extremes. It exhibits moderate turnover times, such that a year's litter production is decomposed in 2–3 years (Larcher 1975).

Microbial decomposition

Microbial decomposition has been studied by Brinson (1977) (see also Triska and Sedell 1976). There are three stages of decomposition:

1. Formation of particulate detritus.
2. Production by saprophytes of immobile humus and the concomitant release of immobile, though soluble, organic compounds (in a relatively short time).
3. Mineralization (mobilization) of humus (a slower process and one that needs further study).

Although the products of stages one and two are not available as nutrients, humic substances and other organics play a vital role in nutrient storage, forming complexes with minerals that enhance uptake by plants. Immobilization through decomposition serves to retain nutrients beyond the period of dormancy to the time of growth.

Ingestion and digestion

Perhaps less widely recognized than microbial decomposition is the nutrient regeneration in soils due to ingestion and digestion of bacteria and fungi by other organisms. Within the soil, metazoan animals account for only 10% of the total metabolism, which leaves the bulk of it (90%) for microorganisms. Thus, we may infer that protozoan digestion and mineralization of bacteria plays a major, vital role in the intrasystem cycling of nutrients. Evidence supports this inference (Pomeroy 1970). In addition, protozoan predation removes cells from overcrowded bacterial populations, which stimulates active bacterial growth and enhances bacterial nitrogen fixation.

Forest trees

Nutrient cycling in forested regions will be discussed in later sections; only some of the roles of forest trees regarding mineral flux are introduced in this section.

Substances released into the soil from roots may provide significant amounts of carbohydrates and the like to the **rhizosphere**, the zone in the immediate vicinity of the plant root. As discussed in Chapter 5, Smith (1976) analyzed root exudates from sugar maple (*Acer saccharum*), yellow birch (*Betula alleghaniensis*), and beech (*Fagus grandifolia*), all of a northern hardwood forest (see Table 5–5). The beech released the most amino acids and organic acids per hectare, and the birch released the most carbohydrates. The cations released were mainly Na, K, and Ca; the anions were chiefly sulfate and chloride.

Smith pointed out that the significance of root exudation is twofold: first, root exudates enhance the growth of the microbial saprophytes and parasites in the rhizosphere; and, second, although root exudates are beneficial to microbial saprophytes and parasites, they may be either beneficial or harmful to individual plants and to the ecosystem. Now a fifth pathway, root exudation, can be added to the four internal cycle pathways between plants, decomposers and soil nutrients shown in Figure 12–3.

The atmospheric contribution of inorganic nitrogen to the watershed of Lake Tahoe, California-Nevada, ranges from 1–2 kg ha^{-1} yr^{-1}. Very little of this drains into the lake, for it is removed efficiently by well-developed stands of conifers. However, perturbation (construction or logging, for instance) reduces the efficiency of nitrate uptake, resulting in an increased export of growth-stimulating nutrients to the lake itself. Thus, there is increased algal growth in the lake, reducing the clarity of the water. Alders are associated with both nitrogen fixation and nitrification, and so significant amounts of nitrate nitrogen are also contributed to the soil water by alder stands in the Lake Tahoe watershed, especially in the fall and early winter (Coats et al. 1976).

In general, trees have not only a very extensive root network, but sometimes they have longer and deeper roots than do shrubs and herbs. Therefore, the portion of the soil column confronted by tree roots is greater, making possible a unique contribution to the ecosystem. Deep-lying minerals are extracted by tree roots, raised into the plant and incorporated into plant biomass. Nutrients, for example, potassium, may be leached from the leaves or bark, and other nutrients, such as calcium, may be returned as litter. Then, decomposition and mineralization add these nutrients to upper soil layers; now these formerly deep-lying nutrients are accessible to the herbaceous and shrub strata of the forest.

MARITIME ECOSYSTEMS

How does nutrient cycling differ from one maritime ecosystem to another? Have conservative mechanisms for retention and internal recycling evolved as a means of obtaining elements in short supply? A Baltic seashore meadow ecosystem exhibits a very tight, conservative mineral nutrient regime (Tyler 1971) in contrast to the loose, open regime of the dune system at Cape Hatteras, North Carolina, for instance. In general, the annual output of nutrients from a terrestrial ecosystem exceeds the meteorological input, and that input is small compared to the nutrient reservoir of the system. However, in dune ecosystems the atmospheric contribution of cations slightly exceeds the calculated output through leaching, and greatly exceeds the nutrients in storage in the system.

Cape Hatteras National Seashore

The dune systems at Cape Hatteras National Seashore, North Carolina (which includes the dunes of Bodie, Ocracoke, and Hatteras Islands), were studied in recent years by van der Valk (1974), with special emphasis on their nutrient cycles. Bodie Island, in the northern part of the National Seashore, is oriented such that it receives the full fury of winter storm winds, which blow mostly from the northeast. The coastline of Ocracoke Island to the south runs in an east-northeast direction, and so is parallel to the winds' path. Also, it is partly protected by Hatteras Island. Bodie Island receives greater cation input than Ocracoke Island, in salt spray and rainfall, probably due to its orientation. The foredune (the most seaward dune) at Bodie is higher than that at Ocracoke Island, and this may also influence salt spray deposition.

The vegetation of the dunes, primarily American beachgrass (*Ammophila breviligulata*) and sea oats (*Uniola paniculata*), forms a sparse cover; coverage is 3–4% for each of the grasses. There are few or no forbs* on the front of the foredune, and perhaps 20 forb species on the back of the foredune. Some contribute as much or more cover than the

*Herbaceous plants other than grasses.

grasses. The principal forb species are horseweed *(Conzya canadensis)*, cudweed *(Gnaphalium obtusifolium)* and goldenrod *(Solidago semper-virens)*.

Nutrient supply

The primary nutrient source in this community is salt spray; there is almost no internal reservoir of nutrients. At Bodie Island, the salt spray contribution of mineral ions exceeds that of rainfall as much as eightfold (Table 12–4). Notice, for instance, that the atmospheric input of cations at Bodie Island in bulk precipitation (rainfall plus salt spray) ranges from 14 kg $ha^{-1} yr^{-1}$ for K, to 433 kg $ha^{-1} yr^{-1}$ for Na, with annual inputs of 35 kg ha^{-1} for Ca and 64 kg ha^{-1} for Mg. By comparison, Ca input from bulk precipitation for an inland Tennessee *Liriodendron* forest was estimated at 10.5 kg $ha^{-1} yr^{-1}$.

What happens to the large supply of nutrients that arrive on the wind? Recall that the foredune is not a truly saline habitat, and that the exchange capacity in soils with a very high sand to clay ratio is poor. Excess cations are quickly leached away.

Output of each cation for a certain depth of soil may be calculated, since

$$S_2 = S_1 + IN - OUT$$

Table 12–4. Average concentration of cations (ppm) and annual input of cations (kg $ha^{-1} yr^{-1}$) in the bulk precipitation (rainfall plus salt spray) and in the rainwater at Bodie Island and Ocracoke Island. (Modified from van der Valk 1974. Copyright 1974 by the Ecological Society of America.)

	K	Na	Ca	Mg
Bodie Island:				
Bulk precipitation (ppm)	1.30	26.00	2.30	3.90
Rainfall (ppm)[a]	0.15	7.16	1.02	1.30
Bulk precipitation (kg $ha^{-1} yr^{-1}$)[b]	14.40	432.60	34.80	64.20
Rainfall (kg $ha^{-1} yr^{-1}$)	2.30	118.00	16.00	21.00
Ocracoke Island:				
Bulk precipitation (ppm)	1.00	21.00	3.40	3.00
Rainfall (ppm)[a]	0.15	4.36	0.82	0.59
Bulk precipitation (kg $ha^{-1} yr^{-1}$)[b]	15.60	319.80	51.60	46.20
Rainfall (kg $ha^{-1} yr^{-1}$)	2.20	65.00	12.00	8.90

[a]Data from Gambell and Fisher 1966.

[b]Estimated from data for May 1971 to January 1973.

where S_2 is the quantity of the cation present at a certain depth at sampling time T_2, S_1 is the quantity of the cation present at the same depth at the previous sampling time T_1, IN is the input of the cation in salt spray or in salt spray leachate during the time period from T_1 to T_2, and OUT is the output of the cation due to leaching by rainwater during the time period from T_1 to T_2. About 160 kg ha^{-1} yr^{-1} of Ca is exported from the 60 cm depth, which is significantly more than the amount of Ca that entered as salt spray. The difference is probably due to leaching of calcium carbonate from shell fragments in the sand. The turnover times for nutrients in dune systems are very short compared to other terrestrial ecosystems: 11–37 days at Bodie Island for K, Na, and Mg. For Ca, a longer time of 32–206 days reflects input from dissolution of shell fragments. Total annual meterological inputs are always higher than exports for the cations K, Na, Ca and Mg.

Limits to growth

Many of the dune perennials have extensive root systems, or rhizome networks, which form clones over large areas. These may pull nutrients from the front of the dune, which receives ample supply of K, to the back, where K may be limiting to vegetation growth. As shown in Figure 12–5, the annual input of Na and Mg at Bodie Island greatly exceeds that present in vegetation.

The sparse vegetation provides almost all the nutrient storage of the system. There is no soil reservoir. The dune system shows virtually no conservative mechanisms for retention and internal cycling of nutrients. Thus, mineral cycles in seashore dunes are less stable than in other terrestrial ecosystems (Jordan and Kline 1972).

Salt Marsh

The salt marsh is similar in some of its attributes to the coastal dune system, and in some aspects it is unique. A marsh is an herb-dominated ecosystem in which the rooting medium is inundated by tidal water for long periods, if not continuously. The substrate is chiefly mineral with typically high humus content. A swamp is similar to a marsh except that the dominant life form is woody plants, not herbs. A marsh or a swamp may have salt, fresh or brackish water.

Nixon et al. (1976) investigated the salt marsh ecosystem which surrounds a 6000 m^2 bay at Bissel Cove, Narragansett Bay, Rhode Island. The single outflow is subject to control, allowing the manipulation of the marsh and the monitoring of various aspects of nutrient cycling. The vegetation at Bissel Cove is primarily cordgrass (*Spartina alterniflora*). *Spartina* marsh is extremely productive, comparable to agricultural crops subject to intensive cultivation. A *Spartina* community

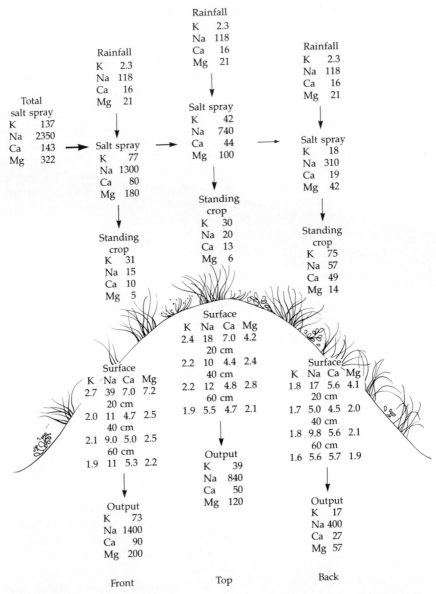

Figure 12-5. Estimated or calculated annual inputs (kg ha^{-1} yr^{-1}) of cations in salt spray and rainwater, and output (kg ha^{-1} yr^{-1}) in drainage water from 60 cm below the surface on the front, top and back of the foredune at Bodie Island. The cation content (kg ha^{-1}) of vegetation at maximum standing crop and average concentrations (kg ha^{-1}) of cations in soil at the surface and 20, 40, and 60 cm below the surface are given in the boxes. Outputs equal salt spray leachates plus rainfall. (After van der Valk 1974. Copyright 1974 by the Ecological Society of America.)

is very stable, but it has little biological structure, is not very diverse, and is often subjected to a stressful environment.

Salt marshes, like coastal dunes, receive most of their nutrients from the sea. Rivers do not, as a rule, fertilize salt marshes and other estuarine communities; their contribution of nutrients is insignificant. Nutrients are concentrated and recycled by salt marshes, quite unlike the functioning of coastal dune mineral cycles. The sediment of the marsh provides a sizable storage component (Figure 12–6) due to the large clay fraction. Although there may be some export, at Bissell Cove during the summer and early fall nutrients cycle intrasystematically and there is little or no export with tidal flow.

The movement of P within a salt marsh system is shown diagrammatically in Figure 12–6. P, in undisturbed nature, is in chronic short supply; it is scarcer than carbon in the lithosphere and it is in the parts per billion range in the hydrosphere. Thus, it is often limiting to plant growth (Deevey 1970). The transfer of P and other nutrients from deep sediments (more than a meter deep) to the water is accomplished in several steps: (1) cordgrass root systems take in P; (2) bacteria degrade the cordgrass; (3) detritus feeders ingest bacteria and return P to the marsh waters. This "pump" role of cordgrass is analogous to the action of deep roots of forest trees, bringing nutrients up from the depths and depositing them in surface layers.

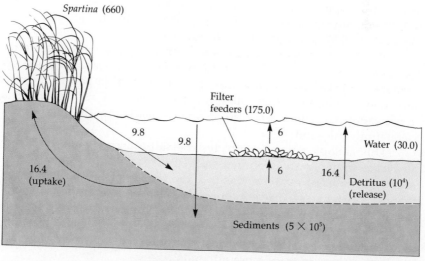

Figure 12–6. The flux of phosphorus in a Georgia salt marsh ecosystem. The arrows indicate transfer of phosphorus and the parentheses indicate storage. Data for the standing stocks are in mg m^{-2} and data for the transfer fluxes are in mg m^{-3} day^{-1}. (Modified from Odum: *Fundamentals of Ecology*, copyright 1971 by the W. B. Saunders Co., Philadelphia, PA.)

In the salt marsh at Bissel Cove, the concentrations of dissolved organic P and N are higher than their inorganic counterparts (in summer and early autumn). This strongly suggests internal cycling. Furthermore, although N fixation is significant, portions of the marsh become anaerobic for part of each day, contributing to rates of denitrification which may equal or exceed N fixation. The C to N ratio is very high, due to cordgrass detritus increasing the N demand on the system. Thus, the internal cycling of N in salt marsh sediments becomes crucial, and may even be more critical than P cycling.

The concentration of available P and N found in salt marsh waters is an order of magnitude higher than that of the adjacent sea or inflowing river water (Pomeroy 1970). The contributions of the large storage compartment in the clay sediments and the internal biological processes which hold and recycle elements provide a nutrient base for a highly productive system.

Barrier Island Forest Ecosystem

The dune forest system on Fire Island, New York, relies primarily on atmospheric input for cations. Fire Island is a barrier island* in the Atlantic Ocean near Long Island, New York. Rainfall exceeds evapotranspiration, and a lens of fresh water is maintained beneath the island at a central depth of perhaps 40 m below sea level. Fresh water moves downward at the center of the lens and upward along the edges, which prevents the intrusion of ocean nutrients into the ground water. Art et al. (1974) investigated this forest ecosystem.

The forest is dominated by American holly (Ilex opaca), sassafras (Sassafras albidum) and eastern service berry (Amelanchier canadensis). As branches of these plants grow through the uppermost part of the canopy, they encounter lethal salt spray. As a result of such pruning, a dense canopy of lateral branches develops just below the interface of vegetation and wind.

This canopy network impedes the salt-carrying capacity of the airstream: minerals are deposited on the foliage, and subsequent leaching by rainfall carries the nutrients to the forest floor. Thus, in a sense, there is a positive feedback loop operating here: the greater the total canopy surface, the greater the quantity of nutrients taken from the atmosphere and deposited on the vegetation. The subsequent leaching and uptake of these nutrients increases productivity, biomass, and leaf area, which in turn traps more minerals.

Given this, what provides the limits to further increase of biomass? The same mineral-laden winds that provide nutrition are also

*A barrier island is an island which hinders the approach to the mainland, or in this case, hinders the approach to Long Island.

periodically intense enough to kill back the twigs of the upper canopy.

The major storage compartments for minerals in this dune forest system are the living biomass, which holds most of the K, Ca, and Mg, and the soil organic matter, which accounts for lesser amounts of these elements. Nutrient uptake is just about balanced by litter fall and by canopy leaching. The system is believed to be at equilibrium, with losses to ground water balanced by atmospheric inputs. Salts are strained from slower-moving portions of the airstream by the finely divided canopy, and precipitation and wind-borne nutrients provide additional nutrition to the forest.

GRASSLANDS

In North America, the grasslands extend in a broad, sweeping belt from Texas north to Saskatchewan, replaced at the margins by mesquite *(Prosopis)*, aspen *(Populus)* or oak *(Quercus)* savannah. Grasslands are characterized by frequent fires, heavy soils and a continental climate with 300–500 mm precipitation annually. Potential evaporation exceeds precipitation. In the **pampa** or grass steppe of South America, precipitation increases to 1000 mm yr^{-1}, but potential evaporation is correspondingly high (Walter 1973) and the precipitation to evaporation ratio is still less than 1. Recall that evapotranspiration is a conservative force, positively correlated with efficient intrasystematic cycling of nutrients; that is, intrasystem cycling is enhanced by increased evapotranspiration.

We will not trace the flux of every nutrient in the grassland ecosystem, but will select one: nitrogen. In grasslands, almost without exception, primary productivity is limited by the available nitrogen supply. Although loss is unlikely during normal functioning, both grasses and decomposers use up soil nitrate and ammonium ions so rapidly that their concentration remains low. When live above-ground biomass reaches 300 kg ha^{-1}, nitrogen becomes severely limiting.

Nitrogen fixation rates, both by symbiotic and by free-living organisms, are low in grasslands (about 1 kg $ha^{-1} yr^{-1}$, compared to 1.5–8 kg $ha^{-1} yr^{-1}$ for fixation by the lichen component of the tropical rain forest). Contribution in precipitation is about 3 kg $ha^{-1} yr^{-1}$. Nitrogen loss may occur through leaching, denitrification or volatilization. Estimates of volatilization range to about 1.5 $ha^{-1} yr^{-1}$; leaching and denitrification require the presence of significant concentrations of nitrate in wet soils, a situation rarely encountered in grasslands.

A generalized biogeochemical cycle for N is shown in Figure 12-7. Let us begin with uptake of nitrogen into plants through living roots as ammonium ion or nitrate ion. The rate of uptake of nitrate nitrogen increases as nitrate concentration increases, but only to a

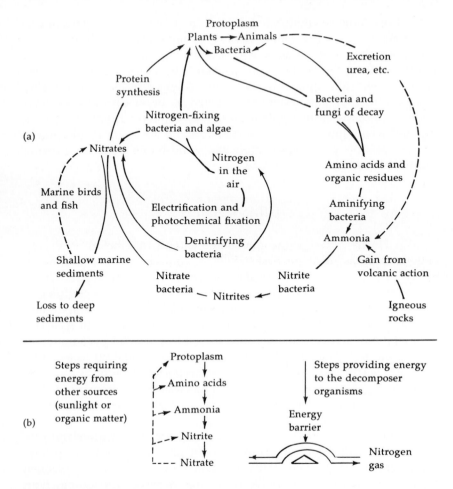

Figure 12–7. Two ways of picturing the nitrogen biogeochemical cycle, which is an example of a relatively perfect, self-regulating cycle with a large gaseous reservoir. In (a) the circulation of nitrogen between organisms and the environment is depicted along with microorganisms which are responsible for key steps. In (b) the same basic steps are arranged in an ascending-descending energy series, with the high energy forms on top to distinguish steps which require energy from those which release energy. (After Odum: *Fundamentals of Ecology,* copyright 1971 by the W. B. Saunders Co., Philadelphia, PA.)

point, leveling off as soil nitrate becomes high. Translocation and distribution within the plant varies seasonally, with the ratio of live shoot N concentration to live root N concentration changing with plant phenology. The ratio of shoot to root N varies from 3.0 to 1.0, remaining high during the first third of growth, then declining. The decline in shoot to root N may be adaptive, pulling N back to living, persisting roots late in the season, before aboveground biomass dies back.

Decomposition and mineralization transfer N back to the various soil compartments. Partitioning of N from decomposing roots and litter into soil organic matter or ammonium is determined by the soil C:N ratio. When this ratio exceeds 50, most of the N in tissue being decomposed will be transformed to stable organic matter. As the C:N ratio in soil declines, more and more N will be mineralized, i.e., converted to ammonium. Mobilization of soil organic N (mineralized to ammonium) provides an annual turnover amounting to about 2% of the substrate. Factors affecting this rate are: (a) the concentration of organic nitrogen, and (b) abiotic factors such as temperature and soil water.

It is apparent that mechanisms have evolved to conserve and recycle vital N within the grassland ecosystem. The natural force that circumvents these mechanisms is fire. It reduces aboveground biomass to mineral ash but allows the escape of volatile elements, including N. However,

> . . . the prairie had two strings to its bow. Fires thinned its grasses, but they thickened its stand of leguminous herbs . . . each carrying its own bacteria housed in nodules. . . . Each nodule pumped nitrogen out of the air into the plant, and then ultimately into the soil. (Leopold 1966)

THE HUBBARD BROOK EXPERIMENTAL FOREST

The study of an entire ecosystem requires many workers of related and widely divergent disciplines, many hours, and an excellent integration system. In part, our understanding of ecosystem function has been enhanced through the use of models, which are essentially of two types: (a) a microcosm, utilizing a closed system with quantifiable inputs and outputs of energy and materials, for instance, a balanced aquarium; and (b) systems analysis using computers to simulate the functioning of a real ecosystem (for example, see Jordan and Kline 1972 or Shugart et al. 1976).

Generalizations and predictions that derive from a microcosm study may not be valid at the ecosystem level. Relationships may not be linear, nor even logarithmic. Extrapolation from exceedingly small, closed systems to, for example, a forest region, may be unwarranted. Simulation also has its drawbacks. Data collection to provide an informational base must be unbiased and painstakingly accurate. Often, simplifying assumptions are made, which of necessity eliminate consideration of short-term fluctuation, of separate categories of producers and consumers, etc.

Ecosystem Analysis

What must be done is analysis of real ecosystems. The small watershed approach of Likens and Bormann (1972; also Bormann and Likens 1967; Bormann et al. 1977; Siccama et al. 1970) combines some of the most attractive features of the microcosm approach with large-scale ecosystem-encompassing measurements. Their criteria for a suitable study site are as follows:

1. the watershed ecosystem is part of a larger, homogeneous biotic and geologic unit;
2. it has an impermeable geologic substrate; and
3. it is characterized by relatively humid conditions, and the input/output budget for nongaseous nutrients can be easily determined from the difference between the meteorologic input (dissolved substances and particulate matter in rain and snow) and the geologic output (dissolved substances and particulate matter in drainage waters).

Let us propose a gradient for nutrient sources: at one extreme is an ecosystem totally dependent on weathering for K, Ca, Na, and Mg; at the other extreme is an ecosystem totally dependent on meteorological input. The forest ecosystem at Hubbard Brook, New Hampshire, is near the first extreme, while the Sunken Forest at Fire Island is near the other. The oak-pine forest at Brookhaven National Laboratory, Long Island, New York, is intermediate in terms of nutrient source. It is suggested that plants at Brookhaven have evolved mechanisms to conserve nutrients and to reduce intrasystem cycling times (see Woodwell et al. 1975, for a discussion of this late successional, young oak-pine forest).

The Hubbard Brook Experimental Forest is located north of Plymouth, New Hampshire, in the lower elevations of the White Mountains. The watersheds studied contain a cool-temperate, meso-phytic, broadleaf deciduous forest at lower altitudes, with some coni-fers higher up. Although the area was logged extensively in the early part of this century, there are some older trees, and the stands are believed to represent a climax association. The physical environment, including edaphological, climatological, geological and hydrological factors, is discussed extensively elsewhere (Likens et al. 1967; Bormann et al. 1970).

The data compiled over the last decade or so form the basis for formulating general principles and concepts about nutrient cycling in this eastern deciduous forest. Experimental manipulation provides additional information which can be used for prediction about the behavior of such systems in response to perturbation, whether the change is natural or is caused by human impact. In this section, we will discuss the nutrient budgets at Hubbard Brook. We will examine two cycles in some detail: one in which there is net loss (the calcium cycle) and one in which there is net gain (the nitrogen cycle).

Nutrient Budgets: Input, Output, and Balance

Nutrient input—geological, biological and atmospheric—can be rather easily estimated. Rainfall, throughfall and dry fallout can all be gauged as the product of the concentration per unit of time and the total precipitation, etc. The Hubbard Brook region is not subject to geological influx, and biological output is assumed to balance biological input, since the area holds no special attraction (e.g., for migratory animals), so that these two sources for nutrients can be disregarded. The climax forest is unlikely to lose many nutrients as wind-borne particulates or aerosols, although an occasional leaf fragment may be lost. Significant losses would most likely be due to geological output, specifically in streamflow. In addition to dissolved nutrients, stream water carries inorganic and organic matter suspended by the turbulence of the stream, and other material which rolls along the bottom of the stream, carried by the current.

The streams at Hubbard Brook are subjected to sensitive hydrologic and chemical measurements, made possible by the geologically impermeable substrate. The assumption is that the streams of the area carry all of the water leaving the system, and that they carry or deposit all of the particulates and dissolved nutrients leaving the watershed. A concrete ponding basin behind a v-notch weir (Figure 12–8) allows for assessment of flow, as well as measurement of chemical concentrations and particulate load.

As the stream flows over the weir, constant monitoring of flow provides baseline data. Periodic sampling measures solution nutrient loss from the system. The chemical output is derived as nutrient concentration (mg l^{-1}) times the volume of water draining from the watershed. The streamflow slows in the ponding basin behind the weir so that as the suspended matter and the bed load are deposited they can be measured also.

As we noted earlier, nutrient budgets are inextricably connected to the hydrologic cycle. Since streamflow nutrient concentration is virtually constant, nutrient loss is high in wet years, and low in drought years. Nutrient budgets developed from these long-term studies at Hubbard Brook show a net gain for nitrate nitrogen and ammonium nitrate, but a net loss for Si, Ca, S, Na, Mg and Al. Input and output for K and Cl is balanced. Net losses are assumed to be possible because of weathering, and so net loss data is used as an estimate of weathering

Figure 12–8. A weir showing the v-notch recording house and ponding basin. (Courtesy of the Northeastern Forest Experiment Station, Forest Service, U.S. Dept. of Agriculture.)

within the system. Annual living and dead biomass accumulation at one of the watersheds is 3930 kg ha^{-1}, which indicates that the system is aggrading and therefore seral.

The Calcium and Nitrogen Cycles

In the Hubbard Brook system (see Figure 12–3), the vegetation plus organic debris contains about 854 kg Ca ha^{-1}. Uptake by vegetation exceeds decomposition by about 9.5 kg ha^{-1} yr^{-1} due to accretion. Sources of Ca are atmospheric inputs (2.2 kg ha^{-1} yr^{-1}) and soil water and exchange surfaces (510 kg ha^{-1}). The estimated 74,200 kg Ca ha^{-1} that is locked up in soil and rocks is unavailable.

The output of Ca (13.9 kg ha^{-1} yr^{-1}) includes 13.7 kg ha^{-1} yr^{-1} carried as solution in stream water. The Ca cycle is effectively closed, with internal cycling of about 1/8 of the Ca pool. Only about 2.3% of the available Ca is lost each year, and this is assumed to come from weathering (11.7 kg ha^{-1} yr^{-1} lost). The system efficiently conserves Ca, recycling it in throughfall, litterfall, decomposition and renewed uptake.

An estimated 20.7 kg N ha^{-1} yr^{-1} enters the system. Of this, 81% is held in long-term storage, partly in litter (46%), with the rest (54%) in living biomass. The bulk of N input (68%) comes from N fixation. About 32% derives from precipitation, and virtually none from weathering (Bormann et al. 1977).

Mineralization contributes inorganic N, of which 1% comes from root exudates. Apparently some 33% of the 119 kg N ha^{-1} used annually by vegetation is drawn from within the plant, used in growth, then withdrawn from leaves and replaced in storage at the end of the growing season. Thus, internal cycling is an important conservation mechanism for the forest N cycle. N tends to be stored primarily in soil organic matter (90%) with 9.5% in the vegetation, and the remaining 0.5% in the soil.

Deforestation

Watershed 2 was deforested in 1965, and regrowth was prevented for three summers by herbicide application. The purpose of this drastic manipulation was to test the resilience of the system, after the severance of uptake transfer of nutrients by vegetation (refer back to Figure 12-3). That is, if one compartment is removed from the system, how will the others function without it?

The deforestation and herbicide application had a great effect on the hydrologic cycle. Brief, intense runoff from the deforested region carried a greatly increased load of particulates. Average annual particulate export on undisturbed sites was about 2.5 metric tons km^{-2}, but was 38 metric tons km^{-2} yr^{-1} from the disturbed site. Usually, the ratio of dissolved substances to particulates averages about 2.3; following manipulation, however, this ratio leaped to greater than 8.0 the first 2 years.

Immediate cessation of transpiration resulted in increased streamflow as follows: in the first year, 35 cm of streamflow, which is 40% above what would be expected in an undisturbed site; in the second year, 27 cm or 28% above expected; and in the third year after deforestation, 24 cm or 26% above expected. This increased streamflow due to lack of transpiration is not surprising because streamflow regularly increases on undisturbed sites in autumn, before the rains begin, in response to synchronous leaf fall.

Since streamflow is related to nutrient loss, the deforestation and herbicide application also had an effect on nutrient loss. Export of ions increased for all major ions, except for ammonium, sulfate and bicarbonate ions. Nitrate, from decaying vegetation, was one of the nutrients whose concentration in streamflow increased greatly. Reflecting the decrease in sulfate concentration and the increase in nitrate concentration, the stream pH went down (5.1 to 4.3), and its character changed from a dilute sulfuric acid to a stronger nitric acid.

Increased export of dissolved organic substances (14–15 times greater than controls) was due to increased erodability, to increased streamflow, and to accelerated chemical decomposition of inorganic materials in the deforested region.

A BOREAL FOREST ECOSYSTEM

Ecologists assume that an equilibrium point is reached in a mature system: energy input/output ratio is unity and the mineral budget is balanced. This certainly seems to hold for the tropical rain forest. Highly productive forests can be maintained in a steady state for thousands of years, unless there is deforestation and subsequent soil depletion. In the jack pine *(Pinus banksiana)* boreal* forest ecosystem, this steady state is reached for even-aged stands somewhere between 30 and 65 years old (Foster and Morrison 1976).

The Environment and the Vegetation

In the Mississagi River area of northern Ontario, Canada, jack pine grows well, in even-aged stands. The growing season in the area is about 175 days, and extends from April to October. The climate is modified continental, with some 860 mm of rainfall per year. The soils are young, of medium texture with poorly developed horizons (entic haplorthod, roughly a humo-ferric podzol; see Chapter 16). The soils are derived from glacial outwash: silt, sand, and stones carried by glacial streams are deposited as the stream water loses velocity.

In addition to jack pine there is a well-developed understory, with a moss floor *(Pleurozium schriberi)* and shrubs and herbs, e.g., dwarf huckleberry *(Vaccinium angustifolium)*, bush honeysuckle *(Diervilla lonicera)* and big-leaf aster *(Aster macrophyllus)*. By Ontario standards, the stands are very productive (Foster and Morrison 1976).

Nutrient Vaults

Does a nutrient storage compartment ever become a vault, locked against internal cycling? Large, old trees become repositories of enormous quantities of organic matter. Yet wood is notoriously low in nutrient concentration, so that the lockup, although not trivial, is not disastrous for the ecosystem. In the boreal jack pine forest the vault is in the litter and humus of the forest floor.

A stand up to about 20 years old is very productive, and organic matter accumulates on the forest floor at about 800 kg ha^{-1} yr^{-1}. After that, the annual increments of organic matter added to the litter decrease to about half, as the growth of the stand slows. Especially apparent is the decline in productivity of stands between 30 and 65 years old.

*Boreal refers to the forest areas of the north temperate zone and Arctic region.

Foster and Morrison (1976) point out that very early in the life of the stand, phosphorus concentration stabilizes at about 12–15 kg ha^{-1} in the aboveground biomass of jack pine. From age 20–30, N and P are immobilized to an increasing degree in microbial biomass. Using incorporation of elements into the forest stand as a measure of growth, we find little to no incorporation in stands 30–65 years old. The flow of minerals from soil to roots to vegetation and back to soil is maintained, but the per cent of nutrients retained in plant biomass is very small. The lack of decomposition locks up N, P, and Mg during this stage in the life of the stand, and so plant growth falters.

How will these sites be returned to high productivity? There are at least two possibilities, but only one favors a vigorous stand of jack pine: wildfire. If fire does not occur, the mature stand, about 80–100 years old, deteriorates, opening the canopy. The increased amount of sunlight reaching the forest floor would perhaps enhance decomposition and mineralization of litter. However, cones of jack pine are late to open (serotinous), and seed release is much more likely following a fire than at any other time.

The favorable aspects of a fire, aside from opening serotinous cones, are as follows: P, K, Ca and Mg (tied up in organic matter) are made available to plants as soluble nutrients in the soil, mineral soil is exposed as a seed bed, and the forest canopy is opened. Beneficial aspects of prescription burning, as well as wildfire, will be more fully discussed in Chapter 15. Note, however, that surface fires of low intensity would provide regeneration of jack pine without a great loss of nutrients to the system.

One negative result of a fire is that N is volatilized, which would be a serious loss to young stands, since N accumulation is greatest in early years. Also, the mineral soils in a boreal forest ecosystem do not have a high exchange capacity and sudden mobilization of nutrients would allow some loss by leaching.

THE ARCTIC TUNDRA

The arctic tundra is thinly covered by grasses, sedges, dwarf shrubs and low perennial herbs supported on very shallow soils over permafrost. Thus, neither vegetation nor soil provides compartments of a reasonable size for nutrient storage. Nutrient content in alpine tundra vegetation may be as low as 17 g m^{-2} (Chepurko 1972), and we may assume a similar low value for arctic tundra plants. Nutrient values in a temperate North American forest (Brookhaven, Hubbard Brook) range from 91 to 147 g m^{-2}, 5 to 8 times that of the alpine tundra.

Jordan and Kline (1972) modeled the phosphorus cycle of the arctic tundra using only three compartments: litter → roots → leaves. The soil was not considered a significant storage compartment. They assume that litter provides the primary storage for P, and that release occurs every 3 or 4 years (Schultz 1969). The system lacks resilience; available P decreases dramatically once the transfer from litter to roots occurs. Roots only buffer the loss about 1 year, and thereafter P content of leaves decreases and does not rebound (Figure 12–9) until the next large pulse is released from litter.

We could further trace the pulse of available P from litter → roots → leaves → lemmings (leaves are forage for lemmings) → predatory birds of the tundra (snowy owl and parasitic jaeger) and the least weasel. A striking correlation is shown if we change the ordinate of Figure 12–9 to represent density of lemmings, rather than phosphorus in forage. Cycles of lemmings at Point Barrow, Alaska, and lemmings and voles in Norway oscillate in a fashion greatly resembling the forage (leaf) cycle modeled in Figure 12–9.

The wide swings in arctic tundra ecosystem functioning can thus be attributed, at least in part, to the lack of large storage compartments for available nutrients.

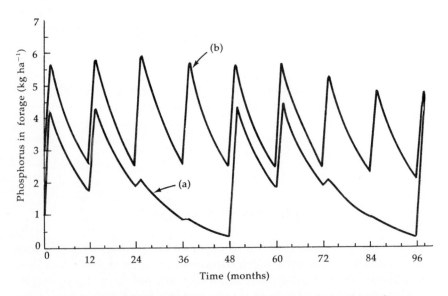

Figure 12–9. Total amount of phosphorus as a function of time (a) in forage (leaves) of a tundra ecosystem, and (b) in forage of an ecosystem differing from the tundra only in that it has a wood compartment equal in size to that of a tropical rain forest. Lines connect monthly values. (After Jordan and Kline 1972. Reproduced, with permission, from the *Annual Review of Ecology and Systematics*, Volume 3. Copyright 1972 by Annual Reviews Inc.)

STABILITY

Stability is a measure of a system's response to some outside perturbation. A **stable** system returns to steady-state conditions in a short-time following some irregularity in either the physical environment or in the flow and cycle of energy and materials. The shorter the recovery time, the higher the stability.

Stability of cycles of single elements may be described in terms of relative stability. Usually, if an element is repeatedly recycled such that its function is great compared to its input, its relative stability is low. In other words, recovery would take a long time following disturbance. Little recycling, compared to input, would imply high relative stability. For example, we infer that the sodium cycle in the coastal dune has a very high relative stability: it would quickly return to steady-state following some interruption.

Over ecological time, populations of organisms accumulate, concentrate and recycle nutrients. Up to a certain point, this accumulation enables succeeding populations to occupy the site. The rate of accumulation of even one element may limit succession. The accumulation process is governed by internal feedback and is, to a certain extent, self-reinforcing. For instance, the climax forest is supported by a very large reserve of nutrients. The forest shields the soil against erosive nutrient loss and enhances evapotranspiration, which reduces stream-flow and thus reduces mineral loss due to leaching.

What factors are involved in the stability of ecosystem nutrient cycles? Evergreenness, size of storage compartments, erosion control, and immobilization of nutrients are properties of ecosystems that serve to conserve and recycle nutrients.

Evergreenness, with progressive leaf fall, may be correlated with site characteristics (Monk 1966). Progressive leaf fall may be a significant conservation mechanism in dry, sterile sites. In a sense, forests on mesic, fertile sites can afford the "luxury" of deciduousness (synchronous leaf fall). If the big pulse of nutrients released in a very short time period cannot be accommodated in the soil storage compartment, the nutrients may be lost through leaching. Furthermore, nutrients are more easily leached from broadleaf litter than from needle litter. Progressive needle fall results in a slowly decomposing litter. Evergreenness may, among other things, be a means of establishing a closed mineral cycle in a xeric site.

The coupling of metabolism and mineral cycling is a stabilizing force within an ecosystem. The bridge between metabolism and mineral cycling is the availability of large storage compartments within the system. Large storage capability tends to buffer nutrient cycles against perturbation. A large standing crop (e.g., trees in tropical rain forest) provides a spacious storage compartment; so do ample exchange sites for cations (e.g., young volcanic soils). The organically bound component in soil provides secure storage for nutrients also. An element's replacing power on soil surfaces is an important feature of nutrient stability. Knowing the magnitude of the replacing power allows prediction of the resistance of that cycle to loss, and so the efficiency of internal cycling of that element can be estimated.

Certain apparently open systems are also stable. Salt marshes seem to have a superabundance of energy supply, nutrients are rarely limiting, and storage compartments are never empty. Although diversity is low, stability remains high. This is probably attributable to a remarkably high metabolic activity level appropriate to the abundance of nutrients available in the system.

But what of stability in terms of response to disturbance? There are two basic ecosystem strategies for ameliorating the effects of perturbation: maintenance of resistance to erosion, and enhanced primary productivity.

The natural undisturbed ecosystem exerts control over erosion. Following some disturbance (e.g., fire, flood, avalanche, drought, or insect attack) the community reestablishes that control. For example, following a fire many standing dead snags provide a measure of protection against the wind, and by the time these topple over, regrowth has begun. Understory shrubs, ferns and herbs may proliferate, further insulating the soil surface. At Hubbard Brook, following deforestation of Watershed 2, herbicide application prevented any new growth. Thus, there could be no biotic control over erodability. Consequently, export of particulates rose sharply the third year after the application of herbicides.

Another effect of disturbance is the mobilization of nitrates. Usually, at least in mature ecosystems, nitrification proceeds very slowly. In the Hubbard Brook deforested region, increased microbial nitrification not only allowed flushing of mobile nitrates from the system, but the alteration of the N cycle also effected an increased net export of other nutrients from the system. It is believed that disturbances such as plowing, logging or burning accelerate nitrification, and thus nutrient loss.

Often a disturbance is followed, for a short time, by an increased nutrient and water supply. Pioneer plants exploit this situation, especially the increased availability of nitrates. Bormann et al. (1974) point out that these are the responses of a besieged ecosystem: enhanced soil water and nutrient availability, enduring resistance to erosion, and opportunistic species that exploit the disturbed habitat. Together, these factors comprise a homeostatic mechanism that propels the system toward stability.

SUMMARY

Materials cycle within the biosphere, but energy flows one way. Biogeochemical cycling includes nutrient cycling, which may be one of two types: gaseous (or perfect), which tends to be regional or global (e.g., CO_2); or sedimentary, which operates primarily within one ecosystem (e.g., Ca). Nutrients may be accumulated with successive seral stages. Nutrients are compartmentalized, occurring in soil and rocks, in soil solution, in the atmosphere, and in biomass. Transfer into or out of an ecosystem may be meteorologic, geologic, or biologic. Cycles may require minutes, decades or millenia to revolve.

Essential nutrients include both macronutrients (N, P, S, K, Mg, and Ca) and micronutrients (Fe, Mn, Zn, Cu, Mo, B, and Cl). Factors involved in nutrient cycling include fire and earth movement and the various aspects of the hydrologic cycle: rainfall, leaching, evapotranspiration, runoff, infiltration, and salt spray. Biotic factors include grazing, decomposition, root exudation, and retrieval of deep-lying nutrients.

Coastal dune ecosystems are open, nonconserving systems, relying on salt spray for nutrient input, with no soil reservoir for nutrients. Salt marshes are extremely productive and exhibit high stability even though diversity is low. Deep-lying nutrients are brought to the soil surface by cordgrass and are thus made available to other organisms. The stability of marshes and their highly efficient nutrient cycling regime is in part attributable to large storage compartments in clay sediments.

The canopy of the forest system at Fire Island filters nutrients from the marine airstream. The forest responds to wind-borne salt with dieback and subsequent branch proliferation below the zone of contact.

Nitrogen commonly limits productivity in grasslands. N is provided by fixation, both from free-living and symbiotic fixation, and lost through volatilization.

The soil provides a large storage compartment in the broadleaf temperate deciduous forest of the eastern United States. The exchange sites are saturated with cations, and streamflow discharge maintains a constant chemical concentration for most major nutrients except K. Experimental deforestation of one watershed resulted in greatly increased runoff and enhanced export of particulates and all major nutrient ions except ammonium, sulfate and bicarbonate. The undisturbed system is stable and conservative, with efficient intrasystem cycling.

Nutrients are immobilized in microbial biomass in the boreal Jack pine ecosystem. Frequent fire maintains the vitality and integrity of that system.

A sort of pulse stability, with wide oscillations on a 3–4 year cycle, characterizes the arctic tundra. This northernmost ecosystem has virtually no nutrient storage compartments analogous to those of other ecosystems.

Stability relies upon resilience: replenishment power of nutrients, large storage compartments, resistance to erosion, and exploitative pioneer species are essential homeostatic mechanisms for ecosystems.

PART IV

ENVIRONMENTAL FACTORS

We have seen in earlier chapters that the population and community attributes of plant species are determined by the response of individuals to the external environment. It is possible that autecological information may, when enough taxa have been characterized, lead to predictability at the community level. In this part, we will consider the environmental factors of light, temperature, fire, soil, and water, including their effect on physiological and community processes.

It should be kept in mind that separate discussions of various factors is artificial; the interaction of all environmental factors determines the response of the plant. Chapter 19 attempts to show how the distribution of major vegetation types in North America is a response to interacting environmental factors.

CHAPTER 13

LIGHT AND
PHOTOSYNTHESIS

Solar radiation is one of the most important qualities of the environment which influences plants. Light is the ultimate source of energy for most organisms; therefore, the variations of light in time and space and the photosynthetic responses to these variations are basic to our understanding of plant ecology. In this chapter we will consider the physical nature of light and look at how it varies spatially and temporally. We will also consider adaptations which reflect the ability of plants to assimilate carbon dioxide by photosynthesis. Plants are able to grow in environments as extreme as a snow bank, a tropical rain forest, and a desert. Survival in these habitats has been made possible by a broad range of adaptations including the photosynthetic and respiratory responses considered in this chapter.

PHYSICAL PROPERTIES OF LIGHT

Global Radiation

There are several forms of radiant energy that are important influences on an exposed leaf (Figure 13–1). The infrared radiation absorbed and emitted by the leaf is considered in Chapter 14. Here we are concerned only with the visible portion of global radiation. Solar radiation may arrive at the surface of a leaf directly or indirectly, as diffuse radiation scattered by the atmosphere (skylight) (S^{sky}) and by clouds (cloudlight) (S^{cloud}), and as radiation reflected from the soil or other objects in the habitat ($r\,S^{direct}$).

The spectral qualities and intensity of radiation (Figure 13–2) vary depending on the distance the radiation travels through the atmosphere and the features of the habitat which absorb, reflect, or transmit the light. The atmosphere effectively removes a portion of direct solar radiation, reducing the intensity from the extraterrestrial level of ~ 2 cal cm^{-2} min^{-1} (the solar constant) to ~ 1.3 cal cm^{-2} min^{-1} at sea level on a clear summer day (Gates 1965). Most radiation received is direct with only a small amount that is rich in the blue wavelengths arriving as skylight. The energy of solar radiation can be measured with devices called pyranometers (see Chapter 14).

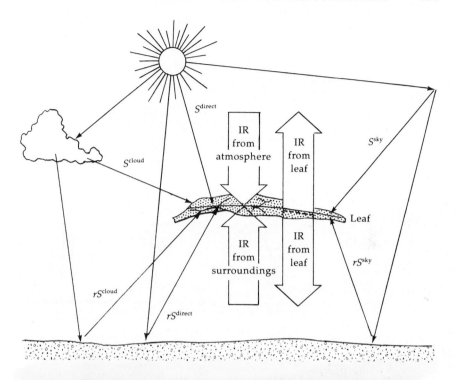

Figure 13-1. Schematic illustration of eight forms of radiant energy incident on an exposed leaf, and the infrared radiation emitted from its two surfaces. (From Introduction to *Biophysical Plant Physiology* by Park S. Nobel. W. H. Freeman and Company. Copyright © 1974.)

The light relationships of an idealized green leaf are diagrammed in Figure 13-3. The amount of radiation absorbed is relatively high at wavelengths shorter than 0.7 μm (700 nm), which includes the visible and ultraviolet wavelengths. The very high energy ultraviolet light (wavelengths shorter than 0.4 μm (400 nm)) can be damaging to biological material; however, water present in leaf cells is very efficient in absorbing these wavelengths, which are not important in the photosynthetic process. High rates of absorption in the visible portion of the spectrum are caused by the presence of chlorophylls, carotenes, and xanthophylls: the pigments in greatest abundance in plant cells. Light transmitted through vegetation is dramatically reduced in the visible region because of these pigments (Figure 13-2).

Photosynthetically Active Radiation (PAR)

The radiation between 0.4 μm and 0.7 μm is the wavelength band absorbed by chlorophyll and therefore these wavelengths are active in

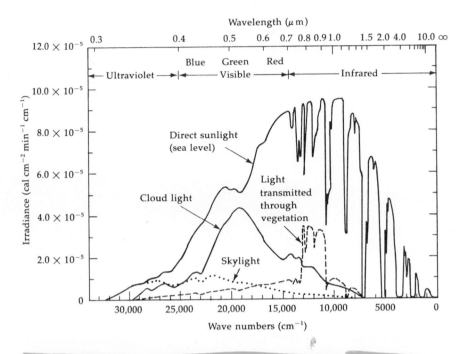

Figure 13–2. Spectral distribution of direct sunlight, skylight, cloudlight, and sunlight penetrating a stand of vegetation. Each curve represents the energy incident on a horizontal surface. (From Gates 1965. By permission of the American Meteorological Society.)

the photosynthetic process. Light in this band has been labeled **photosynthetically active radiation** (PAR). Physicists have shown that radiant energy is transported in discrete bundles called photons. A mole of photons (6.02×10^{23} photons) is referred to as an Einstein (E)—the unit of photosynthetically active radiation. The intensity of PAR is determined by the photon (quantum) flux which, for example, would be about 2000 μE m^{-2} sec^{-1} near midday on a clear day.

Other units of light measurement were used in plant ecological studies prior to the advent of instrumentation to measure only PAR. Common units and their equivalents are listed in Table 13–1. It is not possible to directly convert light values reported in conventional units to PAR in Einsteins, because the broad spectral range of lux or footcandle measurements reports light as perceived by the human eye, which is not identical to PAR. Portable light meters which measure only photosynthetically active radiation are commercially available. Other more inexpensive measures are considered in more detail in Chapter 14.

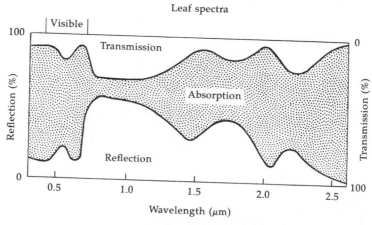

Figure 13–3. Transmission, absorption, and reflection with respect to wavelength for an idealized green leaf. (After Monteith, J. L. *Principles of Environmental Physics*, Edward Arnold, London, 1973.)

VARIATIONS IN THE LIGHT ENVIRONMENT

Latitude and Physiography

The general distribution of global radiation over the surface of the earth is plotted in Figure 13–4. There is a gradual decrease in global radiation over the land areas from latitudes of about 30°N and S toward the poles. This decline is due primarily to the seasonality experienced in temperate regions and poleward. From the equator to 30°N and S, radiation is generally high with limited areas of very high radiation or very low radiation. Radiation over the Amazon Basin is low

Table 13–1. Common units for the measurement of light energy.

Unit name	Units or equivalents	Value at sea level in full sun
langley (ly)	1 cal cm^{-2}	1.3 ly
foot-candle	10.76 lux	10,000 ft-c
lux	1 lumen m^{-2}	107,600 lux
watt	1 joule sec^{-1} or 10^{7}ergs	1,000 watts m^{-2}
Einstein	6.02 × 10^{23}photons	2,000 μE m^{-2} sec^{-1} = 200 nE cm^{-2} sec^{-1}

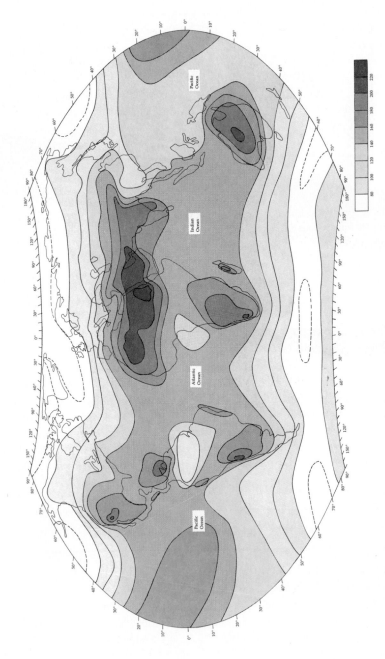

Figure 13–4. World map showing the distribution of solar radiation over the earth's surface in kcal cm^{-1} yr^{-1}. (From Landsberg, H. E., et al. 1966. *World Maps of Climatology.* By permission of Springer-Verlag.)

because of frequent heavy cloud cover and heavy rainfall. Areas of very high solar radiation occur on all the major continents where high pressure systems prevail and cloud cover is very low.

Variations in solar radiation also occur on a finer scale than that just discussed. For example, in the Northern Hemisphere south-facing slopes receive direct solar radiation for longer periods and at greater intensity than do north-facing slopes. The presence of dust, pollutants, or fog reduces the intensity of global radiation in specific physiographic situations and during certain seasons of the year. Also, as the elevation increases, light rays pass through less atmosphere and, as a result, the ultraviolet light is a larger fraction of the total incoming radiation.

Vegetation

Vegetation modifies the light environment in different ways depending on the geometry of the plants, the distribution of biomass, and the spectral properties and orientation of leaves. The modifications due to these factors are variable seasonally because of seasonal differences in the vegetation and in the elevation of the sun.

Hutchinson and Matt (1977) studied the distribution of solar radiation in a tulip poplar forest in Tennessee (Figure 13–5(a)). The year was divided into **phenoseasons**, each of which represents a period when the solar elevation and phenological state of the vegetation create a unique set of influences on the radiation regime within the forest. The light environment was measured as the amount of total solar radiation received per unit time at different levels within the forest. Variations above the canopy were due to the phenological state of the canopy and fluctuations in solar radiation. The highest radiant flux reaches the forest floor between the spring leafless phenoseason and early summer when the canopy is in full leaf (Figure 13–5(b)). It is during this period that sufficient light and adequate temperature allows the spring bloom of herbaceous plants on the forest floor. The gradient in light reduction during the leafless phenoseasons was much less than during the fully leafed phenoseasons at equivalent solar angle (Figure 13–5(a)). Other less obvious variations in leaf orientation and canopy structure modify the extinction rate of light as it passes through the forest canopy.

The extinction of light as it travels through a vegetation canopy depends on the total leaf area index (area of leaves projected on a unit area of ground surface). Campbell (1977) considered the theoretical leaf area which would receive direct sunlight as a function of leaf area

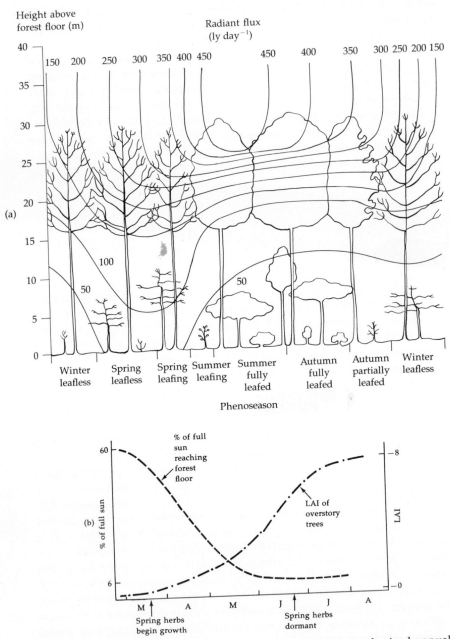

Figure 13-5. Light and phenology in deciduous forests: (a) synthesized annual course of average daily total solar radiation received within and above a tulip poplar forest; (b) seasonal changes in light penetration and leaf area index (LAI) of a temperate deciduous forest. ((a) from Hutchison and Matt 1977. Copyright 1977 by the Ecological Society of America.)

index (LAI) and leaf orientation (Figure 13–6(a)). Very little additional sunlight penetrates a canopy of horizontal leaves with a leaf area index greater than about 3, but when more leaves have a vertical orientation greater light penetration occurs in canopies of higher LAI. Not all leaves within a canopy or even on an individual plant are at the same inclination. Recall from Chapter 7 that forest communities have LAI of 8 or more. Campbell has simulated gross photosynthesis in model canopies assuming that light was the only limiting factor. Figure 13–6(b) shows the results for LAI values of 1, 3, and 5. Notice that high photosynthesis is possible with large LAI values when leaf orientations are random, but at low LAI leaf orientation has little effect.

PHOTOSYNTHESIS

Our understanding of the photosynthetic process and the ecological ramifications of various photosynthetic adaptations has broadened dramatically over the past decade (Burris and Black 1976). Until recently, it was assumed that higher plants performed photosynthesis only by the Calvin pathway. During the 1960s a new photosynthetic process (C_4 photosynthetic pathway) was discovered by Kortshak et al. (1965) in sugar cane and later described in detail by Hatch and Slack (1966). This discovery has stimulated a flurry of research on aspects of the C_4 pathway from biochemistry to the population and community levels. The photosynthetic properties of only a few plants have been characterized but many more remain to be studied. It is possible that additional pathways will be discovered and it is certain that, as research proceeds, we will further modify our concepts of the ecological importance of photosynthetic variations.

Our goal in this section is not to describe the detailed aspects of the various photosynthetic processes, but simply to bring to mind photosynthetic characteristics that are most likely to be of ecological significance. If you wish to have a more complete review of these concepts, most recent textbooks in plant physiology and biochemistry have detailed information.

The photosynthetic process is divided into light reactions and dark reactions. Light reactions convert light energy into chemical energy and are common to all higher land plants. Dark reactions, which do not require light, convert carbon dioxide into sugars and starches. In the light reactions (Figure 13–7) light energy is absorbed by pigment systems within the chloroplasts where it is converted into chemical energy in the form of ATP and NADPH (reduced NADP), which are the sources of energy for the dark reactions. Briefly, the process is as follows: water is oxidized and acts as the source of hydrogen to reduce NADP and the source of oxygen released during photosynthesis. ATP is also formed in this process from ADP and inorganic phosphate. (Note that the following chemical expression is not balanced.)

$$H_2O + NADP^+ + ADP + P \xrightarrow[\text{chloroplast}]{\text{light}} NADPH + O_2 + H^+ + ATP$$

C_3 Photosynthesis and Photorespiration

Prior to the discoveries of Hatch and Slack and Kortshak et al., it was thought that most plants fixed CO_2 only by the pentose phosphate pathway, also called the Calvin cycle. Plants with the Calvin pathway are often referred to as C_3 photosynthesizers because the first stable

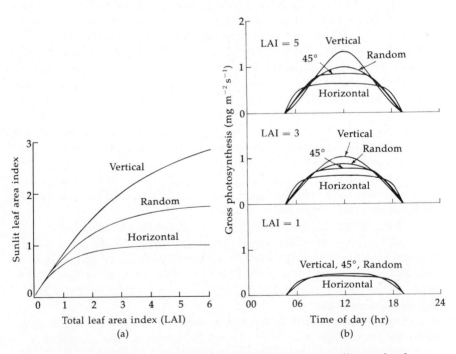

Figure 13-6. (a) Light interception versus leaf area for vertically, randomly, and horizontally oriented leaves. Sunlit leaf area index is found by subtracting the sunlit area below the canopy from 1. (b) Gross photosynthetic rate in model canopies having various leaf areas and leaf distributions as a function of times of day (48°N on 1 June). (From Campbell, G. S. 1977. *An Introduction to Environmental Biophysics*. By permission of Springer-Verlag.)

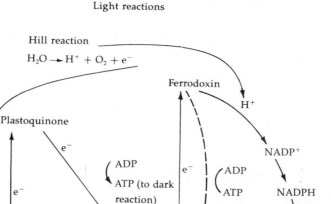

Figure 13-7. Overview of the light reactions of photosynthesis. Note that the chemical expressions are not balanced.

product formed is phosphoglyceric acid (PGA) with a skeleton containing three carbon atoms. A series of reactions follow (Figure 13–8(a)) in which sugars and starch are formed as the products of photosynthesis. The energy necessary for the incorporation of atmospheric CO_2 into photosynthate is obtained from the ATP and reduced NADP formed in the light reactions. Additional ATP from the light reactions provides energy and phosphate to regenerate ribulose biphosphate (RuBP, a 5-carbon sugar), which then can combine with CO_2 to form PGA. This process takes place during the day within the mesophyll cells and cortical cells of stems which contain chloroplasts. Anatomically the leaves of C_3 plants are typically divided into palisade and spongy layers with the chloroplasts widely distributed within the mesophyll (Figure 13–9(a)).

Coupled with the Calvin cycle is a series of reactions which leads to the light stimulated release of CO_2 known as photorespiration (Fig-

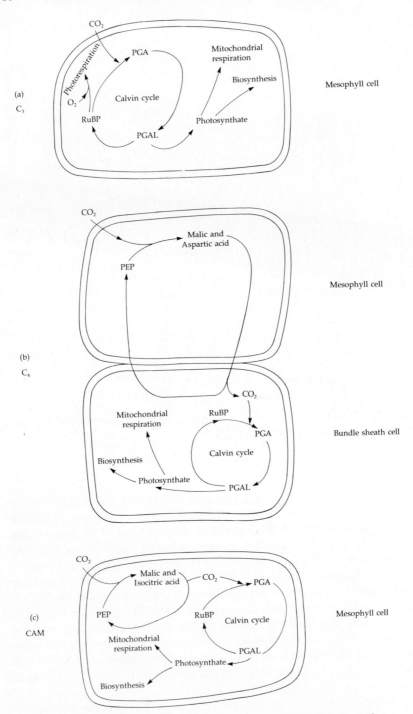

Figure 13–8. Basic dark reactions and product fate of (a) C_3, (b) C_4, and (c) CAM photosynthesis.

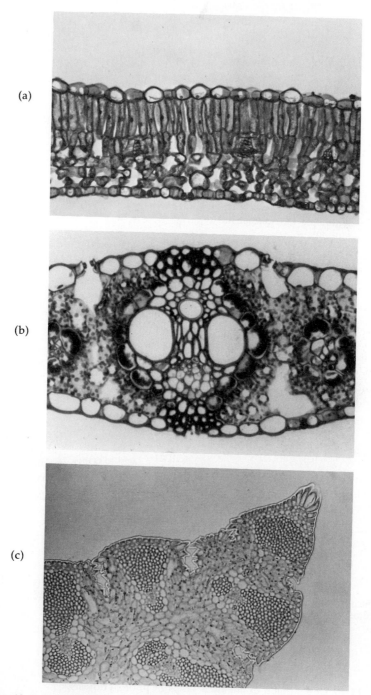

Figure 13–9. Cross sections of (a) C_3, (b) C_4 (Kranz anatomy), and (c) CAM photosynthetic tissues.

ure 13–10(b)). **Photorespiration** is not the same as true respiration, and both processes may go on independently of each other in the same cell at the same time. True respiration (Figure 13–10(c)) oxidizes carbohydrates, such as glucose, and generates energy in the form of ATP and

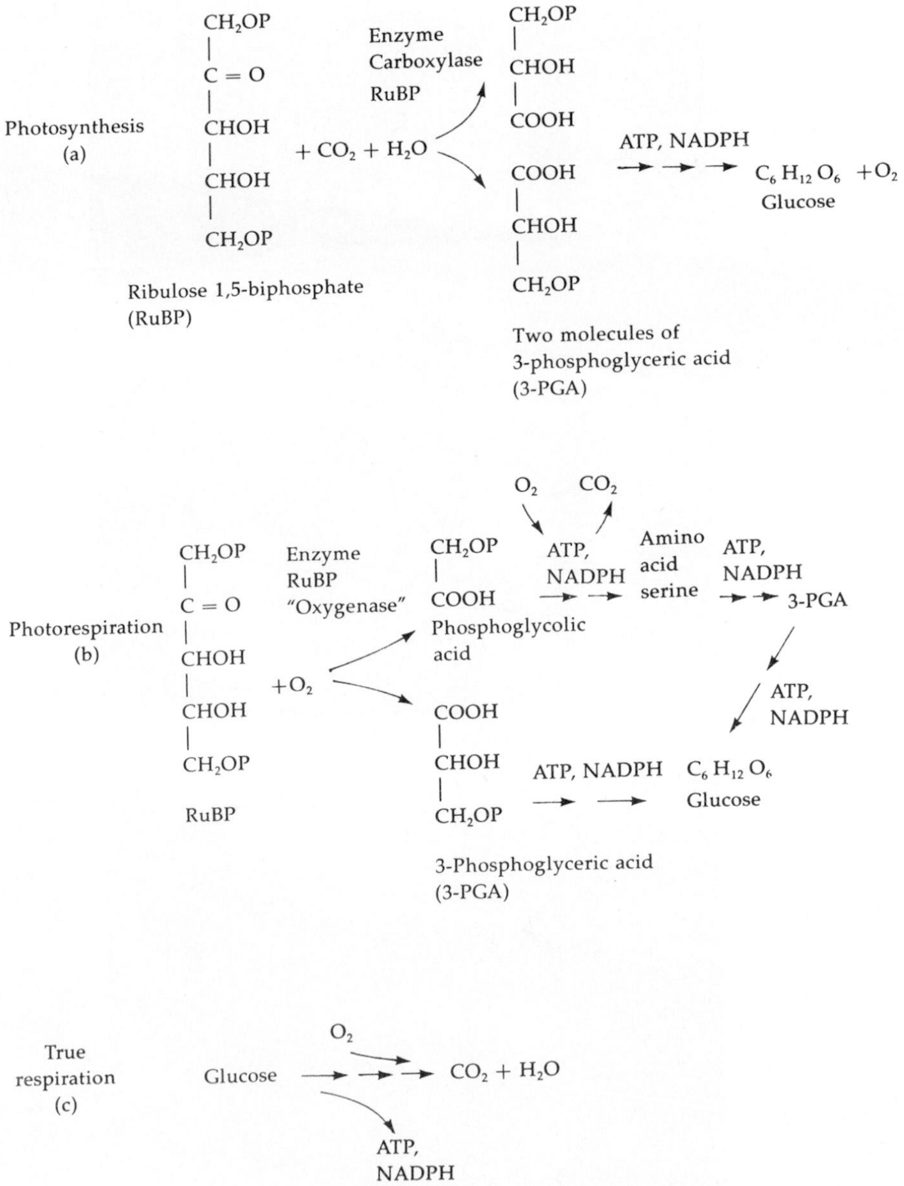

Figure 13-10. Comparison of chemical steps of photosynthesis, photorespiration, and true respiration.

NADPH. Photorespiration oxidizes organic acids, using up oxygen and releasing carbon dioxide, but it uses more energy than it produces. It is wasteful of the plant's resources; it may reduce the rate of net photosynthesis by 30–50% (Goldsworthy 1976; Downes and Hesketh 1968). In the presence of atmospheric levels of oxygen (21%), the enzyme responsible for fixation of carbon dioxide in photosynthesis (RuBP carboxylase) (Figure 13–10(a)) may fix oxygen instead of CO_2 to the receptor molecule of RuBP. Thus, oxygen and carbon dioxide compete for the same enzyme. For convenience, the enzyme is called RuBP carboxylase when it fixes carbon dioxide, and RuBP oxygenase when it fixes oxygen, but it is the same enzyme. When oxygen is fixed, only half as much phosphoglyceric acid is initially produced; the remaining carbohydrate is converted into phosphoglycolic acid. A series of energy-requiring steps converts this to the amino acid glycine and then two molecules of glycine are combined into one molecule of serine, with the loss of carbon dioxide and the consumption of oxygen. Serine can be chemically altered back to PGA with additional consumption of energy. Thus, glucose sugar is produced during photorespiration, as with photosynthesis, but photorespiration requires the expenditure of a great deal more energy.

Another characteristic of C_3 plants relates to their uptake of carbon isotopes. Carbon exists in a stable state in atmospheric carbon dioxide as both $^{13}CO_2$ and $^{12}CO_2$. Carbon with a weight of 12 is much more abundant, and the ratio of $^{13}C/^{12}C$ is a constant. As plants fix carbon dioxide, they discriminate slightly between $^{12}CO_2$ and $^{13}CO_2$, thus changing the $^{13}C/^{12}C$ ratio. The ratio can be measured by burning a sample of plant tissue and subjecting the captured gas to analysis by mass spectrometer. C_3 plants typically reduce the isotope ratio by 2.4–3.4%, which is written as -24 to $-34‰$. A value of $-20‰$ means that the $^{13}C/^{12}C$ ratio is reduced relative to a bicarbonate standard by 20 per thousand (Bender 1971; Smith and Epstein 1971).

C_4 Photosynthesis

Over a dozen families of angiosperms have been shown to fix carbon dioxide by the C_4 dicarboxylic acid pathway. C_4 metabolism may have arisen several times in evolutionary history, as indicated by the conspicuous lack of evolutionary relationship between many C_4 families. Lists of C_4 species have been published in several places (e.g., Downton 1975; Krenzer et al. 1975; Smith and Epstein 1971; Teeri and Stowe 1976; Mulroy and Rundel 1977), and the lists keep growing as more C_4 species are discovered.

In the C_4 pathway, carbon dioxide is fixed by the enzyme PEP carboxylase to form 4-carbon acids such as malate or aspartate in the mesophyll cells (Figure 13-8(b)). These acids are then transported from the mesophyll cell into a bundle sheath cell where CO_2 is released to enter the Calvin cycle. The 3-carbon carrier molecule returns to the mesophyll cell where it is converted to PEP to receive another CO_2 molecule. This mesophyll-bundle sheath shuttle thus concentrates CO_2 at the C_3 fixation site. Photosynthesis proceeds within the bundle sheath cells just as it does in the mesophyll cells of C_3 plants. The special anatomical characters necessary for these processes to take place are referred to as **Kranz anatomy** (Figure 13-9(b)). Kranz anatomy is characterized by the presence of a dense layer of large, thick-walled, chloroplast-containing cells forming a sheath around the vascular bundles and by a lack of differentiation of the mesophyll into a recognizable palisade and spongy layer. Kranz anatomy is easily seen with a hand lens in sections of fresh material. This is a means by which plants with C_4 photosynthesis can be identified in the field.

The oxygen inhibition noted in C_3 plants is not a problem in plants with C_4 photosynthesis. The lack of oxygen inhibition is apparently due to the high concentration of CO_2 at the site of RuBP carboxylase, which prevents RuBP oxygenase activity and allows C_4 plants to reduce the concentration of CO_2 to nearly zero. The CO_2 compensation point (the point at which the concentration of CO_2 is sufficient to support net photosynthesis) has been used to distinguish between C_3 and C_4 plants (Downton and Tregunna 1968).

Another difference between C_3 photosynthesis and C_4 photosynthesis is that C_4 photosynthesis requires two additional ATP molecules to fix each molecule of CO_2. Also, the $^{13}C/^{12}C$ ratio is different in C_4 plants; values range between -10 and $-20\%o$ (Bender 1971). This difference is often used to distinguish between C_3 and C_4 photosynthetic pathways of plants which are dried, of questionable or intermediate anatomy, etc.

Crassulacean Acid Metabolism (CAM)

Many succulent plants in some 15 families (e.g., Agavaceae, Orchidaceae, Crassulaceae, and Cactaceae) exhibit a carbon fixation scheme referred to as crassulacean acid metabolism (CAM). All CAM plants have succulent photosynthetic tissues but do not have specialized bundle sheath cells as do C_4 species (Figure 13-8(c) and 13-9(c)). The stomata of CAM plants are open at night and carbon dioxide is fixed at night by PEP carboxylase, forming malic or isocitric acid. In contrast, the stomata of C_3 and C_4 species are open during the day and CO_2 is fixed during the day. In CAM photosynthesis, acids are accumulated within the vacuoles of cells at night, which results in a pattern of low nighttime pH followed by increasing pH during the day when photosynthesis proceeds by the Calvin cycle. During the day when stomates are closed, CO_2 from the organic acids is fixed by the C_3 cycle. The elevated CO_2 concentrations within the cell probably counteract the inhibitory effect of oxygen by photorespiration (Osmond 1976).

Ratios of ^{13}C and ^{12}C are variable across the entire range of values found in C_3 and C_4 species, making it impossible to identify CAM plants by $^{13}C/^{12}C$ ratios. This is because the ratios vary depending on environment, as will be described later in this chapter. For example, plants from mesic environments have values comparable to those of C_3 species, whereas the same species in an arid environment will have values more like C_4 species (Bender et al. 1973). CAM plants may, at least in artificial situations, shift to a C_3 mode when not stressed by salt, drought, long days, or warm nights.

Table 13–2 summarizes and compares 16 traits of C_3, C_4, and CAM photosynthetic pathways.

ENVIRONMENTAL FACTORS AND PHOTOSYNTHETIC RESPONSE

Gas Exchange

The maximum potential rate of CO_2 uptake and the efficiency with which available CO_2 can be fixed is a quality of great ecological importance. Plants are limited by the concentration of CO_2 at the site of fixation. Therefore, the rate of diffusion of CO_2 through the stomates and into the plant is a critical factor. Since water also leaves the plant when the stomates are open, the balance between water loss and CO_2 uptake is important in determining the relative success of terrestrial plants. The pathway of CO_2 and water diffusion into and out of a leaf may be visualized as a series of steps, any of which can be limiting. The application of this concept to water diffusion is considered in Chapter 18; here we consider the process of CO_2 diffusion.

The steps in the movement of CO_2 can be visualized as a series of resistances and potential gradients defined by a restatement of **Fick's law** for gas diffusion:

$$\Phi = \frac{\Delta c}{\Sigma r}$$

where CO_2 flux (or the flux of any diffusing substance), Φ, depends on the change in concentration from the source to the reaction site (Δc) and the sum of the resistances (Σr) to diffusion. Figure 13–11 depicts the pathway along which CO_2 diffuses from external air to the reaction site within a chloroplast containing cell. The concentrations and resistances dictate the rate of CO_2 assimilation. Therefore, the consideration

Table 13–2. Comparison of photosynthetic pathways. (From Kelly et al. 1976 and Black 1973. Reproduced, with permission, from the *Annual Review of Plant Physiology,* Volumes 24 and 27. Copyright 1973 and 1976 by Annual Reviews Inc.)

No.	Trait	C_3 heliophyte (adapted to a high light environment)	C_4	CAM
1)	taxonomic diversity	very wide: algae to higher planets	no algae, lower vascular plants, or conifers; wide among flowering plants	some species in about 26 families of flowering plants + Welwitchia
2)	typical habitat	no pattern	open, warm, saline (some exceptions)	open, warm, saline (sometimes cool)
3)	leaf anatomy	palisade + spongy parenchyma	no mesophyll differentiation; large bundle sheath; Kranz	no mesophyll differentiation; big cells with large vacuoles
4)	light saturation point	3000–6000 ft-c	8,000–10,000+ ft-c	like C_3 (?)
5)	optimum temperature	20–30°C (lower in tundra)	30–45°C (as for light above, can be lower for C_4 species in different habitats)	30–35°C for CAM mode; lower for C_3 mode
6)	maximum photo-synthetic rate $mg\ dm^{-2}\ hr^{-1}$	30	60	3 (maximum reported = 13)
	$mg\ g^{-1}\ hr^{-1}$	55	100	1 or less
7)	maximum growth rate $(g\ dm^{-2}\ day^{-1})$	1	4	0.02
8)	transpiration ratio $(g\ H_2O\ g^{-1}\ CO_2\ fixed)$	600	300	80
9)	photorespiration	high	low	low
10)	Na required?	no	yes	no (but salts stimulate CAM mode)
11)	fixation path and enzyme	$CO_2 + 5\text{-}C \rightarrow$ 3-C PGA; carboxydismutase	$CO_2 + 3\text{-}C \rightarrow$ 4-C acids; PEP carboxylase	still some debate, possibly just as C_4 but enzyme

Table 13–2, continued

No.	Trait	C₃ heliophyte (adapted to a high light environment)	C₄	CAM
		or also called ribulose biphosphate carboxylase		is light inhibited thus may structurally be different
12)	stomate behavior	open in day, closed at night	open in day, closed at night	closed in day, open at night (unless environment shifts plant to C₃-like mode)
13)	space-time relations	entire Calvin cycle in any mesophyll cell	initial fixation in mesophyll, then transfer of acid to bundle sheath for Calvin cycle	initial fixation at night in any mesophyll cell; storage of acid in vacuole; Calvin cycle during day
14)	effect of environment on pathway	none	none	moist, warm night temperature and long day length put plant in C³ mode
15)	CO_2 compensation point	50 ppm	5 ppm	2 ppm (in dark)
16)	$^{13}C/^{12}C$ ratio	−24 to −34‰	−10 to −20‰	possibly intermediate, although mainly like C₄

of relative values is important in understanding photosynthetic adaptations. **Boundary layer resistance** (r_a) is encountered near the leaf surface where a zone of air may become depleted of CO_2 and through which CO_2 must move by diffusion rather than by turbulent mass transfer as it does further from the surface. **Stomatal and cuticular resistance** (r_l) restricts CO_2 diffusion at the leaf surface. Cuticular resistance has been shown to be so great in most mature leaves that we may assume that essentially all of the CO_2 which enters a leaf diffuses through the stomates. Resistance at the point where CO_2 leaves the gas

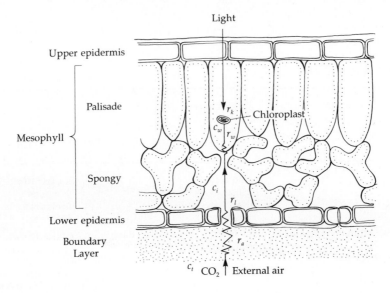

Figure 13-11. Transport pathway, transport resistances, and concentration gradients in a photosynthesizing leaf. See text for definition of terms.

phase in air and is dissolved in the cytoplasm of the photosynthesizing cell is referred to as **liquid phase resistance** (r_w). Limitations placed on CO_2 fixation by the photosynthetic process itself represent the final resistance to CO_2 transport and are called the biochemical resistance (r_k). The sum of r_w and r_k is called mesophyll resistance. It can be equal to or greater than stomatal resistance, depending on the plant and environmental conditions.

The concentration of CO_2 in the atmosphere (c_t), in the intercellular spaces (c_i), and within the mesophyll cells (c_w), in conjunction with the various resistances, determine the rate of CO_2 uptake. In a rapidly photosynthesizing cell intracellular CO_2 concentrations (c_w) are quite low relative to c_t, and so a gradient exists. This gradient becomes greater toward the outside of the leaf where normal CO_2 concentrations are approximately 320 ppm. Carbon dioxide released by photorespiration and mitochondrial respiration adds to the CO_2 concentration within the leaf. The concentration becomes high enough at night to reverse the gradient (except in CAM plants), and CO_2 passes into the atmosphere.

Stomatal aperture (and therefore stomatal resistance) appears to be responsive to light, intercellular CO_2 concentrations and leaf water status. Stomatal aperture responds to these factors by active transport of ions into and out of the guard cells, thereby modifying turgidity by changing osmotic concentrations within the cell. For a detailed account of the mechanisms and causes of stomatal response see Levitt (1974, 1976), Hall et al. (1976), or Raschke (1976).

Different plants have different abilities for utilizing CO_2 at low concentrations. For example, recall that C_4 plants fix CO_2 into organic acids and transport them out of the mesophyll cells. This creates a CO_2 sink in the mesophyll cells so that a steeper concentration gradient can be maintained and C_4 plants can reduce the concentration of CO_2 to near zero, whereas C_3 plants reach their CO_2 compensation point at much higher levels. Figure 13–12 shows the result of studies conducted by the Carnegie Institution of Washington in which C_3 and C_4 plants were subjected to various intercellular CO_2 concentrations (c_i). The ability of the C_4 plant to maintain a positive photosynthetic rate at near zero CO_2 concentration is apparent. It is also interesting to note that at high (~ 700 ppm) intercellular concentrations of CO_2, C_3 and C_4 plants had nearly identical photosynthetic rates. C_4 plants have a higher fixation rate under conditions where intercellular CO_2 concentrations are low. Low intercellular CO_2 concentrations result when stomatal resistance is high. Consequently, C_4 plants have an advantage in arid, warm environments.

Light Utilization

It has been long recognized that some plants are adapted to high light environments (**heliophytes**) and some to low light environments (**sciophytes**). Individuals of the same genotype grown in varying light

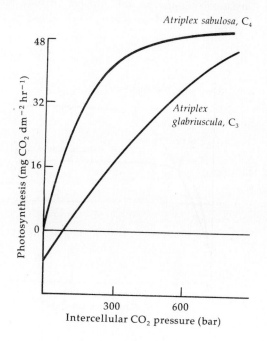

Figure 13–12. Comparison of the photosynthetic reactions of *Atriplex glabriuscula*, a C_3 plant, and *Atriplex sabulosa*, a C_4 plant, to various intercellular CO_2 levels. (From Björkman et al. 1975. Courtesy of the Carnegie Institution of Washington.)

conditions exhibit morphological and physiological differences just as
do individual leaves on a single plant which are differentially exposed
to light. Plants growing in the shade tend to have larger, thinner
leaves, with larger, less dense mesophyll cells, longer internodes and
less overall pubescence; these adaptations increase utilization efficiency, increase available light, or reduce reflection. Figure 13–13
shows a typical response of sun adapted leaves and shade adapted
leaves to increasing incident radiation in the 400–700 nm range (PAR).
Note that shade leaves require less light to reach their CO_2 compensation point and they saturate at lower levels than do sun leaves. The
ability of sciophytes to maintain a positive carbon balance in the shade
is not necessarily due to an ability to photosynthesize more rapidly
than heliophytes in low light. For example, Loach (1967) found that
shade tolerant and shade intolerant species had essentially equivalent
rates of photosynthesis when grown in low light. The key to differential success in shade was a difference in dark respiration rate. Shade
tolerant species had a lower dark respiration rate and therefore a lower
light compensation point, allowing them to maintain a positive carbon
balance (net photosynthesis) even at very low gross photosynthetic
rates. The fact that the dark respiration rate is the key to survival in the
shade has been supported by several other authors (e.g., Logan 1970;
Willmot and Moore 1973).

C$_4$ plants typically have higher light saturation levels than C$_3$ species (Figure 13–13). However, it is now apparent that at least some C$_3$
plants do not saturate even at full sunlight. For example (Figure 13–14),
Camissonia claviformis, a C$_3$ desert annual, was studied by Mooney et al.
(1976) and found to have the highest photosynthetic rates yet mea-

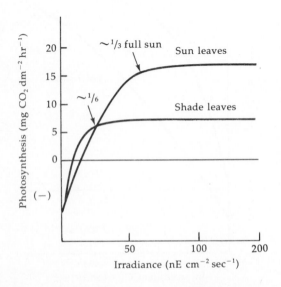

Figure 13–13. Typical light
saturation curves for sun
adapted and shade adapted
leaves of C$_3$ plants. Arrows
indicate saturation
intensity: ∼¹/₃ full sun for
sun leaves, ∼¹/₆ full sun
for shade leaves.

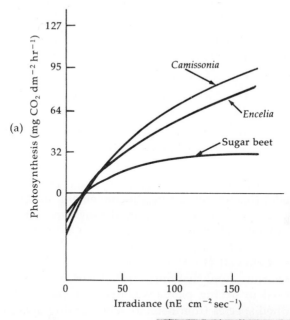

Figure 13-14. (a) Contrasting responses of C₃ species of *Camissonia, Encelia,* and sugar beet to different levels of light. (b) *Camissonia claviformis,* taken in Death Valley. ((a) from "High Photosynthetic Capacity of a Winter Annual in Death Valley," by Mooney, H. A. et al., *Science,* Vol. 194, pp. 322–323, 15 October 1976. Copyright 1976 by the American Association for the Advancement of Science; and from "Leaf Pubescence: Effects on Absorptance and Photosynthesis in a Desert Shrub," by Ehleringer, J. et al., *Science,* Vol. 192, pp. 376–377, 23 April 1976. Copyright 1976 by the American Association for the Advancement of Science. (b) courtesy of Harold Mooney.)

sured in a terrestrial plant. The reason for this very rapid CO_2 fixation rate lies in low stomatal resistance and high levels of RuBP carboxylase. As a result, *Camissonia* responds almost linearly to increasing light up to full sunlight. *Encelia farinosa* is the only other species reported to have a similar capability; it is also a C₃ desert plant (Cunningham and Strain

1969; Ehleringer et al. 1976). It is clear that the C_4 pathway does not, in itself, give plants the ability to have high photosynthetic rates in very high light environments. It is not clear, however, why other C_3 plants have not evolved this capacity.

Water Stress

Several authors (e.g., Boyer 1976; Hsiao 1973) have reviewed the impact of water stress (plant water status when evaporative demands exceed water supply) on the photosynthetic process. There is general agreement that the most immediate effect is an increase in resistance due to stomatal closure, restricting the movement of CO_2 and water. It is difficult to directly measure mesophyll resistance without the interference of stomatal changes. However, there is good evidence that nonstomatal inhibition of photosynthesis does occur in many species. Johnson and Caldwell (1975) measured CO_2 resistance in arctic and alpine species and found that only a small part of the increased resistance measured in drying plants was due to stomatal closure. Bunce (1977) examined species from habitats with varying levels of available water and reported that increases in mesophyll resistance to CO_2 uptake occurred as the leaves dried. Further evidence of the effect of water stress on photosynthesis comes from studies of lichens conducted by Lange and Kappen (1972) on antarctic and desert lichens. Their results show that photosynthetic rates in these plants vary with changes in water availability even though there are no stomates. The actual biochemical relationship between water stress and photosynthesis remains to be determined.

The fact that C_4 and CAM species have the ability to utilize light while restricting water loss leads to the increased numbers of C_4 and CAM species in hot and dry environments. This is because of their higher **water use efficiency,** meaning that less water is necessary to fix a molecule of CO_2 in C_4 and CAM plants than in C_3 species. This does not mean that C_4 plants have a significant competitive advantage over C_3 species in arid environments. Syvertsen et al. (1976) examined the species composition of Chihauhuan Desert communities and found more C_3 species than C_4 or CAM plants, and found that the C_4 and CAM species contributed little to the standing crop. The adaptability of C_4 plants to arid regions is evidenced by the overwhelming predominance of C_4 species in the summer annual populations in the hot desert areas of the southwestern United States. Caldwell et al. (1977) considered the growth and gas exchange characteristics of a C_3 and a C_4 shrub in cold desert regions of Utah and found no advantage for the C_4 species. Even though the C_4 shrub was able to photosynthesize during the summer, it had a lower overall rate of CO_2 fixation during the more moderate springtime and only recuperated its deficit during the drought.

CAM plants respond readily to water levels and will open stomates during the day if ample water is available. Hartsock and Nobel (1976) heavily watered transplanted *Agave deserti* for 12 weeks under laboratory conditions, causing a shift in stomatal movements so that 97% of the CO_2 uptake occurred in the daytime. Artificial watering in the field did not cause a similar change. However, when water is not limiting, desert *Agave* has a postdawn period of reduced stomatal resistance and a concurrent rapid period of CO_2 fixation (Nobel 1976) similar to that in C_3 plants.

Water stress caused by salinity also induces photosynthetic response, at least in *Gasoul crystallinum* (ice plant). Salinity of 200 mM or greater NaCl causes *Gasoul* to switch from daytime to nighttime CO_2 fixation (Figure 13–15). Also, there is an apparent switch from C_3 to CAM as the leaves get older (Winter and Lüttge 1976). *Gasoul* also has nocturnal CO_2 fixation in naturally saline environments in Israel.

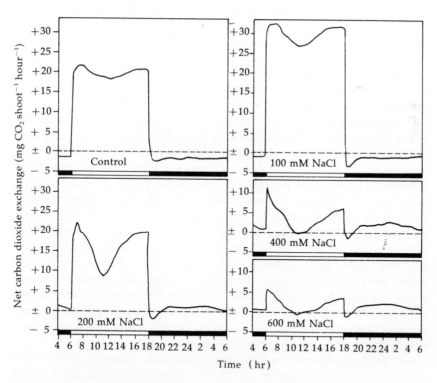

Figure 13–15. Net CO_2 exchange for *Gasoul crystallinum* at various levels of salinity. (From Winter 1975 as redrawn by Winter and Lüttge 1976, Balance between C_3 and CAM pathway of photosynthesis, *Ecological Studies* 19:323–334.)

Temperature Responses

Temperature influences photosynthetic responses in various ways; we have already discussed the ability of C_4 plants to carry on positive photosynthesis at higher temperatures than C_3 plants. The most apparent influence of temperature is the limitation of all enzymatically catalyzed reactions. We would expect that reaction rates would increase gradually to optimum temperatures and then decrease sharply at higher temperatures where enzymes denature. There is, however, much more involved in understanding temperature responses of photosynthesis because plants are able to acclimate to previous temperature regimes. In addition, photosynthesis has ranges of temperature tolerance which are related both to the geographic location of the population and to the age of the tissue.

The very sensitive responses of plants to temperature are illustrated in a study by Fryer and Ledig (1972) in which they measured the photosynthetic temperature optimum of balsam fir seedlings. The seedlings were grown from seeds collected along a 731–1463 m elevational gradient of the White Mountains of New Hampshire. The photosynthetic temperature optimum changed 4.3°C in 500 m of elevation change. This is very close to the change in mean temperature of 3.9°C in 500 m, as recorded by weather stations along the gradient. This rather precise relationship between photosynthesis and temperature is characteristic of habitats with a short growing season.

Where the time for acclimation is greater, we observe acclimation of the photosynthetic apparatus over time. For example, McNaughton (1973) compared Quebec and California ecotypes of *Typha latifolia* and found a narrower range of tolerance for temperature in Quebec populations and a much more restricted capability for acclimating to temperature changes than the California plants. The Quebec ecotype also had an increase in photosynthetic temperature optimum with age, which was not present in the California population. The California plants can shift the temperature optimum depending upon the season and maintain this ability throughout the life of the leaf, whereas the Quebec plants have lower optima in the younger stages and a more precisely fixed response in mature leaves. Pearcy (1976) reported that desert ecotypes of *Atriplex lentiformis* have a much greater capacity for temperature acclimation than the ecotype from a more constant coastal environment. The mechanism of temperature acclimation in desert plants such as *Atriplex lentiformis* (Pearcy and Harrison 1974), *Simmondsia chinensis* (Al-Ani et al. 1972), *Atriplex polycarpa* (Chatterton 1970), and *Larrea tridentata* (Strain and Chase 1966) appears to involve, in part, an adjustment in respiration rate. Instead of continuing to increase with temperature, dark respiration rates tend to be constant with increasing temperature. Under stress situations which restrict photosynthesis, a positive carbon balance can be maintained by the reduction of respiration rates at temperatures in excess of 45°C in some species. Recall that similar adjustments were the mechanism for survival in sciophytes where gross photosynthesis was limited by light. We can generalize that plants from areas with shorter growing seasons or from habitats which are more predictable will have a reduced ability to acclimate to temperature changes and will tend to have a narrower range of tolerance to temperature.

Temperature is also a critical factor regulating CAM photosynthesis. Nighttime CO_2 uptake is dependent upon low temperature. For example, Neales (1973) found that the normal night-day stomatal rhythm of *Agave americana*, a CAM plant, was inverted when night temperatures reached 36°C (Figure 13–16). Kluge (1974) measured no

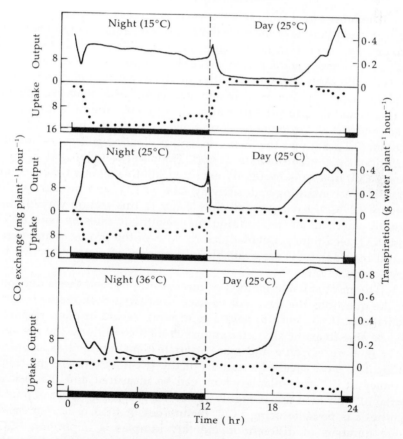

Figure 13–16. Effect of three night temperatures on the patterns of CO_2 exchange (dotted line) and water vapor exchange (solid line) of *Agave americana,* by day and night. (From The effect of night temperature on CO_2 assimilation, transpiration, and water use efficiency in *Agave americana* L. *Australian Journal of Biological Science* 26:705–714. By T. F. Neales 1973.)

night fixation of CO_2 when plants were subjected to 25°C day and 30°C night temperatures. Nobel (1976) supported these findings for *Agave deserti* in which stomatal resistance increased fivefold when leaf temperatures increased from 5°C to 20°C. There is, therefore, strong evidence that the success of CAM plants in water limiting environments is dependent on low night temperatures. This conclusion is further supported by the actual geographic distribution of CAM plants; CAM plants usually do not live in water limiting areas with high night temperatures.

METHODS OF PHOTOSYNTHETIC RESEARCH

The most direct and accurate way of determining photosynthetic rates of land plants is to measure the rate of exchange of CO_2. There are two principal ways that gas exchange rates are determined in terrestrial studies. The most widely used method incorporates an **infrared gas analyzer** (IRGA) to measure the flux of CO_2 to or from a plant or part of a plant sealed in an environmentally controlled chamber. Photosynthesis and dark respiration can be continuously monitored through time, and the environmental conditions within the chamber can be regulated. This allows the measurement of plant responses to individual or multiple environmental factors. The other common way of measuring CO_2 uptake is by briefly exposing photosynthetic tissues to an atmosphere containing radioactively labeled CO_2 and measuring the amount of $^{14}CO_2$ fixed per unit time. Later in this section we will discuss the methods of determining photosynthetic rates by both IRGA and $^{14}CO_2$ techniques. Detailed descriptions of the processes and theoretical considerations are available in a manual of methods edited by Sestak et al. (1971).

The most suitable method of measuring photosynthesis depends on the questions the data will be used to answer. Both methods have limitations. If you wish to have a continuous record of plant response and be able to artificially alter environmental conditions, the gas analyzer is more versatile and more accurate than repeated exposure to $^{14}CO_2$. The most severe limitations of the IRGA system are the small number of different plants which can be measured and the lack of portability for field use. A highly portable and rapid measure of photosynthesis is possible using $^{14}CO_2$ techniques. If field measurement of large numbers of different momentary samples is important, $^{14}CO_2$ techniques are most satisfactory. There are, however, errors inherent in the measurement process which limit its application to comparative studies or to surveys where absolute rates of photosynthesis are not critical. A major source of error relates to the idea of isotope discrimina-

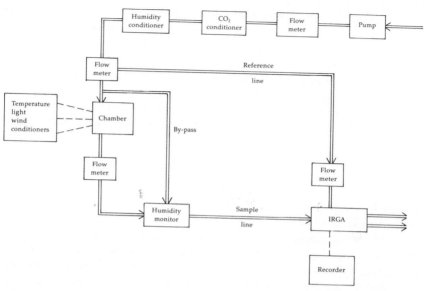

Figure 13–17. Flow diagram showing the major components of an open flow gas exchange system.

tion characteristics noted earlier in our discussion of C_3 and C_4 photosynthesis. Some plants discriminate against $^{14}CO_2$ (Yemm and Bidwell 1969), possibly because of differences in diffusion rates. Whatever the reason, one cannot be sure that $^{14}CO_2$ is absorbed at the same rate as $^{12}CO_2$ under normal conditions. There are additional errors caused by the dilution of $^{14}CO_2$ by $^{12}CO_2$ released by respiration, which modifies the ratio of labeled and unlabeled carbon at the photosynthetic site. Therefore, factors which influence respiration rates will alter measurements obtained by this method. Respiration measurements can only be inferred by measurements of $^{14}CO_2$ dilution by $^{12}CO_2$; such measurements are not as dependable as those obtained by infrared gas analysis.

IRGA Gas Exchange Systems

Gas exchange systems used to monitor CO_2 exchange rates vary widely according to application and, more often, according to the financial resources available. We will discuss a basic open flow design which can be modified for specific purposes (Figure 13–17). The gas is pumped through a series of tubes, usually copper or stainless steel, in which flow rates are carefully monitored. Three flow paths are incorporated so that conditioned gas flows directly to the analyzer, and either through the plant chamber, or through a by-pass and then to the analyzer. The purpose of the by-pass is to assure that all modifications being made to the sample gas are due to plant activity. It is also convenient to have by-pass valves on environmental conditioning components so that any malfunction can be easily identified and corrected.

The plant or leaf cuvette is simply a sealed chamber into which monitoring probes are inserted to constantly measure environmental conditions in the chamber. More sophisticated systems have electronic circuitry which feeds information to the conditioning systems where rates are adjusted to maintain conditions. Some field systems are made so that conditions within the chamber track changes in ambient conditions. The components which condition the air vary widely in sophistication. If a study is being conducted where CO_2 concentrations of the atmosphere are relatively constant, CO_2 conditioning may not be necessary. It is important in all systems, however, to be able to control and measure gas flow rate, photosynthetically active radiation (PAR), temperature, wind speed, and humidity within the chamber. Humidity changes which occur as the air moves through the chamber, when accurately measured, give reliable measurements of transpiration rates.

Additional values such as dark respiration, intercellular CO_2, stomatal and mesophyll resistances, etc., can be measured or calculated from data obtained by manipulating the conditioning components and monitoring plant response.

$^{14}CO_2$ *Absorption Technique*

Shimshi (1969) described an apparatus (Figure 13–18) which can be used in the field to make rapid measurements of photosynthesis. A plexiglass chamber (usually fitted with a circulating fan to reduce boundary layer resistance) is sealed on the surface of a leaf for 10–20 seconds and air containing $^{14}CO_2$ in a known quantity is passed over the leaf. A CO_2 absorbant must be incorporated into the line to prevent releasing $^{14}CO_2$ into the atmosphere. After the leaf has been exposed to the label, a known area of the exposed portion of the leaf is then punched out and immediately killed. Leaf discs are then returned to the laboratory where the amount of $^{14}CO_2$ fixed is determined using liquid scintillation techniques. Sestak et al. (1971) summarized this method and several other techniques of photosynthetic research, and their book should be consulted before attempting to measure photosynthesis.

SUMMARY

Patterns of solar radiation are most significantly affected by latitude, the distribution of atmospheric moisture, and vegetation. Low light intensities are found in high latitudes and in low latitudes where cloud cover is high as in the cloud forests of the Amazon Basin. Very low atmospheric moisture in mid latitudes allows high levels of light to penetrate these arid and semiarid environments. Radiation in the range of 400–700 nm is absorbed by chlorophyll and referred to as photosynthetically active radiation (PAR). High light intensities at the top of the plant canopy decrease rapidly within the canopy; the rate of decrease depends on the season, leaf orientation, wavelengths of light absorbed by leaves, and the phenological state of the vegetation.

There are three photosynthetic pathways in plants, designated as C_3, C_4 and CAM. C_3 plants are most successful and wide-spread, being common in all major climates. Efficiency of C_3 photosynthesis is lowered by photorespiration: the process where O_2 substitutes for CO_2 causing the formation of glycolic acid. C_4 plants are most successful in hot, arid environments because of their low mesophyll resistance and their low CO_2 compensation point which permits them to fix CO_2 despite a high stomatal resistance; the high stomatal resistance reduces water loss. This results in a more efficient use of water by C_4 plants. CAM plants are succulents which have adapted to arid environments

Figure 13–18. $^{14}CO_2$ photosynthesis measuring apparatus: (a) photograph of a typical leaf chamber; (b) diagrammatic representation of the entire apparatus. (Photo compliments of Charles Lambert.)

(a)

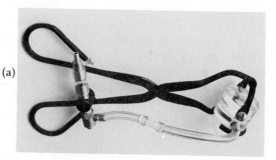

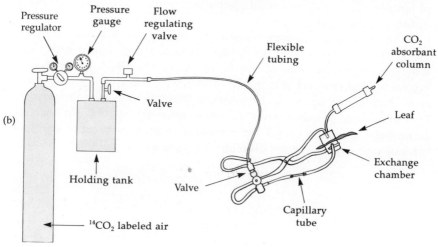

(b)

Pressure regulator

Pressure gauge

Flow regulating valve

Flexible tubing

CO_2 absorbant column

Valve

Leaf

Holding tank

Valve

Exchange chamber

Capillary tube

$^{14}CO_2$ labeled air

by absorbing CO_2 at night when water loss is restricted. CAM plants have the highest water use efficiency but are restricted by very low rates of photosynthesis and growth, and the requirement for cool nighttime temperatures. C_4 species have intermediate adaptations to arid environments; in fact, they are most important in hot, dry habitats. C_3 species, in general, have a lower water use efficiency but individual species are successful in habitats supporting CAM and C_4 species. It is important to keep in mind that these are generalizations and that success in any given environment depends on a wide variety of adaptations other than the photosynthetic pathway (Stowe and Teeri 1978; Eickmeier 1978). Alternate photosynthetic pathways may be a mechanism to partition resources between species and reduce competition by dividing the limited resources in time or space.

Primary limitations are imposed on photosynthesis by resistances to CO_2 uptake, water stress, light, and temperature. The rate of CO_2 uptake is determined by the boundary layer resistance, stomatal resistance, resistance as CO_2 dissolves into the liquid phase within the mesophyll cell, and the rate of fixation by photosynthesis. Different plants have different abilities to use CO_2 in low concentrations. For example, C_4 plants can reduce intercellular CO_2 concentrations to near zero whereas C_3 plants reach CO_2 compensation at much higher levels. Low intensities of light limit photosynthesis and high intensities may surpass the capacity of the photosynthetic apparatus. However, some desert plants have been found in which the photosynthetic apparatus does not saturate even at full light. The most immediate effect of water stress is stomatal closure which increases resistance to CO_2 uptake. There are significant changes in the diurnal pattern of stomatal movements in CAM plants which, under certain conditions, may switch between CAM and C_3 pathways. Plants are able to acclimate to a variety of temperatures and, as a result, show different temperature optima at different seasons of the year.

The two most common ways of measuring gas exchange in plants are systems incorporating an infrared gas analyzer and a technique which measures the rate of uptake of $^{14}CO_2$. The IRGA gas exchange system is expensive but it is the most accurate technique. It can be used, with proper conditioners, to measure plant responses continuously over numerous combinations of environmental factors. Radioactive techniques are most useful for comparative studies because absolute rates of photosynthesis are difficult to obtain by this method.

CHAPTER 14

TEMPERATURE

The two most important environmental factors that influence plant success and distribution are heat and the availability of water. Although it is impossible to discuss one without reference to the other, water availability will be covered in Chapter 17, and our primary emphasis in this chapter will be on heat. You have already been introduced to some of the concepts that we will cover in this chapter through the discussions on light in Chapter 13.

It is important to note that there is only a very narrow range of temperature within which life processes can occur. Most biological activity proceeds within the range of 0° to 50°C. There is little activity above or below this range. At 0°C and below there is immobilization of water, and liquid water is necessary for life processes. Above 50°C the heat is destructive to vital life processes.

Heat is the aggregate internal energy of motion of atoms and molecules of a body. It can be transferred by means of radiation, convection, or conduction. If you enter a small room newly warmed by a small charcoal stove and stand very close to the stove, your face gets hot while your back remains chilled. You are being warmed by **radiation**: energy propagated as an electromagnetic wave, which would pass through a vacuum with the speed of light (3×10^{10} cm sec^{-1}), but would also travel through a medium such as air or water. You will begin to feel some warmth as energy is transferred to you from the heated air of the room by **convection**. If your hand rests on the stove you will experience acute discomfort. Energy has been efficiently transferred by **conduction**, resulting in a minor burn.

In actual use, the term radiation refers both to the energy transferred and to the process of transferring energy. The solar energy that supports the life of this planet comes to us as radiation which is transmitted in a broad spectrum that includes not only visible light, but ultraviolet radiation at one extreme and heat or infrared radiation at the other extreme.

The ozone of the atmosphere, as well as oxygen and nitrogen, is opaque to wavelengths shorter than 0.29 microns, so that most of the ultraviolet and shorter wavelengths do not reach the earth. The longer, or infrared, wavelengths are absorbed and re-emitted by water vapor, liquid water, and carbon dioxide. The spectral distribution of global radiation, which is the sum of direct sunlight and scattered skylight, is shown in Figure 13–2.

It follows that unless the absorbed radiation is reradiated or in some way transferred, the body absorbing it would continue to increase its energy content. Thus, there is a continual exchange of energy between objects and the environment. Direct sunlight is absorbed by plants, rocks, soil, and the air around them. Infrared radiation (heat) is transferred from plants to the air, from soil to plants, and so forth. The radiation energy gains and losses between a young oak seedling and its environment are depicted in Figure 14–1.

Because our eyes do not perceive the longer wavelengths, the radiant energy given off by rocks, trees and the air itself is invisible to us. On a warm spring day (300°K, or 27°C), the young trees, the mature trees, the various rocks, open soils, herbaceous cover, and the various woodland animals radiate energy across a rather broad spectrum but at a maximum wavelength of about 10 microns, far below the visible red wavelengths.

SOLAR ENERGY BUDGET

Let us develop a solar energy budget, and envision the allocation of energy in equation form:

$$S = R + C + G + Ps + LE$$

where S = solar radiation; R = reflected (shortwave, diurnal), or reradiated (longwave, both day and night); C = convective heating of air (called H for heat by some authors); G = heating of solid objects (soil, for instance); Ps = photosynthesis; and LE = latent heat of evaporation.

The amount of energy reaching any given spot on the earth varies seasonally and diurnally, but on the average there are 1.94 gram-calories (or cal) cm^{-2} min^{-1} received as direct solar radiation measured perpendicular to the sun's rays just outside the earth's atmosphere at the mean distance of the earth to the sun. (A gram-calorie is that quantity of energy that is required to raise the temperature of one gram of water from 14.5°C to 15.5°C.) This 1.94 cal cm^{-2} min^{-1} is known as the **solar constant**, and it represents one of the most fundamental and vital components that is known of our environment. Solar radiation (S) is usually expressed in langleys min^{-1}, a unit of **radiant flux**, which is the amount of energy received on a unit surface in a certain time (see Table 13–1). At sea level at high noon, S may be equal to 1.3 ly min^{-1}. On a high mountain on a clear day, S may be as high as 1.7 ly min^{-1}.

Net radiation (N) is equal to solar radiation minus reflectivity and reradiation, or: $N = S - R$. This can be diurnally negative, because at

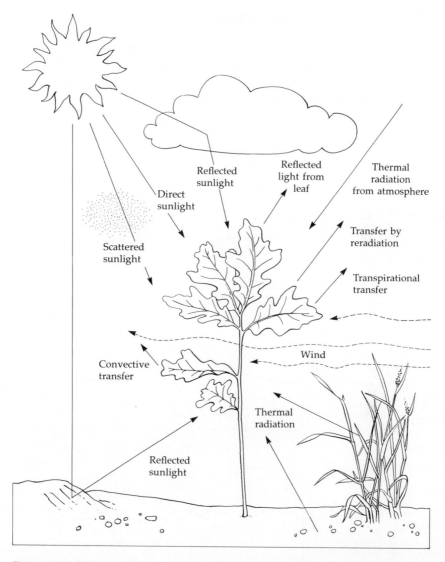

Figure 14-1. Gains and losses of heat and radiation energy between a young oak seedling and its environment. (Reprinted by permission from Heat transfer in plants by D. M. Gates. Copyright © 1965 by Scientific American, Inc. All rights reserved.)

night S is zero, and therefore less than R, and there is a net loss of heat from earth. During the day, S exceeds R, because reflectivity and reradiation are less than solar radiation, and there will be a positive energy balance. During the long winter of the polar regions, N is seasonally negative, and may be so even in more temperate regions. In Siberia, the annual net radiation is negative (Figure 14-2(a)). Or, there may be virtually no period of negative radiation balance, with a large net positive radiation balance; such is the case at Aswan, Egypt, where there is never a negative N (Figure 14–2(b)). In the tropics, there is little seasonal change. The angle of the sun does not vary enough to create seasonal deficits: N is always positive. In fact, diurnal variation exceeds the seasonal variation of daily temperature means (Figure 14–2(c)).

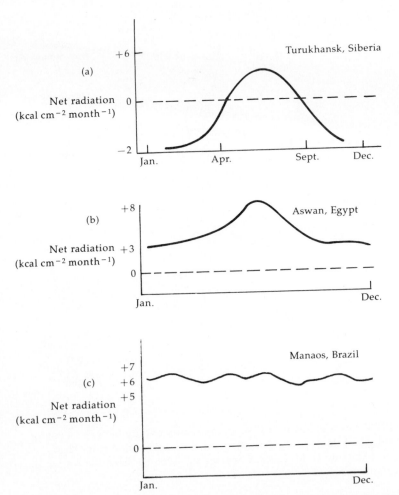

Figure 14–2. (a) Net radiation for a weather station in Siberia. Note that the annual average is negative. (b) Net radiation for a weather station at Aswan, Egypt. Note that there is virtually no period of negative radiation balance. (c) Net radiation for a tropical weather station. Note that there is very little seasonal flux in the radiation balance. (From Billings 1970. By permission of Wadsworth Publishing Co., Belmont, CA.)

Allocation of Energy within the Community: A Wet Mountain Meadow

The allocation of energy within the solar budget varies over time, and varies from one community to another. How is the solar energy divided within a specific community? Let us first consider a wet mountain meadow. The maximum S incident upon a wet mountain meadow is $+1.3$ ly min^{-1}. Meadow vegetation has an **albedo** of 0.7. Albedo refers to the reflectivity of a material summed over all wavelengths for a sunlit surface. S times albedo equals R, and so in this example $R = (0.7)(1.3$ ly min$^{-1}) = 0.91$ ly min^{-1}. The rest of the solar radiation is apportioned among G, C, Ps and LE.

Mountain meadows are often densely clothed with herbaceous vegetation, and at 100% cover G will be rather low. For various physical reasons, C is also low. Only 1–2% of incident light is usually converted into chemical energy, and photosynthesis here will be lower than usual due to low temperatures. Although Ps is low, LE will be high, simply due to the wetness of this mountain meadow system.

Allocation of Energy within the Community: The Mojave Desert Region

A contrasting situation is seen in an arid area such as the Mojave Desert. Again, S_{max} is 1.3 ly min^{-1}. R will be higher than in the mountain meadow.

In very arid regions such as the Mojave Desert, plants are very widely distributed, and the open expanses of exposed soil absorb a relatively high fraction of S. Thus, G will be much higher than in the wet mountain meadow system and photosynthesis will be very low, less than 0.1% of S. C will be high, as the denser air absorbs radiant energy from sunlight. There is very little liquid water available, and LE will be correspondingly low.

ENVIRONMENTAL INFLUENCES ON TEMPERATURE

Insolation (exposure to solar radiation) varies temporally and spatially. The primary factors that influence spatial variations in temperature are latitude, altitude and proximity to water. Topography, cloud cover, vegetation, and slope aspect are contributing factors to temperature variation. The obvious and immediate temporal differences derive from the earth's rotation. As we noted earlier, the darkened half of the earth is losing heat during the night (reradiating energy). During the day, radiant energy from sunlight usually replaces what was lost, so there is an energy balance.

Latitude

Apart from the obvious diurnal differences in temperature, the primary influence upon temporal variation is latitude. Both latitudinal and altitudinal differences are spatial differences. The earth moves about the sun in an elliptical orbit, so that there is some variation in the earth-sun distance. However, it is the tilt of the earth that gives us our seasons. There are periods of equability, when sunlight shines from pole to pole, and day and night are virtually equal in length. These periods, when the earth's polar axis is perpendicular to the radius between the earth and the sun, are called vernal and autumnal (spring and fall) equinoxes. Alternatively, when the earth's polar axis is in a plane parallel to the earth-sun radius, one pole will be in virtual 24 hour darkness, the other bathed both day and night in sunlight. In the Northern Hemisphere, our longest night, or winter solstice, is in December, and our longest day, or summer solstice, is in June. A comparison of solar radiation received throughout the year at latitudes ranging from 0° to 80° shows the greatest equability at the lowest latitudes, and the greatest range and the highest energy received in a day, at the highest latitudes (Figure 14–3; refer back to Figure 13–4).

Altitude

In general, temperature will be lower at higher altitudes. This inverse correlation of altitude with temperature creates a gradient or **lapse rate**, usually given as about 10°C per 1000 m of elevation. There are several lapse rates, and the one given here is a **dry adiabatic** lapse rate, which means the cooling of air as it rises without condensation and without cloud formation. Lowry (1969) discusses lapse rates and their significance. There is less atmospheric pressure at the higher elevations, so air there expands and loses energy. Conversely, coming downslope, air is more compressed and therefore warmer. Thus, in addition to latitudinal differences, there will be, at any given latitude, elevational differences in temperature (see Figure 19–12).

Topography

The decline in air temperature due to increased elevation is not constant, especially in dissected topography. The situation is complicated by various physical phenomena, including temperature inversion. During the night, the soil gives up heat to the air, sometimes so quickly that the soil may become cooled below the temperature of the overlying layers of air. These layers return heat by conduction to the

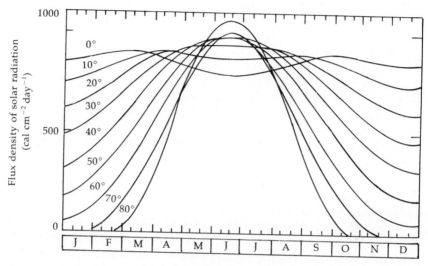

Figure 14–3. Daily total of the undepleted solar radiation received on a horizontal surface as a function of latitude and time of year. (From Gates, *Energy Exchange in the Biosphere,* Harper and Row, Publishers, Inc., New York, 1962.)

soil, and may become cooler than the overlying strata, resulting in a **temperature inversion**. That is, the normal vertical temperature gradient is inverted. However, the warm air capping the cold eventually is cooled (usually by 300 m in elevation) and the normal lapse rate is resumed. There is thus an upslope thermal belt warmer than the lower-lying regions.

Furthermore, cold air is denser than warm air, and drains into ravines and valleys at night, such that these low-lying canyon bottoms and river valley floors are colder than the slopes above them, augmenting the thermal belt effect. The flow of air may be reversed in the daytime as warmed air moves back upslope. Application of this principle of cold air drainage to desert regions is discussed in Chapter 19 and illustrated in Figure 19–39.

There are often tiny basins or frost pockets in which cold air accumulates and remains due to feeble air movement. A further complication of temperature variation in mountainous regions is the width of ravines and valleys. Narrow mountain valleys or canyons receive proportionately less sunlight. The walls therefore reflect less heat into the canyon and it will be colder and damper than the surrounding area. By contrast, broad, open mountain valleys build up high temperatures due to the accumulation of heat reflected from the wide sides, and they are consequently hotter and drier than surrounding regions.

Proximity to Water

The temperature of a body of water, especially a large lake or ocean, is much more stable than that of a comparable land mass. This is due to the high specific heat of water, its ability to lose energy by evaporation, its high reflectivity, and the contribution of vertical mixing to stability of temperature. Therefore, proximity to a large lake or ocean buffers the climate for adjacent land areas. The water gives up energy during the winter, warming the nearby area; thus, its average winter temperature is higher than a comparable area at the same latitude further inland. The inland region is said to have a continental climate. Conversely, summer temperature will also be more moderate under a marine climatic region as water absorbs energy from the surrounding air. Thus, the amplitude of temperature oscillations in an extreme continental area is greater than that of a maritime region. The effect of a maritime climatic regime may be considered with regard to dates of last and first killing frost. These data are especially important to agriculturalists and horticulturalists, as well as plant ecologists. Miller and Thompson (1975) have shown in map form the average dates of the last killing frost in spring and the average dates of the first killing frost in autumn for the continental United States (Figure 14–4).

Let us now combine considerations of diurnal variation with the effect of proximity to water. Note the narrow range of temperatures and the minor seasonal differences at a maritime station, San Francisco, California, compared with a more continental station, El Paso, Texas, shown in Figure 14–5.

Cloud Cover

The influence of cloud cover is twofold. Water vapor is opaque to certain wavelengths of solar radiation. A place frequently covered by clouds will not be warmed nearly as much as an area under clear skies. There is, however, some reflection of sunlight from clouds to the ground, and some reradiation heat from vegetation, soil surface, and rocks, so that cloud cover may augment direct sunlight and impede heat loss from a land surface. For instance, direct sunlight and sunlight reflected from clouds may simultaneously warm an alpine slope such that radiation values are as high as 2.2 ly min^{-1}, which is in excess of the solar constant. The considerations mentioned regarding the moderating influence of liquid water apply also to water vapor. We cannot overemphasize the incredibly important role of moisture in the atmosphere in holding and reradiating energy. Temperature variations in the tropics vary as little as 2°C on cloudy days, but can vary as much as 9°C on sunny days. Without the moderating influence of cloud cover,

(a)

(b)

Figure 14–4. (a) Average dates of last killing frost in spring, and (b) average dates of first killing frost in autumn. (From *Elements of Meteorology,* 2nd ed. by Miller and Thompson 1975. Reprinted by permission of Charles Merrill Publishing Co.)

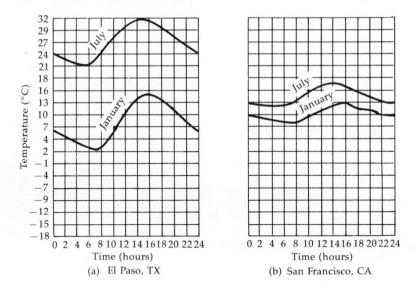

Figure 14-5. Diurnal temperature variation at (a) a continental weather station and (b) a maritime station. (From *Elements of Meteorology,* 2nd ed. by Miller and Thompson 1975. Reprinted by permission of Charles Merrill Publishing Co.)

deserts experience great diurnal extremes of temperature. Also, frosts are much more likely on clear nights. In general, moisture in the atmosphere has a stabilizing influence on temperature variations.

Plant Cover

We discussed the influence of vegetation on the light environment in Chapter 13: the modification of temperature by plant cover is both significant and complex. Shaded ground is, of course, cooler than open area, if there is little exchange of air. Vegetation interrupts the laminar flow of air, impeding heat exchange by convection. The sensual impact of a forest is usually one of moist coolness. This is attributable first to the reduction of soil heating due to shading: the soil absorbs energy from the warmer air above it, cooling the forest air. Secondly, because the forest vegetation transpires as photosynthesis proceeds, the forest microclimate will be more humid than adjacent clearings, and the quantity of energy needed to raise the air temperature is increased. The reradiation of heat is impeded by plant cover, so that the nocturnal forest air and soil are warmer than those of nearby clearings. Temperature oscillations are thus damped both day and night by heavy, continuous plant cover (Table 14-1). This situation may be changed, or even reversed, if the plants are widely dispersed.

Slope Aspect

Due to the inclination of the sun, a south-facing slope in northern temperate latitudes will always experience greater total insolation than will the north-facing slope of the same region. Thus, there will be warmer air and soil, less moisture, and sparser vegetation, in general, on the south-facing slope. North-south slope aspect difference is often manifested in very different plant cover. In central California, for instance, a north-facing slope will be oak woodland (blue oak *(Quercus douglasii)* for instance), and the south-facing slope of the same ravine will be covered by chaparral (chamise *(Adenostoma fasciculatum)* for instance; Figure 14–6).

Figure 14–7 illustrates the variation in energy reception on sloping surfaces as a function of the time of day. Figure 14–7(a) depicts energy received at winter solstice, 22 December. Note that north facing slopes steeper than 22.5° receive no direct insolation on that date. Figure 14–7(b) depicts energy reception at the vernal equinox. The north vertical receives no direct radiation, but the south slope at 45° intercepts almost as much as it receives at the summer solstice, which is depicted in Figure 14–7(c). As we would predict, the total energy received is the greatest for the community at the summer solstice, 22 June (see also Figure 19–11).

The study of the relationship between climatic factors and periodic phenomena is referred to as **phenology**. A significant climatic factor is temperature. Phenological events concerned with reproductive activity in plants may be retarded by several days on a north-facing slope with respect to plants of the same species on an adjacent

Table 14–1. Comparison of certain habitat factors in a virgin pine forest and in an adjacent clearing in northern Idaho. Data for the month of August. (From Larsen 1922. By permission of the Ecological Society of America.)

Factor		Forest	Clearing
Air temperature (°C)	Maximum	25.9	30.0
	Minimum	7.4	3.9
	Range	18.5	26.1
Mean relative humidity at 5 P.M. (%)		38.8	35.2
Mean daily evaporation[a] (ml)		14.1	36.1
Mean soil temperature at 15 cm (°C)		12.8	17.0
Mean soil moisture at 15 cm (%)		32.0	43.2

[a]Livingston atmometer mounted 15 cm above the ground.

Figure 14–6. Slope aspect difference.

south-facing slope. Jackson (1966) found significant correlation between air temperature sums and flowering dates. The increased insolation received by south-facing slopes, as compared to nearby north-facing slopes, results in warmer temperatures earlier in spring and therefore, advanced flowering (Figure 14–8).

We might relate time range of flowering dates to latitude (northward) and to elevation (upward). For instance, one day of retardation or advance of flowering might be equivalent to more than 27 km in distance northward, or more than 30 m in elevation. Jackson found flowering variation of an average of 6.0 days between a north-facing slope of a large gorge and the south-facing slope only 46 m distant. Across a small east-west gorge, plants of the same species blooming less than 8 m apart, but on opposite sides, evidenced a delay of 2.8 days on the north-facing slope as compared to the south-facing one. This could be assigned a difference comparable to 80 km (northward) or 85 m (upward). She found a mean range of flowering dates of 7.2 days for all species studied. This may be interpreted as equivalent to a geographic distance of 201 km northward and 219 m upward (Hopkins 1938), or as 217 km northward and 168 m upward (Jeffree 1960).

TEMPERATURE-MEDIATED PLANT RESPONSES

We have seen how energy is allocated within the community. We have recognized that a balance is needed if a plant, or even a leaf, is to live. Outgoing energy must equal incoming energy, except for that

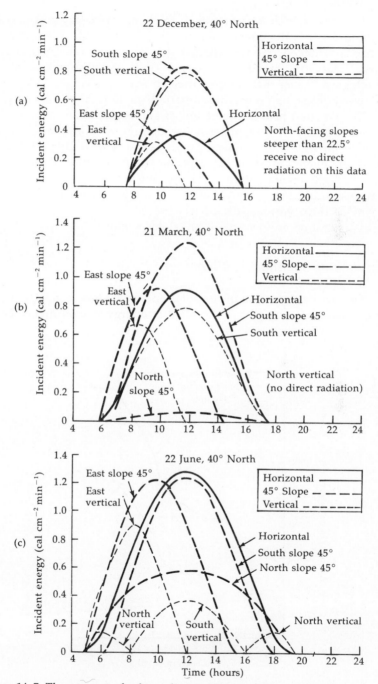

Figure 14–7. The amount of solar radiation incident upon sloping surfaces as a function of time of day at latitude 40°N for (a) the winter solstice, (b) the vernal equinox, and (c) the summer solstice. (From Gates, *Man and His Environment: Climate,* Harper and Row, Publishers, Inc., New York, 1972.)

radiant energy which is transformed to chemical energy and can thus be stored. How is the energy that reaches the plant dealt with so that balance is achieved? We will discuss the concept of energy allocation by the plant in broad terms, and then examine some of the phenological events that are governed in part by changes in temperature: events associated with thermoperiodism, dormancy, stratification, vernalization, and heat pretreatment or **sumorization**.

The activity of a plant must be adaptive to the plant or it will perish, even before reproduction can occur. The timing of various phenologic events must be in response to environmental cues: light duration, light quality, light intensity, moisture availability, various chemicals in solution, gravity, touch, and temperature. Plant responses to many of these various stimuli are discussed elsewhere in this book and in general botany and plant physiology books.

Allocation of S within the Plant

How is the right side of the solar energy budget, e.g., R, C, G, Ps, and LE, dealt with by plants? Recall that energy is continually being

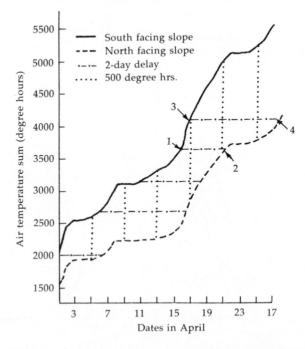

Figure 14–8. Comparison of variations in the lag in air temperature sums due to slope differences in a large gorge during April. (From Jackson 1966. Copyright 1966 by the Ecological Society of America.)

exchanged between a plant and its environment (Figure 14–1). Incident radiation from a variety of sources will be absorbed by a leaf: direct sunlight, scattered skylight, reflected sunlight, and infrared wavelengths. If the air is cooler than a leaf, energy moves away from the leaf. If the air is warmer, the leaf will take on energy.

Cooling results from the evaporation of water *(LE)*. The vaporization of a gram of water requires about 580 cal of energy, a process which cools the leaf much as an evaporative cooler cools a room. There may be a limited water supply and desert plants, for example, cannot depend upon unlimited transpiration for balancing the energy budget. Laminar flow of air removes heat from the leaf *(C)* if its temperature is warmer than that of the air. The transfer of heat is directly proportional to the square root of the wind velocity and inversely proportional to the square root of the leaf width. In other words, the larger the leaf, the less effective the cooling. At the same air temperature and wind velocity, a small leaf will lose heat more effectively than a larger leaf. Small leaves, dissected leaves, and lobed leaves are characteristic of plants of arid regions.

Either a very simple system or a stressed one, such as a desert plant community, can provide a model system of heat transfer. Daytime temperatures of desert soils and the air immediately above them are often a great deal higher than that of the air at the standard height above the ground for measuring temperature (1.5 m). For instance, the highest standard air temperature ever measured was 57.8°C (Azizia, Tunisia, on 13 September 1922, and San Luis, Mexico, on 11 August 1933). However, air next to soil surfaces (dark soil, in full summer sun) may exceed 70°C (Gates 1972). As the spring season gives way to early summer, some small desert annuals assume a basket shape, as the rosetted leaves are lifted into the cooler air just a few cm above the soil surface. This adaptation of leaves of desert plants increases chances for heat loss and reduces absorption of heat from the soil *(G)*. Raising leaves may also change the angle of the leaf relative to the sun and thus reduce S.

Desert plants must deal with strong insolation, often in still air. Insolation *(S)* may raise leaf temperature above that of the surrounding air. If convection *(C)* and evaporation *(LE)* cannot transfer energy as rapidly as it is absorbed, then the leaf will store heat and may then have a temperature of 10°C, or even 15–20°C, above the ambient temperature. Heat exchange is enhanced by winds, which may remove air to within a few mm of the leaf surface. Conversely, heat exchange is hampered in still air, although the layers of warm air above the leaf will tend to rise above cooler air, generating a turbulence which also hastens cooling. Loss of leaf area, and therefore reduction of leaf exposure to insolation *(S)*, is achieved in a variety of ways. Most members of the Cactaceae, for instance, have leaves reduced to simple spines. Barrel

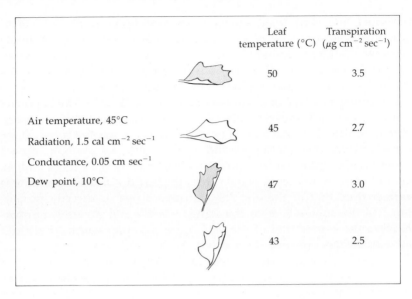

	Leaf temperature (°C)	Transpiration (μg cm^{-2} sec^{-1})
	50	3.5
Air temperature, 45°C Radiation, 1.5 cal cm^{-2} sec^{-1}	45	2.7
Conductance, 0.05 cm sec^{-1} Dew point, 10°C	47	3.0
	43	2.5

Figure 14–9. Leaf temperature and transpiration of leaves of differing absorptances and angles as determined by the energy equation. The wavelength absorptance (ratio of absorbed to incident radiation) is 0.5 for a typical desert holly *(Atriplex hymenelytra)* leaf (shaded) and 0.25 for a summer desert holly leaf (white). Leaf conductance to water vapor transfer was measured on summer leaves (see Chapter 18 for details regarding conductance). Environmental conditions are those characteristic of Death Valley at midday during the summer. Leaf orientation is horizontal or at 70° as indicated. Leaf size is 3.9 cm^{-2}. (From Mooney, H., J. Ehleringer, and O. Björkman. 1977. The energy balance of leaves of the evergreen desert shrub *Atriplex hymenelytra*. *Oecologia* 29:301–310.)

cactus *(Ferocactus acanthodes)* has central spines of 7.5–15 cm and inner radial spines of 4–9 cm in length, which rise much as leaves would. A different method for avoiding heat from the ground is seen in members of the Fouquieriaceae, a family with only one genus, and with several species found in Mexico and the southwestern United States. Ocotillo *(Fouquieria splendens)* (Figure 19–41) loses its leaves seasonally in response to drought stress, replacing its leaves as many as five times in a single year. Vertical leaf orientation, an adaptation often exhibited by desert shrubs, is another way to avoid excessive *S*.

Reflectivity *(R)* is enhanced by a surface with high albedo. Possession of a white or light colored leaf surface increases *R*. Desert holly *(Atriplex hymenyletra)* in the Chenopodiaceae provides a good example of how one desert shrub species is adapted to increase reflectivity by different means and thereby reduce heat load and drought stress. Unlike the cacti, which have virtually no leaves that are photosynthetic, or ocotillo, which is drought deciduous, desert holly is an evergreen perennial of the hot desert. Mooney et al. (1977) found that the adaptation of this plant is to change the characteristics of the leaves during the year. Those leaves produced during the cool season are nearly twice the size of the summer leaves. Furthermore, young leaves are covered with expanded, hydrated salt bladders which collapse as the summer progresses. The salt solution contained in the bladders crystallizes, giving the leaf a white appearance and increasing the reflectance (R) of the leaf surface. Leaf reflectance (R) (at 550 nm) is inversely correlated with leaf water content. Leaves with 60% leaf moisture exhibit a reflectance above 75% at some wavelengths. Also, at all wavelengths their reflectance is greater than that of leaves with 89% leaf moisture.

The leaves of desert holly are held at a steep angle. Field measurements of 50 leaves yielded a mean angle of 70° from the horizontal, with random orientation to the azimuth. Although the angle would seem to be more adaptive in summer than during the very active spring period, it does not change seasonally. This tough shrub is a C_4 plant, but it photosaturates at a rather low light level. Thus, photosynthesis is not reduced, even in spring, by the steep angle. The angle does provide protection by reducing heat load and by allowing greater interception of light during the early morning and late afternoon, which are the times when relative humidity is most favorable (Figure 14-9).

Angle of leaf, reduced leaf size, and high reflectance from the leaf surface enhance the capacity of this desert shrub to remain photosynthetically active even during the hottest summer months. In other words, these adaptations allow the luxury of evergreenness.

Thermoperiodism

There are many plants which require a day-night temperature difference for optimal growth. For instance, red fir *(Abies magnifica)* and Jeffrey pine *(Pinus jeffreyi)* respond positively to a **thermoperiod** of 13°C, that is, a diurnal difference (between day and night) of 13°C. A positive response to such a temperature regime is termed **thermoperiodism**. Is thermoperiodic response of value in those conifers along the coast, where diurnal temperature fluctuations are unlikely to be severe? Hellmers (1966) investigated thermoperiodism in coast redwood *(Sequoia sempervirens)* seedlings, and determined that a thermoperiod of 4°C elicits a response; however, that response was not significantly different from the growth of the plant under a constant temperature regime (Figure 14-10).

The coast redwood does not form dormant buds, but can continue to increase in height in response to favorable light and temperature

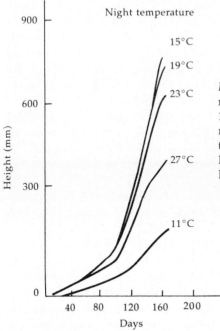

Figure 14-10. A comparison of redwood seedling growth with a 19°C day temperature and a range of night temperatures. There were eight trees in each treatment. (From Hellmers 1966, *Forest Science*, Vol. 12, No. 3, pp. 276–283.)

regime. (Coast redwoods are generally accepted to be the world's tallest trees.) This is in contrast to trees that occupy inland, high elevation sites, such as Jeffrey pine and red fir. These species do set terminal buds, and this restricts upward growth, even though the plant may continue to acquire new biomass. A thermoperiod requirement is adaptive in the pines and firs, since they live in areas of great diurnal temperature fluctuation. It is appropriate that the coast redwood does not require a thermoperiod, since it does not live in an area with a large diurnal temperature difference.

Thermoperiodism in woody plants has a counterpart in seed response to a fluctuating diurnal temperature regime. Figure 14–11 shows the effect of varying temperatures on seed germination for three species (Grime and Thompson 1976*). *Carex otrubae,* a sedge, requires no temperature fluctuation for germination, but the other two species show a positive response to temperature depression during the dark period. Sorrel *(Rumex sanguineus)* achieved 50% germination at a thermoperiod of 2.5°C, and yellow-cress *(Rorippa islandica)* at 9°C.

Dormancy

Dormancy is a period of inactivity for seeds or plant organs which can be broken only when certain environmental requirements have

*This report describes a useful compact germination incubator designed to provide controlled fluctuations in temperature automatically.

been met. For instance, a seed that is supplied with moisture, oxygen, and the proper photoperiod which still does not germinate is said to be dormant. A key feature in the maintenance of seed dormancy is the need for oxygen. Plant tissues are predominantly aerobic, and the exclusion of a large molecule such as oxygen can therefore help to maintain dormancy in seeds. However, the morphology of a dormant bud does not provide for oxygen exclusion. Thus, the correlation between the maintenance of an anaerobic environment and the maintenance of dormancy may not apply to buds.

Dormancy is a mechanism evolved by plants to escape periods of unfavorable weather, and to allow less competitive species to last through periods of high competition. During times of a short photoperiod and a lack of heat or sunlight or liquid water, growth—whether of an established plant or a new one—is very expensive, if not impossible. The protection afforded by the dormant state, especially in

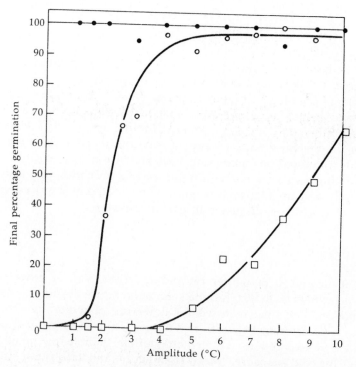

Figure 14–11. The germination response to various amplitudes of diurnal temperature fluctuations in seeds of Carex otrubae (●), Rumex sanguineus (○), and Rorippa islandica (□). The depression in temperature coincided with a dark period of 6 hours. During the photoperiod of 18 hours all seeds experienced a constant temperature of 22°C. (From Grime and Thompson 1976. Annals of Botany 40:795–799.)

extra-tropical regions, is very significant to the plant, and thus dormancy is a very adaptive feature. The dormancy of apical buds of many woody plants in temperate climates is initiated by short days interacting with cold temperature; this is a photoperiodic response mediated hormonally.

Environmental cues, such as lengthening photoperiod and increasing temperatures, trigger the resumption of activity in more favorable times. The correct interpretation of these cues is vital to the dormant plant or seed. If a dormant beech, for instance, began to leaf out in response to unseasonally warm weather in early November, subsequent frost would kill or injure the new soft tissues, at great cost to the plant. The environmental stimulus that enables the plant to perceive the passage of winter is cold temperature. Degradation of inhibitors and concomitant biosynthesis of growth-stimulating substances take place during the cold period and early spring, and the seed or plant is then prepared to respond to more favorable weather in an appropriate manner.

Stratification

Stratification, the layering of seeds in a moist medium which is then kept at a low temperature, is known to enhance germination. Most plants growing at higher latitudes possess seeds which require cold temperatures for germination. That is, the seeds are dormant prior to exposure to cold. Actual germination following release of dormancy may be in response to a variety of environmental cues, including changing photoperiod, increasing heat and water availability, and a changing hormonal regime within the seed. Germination may also rely upon a physical or chemical disruption of the seed coat, a process called scarification which is discussed in part in Chapter 15.

Vernalization

The flowering of winter rye (Secale cereale) and of other cereal plants is influenced by exposure to cold temperature during the time of germination. The time required for flowering of winter rye may be reduced by half, from 14 weeks to 7 weeks, if the seeds are planted in autumn and exposed to winter cold, rather than being planted in spring. This cold exposure, copied from nature and used commercially, is called **vernalization**, from the Latin word for spring, vernus. Seeds of winter strains of certain cereal grasses, if kept near freezing temperatures during germination, will flower the same summer even though not planted until late spring. We repeat an often-stated, but very important qualifier: even with vernalization, the appropriate photoperiod is necessary for flowering to be initiated.

Sumorization

Heat cracking is necessary to rupture the seed coat of some fire adapted plants that possess resistant seeds. But what is the influence of temperatures below those of a fire? We mentioned the high temperatures of desert soils. The seeds of most desert annuals are subject to this heat soon after dissemination, and it probably promotes seed maturation in these ephemerals. Capon and Van Asdall (1966) found that germination of eight species of desert annuals native to the Mojave and Sonoran Deserts was enhanced by heat pretreatment. This treatment is called **sumorization**, after the Anglo-Saxon root *sumor*, for summer.

Maximum germination for these species manipulated by Capon and Van Asdall was reached by the fifth week of storage at 50°C (Table 14–2). Untreated seeds, and those stored at 4°C, showed poor germination percentages, especially the three Sonoran species tested. Of these, none showed more than 12% germination without heat pretreatment. Seeds of all species, including the plantain *(Plantago insularis)*, failed to germinate after storage at 75°C.

THE MEASUREMENT OF RADIATION

Temperature may be ascertained using a thermocouple, a mercury or alcohol thermometer, or various other devices. An inexpensive chemical integrator of average temperature over a long period of time is sucrose. The rate at which a sucrose solution is hydrolized (inverted) to glucose and fructose is a function of temperature. About 10–15 ml of buffered sucrose solution is sealed in a tube and left at a desired height aboveground or belowground at a field site. The amount of sucrose that was inverted during the field exposure period can be determined by measuring the optical rotation of the solution with a polarimeter. The

Table 14–2. Percentage germination of seeds of several species of desert annuals in response to temperature pretreatment. (From Capon and Van Asdall 1966. Copyright 1966 by the Ecological Society of America.)

	Species	Un-treated	Storage at 20°C for		Storage at 50°C for					
			4 weeks	8 weeks	1 week	2 weeks	3 weeks	4 weeks	5 weeks	10 weeks
Mojave Desert spp.	*Coreopsis bigelovii*	9	14	30	26	16	24	20	32	2
	Eriophyllum wallacei	5	16	24	32	48	30	18	36	2
	Euphorbia polycarpa	1	1	2	2	10	12	4	2	0
	Geraea canescens	14	14	16	60	74	32	34	48	4
	Salvia columbariae	6	4	14	80	26	28	14	16	10
Sonoran Desert spp.	*Lepidium lasiocarpum*	0	2	4	53	90	72	61	54	40
	Sisymbrium altissimum	10	10	12	12	13	92	73	71	48
	Streptanthus arizonicus	0	3	8	21	23	29	50	50	30
	Plantago insularis (Not leached)	0	0	0	0	0	0	0	0	0
	Plantago insularis (Leached)	100	—	—	100	—	—	—	—	—

method has been described in detail by Berthet (1960, in French) and by Lee (1969).

Through a global system of weather stations we have a reasonably accurate picture of the world's climate. We can also measure radiation, the source of heat, but measurement of energy flux has rarely been done, and only recently has the instrumentation been developed to support global-scale quantification of the energy environment.

Radiation can be measured by devices called **radiometers**, which integrate all incoming wavelengths and measure in energy units. **Pyranometers** are used to measure direct and diffuse sunlight. They are sensitive to 360–2500 nm, which includes some ultraviolet light (<450 nm), all visible light, and considerable infrared light (>700 nm). Filters can be added to a pyranometer, restricting its sensitivity to only the visible light, or all but a portion of the visible spectrum can be filtered out, if that is needed.

An Eppley black and white pyranometer (Figure 14–12) utilizes the temperature differential between hot and cold areas to monitor radiation intensity. However, it does not separate direct rays from indirect rays. On a clear day, direct sunlight accounts for approximately 85% of the energy received. On an overcast day, up to 100% of the incoming energy is due to diffuse, indirect sunlight. If we wish to monitor only diffuse sunlight, a shading device called an occulting ring, or shadow band, would be used with the pyranometer. The Eppley Precision Spectral Pyranometer (Figure 14–13) is believed to be the most accurate instrument that is produced commercially for mea-

Figure 14–12. The Eppley pyranometer. The sensing element under the glass dome is a thermopile with hot (black) and cold (white) areas.

Figure 14-13. The Eppley Precision Spectral Pyranometer. This instrument utilizes a plated, wirewound thermopile which is temperature compensated to be virtually unaffected by ambient temperature, and is able to withstand severe mechanical vibration and shock.

surement of global sun and sky radiation. It may also be used with a shading device, to screen out either the sun or the diffuse sky component.

An instrument called a **pyrheliometer** measures only direct solar radiation. It consists of a sort of tube and a sensor to track the sun such that only solar radiation of normal incidence enters the monitoring device, the tube. The output can be calibrated to read in ly min^{-1}. Without the use of the pyrheliometer, the intensity of direct beam (direct sunlight) may be estimated, using data obtained with the pyranometer alone, and with the pyranometer combined with the occulting ring:

$$R_S \text{ (direct beam)} = R_S \text{ (total global)} - R_S \text{ (diffuse)}$$

The monitoring of net radiation is of special interest to plant ecologists. Ideally, the net radiometer absorbs all incoming (downward) radiation (I_R), and all outgoing (upward radiation (O_R). The difference between I_R and O_R gives the net radiation, which is the energy available in the community for work. (For current information on meteorological instrumentation see also Wang 1979.)

CLIMATE CLASSIFICATION

Climatology and climate classification have evolved as more knowledge has accumulated and as communication has improved. A semiquantitative classification system was attempted by Hinds during the first half of the nineteenth century. He considered both temperature and moisture, and included variability, or stability, in his scheme. However, his system is quantitative only for temperature, and is qualitative for both precipitation and variability (Gates 1972).

Wladimir Köppen, a climatologist, developed and published a climate classification in 1918, which he subsequently revised in 1923. He divided climates as follows:

A   Tropical rainy climates
B   Dry climates
C   Warm temperate rainy climates
D   Cold snow forest climates
E   Polar climates

These climate classes were then subdivided into climate provinces. Köppen's eleven climate provinces and their characteristic vegetation are shown in Table 14-3. Although the system is widely used, it is severely limited in application because it has such a narrow base (a single climatic parameter, that of temperature).

The use of a **precipitation-effectiveness index** in defining climatic regions was incorporated into the **humidity provinces** of C. Warren Thornthwaite (1931). These humidity provinces are shown in Table 14-4. The index was derived by dividing the mean monthly precipitation (P) by the mean monthly evaporation (E) of a free water surface, and calculated as follows:

$$P\text{-}E \text{ index} = \sum_{n=1}^{12} 10\left(\frac{P}{E}\right)_n$$

Table 14-3. Types of climates as classified by Köppen. (From Gates, *Man and His Environment: Climate*, Harper and Row, Publishers, Inc., New York, 1972.)

Main type	Climate province	Symbol	Temperature	Rainfall	Vegetation
A Tropical rain	Rain forest	Af	Coldest month >18°C Annual variation <3°C	Heavy rain Min. 6 cm/month	Dense forest; heavy undergrowth
	Savanna	Aw	Coldest month >18°C Annual variation <12°C	Dry season with <6 cm/month	Open grasslands; scattered trees
B Dry	Steppe	BS	Variable winters cold	Rain <50 cm/year	Short grass
	Desert	BW	High summer temp.	Very dry	Sparse or none
C Warm, humid (Temperate)	Mediterranean	Cf	Coldest month between 18°C and −3°C Warmest month >10°C	Ample rain throughout year	Cotton, wheat, and corn
	Winter dry	Cw	Coldest month between 18°C and −3°C Warmest month >10°C	Wet summer	Forest
	Summer dry	Cs		Dry summer	
D Cold, humid	Cold with no dry period	Df	Mean temp. of summer is 10°C, of winter −3°C	Rain even throughout year	Canada
	Dry winter	Dw			N.E. Asia only
E Polar	Tundra	ET	Mean temp. of warmest month <10°C	Frozen soil	Moss and small shrubs
	Ice cap	EF	Mean temp. of warmest month <0°C		

Table 14–4. Humidity provinces of Thornthwaite. (From Thornthwaite 1931. The climates of North America; according to a new classification. *Geographical Review* 21:633–655.)

Humidity province	Characteristic vegetation	P-E index
A Wet	Rain forest	128 and above
B Humid	Forest	64–127
C Subhumid	Grassland	32–63
D Semiarid	Steppe	16–31
E Arid	Desert	less than 16

Not every station has pan evaporation data available, and so to resolve this deficiency, Thornthwaite devised this alternate, if somewhat unwieldy, formula:

$$P\text{-}E \text{ index} = \sum_{n=1}^{12} 115 \left(\frac{P}{T - 10} \right)_n^{10/9}$$

As we would expect, a lower P-E index is less favorable for vegetative growth, and a higher P-E index favors vegetative growth. In regions of comparable edaphic conditions, thermal conditions, and human influence, Thornthwaite found that the amount of vegetative growth is a function of the P-E index (Gates 1972).

The term **life zone** was applied to a series of regions defined by C. Hart Merriam in 1895 (Chapter 2). Although he based his proposal on temperature differences, his life zones are currently based primarily on the distribution of birds and mammals, and the original temperature criteria have been abandoned. Clements also developed a system of region classification, but his divisions were based mainly on climax vegetation, including some parameters of climate (Chapter 2). Presently, in both of these systems, the divisions are considered subjective, taxonomic and lacking in a sound base of systematization.

Holdridge (1967) proposed categorizing defined groupings of associations into life zones, based solely on three major climatic factors: heat, precipitation, and moisture, recognizing that the third is dependent upon the first two. A life zone, according to Holdridge, is a group of associations related through these major climatic factors. Holdridge's sytem utilizes the measurement of only the heat that affects plant growth, which is called **biotemperature**. Biotemperature is an average of all the daily temperatures that fall between 0° and 30°C relative to the total period. All temperatures lower than 0° are considered as 0° for calculation purposes, and those above 30°C are eliminated from consideration. Holdridge made several important contributions, and two are especially worthy of mention. The first is the discovery that there is a logarithmic progression in the steps of increasing temperature and precipitation that affect vegetation (Figure 14–14). The second is the discovery of the factor 58.93. Potential evapotranspiration for a given station can be obtained, and rather accurately, by merely multiplying the normal annual biotemperature of the station (in degrees C) by 58.93. Recall the very complex Thornthwaite method of obtaining the precipitation effectiveness index. The results that Holdridge easily and quickly obtained are comparable in most cases.

The depiction of world life zones (Figure 14–14) is derived from consideration of biotemperature, latitude, altitude, precipitation and potential evapotranspiration/precipitation. This last parameter, a ratio, is obtained by dividing the mean annual potential evapotranspiration in mm by the mean annual precipitation in mm. This is considered a standard of humidity (and moisture availability) which can be used for comparison from one life zone to another.

A simpler depiction of life zones is diagrammed in Figure 14–15, showing the relative positions of the latitudinal regions, basal life zones, and altitudinal belts.

As an example of how to use Holdridge's system, consider Verkhoyansk, in northeastern Siberia. Monthly mean temperatures (in degrees C) from January to December are as follows: -50.0, -44.0, -32.2, -15.5, 0, $+12.2$, $+13.3$, $+9.4$, $+1.7$, -15.5, -37.2, and -47.8. Disregarding the magnitude of the negative temperatures and recording them as 0s, the sum for the year is $+36.6°C$. This total, divided by 12, gives a mean annual biotemperature of 3.05°C. At Verkhoyansk the mean annual precipitation is 135 mm. We find the juncture of these two parameters, and place Verkhoyansk in the boreal dry scrub life zone (Figure 14–14). Although we might derive a more precise annual biotemperature by adding up daily biotemperatures and dividing by 365, these figures are useful and place the station within a life zone.

By contrast, consider a station where the mean annual biotemperature is 8°C and the mean annual precipitation is 350 mm. This falls into a steppe hexagon, the cool, temperate latitudinal region, and the montane altitudinal belt. A more precise determination is given by knowing elevation. At 200 m (see Figure 14–15) the station is found in a basal life zone, and is correctly called cool temperate steppe. However, what of a station with the same parameters but at 1200 m elevation? To calculate the basal belt correctly, use a lapse rate of 6°C 1000 m^{-1}: $8°C + (1200 \text{ m}/1000 \text{ m})(6°C) = 15.2°C$. This means that 8°C at 1200 m is in the same basal belt as 15.2°C at sea level; according to Figure 14–15, the basal belt for 15.2°C at sea level is warm temperate. The station at 1200 m then falls correctly into the warm temperate montane steppe life zone.

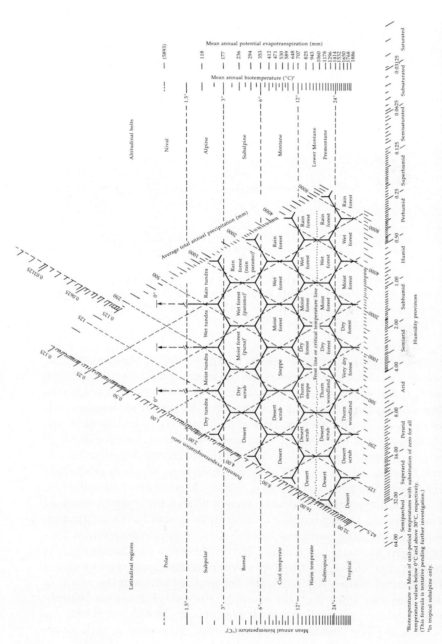

Figure 14–14. Diagram for the classification of world life zones or plant formations. (From Holdridge 1967. *Life Zone Ecology.* Rev. ed. San José, Costa Rica: Tropical Science Center.)

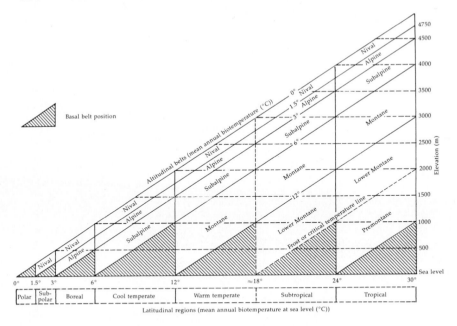

Figure 14-15. Approximate guideline positions of latitudinal regions and altitudinal belts of Holdridge's life zone system. (Based on an average lapse rate of 6°C 1000 m^{-1}.) (From Holdridge 1967. *Life Zone Ecology.* Rev. ed. San José, Costa Rica: Tropical Science Center.)

The Holdridge system seems complex and confusing at first. However, the climatic parameters are the most significant to vegetative growth, and the resultant categorization does reflect the natural distribution of vegetation. It does not consider, however, secondary variables, such as edaphic factors, aspect, or proximity to bodies of water, but is most useful for vegetation of zonal soils, with zonal climate. It does not account for the tall grass prairie of North America, for instance. Even so, the use of the physiognomic type which is most characteristic of a life zone as the name of the life zone chart is useful.

The nature of the Holdridge system allows for categorization of traditional units (associations) but is applicable also to the continuum viewpoint providing (with the basic three-axis graph) geographic limits in regard to the major climatic factors. Lindsey and Sawyer (1971) have applied Holdridge's three-axis sytem to the forest regions in the eastern United States (Figure 14-16). From examination of the graph as drawn, one may infer that, for example, the oak-pine region is probably transitional or seral, for it is overlapped almost completely by other figures. The hemlock-hardwoods type, however, is little overlapped, and may be viewed as a climatic climax.

Climate for a particular station may be depicted as a **hythergraph,** or climograph: a plotting of mean monthly temperature against mean monthly precipitation (Smith 1940). The twelve points derived are connected to form a distinctive polygon (Figure 14–17) which differs from polygons for stations under different climatic regimes. A little practice allows the immediate recognition of characteristic polygons, and the mental association of the vegetation found under the climatic regime symbolized by that polygon. For instance, the hythergraph for a deciduous forest differs dramatically from the hythergraph for a boreal forest (Figure 14–17; see also Figure 19–30).

A more complex graphic system is the climatic diagram depictions of Walter (Figure 14–18). There is a wealth of information conveyed at a glance: rainfall, mean annual temperature, mean annual precipitation, mean daily minimum of the coldest month, and so forth. Included also are the elevation of the station depicted and the duration of observation in years. The visual image provided by climate diagrams allows for rapid interpretation and comparison of climatic regimes.

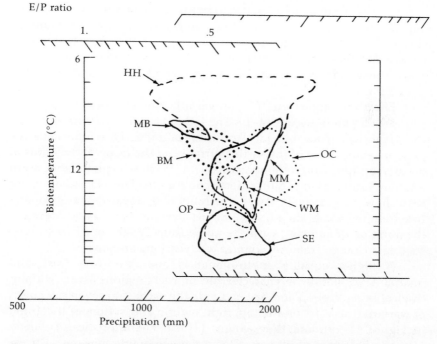

Figure 14–16. Three-axis graph from plotting all stations to show climatic amplitudes of the forest regions. HH = hemlock-hardwoods; MB = maple-basswood; BM = beech maple; MM = mixed mesophytic; OC = oak-chestnut; OP = oak-pine; WM = western mesophytic; and SE = southeastern evergreen. (From Lindsey and Sawyer 1971. By permission of the Indiana Academy of Sciences.)

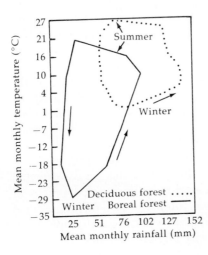

Figure 14–17. A composite climograph, which permits comparison of two forest climates. Monthly means of temperature and rainfall have been plotted against each other and the points connected by lines. Those polygons with less than 12 angles have points along straight lines. (From *The Study of Plant Communities* by H. J. Oosting. W. H. Freeman and Company. Copyright © 1956.)

These various systems, in essence, provide a prediction of vegetation based on two climatic parameters: temperature and moisture. The system of Axelrod and Bailey (1969) predicts the temperature based on vegetation. They suggest that two temperatures—the mean annual temperature *(T)*, and the mean annual temperature range *(A)*—very closely delineate the thermal regime for living plant communities. These considerations are applied to a paleotemperature analysis of Tertiary flora.

Effective temperature *(ET)*, or warmth, and an index of equability *(M)** are shown in Figure 14–19. The index $M = 100$ is represented on the graph as a point, an ideal centrum of 14°C. This represents the average temperature of the earth's surface and the mean of the limits of mean monthly winter temperatures for tropical regions and mean monthly summer temperatures for arctic regions. The arcs radiating out then show progressively lower equability *(M)*, departures of temperature from the ideal. The rays that depict warmth *(ET)* vary according to the duration of summer, with the range from *ET* 45.5 to *ET* 64.4 representing a change from no summer to a yearlong summer.

A highly equable environment is one free from frost, and luxuriant vegetation is characteristic of such regions, even without tropical heat. Axelrod and Bailey divide the earth into seven categories of warmth (these are on the top, right margin and bottom of the graph in Figure 14–19), and they assign the leaf characteristics of woody plants to be expected in each category, given ample moisture. For example, leaves in hot climates tend to be large, with entire margins. In

*For a derivation of *ET* and *M*, see Axelrod and Bailey 1969, p. 174 and 179–180.

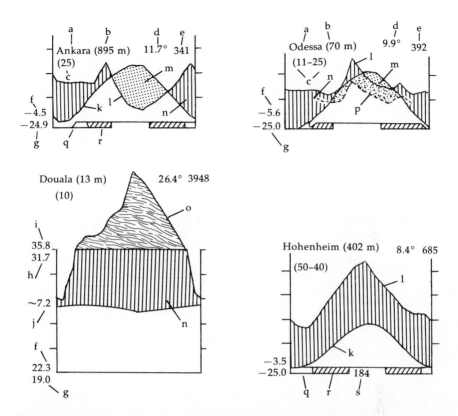

Figure 14–18. Key to the climatic diagrams. Abscissa: Months (N. Hemisphere January–December, S. Hemisphere July–June); Ordinate: one division = 10°C or 20 mm rain. a = station, b = height above sea level, c = duration of observations in years (of two figures the first indicates temperature, the second precipitation), d = mean annual temperature in °C, e = mean annual precipitation in mm, f = mean daily minimum of the coldest month, g = lowest temperature recorded, h = mean daily maximum of the warmest month, i = highest temperature recorded, j = mean daily temperature variations, k = curve of mean monthly temperature, l = curve of mean monthly precipitation, m = relative period of drought (dotted), n = relative humid season (vertical shading), o = mean monthly rain >100 mm (black scale reduced to 1/10), p = reduced supplementary precipitation curve (10°C = 30 mm) and above it (spots) dry period, q = months with mean daily minimum below 0°C (white) = cold season, r = months with absolute minimum below 0°C (diagonal shading) = late or early frosts occur, s = mean duration of frost-free period in days. Some values are missing, where no data are available for the stations concerned (h–j are only given for diurnal types of climate). (From Walter, H. 1973. *Vegetation of the Earth in Relation to Climate and the Eco-physiological Conditions.* By permission of Springer-Verlag.)

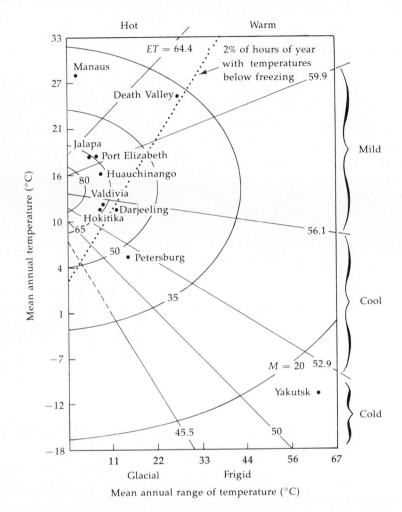

Figure 14–19. Isothermal relations weighted by biological considerations are best expressed by: (a) the rays of effective temperature *(ET)* that connote warmth and its duration; and (b) the arcs of equability *(M)* that measure departures of temperature from an ideal centrum of 14°C. (From Axelrod and Bailey 1969. Paleotemperature Analysis of Tertiary Floras. *Palaeogeography, Palaeoclimatology, Palaeoecology* 6:163–195.)

cool climates, leaves are typically small and serrate in the Northern Hemisphere, or small and entire in the Southern Hemisphere. In cold climates in the Northern Hemisphere needle leaves dominate.

At the right border, the arcs describe a range of variation. For example, Yakuktsk, with *A* = 55°C and *T* = −12°C, has an equability index *(M)* of less than 20 (not a very equable place). Even worse, *M* = 0, or minimum equability, is assigned to Antarctica. Towards the left of the graph, seasonal variations become vanishingly small. For instance, Bogota, in the Andean Highlands, has an index of *M* 92, which approaches ideal equability.

Warmth *(ET)* refers to temperature at the beginning and end of a period free from frost; index of equability *(M)* represents magnitude and frequency of extremes of temperature. Axelrod and Bailey (1969) found a "greater universality in plant-temperature relations" in terms of *ET* and *M* than with other thermal terms.

The usefulness of this equability system has been limited in application, primarily to paleobotanical work. It is, however, a very important and meaningful way of considering morphological response of plants to warmth and to equability.

The aims in plant ecology have been: 1) qualitative descriptions of vegetation; 2) quantitative descriptions of vegetation; and 3) developing predictions about form and function of vegetation. We have clearly passed the first step, the outlines of the second have been accomplished but much work remains, and the third is now being attempted. Ecologists are devising ways of quantifying ecological data, to allow for statistical analysis by computers as well as increasing our information storage capacity. Predictability is the logical next step. The symbolization devised by Axelrod, Holdridge, and others is part of the progress toward that end.

SUMMARY

For plants, a very important environmental factor is heat. Little biological activity occurs below 0°C or above 50°C. Heat may be transferred by radiation, convection, or conduction. Energy that fuels this planet comes to us as solar radiation, including ultraviolet light, visible light, and infrared light. Earth is shielded from most of the UV and IR wavelengths by its atmosphere.

All objects absorb and transmit or reflect radiation. Energy gains and losses must balance within a leaf, an organism, and the biosphere itself.

A solar energy budget may be symbolized as

$$S = R + C + G + Ps + LE$$

where S = solar radiation; R = reflected or reradiated; C = convection; G = heating of solid objects; Ps = photosynthesis; and LE = latent heat of evaporation. The solar constant (energy reaching earth's outer atmosphere) is 1.94 ly min^{-1}. Locally, radiation balance may be diurnally, or seasonally, negative, or may be always positive, as in some tropical regions. Energy allocation within a wet mountain meadow differs dramatically from that of, for instance, a desert region, due to plant cover, amount of open soil, and standing water.

Factors that influence variation in temperature include latitude, altitude, topography, proximity to water, cloud cover, vegetation, and slope aspect. Mean annual temperatures decline poleward, but diurnal and seasonal flux increase. Air cools at a certain lapse rate as it moves upslope. Plant cover tends to dampen temperature oscillations locally. South-facing slopes will be warmer, in general, than nearby north-facing slopes in northern temperate latitudes, the situation being reversed in southern temperate latitudes. Phenological events may be retarded on the cooler slope with respect to the warmer.

A plant allocates energy to maintain a balance, losing heat by convection, reradiation, reflectivity, and evaporation of water. Small leaves aid in loss by convection; white or light colored leaves enhance reflectivity. Thermoperiodism, a requirement for a day-night temperature differential for optimal growth, characterizes certain plants. Some seeds also exhibit a need for temperature fluctuation for germination. Onset of dormancy is triggered, in part, by cold temperature, and cold is used as an environmental cue by plants to perceive the passage of winter. Exposure to cold enhances germination, as well as early flowering, in some seeds and plants, e.g., lupines and cereal grains, respectively. Heat pretreatment enhances germination of some desert annuals.

Radiation is monitored by means of pyranometers and pyrheliometers.

Climate classification schemes have been devised by many authors. For instance, the climate provinces of Köppen, the humidity provinces of Thornthwaite, and the life zones of Merriam have been useful. A new type of life zone, based upon precipitation, moisture and heat (but only the heat that affects plant growth, biotemperature) was developed by Holdridge. Hythergraphs and climatic diagrams were devised by Smith and Walter, respectively, to present the climate graphically for a given station. Axelrod proposed an equability system that allows prediction of vegetation type based on mean annual temperature and mean annual temperature range. All of these systems are attempts to make ecological phenomena predictable, an important and essential goal.

CHAPTER 15

FIRE

Fire must be included as an evolutionary force in the development of communities, particularly those subject to lightning, as pointed out by Komarek (1964):

> Lightning is of such frequency and magnitude that there are not many lands, if any on earth, that at some time or other have not been subjected to reoccurring lightning-caused fires. These have occurred at frequent enough intervals to have lasting effect on plant and animal communities.

Lightning is the major cause of natural fires in almost every community. The earth is bombarded by lightning daily and has been throughout its history. There are an average of 100 lightning strikes to the earth every second, 24 hours a day, 365 days a year (Komarek 1964). This totals over 3 billion strikes per year. The energy of these strikes is impressive; some may range well into hundreds of thousands of volts, with a current of as much as 340,000 amperes. In a one hour period at midnight in 1966, some 2379 lightning thunderstorms passed over 275,000 square miles of the central United States, according to Komarek (1968). He points out that this represents only one frontal activity, and that such fronts sweep North America every seven to ten days in the summertime.

The influence of fire dominates the history of forests, as well as grasslands and the xeric shrub communities (chaparral, fynbos, maqui, matorral) of the mediterranean climate regions of the world. With the possible exceptions of the wettest or coldest regions of the earth, great tracts of land have been subject to periodic fires for millenia (Komarek 1963; Spurr and Barnes 1973). Fires may be caused by lightning, by sparks from falling rocks, by volcanic activity, by spontaneous combustion, and by human activity. Of these, only the first and the last may develop any periodicity and thus act as a directional and consistent evolutionary force in the community. In the Old World and, to a lesser extent, the New World, periodic burning by humans has played a part in the evolution of the biota. Also, "Lightning fires are an integral part of our environment and though they may vary in number in both time and space they are rhythmically in tune with global weather patterns" (Komarek 1967).

HUMAN HISTORY

Humans have used fire as a manipulative tool for at least 0.5 million years (Stewart 1963). Even before we learned to produce fire, the human family used and carried fire from place to place. We can infer that grassland and forest fires may have been inadvertently caused by aborigines, whose practice was to leave a fire banked at their home base. This was to avoid calamity in case the fire being carried went out. Primitive humans living today are keen observers of minutiae, as were our human and prehuman ancestors, undoubtedly. They observed the consequences of fire: game is driven, though casually and without panic (Komarek (1969) suggests that animals have no innate fear of fire); the capture of insects, rodents and reptiles is facilitated; visibility is enhanced; travel is easier; and forage is renewed. What had been serendipitous accidents surely became deliberate, purposeful arson. "Pre-man primates as well as early man inhabited fire environments and had to be 'fire-selected' . . . to live in such environments" (Komarek 1967). Undoubtedly, humans had learned to produce fire by 10,000–20,000 years ago (Stewart 1963). In the tropics, grasslands can be maintained by fire at the expense of trees if there is a periodic dry season. Thick forests held little of use for stone age peoples; grasslands and savannah were of far greater value.

In western North America, fire was used by native peoples to aid in the production of wild seeds, tubers, berries, nuts, and even wild tobacco. In California, winter ranges for deer were burned in early summer to prevent intense wildfire in late summer, which would have destroyed the primary forage, wedgeleaf ceanothus (*Ceanothus cuneatus*). The Indian practice of burning forested areas was recorded by various early explorers and naturalists, such as Galen Clark, guardian of Yosemite, Dr. L. H. Bunnell, a member of the 1851 Yosemite discovery party, and Joaquin Miller, who wrote in 1887 (Biswell 1974).

We have come to accept the views of Komarek, Biswell, and others regarding the natural and critical role that fire plays, and has played, in the maintenance of communities. The role of fire is studied in virtually every country, and certainly in those countries in temperate and tropical latitudes. Whether to maintain a wilderness, to increase forest production, to enhance the habitat for game, or to increase grassland productivity, the naturalist, scientist, and manager seeks to understand the vital role that fire plays in the ecosystem.

CLASSES OF FIRE

The various types of fires are divided into three main groups, based on location and intensity: surface, crown, and ground fires. Fire behavior and fire intensity are affected by meteorological conditions, by fuel loading, and by fuel moisture and soil moisture.

Surface Fires

Surface fires are the least destructive fires, whether in a grassland or in a giant sequoia forest. They are relatively cool, fast moving fires. These fires do not build up high temperatures at the plant and ground levels because they are usually fed by lightweight fuels that are quickly converted to ash. Consequently, the basal portions, root stocks, and tubers are not harmed in grasslands and shrublands. In forested regions, cones may be opened, bark and needles scorched, and seedlings killed, but few mature trees will be severely damaged.

Crown Fires

Crown fires occur in the upper parts of trees and so, by definition, take place only in forest situations. They may result from lightning storms or from intense surface fires fueled by heavy accumulations of litter and debris. This type of fire never occurs in the absence of surface fire. From the human viewpoint, crown fires may be devastating, killing and consuming mature trees and dropping boles and branches to ignite and further spread the fire at lower levels (Phillips 1974). However, in some forests, a clean burn is necessary to open the canopy and thus allow seedling survival in full sunlight. For example, in Canada and the Lake States, nearly pure stands of jack pine *(Pinus banksiana)* often burn during hot dry periods, spectacularly crowning, generating fire storms, and killing all aboveground vegetation. Periodic fires of such intensity maintain the jack pine in vigorous growth. This requirement for intense fires is thought to be also true of the bishop pine *(Pinus muricata)* and of the knobcone pine *(Pinus attenuata)* in California (Aho 1975; Kuhlman 1977). It is paradoxical that the pure stand must be destroyed in order to be maintained in vigorous health.

The occurrence of fire or its absence results in certain changes in the environment. Our goal should be understanding, as best as we can, why such changes come about, and in what manner fire is responsible for them.

Ground Fires

Ground fires are termed "retrogressive agents" by Vogl (1974). In grasslands, ground fires consume soils down to mineral substrate, creating depressions that then sometimes become ponds. Although of infrequent occurrence, ground fires can be very destructive, not only of roots, tubers and rhizomes, but of the organic matter in the soil itself. Thus, recovery of the plant community may take tens to thousands of years following such a catastrophe.

Ground fires are especially destructive to peat bogs, and usually occur after drainage of the bog or after several drought years. Many thousands of years of peat accumulation may be completely removed by a ground fire, with the area subsequently reverting to a swamp. Many of the forest sites in bogs on recently glaciated parts of the northeastern and lake states of the United States have been lost to ground fire (Spurr and Barnes 1973).

However, light, dry soils may provide layers of dry organic matter—peat, fern rhizomes, and roots of other plants—that carry the fire easily. Further, ground fires tend to persist. They are shielded from environmental influences, especially human efforts at extinction, and will continue as long as the fuel remains and there is sufficient energy available to maintain a fire. Thus, when conditions are favorable, a ground fire can serve as a re-ignition source for surface fire (Vogl 1974).

FIRE EFFECTS ON SOIL

In general, fire has the following effects on soil: soil temperatures are higher, some soil nutrients and organic matter are lost, water holding capacity and soil wettability are changed, and populations of soil microbes are altered. We will consider the impact of fire on soil as it affects chemical aspects of soil, physical aspects of soil, and the biotic components of soil.

The effect that fire has on the soil and on soil organisms depends on the intensity of the fire. This intensity, in turn, depends on the heat yield, the availability of fuel, and the rate of spread of the fire. Wakimoto (1977) has symbolized this relationship as

$$I = H \times W \times R$$

where I = intensity (kcal sec^{-1} m^{-1}), H = heat yield (kcal g^{-1}), W = available fuel (g m^{-2}), and R = rate of spread (m sec^{-1}). The chemical constitution and various physical characteristics of the fuel influence its heat yield. For example, the pine needle litter of a yellow pine (Pinus ponderosa) forest has a high surface to volume ratio and is not compacted, so an abundance of oxygen is available. This type of fuel carries a fire readily, but will produce a relatively cool fire.

Western coniferous forests may produce as much litter as 145 kg ha^{-1} yr^{-1} (Hartesveldt and Harvey 1967). Thus, in part, productivity determines the available fuel. Rates of decomposition of litter and fire frequency are the other determiners of fuel availability in the natural environment. The rate of spread of the fire depends on wind speed, nature of the fuel, humidity, and temperature. In Figure 15-1, notice that the primary determinants of fire behavior are fuel, weather, local landforms, and the ignition pattern. Enlightened management of a fire-prone ecosystem requires some estimate of available fuel. Quanti-

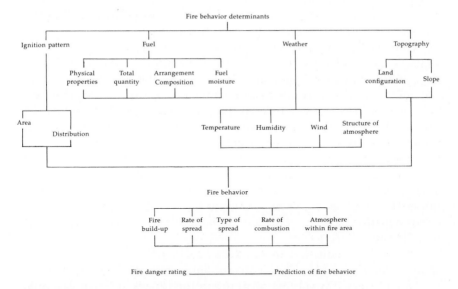

Figure 15–1. Determinants of fire behavior. (From Wakimoto 1977*b*.)

tative assessment of available fuel in almost any ecosystem requires a lot of time and money. Wakimoto (1977*b*) has developed a method of fuel inventory for chaparral which utilizes linear regression techniques that may, with use, reduce some of the expense. He found that cover (crown area) appears to hold the most promise as a single, easily measured variable which can be used as a basis for predicting total biomass.

Fire Effects on Organic Matter

The heating of soil is of less significance than the action of the fire on the organic matter in the soil. A very intense burn will destroy all the organic matter at the soil surface. Some destructive distillations will be effected at levels within the soil column. At temperatures of 200–300°C, 85% of the organic substances will be destroyed. Nitrogen and potassium loss is a function of temperature. Nitrogen and potassium compounds are subject to volatilization, and will be released and lost by distillation at temperatures of 100–200°C. Although there is little loss of nitrogen up to 200°C, as much as 60% may be lost at 700°C. Other nutrients such as calcium, sodium, and magnesium will be released and deposited on the soil even at moderate temperatures, and rapidly recycled into the biota (DeBano et al. 1977). One consequence of the loss of organic matter by burning is the concomitant loss of the high cation exchange capacity that characterizes organic matter, thus removing a holding substrate for vital nutrients.

Bulk density is the weight of the soil solids per unit volume of total soil. The units are given in g cc^{-1}. Although pore space is a part of the volume, the soil is dried in an oven before weighing so that soil water is not a part of the weight. Organic matter decreases bulk density, both because it is lighter than a corresponding volume of mineral soil, and because it gives increased aggregate stability to soil. The heat of an intense fire may break down these aggregates, leading to a loss of soil structure, and to lowered water infiltration. It has been shown that, in prescribed burning in the southern Sierra Nevada, the soil carbon is not lost during burning. The carbon content is inversely correlated with the bulk density of the soil: the higher the carbon, the lower the bulk density. And, the bulk density is not changed significantly by prescribed burning. However, a ground fire, in which almost all of the organic matter is destroyed, would change both carbon content and bulk density (Agee 1973).

There is a reduction in the thickness of the humus layer due to compaction after a fire. This is measurable, and especially evident, at the base of trees. Viro (1974), working in the forests of Fenno-Scandia, found that the reduction in thickness of the humus layer after fires ranged up to 20 cm, although the average reduction was only 1.6 cm. The humus layer was thinnest in the 8–20 years following the fire.

Chemical Considerations

In general, soil pH will be higher after a fire than before. The degree of change is based on the nature of the cations released by burning, and of course on the acidity of the soil before the fire. Forest soils often have a low pH, due to the high acidity of the litter. Thus, the magnitude of change when soluble basic cations are released by burning will be greater in forest soils than in chaparral soils (DeBano et al. 1977). The activity of free-living nitrogen-fixing bacteria is enhanced, both by the addition of nutrients by fire, and by the higher pH due to release of mineral bases in the soluble ash.

Physical Effects of Fire on Soil

Soil acts as an effective insulator

Even the ash layer that results from the combustion of the litter can insulate the soil. Heat energy may be transferred downward by conduction, convection, and vapor flux (DeBano et al. 1977). The rate of this transfer is affected most by the levels of soil moisture, although there are many other factors involved. Temperatures will not rise above 100°C in a given layer until all the water has been evaporated, and this evaporation requires a great deal of energy, slowing the transfer of heat downward. Some typical temperatures in chaparral, in intense, moderate, and light burns, are shown in Figure 15–2. Note the dramatic attenuation, even in intense burns, as little as 2.5 cm below the soil surface. Even under a hot fire, temperatures do not exceed 200°C 2.5 cm down, and are below 100°C under a light burn 2.5 cm down. In other words, a light surface fire would heat only the top fraction of a cm of mineral soil to near the boiling point.

In Redwood Mountain Grove, Kings Canyon National Park, temperature determinations were made by use of Templaq© in burn piles (Hartesveldt and Harvey 1967). Templaq© consists of strips of a special paint that fuses at a predetermined temperature. The recorded temperatures are presented in Figure 15–3. Notice that, even near the center of large burn piles, the temperature is less than 66°C at 25 cm depth, and less than 177°C at 12.7 cm depth. Grassland fires tend to be cooler than forest fires, with soil surface temperatures in grassland surface fires reaching temperatures of 120°C, and the air temperatures rising to 600–700°C. Air temperatures in forest fires may reach greater than 1100°C. In contrast to the Redwood Mountain Grove results, soil temperatures in grassland fires may rise only 10–15°C above ambient in soils only 1–5 cm down, and only for short periods of time (Daubenmire 1968). The nature and availability of fuel probably accounts for observed differences.

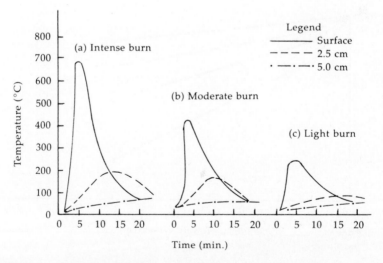

Figure 15–2. Typical temperatures at the surface and downward in the soil during (a) an intense, (b) a moderate, and (c) a light burn in chaparral. (From DeBano et al. 1977. Fire's effect on physical and chemical properties of chaparral soils. In *Proc. Symp. Env. Cons. Fire and Fuel Mangmt. in Medit. Ecosyst.*, pp. 65–74.)

Sweeney (1956), working in California chamise chaparral *(Adeno-stoma fasciculatum)*, found the highest temperature to be 650°C at the surface, 335°C at 1.25 cm depth, and 154°C at 3.8 cm depth during a burn. Beneath the soil during a burn in a mixed chaparral community, the temperatures recorded were even lower: 530°C at the surface and 198°C at 1.9 cm depth.

Soil moisture and fuel moisture

Soil moisture levels are affected by fire, but not in a simple fashion. In certain communities, living moss and humus can take in large amounts of water, effectively preventing rain water from reaching tree roots. The removal of that layer by fire should allow percolation of more water. However, the layer also prevents evaporation, so that water holding capacity is lower and evaporation is higher on burned plots than on unburned plots (Viro 1974). Burned sites also have higher temperatures than unburned sites due to the increased absorption of sunlight by blackened soil surfaces. This is especially true in grasslands. The higher temperature initially increases evaporation, but quick drying breaks capillary connections, and thus reduces evaporation.

The influence of fire on water runoff is temperature dependent. Following extremely hot fires which completely volatilize hydropho-

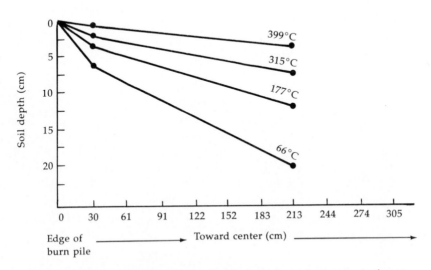

Figure 15–3. Temperatures recorded in a pile burn in a giant sequoia forest. Note the insulating value of the soil, even in the areas that approach the center of the burn. (From Hartesveldt and Harvey 1967. The fire ecology of sequoia regeneration. In *Proc. Tall Timbers Fire Ecol. Conf.*, pp. 65–78. By permission of the Tall Timbers Research Station, Tallahassee, FL.)

bic substances and thus increase wettability (see next section), there may be greater infiltration of water into the soil. However, after less intense fires, runoff is increased, due to lowered infiltration. Increase in the sediment yield is usually observed too, and this may correlate with, for instance, the loss of soil structure just mentioned, increased nonwettability, or both of these. Increased water runoff and sediment yields are significantly correlated with the decrease in the forest floor depth (or litter) that is observed following a forest fire. Although heavy and aerial fuels (attached branches, epiphytes, etc.) are often not affected, the weight, depth, and water holding capacity of fine surface fuels are reduced by fire (Agee 1973).

Fuel moisture is estimated by means of **fuel sticks**. These are wooden rods which when oven-dry weigh exactly 100 g. They are reweighed in the field, and the assumption is made that additional weight (over the original weight, expressed as a percentage) is an approximation of the fuel moisture levels in available fuels on the site.

Soil nonwettability

In certain communities, hydrophobic substances prevent the movement of water into the soil. This condition is called **nonwettability** and is characteristic of, for instance, ponderosa pine *(Pinus ponderosa)*/incense cedar *(Calocedrus decurrens)* forest, white fir *(Abies concolor)*/giant sequoia *(Sequioadendron giganteum)* forest, and the chaparral community of California. Hydrophobic substances which reside in decomposing plant parts accumulate on the soil surface and in the upper part of the soil column in the years between burns. Temperature gradients, established during fires (Figure 15–3) enhance the translocation of these substances into the lower soil layers (Figure 15–4). The nonwettable layer is thus moved down lower in the soil column, where it decreases water infiltration. The depth of the layer, and thus the depth of water infiltration, is determined by fire intensity, fuel availability, and physical condition of the soil (DeBano et al. 1977). Notice (Figure 15–4) the movement of the hydrophobic substances into the soil during a fire. The water repellent zone moves downward, such that a wettable layer exists both above and below it. Water can thus move into the upper soil layer, and perhaps cause sheet erosion. This phenomenon is sometimes seen in chaparral of southern California following a fire and subsequent rainfall.

Effects of Fire on Soil Biota

Ahlgren (1974) pointed out that the ecology of soil organisms following a fire is not well understood. Even so, numerous studies have been made of soil microflora and fauna in reference to fire (Dunn

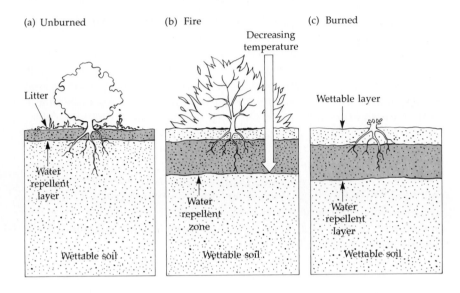

Figure 15–4. Water repellency before, during, and after fire. (a) Before fire, the hydrophobic substances accumulate in the litter and mineral soil immediately beneath it; (b) fire burns the vegetation and litter, causing the hydrophobic substances to move downward along temperature gradients; (c) after fire a water repellent layer is located below and parallel to the soil surface on the burned area. (From DeBano et al. 1977. Fire's effect on physical and chemical properties of chaparral soils. In *Proc. Symp. Env. Cons. Fire and Fuel Mangmt. in Medit. Ecosyst.*, pp. 65–74.)

and DeBano 1977; Ahlgren 1974; Parmeter 1977; Wright and Tarrant 1957). The effect of fire is usually to reduce fungi, but to increase populations of soil bacteria and actinomycetes. These changes are due not only to the heat of the fire directly, but more importantly to changes in the post-burn environment. As we would expect, any changes in the microbiota are most obvious in the very uppermost soil levels.

The diminution of populations of heterotrophs may be achieved by the removal of their food base. As Parmeter (1977) suggests, we may infer that pathogens will be destroyed by fire, if these pathogens sporulate on litter and if that litter is consumed by fire. For example, damping-off fungi, seedling root rot fungi, and organisms that decay seeds will be removed from thoroughly heated seedbeds. The brown spot disease pathogen (*Septoria acicola*) of longleaf pine (*Pinus palustris*) is controlled by burning. (We will discuss the biology of longleaf pine in more detail in a later section of this chapter.) Thus, the success of seedlings in a burned-over area may be attributable in part to the removal of soil pathogens or their substrates (Parmeter 1977).

The other face of this coin is the possible stimulation of sporulation and growth of fungi, especially of ascomycetes, due to fire (Ahlgren 1974). There are a number of early post-fire species, mostly **pyrophilous** (fire-loving) discomycetes such as *Pyronema*. The succession of fungi on burned lands has been studied, and is thought to be analogous to succession of higher plants on disturbed lands (Ahlgren 1974).

In some cases, fire actually increases disease: heart rot of aspen and pine, and root rot of Scots pine and mugo pine may be stimulated by fire. Trees that are weakened by fire may become highly susceptible to diseases, as well as to insect entry.

It has been reported that "fire has no effect on the bacterial populations, that fire completely destroys it, or that fire results in a striking increase in bacteria" (Ahlgren 1974). Methods of study, intensity of fires that are studied, edaphic factors, and soil moisture of the site are all variable from one locality to another, or even within the same area but dealing with different burns. This variation accounts for some of the diversity of opinion regarding the effect of fire upon bacterial populations.

Some generalizations can be drawn, however. Bacterial populations are, in general, favored over fungi in burned sites, due to the increased pH. And, burning enhances the growth of nitrogen-fixing species of the bacterial genera *Azotobacter* and *Clostridium* in coniferous forest soils.

Ahlgren (1974) noted that actinomycetes are less sensitive to heat and desiccation than are bacteria. For instance, the ratio of bacteria to actinomycetes was reduced immediately following a prescription burn in jack pine *(Pinus banksiana)* in northern Minnesota. This ratio was increased sharply following the first rain. Actinomycetes population growth may be enhanced in the first few post-burn years, especially in the lower levels (Table 15–1), because of the enrichment due to soluble minerals being moved downward from surface ash.

Nitrogen losses due to volatilization of nitrates, ammonia, and amino acids (as well as through alterations in forms of both organic and inorganic nitrogen), may be recovered, both on grasslands and in the chaparral, by early dominance of the post-burn community by herbaceous legumes. Examples of these legumes are such nodulated plants as *Trifolium, Lotus,* and *Lupinus,* which dominate post-burn regions in southern California. Nitrifying bacteria, such as *Nitrosomonas* and *Nitrobacter,* are very sensitive to fire. Their populations are completely destroyed at 140°C, and evidence great mortality even at 100°C. Following a burn, the recovery of such nitrifying bacteria is slow (Dunn and DeBano 1977).

Table 15–1. Ratio of bacteria to actinomycetes on moderate and severe jack pine *(Pinus banksiana)* burns. (From Ahlgren 1974. The effect of fire on soil organisms. In *Fire and Ecosystems,* ed. T. T. Kozlowski and C. E. Ahlgren, ch. 3, by permission of Academic Press.)

	Level 1[a]			Level 2[a]		
	Burn	Cut	Control	Burn	Cut	Control
1961 (moderate burn)						
Pre-burn	0.74	0.63	0.93	0.66	2.68	0.42
2 days after fire	0.50	0.74	1.01	1.86	2.91	1.28
9 days after fire	1.52	1.11	0.91	1.03	4.10	1.67
16 days after fire	1.20	1.03	0.81	1.49	5.00	1.03
30 days after fire	12.14	1.50	1.05	1.51	2.30	8.76
1962 (second post-fire year)						
June 15	0.77	1.18	2.29	1.48	1.80	4.11
July 15	1.30	1.32	1.88	4.32	2.87	5.50
August 15	0.55	0.33	2.51	2.25	1.73	1.03
1963 (third post-fire year)						
June 15	1.83		1.89	2.04		1.59
July 15	9.00		2.20	1.67		4.44
August 15	12.90		7.89	1.32		2.39
1963 (severe burn)						
Pre-burn	9.16			4.00		
2 days after fire	2.33			2.91		
9 days after fire	26.33			2.26		
16 days after fire	22.80			2.37		

[a]Level 1 is the top 3.8 cm of soil, including burned humus; level 2 is 3.8–5.1 cm into the mineral soil.

FORESTS OF THE SOUTHEASTERN UNITED STATES

The region of the United States that stretches from the Appalachian Mountains to the Atlantic is very old geologically. When the northern parts of the continent were glaciated, this southeastern portion became a refuge for many species of plants and animals. Fire has been a major factor in the development of many communities in this area.

The complex patterns of weather, influenced by topography, and by proximity to the ocean, have created a banded mosaic of communities in the southeastern United States. The frequency, size, and intensity of lightning fires varies accordingly. Komarek (1968) has divided the Southeast into two provisional lightning fire bioclimatic regions. These are defined by the variations in fire weather and other meteorological phenomena, and on the basis of certain prominent forest types. The two regions are the eastern deciduous forest (the region of the Appalachian Mountains characterized by deciduous hardwood forests, located from the Appalachian highlands westward) and the southern pine forest. Space does not permit the close examination of all the communities of these regions, but we will look at one in more detail: the longleaf pine community.

The Southern Pine Forest Lightning Fire Bioclimatic Region

This region includes two biogeographic regions: the coastal plain forest and the coastal plain pine savannas and grasslands. In the former are found loblolly, pond, sand, slash, and longleaf pine forests (*Pinus taeda, P. serotina, P. clausa, P. elliotii*, and *P. palustris*). In the latter are found pine-bluestem (*Andropogon* spp.), pine-wiregrass (*Aristida stricta, Sporobolus curtissi*), pine-cane savannas, the "Red Hills" savanna, and the coastal marshes and prairies (see Komarek 1974 for more details).

The longleaf pine-bluestem savanna has a grass and forb understory that is actually a relic of a prairie. The open, parklike forestscapes of pre-European settlement were, and are, maintained by fire. Both the forest and the herb layer are maintained by frequent fire. There are occasional broadleaved hardwood trees and shrubs, but the grassland understory includes little bluestem (*Andropogon scoparius*) and slender bluestem (*A. tener*), and is a community rich in herbaceous plants, including some 200 species of grasses.

Longleaf pine (*Pinus palustris*) and the role of fire in its ecology has been very well studied. Komarek (1974) pointed out that two reviews cite over 1300 articles on this one species, with over 200 references to the fire ecology of longleaf pine. It seems to be a species almost ideally suited to frequent fires. Cooper (1961) and others, e.g., Spurr and Barnes (1973), described the life history of longleaf pine.

Following germination, most pines grow rather uniformly. Longleaf pines reach a height of several centimeters in just a few weeks, but then upward growth ceases. The young seedling sprouts a pompon of long, slender, and drooping needles, which surround the stem apex (Figure 15–5). This "grass stage" endures for three to seven years, and during that time the plant allocates most of its energy to the production of a sturdy root system, with abundant stored food.

The pines of the Southeast, with few exceptions, require mineral soil as seedbeds, and most of the seedlings are shade intolerant. In addition, longleaf pine is very susceptible to a fungal disease called brown spot disease. The causal agent, brown spot fungus (*Septoria acicola*), matures during the summer months. Multitudes of spores are released to the autumn rains, and they are splashed on low-growing plants, including seedlings of various kinds. Control of the fungus is gained by burning the infected needles before new growth resumes the next spring. The dry needles are burned, as are the aboveground portions of competing hardwoods and herbaceous plants. But, the "grass stage" of longleaf pine protects the meristematic regions from heat damage, and the young seedling lives.

If the young tree can survive a full year before fire comes, it is then very tolerant of fire for another one to six years. Of the pines that Biswell (1958) studied (longleaf and slash in Georgia and ponderosa and sugar pine in California *(Pinus taeda, P. elliotii, P. ponderosa, P. lambertiana)*, the longleaf is the most fire tolerant. There is a sensitive period, once the seedling resumes upward growth (terminating the grass stage), until it is about 2 m tall. The mature pines possess the heavy, protective bark necessary to prevent damage to sensitive growing tissue by fire, and, in fact, even young saplings are rather well insulated with corky bark.

Fire, then, functions in several ways to promote longleaf pine forests: it clears the mineral soil for seed germination and seedling growth; it reduces competition from hardwoods and herbaceous plants; it eliminates brown spot blight; and, we infer, it opens the forest canopy to allow growth of these shade intolerant young trees.

Prescribed Burning

We have referred to prescription burning in previous sections of this chapter. A formal definition is given by Biswell (1977) as follows:

Figure 15-5. The grass stage of longleaf pine. The long, slender needles protect the perennating bud against fire.

Prescribed burning is the skillful application of fire to natural fuels under conditions of weather, fuel moisture, soil moisture, etc., that will allow confinement of the fire to a pre-determined area and at the same time will produce the intensity of heat and rate of spread required to accomplish certain planned benefits to one or more objectives of silviculture, wildlife management, grazing, hazard reduction, etc.

In forestry, the main purpose of prescribed burning is to enhance the growth of certain trees, by manipulating the major factors that influence that growth: light, temperature, water, and mineral nutrients. Foresters may desire a change in the relative abundance of species in a forested region, such that desirable species are favored and competing species are eliminated or at least put at a disadvantage. Facilitation of the renewal of valued species and the provision of optimum conditions for growth are the aims of prescribed burning in the forest.

Beginning in the 1920s, burning was used for range improvement, for upland game management, and ultimately was applied to forestry. By the 1940s, prescription burning was regularly applied as a management tool in longleaf pine *(Pinus palustris)* forests and slash pine forests *(P. elliotii)*. By the late 1950s, fire was also used as a silvicultural tool for loblolly pine *(Pinus taeda)* and the shortleaf types.

For prescribed burning to be successful, there must be enough fuel to carry a fire and, ideally, the fuel should be a "material that burns readily and uniformly soon after a rain" (Biswell 1958). This allows a surface fire, but, since heavier fuels that might contribute to a crown fire would still be too wet to burn, prevents a holocaust. In Georgia, grasses (pineland threeawn *(Aristida stricta)*, Curtiss dropseed *(Sporobolus curtissi)*, and *Andropogon* spp.) and some pine needles carry the fire. The loose, well-aerated, high surface-to-volume ratio fuels will dry quickly enough to burn in one to two days after a rain.

GRASSLAND FIRE

Both grassland climates and grassland vegetation favor fire. These arid and semiarid regions are usually subject to prolonged drought. The periods of lowest moisture availability correspond temporally with those of the highest temperatures. The prevailing growth form of grassland species is hemicryptophyte (Figure 1–3). We will consider mainly the prairie of North America, which begins south of 55°N latitude, and continues to south of 30°N latitude. Both grasses and forbs are persistent under the stress of recurring fires. Shrub tops, however, are kept in juvenile condition by annual burning (Figure 15–6).

Figure 15-6. Maintenance of grassland by fire. Mesquite and other shrubs are a minor component of grassland being restrained by periodic fire. (a) In the absence of fire, shrubs gradually increase and replace some grasses. (b,c) Fire consumes both shrubs and grasses; however, the hemicryptophytic nature of the grass life form protects the perennating bud, whereas the fully exposed buds of the shrubs are destroyed. (d) Grasses are capable of producing large crops of seeds much more rapidly than the shrubs. Shrubs, therefore, lose several years' growth following fire and are again vulnerable when fire reoccurs. (Reprinted by permission from The ecology of fire by C. F. Cooper. Copyright © 1961 by Scientific American, Inc. All rights reserved.)

Causes of Fire in Grasslands

Fire requires ample fuel, abundant oxygen, and an ignition source. Before the time of overgrazing, fences, roads and the like, there were miles of continuous, contiguous, flammable vegetation on the prairie, which allowed fire to sweep across, virtually unimpeded by structures or firebreaks of any kind. One series of thunderstorms can travel great distances igniting fires in its path. Under primeval conditions, these fires could simply spread until they burned out, either because the fuel was gone, or because the weather changed (Komarek 1965). Fires caused by lightning are primarily summer fires, and probably the most frequent type of fires in the grasslands.

Early and modern humans also cause fires, and humans, along with lightning, have the capacity for a rhythmic periodicity in the fires they cause. The intentional burning of grasslands, probably the first land management tool, would have been rapidly rewarded. After a few days of good regrowth practically any burned grassland will act as a powerful attractant for animal life. Actually, deliberate burning serves a second function, perhaps more meaningful to us today: reduced danger of wildfire, which spreads quickly and destructively over large areas that have been protected from fire for long period (Lemon 1968).

Interaction of Fire and Vegetation

There are benefits that accrue to those grassland species that survive fire. Silica content of the grasses increases, which enhances resistance to decay, and thereby provides more fuel for the next fire. The more immediate advantage is the increased protection afforded the seeds which remain above the hottest part of the fire on fire resistant seed stalks, thus ensuring survival of the seeds (Vogl 1974). Komarek (1965) pointed out that there are many species of fire-adapted grasses which have very hard stems with a high silica content among the genera *Aristida, Andropogon, Imperata,* and *Setaria.* Tooth adaptations of the fauna of the grasslands are determined by the silica content of the flora. Thus, mammals possessing ever-growing teeth, with ridges that are actually sharpened by wear, came to dominate the grassland community.

The corollary to the attribute of decay-resistant foliage is the immobilization of nutrients in plant tissue. Apparently, grasslands have become dependent on frequent fire as the agent of soil nutrient renewal, releasing the nutrients from plant tissue and recycling them into new biomass (Mutch 1970).

Anderson (1965), writing about Kansas prairie forbs, discussed the benefits of fire for the bluestem (*Andropogon* sp.) range. The indirect benefits of a late spring fire are these: (a) it prevents early annuals from reseeding, (b) it removes the green tops of weedy perennials, and (c) thus it reduces competition. There are also direct benefits: scarification* of certain resistant seeds, so that germination is possible, and the possibility that certain insect species, like certain plant species, may respond differentially to range burning. We should perhaps emphasize, again, that adaptations of plant species are to *periodic* burning, which can only be initiated by one of two agents: humans or lightning.

Grassland Fires and Evolution

The bulk of the lightning-caused fires occur during the reproductive periods of the inhabitants of the grasslands, both plant and animal. Stress, such as that provided by the fire and storm, provides a natural setting in which selection can operate most effectively and quickly. That stress becomes the vehicle of selection, allowing the survival of only those organisms whose responses to fire are adaptive. Lightning and fire are both mutagens, and so the possibility for very sudden change exists.

Mechanisms for Survival

After a fire, one of the following usually occurs, which aids in the survival of grassland plants: (1) the plant resumes life from a protected (perennating) bud at the soil surface or (2) there is increased reproduction from seeds.

Hemicryptophytes

On the prairie, most of the plants are **hemicryptophytes**. This means that, at least once a year, the aboveground portion of the plant dies back. Prairies are windy places, and the winds tend to dry the aerial portions of the plants. In addition, the plants have a low growth form, and this allows sunlight penetration which hastens drying. Thus, when the vegetation is dormant, there is often abundant flammable material at the ground level. In addition, the plants have low moisture content, even before drying. We have, then, ideal conditions for a fire: dry, uncompacted, fine fuels, with plenty of available oxygen.

These fine fuels will be ashed quickly, and very high temperatures do not build at ground level. Rather, the flames pass very quickly through the vegetation, the highest temperatures occur at the top of the flames, and the heat is dissipated by the prairie wind. Soil surface

---

*Scarification is used here in the sense of any agent that ruptures the integrity of the seed coat, leaving it permeable to water and oxygen.

temperatures in grassland fires reach only about 120°C, and in soil 1–5 cm deep, there is a temperature increase of only 10–15°C above ambient soil temperatures. There is evidence that 60°C is the lethal temperature for the shoots of most land plants. However, leaf meristems (perennating buds) of many grasses are 40 mm or so below the soil surface. For example, pineland threeawn (Aristida stricta) and Curtis dropseed (Sporobolus curtissi) have such buried meristems, which are additionally protected by closely packed, persistent leaf sheaths, which tend to exclude oxygen, and therefore do not burn. The living portion of the plant is virtually unharmed by the fire and will resume growth in the favorable time of the year in a nutrient regime actually enhanced by fire (Daubenmire 1968). As we noted previously, the decay-resistant aboveground plant parts contain needed minerals that are recycled by fire.

Increased reproduction from seeds

Enhanced reproduction may occur as a direct response to fire: certain grassland species have resistant seeds that require scarification before germination will occur. Legumes, such as species of Astragalus and Trifolium, are known to produce seeds that require scarification before seed dormancy will be broken. Bermuda grass (Cynodon dactylon) exhibits enhanced reproduction from seed following a fire because of increased seed set. However, many grasses do not possess fire-resistant seeds. Table 15–2 summarizes heat tolerances of seeds of various grass species. Note that the survival times are brief, but that tolerances range well into the temperatures attained by surface fires at the soil surface in grasslands (Daubenmire 1968).

Table 15–2. Heat tolerances of seeds of various grasses. (From Daubenmire 1968b. Ecology of fire in grasslands. Advances in Ecological Research 5:209–266. Copyright by Academic Press Inc. (London) LTD.)

Taxon	Common name	Maximum temperature with survival	Duration of heating
Avena fatua	wild oat	104–116°C	5 min.
Bromus mollis	soft chess	116–127°C	5 min.
Bromus rigidus	ripgut	93–104°C	5 min.
Stipa pulchra	needle grass	116–127°C	5 min.
Bromus mollis	soft chess	>200°C	2 min.

Management

There is good evidence of increasing dominance of woody plants in some grasslands. This may be an invasion, or may simply reflect changes in the density of shrubs already present. The change may be attributed to, for instance, the absence of hot fires, increased water availability (developed for livestock use), reduced fuel availability, and reduced herbaceous competition for shrub seedlings. The latter two factors are due to grazing and overgrazing (Vogl 1972; Box 1967). In California, some 3,000,000 ha of woodland-grassland is used for grazing. Since 1948, ranchers have prescribed burned an average of 40,000 ha yr^{-1} to destroy invading woody plants and to enhance the growth of palatable grasses (Biswell 1963).

Six hundred thousand ha of National Forest land has been judged suitable for conversion to grassland through manipulation, whether by fire, mulching, chipping, or burying, and some 380,000 ha has been judged suitable for conversion to timberland. The conversion to grassland is being attempted in a series of steps, including preparation for prescription burning, seeding, and, finally, control of brush sprouts and seedlings (Doman 1967).

MEDITERRANEAN CLIMATIC REGIONS

The mediterranean climatic regime is found between 30° and 45° north and south latitudes, on the west coast of continents, and in the Mediterranean Sea region. The mediterranean climate is defined by four characteristics: (1) moderate precipitation concentrated in the cool winter months, and summer drought; (2) in summer, long periods of sunny days and cloudless skies; (3) marine-moderated atmosphere influence year-round; and (4) warm-to-hot summers and mild winters (McCutchan 1977).

Within that regime certain areas are occupied by **sclerophyllous** (hard-leaved) plants and by thin, stony soils with little organic matter; these areas also have less than 750 mm of precipitation annually. In California, such sites are occupied by a dense scrub called chaparral (originally a Basque word for scrub oak, *chabarro*, spelled by the Spanish *chaparro*, and used by them to designate the mediterranean type scrub of California) (Figure 15–7). In corresponding regions elsewhere are the *matorral* of Chile, the *fynbos* of the Cape of Good Hope, the *phrygana* of Greece, the *garrigue* or *garigue* of dry, calcareous soils of France, known as *macchia* or *macquis* on siliceous soils, and the *tomillares* of Spain (Biswell 1974). In each area the physiognomy of the vegetation is very similar, as are the adaptations to fire. For convenience, we will concentrate on the chaparral of California, and then consider some higher elevation coniferous forests in the same climatic regime.

Figure 15-7. (a) Typical chamise chapparal community. (b) The shrubs are dense, compact, and finely branched.

(a)

(b)

Mutch (1970) hypothesized that communities maintained by fire possess vegetation that is more flammable than communities not adapted for fire. The woody plants of the chaparral exhibit such flammability, and are characterized as follows: reduced leaves with sclerophyllous adaptations; often leaves that are vertically oriented; intricate branching pattern, resulting in a large surface-to-volume ratio; periodic die-back, resulting in highly flammable twigs; and possession of volatile oils (see Philpot 1977 for more details). These flammable properties contribute to the maintenance of fire-dependent communities. Those species that possess special reproductive mechanisms that enable them to survive fire, or that possess anatomical mechanisms such as thick bark and epicormic sprouting (sprouting from the stem), may also have flammable characteristics that enhance fire spread.

In the Mediterranean region, centuries of manipulation—deliberate burning, overgrazing—have modified the natural situation greatly. Certainly the chaparral of California has also had the influence of human manipulation. We have already noted the California Indians' practice of burning browsing areas in early summer to prevent destructive late summer wildfires. There were drastic changes made by miners, sheepmen, and loggers in some areas of the state. The differences between the Mediterranean region and California are probably of degree, and not of kind. Both of these areas also have a long history of volcanism, which would provide fires, but not on a periodic basis.

The Fire Climax Community

The concept of a fire climax community does not imply a single, fixed community, nor even a certain flora varying in dominants. The key to which fire climax community will be maintained is fire frequency. For example, if fires occur every 5–10 years, shrubs will be selected against, and grassland will be favored (Humphrey 1974; Vogl 1974). In general, fire is more destructive of woody plants than of grasses. In the chaparral, most species begin seed production within 3–5 years after a fire, and brushlands rarely burn more frequently than that (Biswell 1974).

Fire Adapted Plants

Exactly what is meant by the phrase, "fire adapted plants"? Gill (1977) listed the adaptive traits for mediterranean regions as follows: (a) fire-induced flowering; (b) bud protection and sprouting subsequent to fire; (c) in-soil seed storage and fire stimulated germination; and (d) on-the-plant seed storage and fire stimulated dispersal. Other authors (Lotan 1974; Harvey et al. 1977) have commented on the adaptiveness of such traits as thick bark, evanescent branches, rapid growth, and early maturity.

Fire-induced flowering

The phenomenon of increased flowering, and concomitantly enhanced seed-set, would appear to be advantageous in two ways: (1) those plants that do produce increased numbers of seeds would be at a competitive advantage over those plants not so adapted; and (2) the seedbed, which receives more light due to canopy removal and has a deposit of soluble mineral ash nutrients, provides an additional opportunity for enhanced growth and reproduction, giving further advantage to those plants that can quickly exploit this resource.

Several of the bulb-forming perennials, which may have been present but not numerous in the chaparral between burns, exhibit enhanced flowering following a fire. These include death camas *(Zygadenus fremontii)*, soap plant *(Chlorogalum pomeridianum)*, purple-head brodiaea *(Dichelostemma pulchella)*, and several species of mariposa lily *(Calochortus spp.)* (Muller et al. 1968). In other mediterranean regions geophytes also exhibit this trait. Bermuda grass *(Cynodon dactylon)* and fireweed *(Epilobium angustifolium)* are known to exhibit enhanced flowering in response to fire.

The seeds of grasses do not require scarification for germination, and the annual grasses will usually not predominate in the first postburn year, but only appear around rocks and in other protected regions. However, due to enhanced seed set, by the third year grasses will be very abundant, even dominant (Biswell 1974, Table 15–5).

Bud protection and subsequent resprouting

Buds near the base of the shrub may survive a fire and be released from dormancy by the death of shoots above them. Sprouting results as a hormonal response or with the onset of winter rains (Figure 15–8). The sprouting may be from the stem (epicormic) or may be from a basal burl or **lignotuber**. Protection for the stem bud is provided by the bark. Buds of lignotubers are protected by the soil.

In the genus *Arctostaphylos*, there are several examples of species that possess basal burls. The ubiquitous chamise *(Adenostoma fascicu-*

Figure 15–8. Stump sprouting. Latent axillary buds are released from inhibition when fire removes the apical meristem.

latum) also possesses a well-developed lignotuber. There are many other species which sprout following a fire, as detailed by Biswell (1974).

In-soil seed storage and fire stimulated germination

There are some chaparral species whose seeds remain dormant in the soil for many years if there is no fire. For example, it is believed that the seeds of snowbrush (*Ceanothus velutinus*) might remain viable in forest litter for up to 575 years (Zavitkovski and Newton 1968). Various authors discuss the use of heat treatment, or some other form of seed cracking, to induce germination of resistant seeds (Biswell 1974; Sweeney 1956 and 1967; Quick 1959; Sampson 1944). Table 15–3 shows the response of seeds of deerbrush and mountain whitethorn (*Ceanothus integerrimus, C. cordulatus*) to treatment with hot water. Chamise (*Adenostoma fasciculatum*), a sprouter which depends on fires every 15–20 years for optimum health and stability, also responds to fire with phenomenal germination of stored seeds: after a burn, as many as 3000 chamise seedlings m^{-2} may be counted. Seeds of both chamise and mountain whitethorn withstand dry oven temperatures of 138–149°C, the higher temperatures providing the trauma required to rupture the seed coat (Biswell 1974).

At Redwood Mountain, in Sequoia National Park, prescribed burning on three plots yielded the data shown in Table 15–4. The plot designated as Burn #1 was the lightest fire and Burn #3 was the hottest. The greater numbers of shrub seedlings are in areas of lighter burns, which provide heat for cracking but do not kill the seeds. Hartesveldt and Harvey (1967) suggest several hypotheses for higher rates of sequoia seedling survival in areas of very hot burns, including the destruction of pathogens by the fire and the removal of competi-

Table 15–3. Germination of heat-treated seeds of deerbrush and mountain whitethorn. (From Biswell 1974. The effects of fire on chaparral. In *Fire and Ecosystems*, ed. T. T. Kozlowski and C. E. Ahlgren, ch. 10, by permission of Academic Press.)

Taxon	Treatment length	Germination
Ceanothus integerrimus	Boiled 20 min	12%
Ceanothus cordulatus	Boiled 25 min	25%
Ceanothus cordulatus	Boiled 30 min	0%

Table 15–4. Sequoia and deerbrush seedling response to 1969 burning at Redwood Mountain in Sequoia National Park. (Recalculated from Kilgore and Biswell 1971. Reprinted by permission of *California Agriculture Magazine*, University of California.)

Plot No.	Size (hectares)	Mature sequoia[a]		Sequoia seedlings		Deerbrush seedlings	
		no. per plot	no. per hectare	no. per transect	no. per hectare	no. per transect	no. per hectare
Burn #1	1.52	11	7.2	138	18,567	120	16,145
Burn #2	2.47	28	11.3	337	45,343	53	7,131
Burn #3	2.53	58	22.9	737	99,163	4	539
Totals	6.52	97		1,212		177	
Means			13.8		54,357		7,938
Control	2.14	31	14.5	0	0	0	0

[a]Trees more than 183 cm dbh.

tion. A control plot did not show a single shrub or sequoia seedling (Kilgore and Biswell 1971).

Plants with very resistant seeds will often appear in great numbers following a fire (Sweeney 1956; Armstrong 1977). Examples of these fire followers are shown in Table 15–5. Notice the decline of many of the herbs in the post-burn years. Many of these same plants will germinate from seed, if a light fire burns over a seedbed, mimicking a fire in nature. Otherwise, they lie dormant until some scarifying agent provides the vehicle for resumption of activity.

Thus, chaparral shrubs exhibit two responses to fire: (1) seeders die and come back as seedlings, and (2) sprouters remain alive and come back as sprouts from burls or root crowns (Figure 15–9). Long fire-free periods favor seeders. The resistant seeds, produced in greater or lesser amounts each year, are then stored in the soil. They remain dormant in the soil until scarification (fire is the most efficient method) induces germination in a nearly optimal environment. During the periods between fires, seedlings grow to maturity within 5–8 years (Biswell 1974), produce abundant seed, and the stage is set for another cycle.

This division is not a strict dichotomy, for some sprouters are also seeders, though some species are seeders only. There are also plants that sometimes sprout, and sometimes do not, including bitterbrush (*Purshia tridentata*) and mesquite (*Prosopis velutina*). There are also degrees of sprouting response, governed in part by the time or season of burning (Humphrey 1974). Sprouters have distinct advantages. Those

Table 15-5. Density of certain herbs which appeared 1–4 years after a burn of chaparral in the Highland Springs area of California. Density is given as number of plants per 28 m^2. (From J. R. Sweeney, 1956, University of California Publications in Botany 28:143–206. By permission of the University of California Press, Berkeley, CA.)

Category	Species	Year 1	Year 2	Year 3	Year 4
Species which peak the first year	*Mimulus bolanderi*	23	3	7	3
	Oenothera micrantha	19	3	4	2
	Antirrhinum vexillocalyculatum	15	4	6	0
	Emmenanthe penduliflora	280	0	3	3
	Silene antirrhina	25	5	12	10
	Malacothrix floccifera	47	8	14	23
	Mentzelia dispersa	31	11	19	14
	Mimulus rattanii	16	0	0	0
Species which peak the second or third year	*Barbarea americana*	9	21	34	17
	Cryptantha torreyana	19	62	844	127
	Gilia capitata	14	7000	0	3
	Lotus humistratus	132	738	33	17
Species which peak the fourth year or later	*Bromus mollis*	3	43	217	498
	Bromus rubens	19	23	147	1499
	Festuca megalura	20	24	417	737
Species which remain constant	*Chlorogalum pomeridianum*	4	5	3	3
	Penstemon heterophyllus	1	1	1	1

with lignotubers quickly resume activity following a burn. Even before seasonal rains, chamise *(Adenostoma fasciculatum)* and manzanita *(Arctostaphylos* spp.) will send up sprouts several cm tall by drawing from stored nutrients and water in the basal burl. During the first post-burn season, greenleaf manzanita *(A. patula)* will send up luxuriant growth, exploiting mineral and moisture resources via an already established root system. This chaparral plant with both resistant seeds and a basal burl seems to have the best of both worlds.

Sweeney (1977) pointed out the absolute dependence of some of the chaparral seeders on fire and suggested that, although the seeds are long-lived, there is some finite life span. When that time is exceeded because of fire suppression for decades, that population will become locally or even regionally extinct. There is a complicating factor, when burns are prescribed in habitats where fire followers are suspected. These burns are usually done in the cool, damp weather of the late fall, when the risk of fire escape is minimal. But, the rate at which the seeds succumb to the heat of the fire is directly related to moisture content: as moisture goes up, viability and resistance to heat goes down, and, ironically, some of the populations of fire followers may be lost to fire.

On-the-plant seed storage and fire stimulated dispersal

There are coniferous plants that retain viable seeds for many years, or even decades. In the condition called **serotiny** (literally, late to open), the cones of some conifers remain on the tree and remain closed until mechanical breakage severs the vascular connection with the parent plant, or remain closed by resin until the heat of a fire opens them. Serotiny is characteristic of many different plants such as jack pine *(Pinus banksiana)*, some subspecies of lodgepole pine *(P. contorta)*, bishop pine *(P. muricata)*, knobcone pine *(P. attenuata)*, pitch pine *(P.*

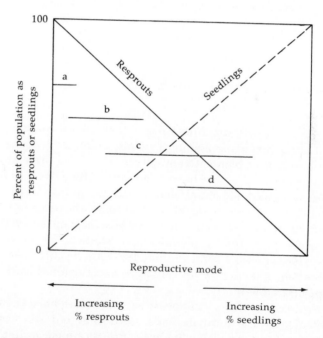

Figure 15–9. Reproductive options available to chaparral shrubs. Plants which resprout after a fire but establish very few seedlings would occupy region (a), farthest to the left. Plants that are mostly seeders are indicated by region (d). The majority of manzanita *(Arctostaphylos)* and *Ceanothus* species do not occupy a region along the abscissa, but rather a single point: they are obligate seeders. (From Keeley, J. E. 1977. Fire-dependent strategies in *Arctostaphylos* and *Ceanothus*. In *Proc. Symp. Env. Cons. Fire and Fuel Mangmt. in Medit. Ecosyst.*, pp. 391–396.)

rigida), Monterey cypress (Cupressus macrocarpa), and giant sequoia (Sequoiadendron giganteum). Some subspecies of lodgepole pine require temperatures of 45–50°C (or even higher) to break the resinous bond between the cone scales. Summer soil temperatures may suffice to open low-hanging cones. Knobcone pine, however, requires 200°C for opening the cones. By contrast, giant sequoia cones remain closed due to cone scale turgidity, and open in response to the severity of vascular connections (Lotan 1974; Vogl 1973; Harvey et al. 1977).

Serotiny is rarely exhibited by every tree in a stand. However, in regions frequently disrupted by fire a high percentage of serotinous-coned trees would be expected. In an even-aged stand of lodgepole pine of recent fire origin (which was contiguous with an uneven-aged, undisturbed stand), 58% of the trees bore serotinous cones. The uneven-aged stand showed only 38% serotiny. Frequent fire must play a very dramatic role in selection. The genetics of serotiny is not well understood, but we can infer that a stand of trees established from fire-opened, serotinous cones would show a high degree of serotiny (Lotan 1974).

The Post-burn Chaparral Community

In the first few years after a fire in a chaparral community, there will be a noteworthy succession of plants, dominated in the first few years by herbaceous plants. Often, many or most of these are annuals not seen except in the post-burn community: they are fire followers. However, it should be noted that seedlings and sprouts of woody plants will be present, even the first season following the fire. This situation has given rise to the term **autosuccession** (or cyclic succession; see Chapter 10), a phenomenon studied by many authors (Sweeney 1967; Biswell 1974; Philpot 1977; Keeley 1977). By the sixth year following fire, the shrub canopy is once again closed, and few herbs appear below it.

Why do the annuals disappear? Several causes have been hypothesized. Many of the annuals have seeds that require scarification, which is provided most effectively by fire, as noted earlier. In addition, shrubs, present from the early post-burn time, overtop the low-growing herbaceous plants, reducing the light intensity at the soil level. Photosynthesis is reduced in the lowered light regime, and the annuals do not survive. Also, it is thought that many of the shrubs are producers of allelopathic chemicals that inhibit the growth of other plants (see Chapter 5).

GIANT SEQUOIA FORESTS

The forests of the Sierra Nevada, especially those that are giant sequoia forest, were characterized primarily by surface fire prior to the time of fire suppression. The fire ecology of giant sequoia *(Sequoiadendron giganteum)* has been studied extensively (Biswell 1977; Hartesveldt 1964; Kilgore 1972; Agee 1973; Hartesveldt and Harvey 1967; Harvey et al. 1977). These studies indicate that fire is essential to the continued vigor of the sequoia-mixed conifer forest, and that it plays these primary roles: (a) reduction of fuel, decreasing wildfire hazard; (b) recycling of nutrients; (c) destruction of pathogens such as damping-off fungi; (d) maintenance and development of a mosaic of vegetation age classes and types, which in turn provides (e) enhancement for wildlife; (f) modification of insect life in the forest; (g) preparation of a suitable seedbed on mineral soil in a favorable light regime; (h) maintenance of a subclimax community; and (i) opening of serotinous cones. Although fire suppression was incorporated into National Park policy as early as 1886, there has been a re-introduction of fire as let-burn and prescription burns in many National Parks, including Sequoia-Kings Canyon National Park.

The giant sequoia groves are not pure stands of the big trees, but associations of conifers that include white fir *(Abies concolor)*, incense cedar *(Calocedrus decurrens)*, and sugar pine *(Pinus lambertiana)*. These occur at mid-elevation on the western slope of the Sierra Nevada of California.

The giant sequoia is a fire adapted species and possesses characteristics that enhance its ability to survive fires. The mature trees are well adapted to resist surface fire damage. The thick, fibrous bark protects the living tree. This protection is enhanced by the possession of **evanescent** lower branches: as they are shaded by the forest canopy, they drop off, and do not remain as a "fire ladder" which would otherwise give fires access to the tree's crown. Rapid growth is another fire adapted feature of the big trees. The young saplings grow quickly up into the light, outgrowing the competition, and raising the canopy above the level of surface fires.

Many, if not most, of the mature trees bear fire scars, and many have scars of many fires. It has been shown that between 1760 and 1900 fires in the Mariposa Grove of Yosemite National Park averaged one every seven to eight years. In fact, up to 85 to 95% of the tree can be burned without resulting in the tree's death (Hartesveldt 1964). A giant sequoia stands near the entrance station of Sequoia-Kings Canyon National Park, showing evidence of heavy burning, but no indication of loss of vigor (Figure 15–10).

Figure 15-10. Giant sequoia at the entrance to Sequoia-Kings Canyon National Park. The fire scar covers more than one-third of the base of the tree and extends nearly 30 m up the trunk, and yet the tree is vigorous and sturdy.

Cone production in giant sequoias is phenomenal. At any one time, a mature tree might have 11,000 cones, 7000 of which would be green, photosynthetic, and closed (serotinous) (Figure 15–11). However, the remaining 4000 cones would be open and without seeds. Cone production begins at a surprisingly early age (around 8 years), and the cones mature within 2 years. The seeds remain viable for as long as 22 years. However, the seeds are tiny (Figure 15–12): it would take over 41,000 to weigh a kilogram. Food reserves are insufficient to allow for germination atop a thick layer of litter. Notice the comparison of white fir seeds and giant sequoia seeds (Figure 15–12). The greater amount of stored food in the former should allow for greater root elongation, through forest litter, to reach mineral soil.

Though the reproduction of giant sequoia is fire-dependent, there is an almost continual, if slight, shedding of seeds, due to the activities of various animals. These seeds will germinate, given a suitably moist medium. And, there are seedbeds available from time to time: root pits of fallen trees; sandbars in streams; snow avalanche paths; and areas cleared by human activity. Thus, at least in theory, there would always be several age classes of giant sequoia. However, in the upper Mariposa Grove, Hartesveldt (1964) counted no more than thirty sequoias that survived past the seedling stage between 1934 and 1964.

Seedling mortality is high in the first year, often more than 86% (Hartesveldt 1964; Harvey et al. 1977). Desiccation is the primary cause of death. Heat canker, damping-off fungi, insect damage, and falling debris can also cause the seedling to die. The combination of environmental disturbance and enrichment that seems to provide optimum conditions for seedling survival is derived from periodic fire in the forest: penetration of sunlight to mineral soil; reduction of competition; reduction of pathogens; and the opening of serotinous cones, allowing as many as 200 seeds per cone to fall to the receptive soil below.

The large size and the longevity of the trees enable them to dominate a forest and to linger on as relict individuals long after the climax forest (shade-tolerant white fir-incense cedar) has begun to dominate the understory. These climax species are shorter and denser than giant sequoia, and they may serve as fire ladders to carry a fire to the crown of a giant sequoia. Fire suppression favors the climax species. Frequent fires remove them, both from competition and as a fire hazard, and favor the maintenance of giant sequoia.

Figure 15-11. A closed cone of the giant sequoia. This cone was cut by a chickaree, in Giant Forest, Sequoia National Park. The fleshy scales are eaten by the rodent, and seeds may be dispersed in the process.

Figure 15-12. Comparison of a white fir seed (left) and a giant sequoia seed (right). White fir is shade tolerant, and the greater amount of stored nutrients in the seed allows the root of the seedling to grow through the litter of the forest floor to reach mineral soil; at least the root can grow farther than would be allowed by the much smaller amount of stored nutrients in the giant sequoia seed.

SUMMARY

Fire has undoubtedly occurred at frequent enough intervals, in most communities, to be a major force in the development of plant and animal communities. Forest, grasslands, chaparral, and other communities of the mediterranean regions show the influence of fire. Of all the causes of fire, only lightning and human activity have the capacity for periodicity, and therefore exert a consistent effect upon the evolution of organisms.

Humans have used fire to manipulate the environment to their own advantage since before history began. Native peoples in various lands have used fire to enhance hunting, improve visibility, provide forage, etc.

Fires may be divided into three classes: surface fire, which is rather low in intensity and is not destructive of overstory vegetation; crown fire, which is a spectacular blaze that consumes whole trees, opening the canopy completely; and ground fire, which is termed a retrogressive agent because of the catastrophic effect it has on a community.

The effects of fire on the soil include some loss of soil nutrients due to volatilization, changes in water-holding capacity by reducing organic content, and impact on soil microbiota. Fire behavior is determined by character of fuel, weather patterns, humidity, and temperature. Fire destroys some organic matter, vaporizes some, and ashes the rest. In addition, soil structure is changed by fire. Soil pH will usually be higher following a burn. Soil insulates against the heat of a fire, such that even a few centimeters down, the temperature does not exceed 100°C under a light burn.

Water runoff is increased by burning, as is sediment yield. Soil wettability is increased by hot fires. Fungal diseases may be controlled by burning. Bacterial populations are often favored over most fungi. For example, nitrogen losses due to volatilization are counterbalanced by the enhancement of the regime for root-nodule bacteria in certain plants.

Lightning fires have been a common occurrence for millenia in the southeastern United States, and some coniferous forests there depend on fire for maintenance. The longleaf pine is most highly fire resistant, and fire dependent for its maintenance. Fire is used in the Southeast in many pine forests and pine savannas as a silvicultural tool. In general, the effects of fire are to reduce competition, to prepare a mineral seedbed, to reduce fungal disease, and to open the forest canopy.

The grasslands are maintained by fire. The adaptations of grassland plants, most of which are hemicryptophytes, allow them to survive, and even thrive in the presence of periodic fire. Fire recycles nutrients and reduces shrubs to the juvenile stage. The fire environment, including lightning and heat, provides a very powerful selective force in the grassland. Mechanisms for survival in grasslands include increased reproduction from seeds, as well as resprouting from perennating buds protected at the soil's surface.

The mediterranean regions support vegetation that is inherently flammable, and usually well-adapted to fire. The adaptations include: fire-induced flowering; bud protection and sprouting subsequent to fire; in-soil seed storage and fire stimulated germination; and on-the-plant seed storage and fire stimulated dispersal.

The giant sequoia of the Sierra Nevada provides an excellent example of adaptation to fire. With its serotinous cones, small seeds, fire-resistant bark, and long life, this impressive tree actually requires periodic fire, or it will be succeeded ecologically by a white fir-incense cedar forest.

CHAPTER 16

SOIL

The soil is the common ground between the living and the nonliving world. We could have logically begun this book with a foundational chapter on soils. The importance of the medium for growth cannot be overstated. However, plant ecology is much like any other science: we could begin almost anywhere, and work through a cyclical progression of information with effectiveness. We can now use the knowledge gained from other chapters to support our discussion of soils as they influence plant distribution. We will consider soil development, profiles, texture, and structure. Following a brief discussion on soil physics and chemistry, we will consider the taxonomy of soils.

There are four separate components of soil: mineral grains, organic matter, water, and air. The mineral substrate provides anchorage, pore space for storage of water and air, and nutrients on an exchange basis. Organic matter refers to the plant and animal residues in various stages of decomposition, as well as to the cells, tissues, and exudates of soil organisms. Organic matter enhances the intrasystem cycling of nutrients and improves soil structure, pore space, and water storage, among other things. Soil water is the solvent medium for nutrients needed by growing plants, maintaining equilibria among the cations and anions adsorbed* on soil particles, the living plants, and the soil water (soil solution) itself. Soil air contains oxygen for cellular work, carbon dioxide that augments further mineral weathering and biotropic processes, and atmospheric nitrogen for nitrogen-fixing soil organisms.

Soil, along with climate, is a primary agent in plant selection through evolutionary change. The term soil comes from the Latin word for floor, *solum*. A soil scientist, or **pedologist**, views soil as "a natural product formed from weathered rock by the action of climate and living organisms" (Thompson and Troeh 1973). The foregoing statement implies that soil is mainly composed of mineral materials, and for the most part, it is. However, pedologists do not overlook the fact that some soils are formed mainly from the weathering of accumulating

---

*Adsorbed refers to a material being attracted and held to, whereas absorbed refers to a material being taken in and probably held more tightly than material adsorbed.

organic debris (organic soils). We will use an edaphological definition of soil, considering it primarily as a medium for growing plants: soil is a mixture of mineral and organic materials, capable of supporting plant life. Implicit in both views is the interaction of soil and life: each one exists because of, and is formed at least in part by, the other.

The materials in the upper fraction of the earth's crust are exposed to the weathering agents more intensely than are the underlying strata. Three levels can be seen in the section of an exposed hillside diagrammed in Figure 16–1, which has loose material resting on underlying rocks. The foundation is termed **bedrock**. Above that is the **regolith**, the unconsolidated mantle of weathered rock and soil material, of which the upper 1–2 meters or less of highly weathered, biochemically altered material is termed **soil**.

THE SOIL CYCLE
AND SOIL DEVELOPMENT

Although the process of soil development may seem to be linear and formational in terms of human time, in geologic time the process is

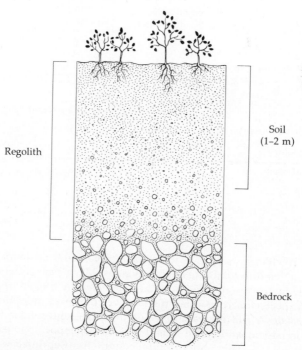

Regolith

Soil
(1–2 m)

Bedrock

Figure 16–1. A vertical section of soil, showing relative positions of bedrock, regolith (unconsolidated material), and soil (highly weathered portion of regolith).

part of a cycle. The processes that take place in the cycle of soil development include sedimentation or deposition of fragments from clay size up, lithification (stone formation), metamorphism and melting, crystallization, volcanism, and the agents of erosion and transport. The process of forming soil may be conveniently divided into two phases. The first is rock weathering and the second is soil formation. Whether the substrate is a granitic dome, a lava deposit, or a limestone cliff, there are certain weathering phenomena that are common to all rock substrates, and we will examine a few of them. Weathering pathways are shown in Figure 16–2. Notice that a variety of materials are produced, but that common pathways yield the same minerals, whether derived from sedimentary, igneous, or metamorphic rocks.

Mechanical Factors

Mechanical erosion, or disintegration, may be attributed to the expansion and contraction of rocks in response to thermal flux. The outer layers are heated differentially compared to the inner rock, creat-

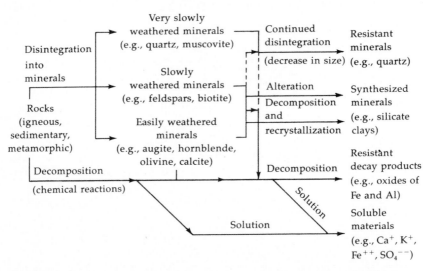

Figure 16–2. Weathering pathways which take place under moderately acid conditions common in humid temperate regions. Major paths of weathering are indicated by the heavier arrows, minor pathways by broken lines. As one would expect, climate modifies the exact relationships. In arid regions physical breakdown (disintegration) would dominate, and soluble ions would not be lost in large quantities. In humid regions decomposition becomes more important, especially under tropical conditions. (Reprinted with permission of Macmillan Publishing Co., Inc. from *The Nature and Properties of Soils*, 8th Ed. by Nyle C. Brady. Copyright 1974 by Macmillan Publishing Co., Inc.)

Figure 16–3. Cracks in granite caused by the force exerted by the freezing of water which seeped into minute, hairline fractures.

ing stress that is relieved by cracking and spreading. The fissures and crevices thus formed, along with those caused by tectonic activity and gravitational tensions, will be filled with water. Expansion caused by the freezing of this water is impressive: normal hydrostatic pressure amounts to more than 10,000 kg cm^{-2}. Ice floats on liquid water, and the water at the air/liquid interface is cooled more rapidly than that within the rock, so that the ice acts as a wedge, forcing still wider cracks into the rock (Figure 16–3). The stresses generated by the freezing of water are also significant after initial fragmentation of rocks. The regolith, including the soil at the top, continues to be altered physically by this force.

Exfoliation, the peeling off of layers of rock, is attributed to the differential expansion just mentioned and also to the expansion allowed when surface material is removed by erosion. Igneous intrusions which crystallize far beneath the surface are, in a sense, "spring-loaded." The removal of the pressure of overlying material allows exfoliation, and so layers of rock are released much as the modified leaves peel away from an onion bulb. Half Dome, in Yosemite National Park in California, is an example of exfoliation.

Erosion by wind and water, and deposition by these same forces, are significant to rock weathering. The force exerted by plant roots must also be counted as a very important physical force, even though plants exert biological and chemical influences as well, and also influence subsequent soil formation. Moving ice, or glaciers, transport tremendous amounts of material. In addition, the abrasive materials carried by the ice polish and further reduce the rocks and rock fragments. Alluvial (stream-carried), glacial, aeolian (wind-carried), and lacustrine (lake deposited) materials form the basis for parent materials of many kinds of soils. Streams carry an enormous amount of material into lakes or into rivers which ultimately empty into the sea. The Mississippi River is estimated to discharge in excess of 660,000 kg of soil into the Gulf of Mexico each year. The size of alluvium may range up to house-size boulders (Figure 16–4), depending on the volume and velocity of the water. By contrast, aeolian materials are fine, usually dust-like particles which abrade, polish, and further erode rocks in their passing.

Chemical Factors

Chemical processes of weathering (decomposition) are active almost as soon as disintegration has begun. Included under chemical processes are a number of rather simple reactions. Oxidation, especially of iron-carrying rocks, is a very obvious form of chemical weathering. Carbonation, produced by the action of a weak concentration of hydrogen ions in percolating waters, and attack by other acids including organic acids, causes decomposition. Some of the latter may, however, initiate processes leading to recrystallization of clays, and thus are synthetic in effect. The most significant chemical weathering is exerted by water, in one of three ways: solution, the solvent action of water upon the mineral substrate; hydrolysis, the cleavage of a mineral by water; and hydration, the rigid attachment of H^+ and OH^- ions to the material being decomposed. The products of some of these processes, such

Figure 16–4. Alluvium forming a fan originating at the mouth of a desert canyon. The sparseness of the vegetation allows a clear view of the sizes of material moved; some of the "particles" are huge boulders. (Courtesy of Dr. John S. Shelton.)

as oxidation and hydration, usually have a larger volume than the initial minerals, and, if not displaced, contribute to the mechanical forces involved in exfoliation. These processes are all influenced by climate, and are very effective in tropical regions.

Soil Formation

Soil formation *(S)* is characterized by Jenny (1941) as dependent on a set of independent variables as follows: climate *(c)*, organisms *(o)*, topography *(r)*, rock type or parent material *(p)*, and time *(t)*. Jenny's classic equation is as follows:

$$S = f^{(cl, o, r, p, t \ldots)}$$

He defines organisms as the biotic potential of a site, not just the existing organisms which are indeed interdependent or dependent factors. The biotic potential includes consideration of soil organisms, litter, vegetation type, etc. The soil macrobiota and microbiota interact with the growth medium, altering its capacity to support life. The amount of litter deposited on the soil is determined by the productivity of the community, and the rate of decay or removal of that litter is determined by climate, by the nature of the litter, and by fire frequency. Also, the incorporation of litter-derived materials into the soil, the maintenance of soil porosity and permeability, and ultimately the soil fertility, all rely at least in part on the activities of soil organisms. Vegetation and climate determine what kind of soil is formed. Podzol soils develop beneath coniferous forests in cool, moist climates; podzol soils have a prominent ashy gray upper horizon. Chernozem ("black earth") soils develop under grasslands in temperate regions. As previously noted, a warm, moist environment hastens both disintegration and decomposition. The concepts relating to plant community succession developed in Chapter 10 also apply to soil succession.

So, parent materials interacting with soil biota and climate produce a certain soil, with its characteristic features. Another very important factor in soil formation is time: time for large mineral grains to be disintegrated, decomposed, and changed to secondary minerals; time for finer and finer mineral grains to accumulate; and time for organics and humus to become a significant fraction of the total soil. But, also, time for finer grains and soluble anions and cations to be leached away (Chapter 12).

The composition of parent material largely determines the chemical make-up of its derived soils. For instance, acid soils derive from (a) aluminum-rich bauxites, (b) silica-rich substrates such as sands, slates, and diatomaceous earth, (c) mine tailings of lead and zinc deposits, (d) sulfide-rich, hydrothermally altered volcanic rocks and ash, and (e) estuarine deposits. Soils with poor fertility may develop from ultrabasic (ultramafic) rock types rich in silicates of iron and magnesium. For example, serpentine develops from igneous rocks such as peridotite and dunite, which contain silicates of iron and magnesium. Serpentine soils are high in magnesium and low in calcium (Table 16–1). They may also be high in nickel and chromium and deficient in nitrogen and phosphorus; thus, they are not very fertile.

Soil may form from transported material or *in situ* parent material (residual soil). Topography influences the stability of a soil body in a variety of ways, including its effect on rainfall patterns and the steepness of slopes. Resistant parent materials may result in steep slopes; parent rocks or materials that erode easily produce wide valleys, rounded hills, and often deep soils. Steep slopes may continually expose new surfaces to weathering and soil formation, as alluvium or talus (material moved by gravity) is moved downslope; resident soils, if present at all, are young and shallow. Gentle slopes restrict transport, and often enhance development of deep, strongly developed soils.

SOIL PROFILES

Soil Profile Development

Soils, especially deep soils of temperate latitudes, develop very characteristic layers or horizons, more or less differentiated and discernible from each other if a pit or trench is dug into the soil. A section encompassing all the horizons of the soil and extending into its parent material is called a **soil profile** (Figure 16–5). The rock weathering that we discussed earlier is mainly a physical process. Soil formation, by

Table 16–1. Comparison of certain soil nutrients in serpentine and nonserpentine soils. (In part from Walker 1954. Copyright 1954 by the Ecological Society of America.)

Nutrient	Nonserpentine soil	Serpentine soil
Replaceable Ca (meq 100 g^{-1})	1.56	0.21
Replaceable Mg (meq 100 g^{-1})	0.71	1.89
Typical Ca/Mg ratio	3.0	0.3
Percent N (total, all forms)	0.14	0.01
Percent P (total, all forms)	0.13	0.002
K (meq 100 g^{-1})	0.83	0.22
Ni (meq 100 g^{-1})	0.005	0.040
Cr (meq 100 g^{-1})	0.0001	0.011

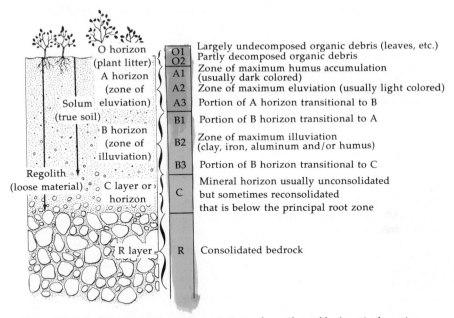

Figure 16-5. A diagrammatic representation of a soil profile (vertical section of the soil) showing soil horizon nomenclature and symbols for mineral soils. Additional subdivisions of the O horizon, not shown, are used in organic soil. A and B horizons taken together are referred to as the **solum** or true soil. (Reprinted by permission of McGraw-Hill Book Co. from *Soils and Soil Fertility* by Thompson and Troeh, 1973.)

contrast, is largely biochemical weathering and the gradual mixing of organic material with inorganic material.

As the upper layers of soil are exposed to the roots of plants, to animal activities, to the activities of both soil microflora and microfauna, and to accelerated chemical weathering, mineral grains are altered. Primary or original minerals are transformed into **secondary minerals**. Some primary and secondary minerals are shown in Table 16-2. Recall from Figure 16-2 that quartz is a very resistant primary mineral. Primary minerals are found in abundance in igneous and metamorphic rocks and secondary minerals are often found in sedimentary rocks, as we would expect.

The presence of secondary minerals and abundant organics, and changes due to interactions between soil and soil biota result in differentiation of soil horizons. Two of the processes are significant enough that they provide names for the principle (master) horizons of mineral soils in which they occur. The A horizon includes the surface soil in which surface or near-surface organic matter accumulates. It is also referred to as the **zone of eluviation**, due to the movement of materials downward from the A horizon, mostly by leaching. The B horizon

Table 16-2. The more important original and secondary minerals found in soils. The original minerals are also found abundantly in igneous and metamorphic rocks. Secondary minerals are commonly found in sedimentary rocks. (Reprinted with permission of Macmillan Publishing Co., Inc. from *The Nature and Properties of Soils,* by Nyle C. Brady. Copyright 1974 by Macmillan Publishing Co., Inc.)

Name	Formula
Original minerals:	
Quartz	SiO_2
Microcline	
Orthoclase	$KAlSi_3O_8$
Na plagioclase	$NaAlSi_3O_8$
Ca plagioclase	$CaAl_2Si_2O_8$
Muscovite	$KAl_3Si_3O_{10}(OH)_2$
Biotite	$KAl(Mg, Fe)_3Si_3O_{10}(OH)_2$
Hornblende[a]	$Ca_2Al_2Mg_2Fe_3, Si_6O_{22}(OH)_2$
Augite[a]	$Ca_2(Al, Fe)_4(Mg, Fe)_4Si_6O_{24}$
Secondary minerals:	
Calcite	$CaCO_3$
Dolomite	$CaMg(CO_3)_2$
Gypsum	$CaSO_4 \cdot 2H_2O$
Apatite	$Ca_5(PO_4)_3 \cdot (Cl, F)$
Limonite	$Fe_2O_3 \cdot 3H_2O$
Hematite	Fe_2O_3
Gibbsite	$Al_2O_3 \cdot 3H_2O$
Clay minerals	Al silicates

[a]These are approximate formulae only since these minerals are so variable in their composition.

includes the subsurface part of the soil, or subsoil, which may be a **zone of illuviation**, where the materials lost from A are accumulated. The illuvial materials may include iron and aluminum compounds, clays, and humus. Humus is the dark brown or blackish stable fraction left over after most of the organic residues have been decomposed.* Alternatively, the B horizon may be a **zone of alteration** where little material accumulates from A but significant changes in structure or secondary mineral formation take place.

Soil horizon nomenclature is shown in Figure 16-5. The O (organic) horizon and its subdivisions may be largely or totally lacking in certain desert soils, and may be the only horizons existing in organic soils. The R layer underlies many, but not all, upland or residual soils,

---

*The formation of humus is a complex topic that we will discuss in a later section.

where it may displace the C horizon. A young soil may have A and C horizons, but no B horizon. This lack of a B horizon results when there has been no translocation or eluviation of clay from the A horizon downward; such a profile is characteristic of the chernozem or "black earth" profiles in steppe regions.

Soils can be identified in the field based on the differentiation of the horizons, the texture, the color of the principal materials, and even which horizons (or portions) are lacking. A field sheet for recording soil characteristics is shown in Figure 16–6, and Figure 16–7 is a photograph of the soil described in the field sheet. Note the site characteristics that are recorded: elevation, slope, aspect, erosion, and the like. This soil is an Inceptisol, identified in part by its lack of a textural B horizon. There is no C horizon, and the lower B horizon rests directly on the R horizon, or bedrock.

Zonal, Azonal, and Intrazonal Soils

Although we will discuss the classification of soils in detail in a later section, it is appropriate to mention here the three orders in the 1938 system of soil classification (Baldwin et al. 1938). These include: zonal soils, whose characteristics and profiles are mostly influenced by climate and vegetation (e.g., desert soils, lateritic (red tropical) soils); intrazonal soils, which exhibit unique soil profiles due to unusual parent material or topography (e.g., saline soils, bog); and azonal soils, which have little or no profile development (e.g., recent alluvial soils, rapidly eroding soils). The appearance of a soil profile largely determines the fit of a soil into a soil classification scheme.

PHYSICAL PROPERTIES OF SOIL

Soil Texture

Soil texture is defined by the percent by weight of sand, silt, and clay. That is, only mineral components are considered in determining soil texture, even though soil is also composed of organic matter, water, and air, with organic materials contributing up to 5% or more of the total. The United States Department of Agriculture classification of soil particles places the upper limit for silt at 0.05 mm in diameter and the lower limit at 0.002 mm. Particles greater than 0.05 mm in diameter are considered sand, and particles less than 0.002 mm in diameter are considered clay (Brady 1974). Although almost any soil contains pebbles and cobbles that are larger than the coarsest sand (2 mm diameter), such coarse fragments play but a minor role in soil function, unless the bulk of the soil is composed of these fragments. In the field, a good

FIELD SHEET FOR RECORDING SOIL CHARACTERISTICS

No. 8

Soil Type _Henneke variant_
Location _9 km S of Coloma NE ¼ SW ¼ Sec 1 T 10 N, R9E   MDBM_
Geographical Landscape _Upland_
Elevation _326 m_   Slope _20%_   Aspect _S - SW_   Erosion _slight - moderate_
Groundwater _deep_   Drainage _mod. well_   Alkali _none_
Mode of Formation _Primary serpentinitic_   Parent Material _Serpentine_
Climate MAT _14°C_ : Jan _4.5°C_, Jul _23°C_   MAP ≅ _89 cm_   FFD _223 @ 0°C_
Natural Cover _Chamise, Digger Pine, Ceanothus_   Soil Region _VIII_
Profile Group _VII - II_   Higher Categories _loamy - skeletal, serpentinitic,_
Genetically Related Soil Series _Del Piedra (an alfisol);_ _thermic, Typic Xerochrept_
Dukabella (higher, timbered with Jeffrey pine)

PROFILE SKETCH		COLOR	TEXTURE	STRUC- TURE	CONSISTENCE	REAC- TION	MISC: Roots, Pores, Clay films, Concretions
A 11 0-13 cm	a s	7.5YR 4/4	g l	1 f gr	so, vfr; so, po	7.3	3 vf, 3f, 2m, 2co rts 3 vf, 3f, 2m, 1 co pores
A12 13-20 cm	A w	5YR 4/6	g l	2 m gr	so, vfr; so, po		2 vf, 2f, 1m, 1co rts 2 m, 1 co, 3vf, 1f pores
B1 20-28 cm	C w	5YR 4/6	g l	1 c abk → 2 c gr	so, vfr; ss, ps		1 vf, 1f, 2m, 2co rts 2 vf, 2f, 2 m pores
B 21 28-51 cm	c w	5YR 3/4		1 m abk → 2 c gr	so, fr; ss, ps		1 f, 1m, 1co rts 2 fi, t; 2vf i pores
B 22 51-68 cm	c i	5YR 4/4		1 m sbk → 2 c gr	sh, fr; ss, ps		1f, 1m, 2 co rts 2f, 2vf pores
R							

Natural Land Division _E 8 - 3m_
Soil Rating (Storie index) _11_   Soil Grade _5 (6)_

Figure 16-6. Field sheet for recording soil characteristics. a s = abrupt smooth; a w = abrupt wavy; c w = clear wavy; c i = clear irregular; 7.5 YR = Yellow-Red hue and 4/4 = value and chroma (in Munsell notation); g l = gravelly loam; f = fine; m = medium; c = coarse; gr = granular; abk = angular blocky; sbk = subangular blocky; so = soft; ss = slightly sticky; sh = slightly hard; vfr = very friable; po = non-plastic; ps = slightly plastic; fr = friable; vf = very fine; co = coarse; rts = roots; t = tubular pores; i = interstitial pores. For structure, 1 = weak and 2 = moderate. For roots and pores, 1 = few, 2 = common, and 3 = many.

estimate can be made of texture by dampening and handling a small amount of the soil. The particles that feel grainy and gritty are sand; those that are silky, like moist talcum powder, are silt. If the sample is sticky and if a self-supporting, flexible ribbon can be extruded between the thumb and finger, there is a high percentage of clay. If it is nonsticky or only slightly sticky, and no ribbon can be extruded, there is a low percentage of clay. Different soil textures can be recognized by various combinations of the foregoing.

In general, the structural or skeletal support of the soil is provided by the sand fraction, or at least by the largest soil particles, and the finer particles help to store nutrients and to bind particles together into aggregates. Size classes are assigned arbitrarily, recognizing that minerals (and the building blocks of minerals, such as silicon tetrahedra and iron octahedra) form a continuum of sizes, ranging from single tetrahedra to large and very complex mineral crystals. Material that is without identifiable structure in the fine clay fraction is called **allophane**.

If mineral grains are not spherical, then their diameter is taken to be an average of their maximum and minimum dimensions. Sand grains tend to range from blocky and irregular to subround in shape, and silt particles are similar but smaller. Clay particles are very dissimilar to sand grains, and their plate-like shape is reminiscent of mica (Figure 16–8).

A more precise determination of soil texture can be made in the lab. Particles larger than 2 mm are sieved out of a given sample, which is then treated to oxidize and remove the organic matter (usually with hydrogen peroxide: this also has the effect of rupturing the binding of clays by organic matter). The sample is then dried and weighed. The

Figure 16–7. Soil profile of Henneke variant. The soil is classified as loamy-skeletal, serpentinitic, thermic, Typic Xerochrept, an Inceptisol derived from serpentinitic materials.

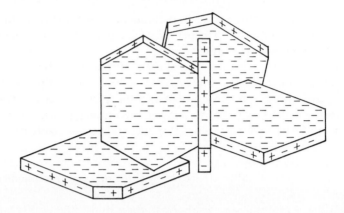

Figure 16–8. An idealized group of clay particles representing the **card house** effect in soil structure in which positive charges on edge positions are attracted by negative charges on the broad surfaces. The negative charges provide exchange sites for cations, and are thus a storage facility. (Reprinted by permission of McGraw-Hill Book Co. from *Soils and Soil Fertility* by Thompson and Troeh, 1973.)

remaining materials are carefully flushed and shaken through a series or nest of sieves which allow particles 1 mm, 0.5 mm, 0.25 mm, 0.1 mm, and 0.05 mm to pass through. The total weight of sand and its distribution can be determined by this process. The fraction passing through the sieves contains silt and clay together, and these can be separated by using a hydrometer or pipette method of analysis, based on differential rates of settling by silt and clay.*

Once the percentage by weight of separate components has been determined, the use of a soil texture triangle (Figure 16–9) facilitates textural class determination. Assume, for instance, that our soil contains 30% clay, 40% silt, and 30% sand. This soil would be called clay loam. Make note of two features: the designation is an area, not just a point on the triangle; and, there are four terms commonly applied to soil texture: clay, silt, sand, and loam. Loam designates a mixture of sand, silt, and clay that exhibits the properties of each fraction about equally. Thus, our sample exhibits the clay fraction more strongly than does loam, but is still a loamy soil, hence "clay loam." Table 16–3 lists the general terms used by the United States Department of Agriculture to describe soil texture.

To understand the function of these various soil fractions, we must comprehend two more physical properties of soil: porosity and permeability. Porosity refers to the total space that is not occupied by soil particles. Permeability deals with the ease of passage through soil

*For a more complete discussion, consult Thompson and Troeh 1973.

of gases, liquids, and plant roots. In general, finer textured soils have greater porosity, with a pore space of perhaps 40–60% of the total, and sandy soils have lesser porosity, with perhaps 35–50% pore space.

Water moves into and drains out of a sandy soil with much greater ease than for a finer textured soil, which means that sandy soils are more permeable than fine textured soils. This is because water has strong adhesion and cohesion capacity. Water molecules form a film **(hygroscopic water)** about each clay particle, and easily bridge the gap

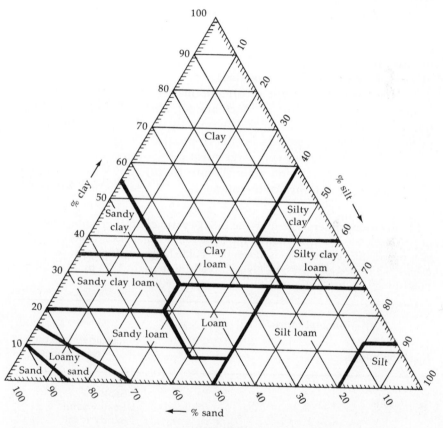

Figure 16–9. Relationship between the particle size distribution of a soil and its class name. In using the diagram the points corresponding to the percentages of silt and clay present in the soil are located on the silt and clay lines, respectively. Lines are then projected inward from % silt parallel to the clay side of the triangle and inward from % clay parallel to the sand side. Note that the third particle size (in this case, sand) is defined by the two projected lines. The name of the compartment in which the lines intersect is the texture class of the soil in question. (Reprinted by permission of McGraw-Hill Book Co. from *Soils and Soil Fertility* by Thompson and Troeh, 1973.)

between particles; the cohesion between water molecules is sufficient to maintain the bridge. Such soils are nutritionally rich, but can be a problem agriculturally, due to low permeability and poor aeration. Hygroscopic water also forms a tight film about sand grains, but the surface-to-volume ratio is very low, there are fewer grains than in clays, and the large pore size does not allow efficient cohesion of water molecules, which therefore drain away under the influence of gravity.

Soil water is a system under tension. The tension is expressed in atmospheres of pressure (1 atm = 1.034 kg cm^{-2}), in bars (1 bar = 10^6 dyne cm^{-2}, or 0.978 atm), or, more commonly now, in MPa (1 bar = 0.1 MPa). Soil water is under tension because water, like any gas or solute, moves down a chemical concentration gradient. The tendency of water to move is called **water potential** (see Chapter 17 for a full discussion of water potential and soil moisture).

Clay particles and bits of organic matter within the soil are referred to as **micelles**, most of which are polyanions. Micelles are negatively charged on their surfaces. The capacity of micelles to provide the primary soil nutrient storage results from (a) the negative charges on the micelles and (b) their very large surface-to-volume ratio. The hypothetical dicing of a mineral grain (Figure 16–10) allows us to envision the difference in the surface-to-volume ratio of a sand grain with sides

Table 16–3. General terms used by the U.S. Department of Agriculture to describe soil texture in relation to the basic soil textural class names. (Reprinted with permission of Macmillan Publishing Co., Inc. from *The Nature and Properties of Soils,* by Nyle C. Brady. Copyright 1974 by Macmillan Publishing Co., Inc.)

Common names	General terms	Basic soil textural class names
	Texture	
Sandy soils	Coarse	Sand Loamy sand
Loamy soils	Moderately coarse	Sandy loam Fine sandy loam
	Medium	Very fine sandy loam Loam Silt loam Silt
	Moderately fine	Clay loam Sandy clay loam Silty clay loam
Clayey soils	Fine	Sandy clay Silty clay Clay

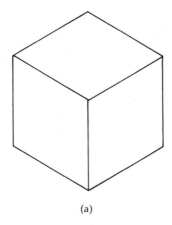
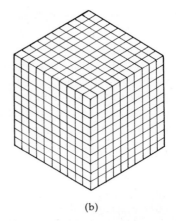

(a)   (b)

Figure 16–10. (a) A block with sides 1 mm in length. (b) The same block sliced vertically and horizontally into blocks with sides 0.1 mm in length. The block with 1 mm sides has 6 mm^2 of surface area. The block sliced into smaller blocks with 0.1 mm sides has resulted in 1000 blocks with 60 mm^2 of total surface area exposed. Slicing the block into smaller blocks with 0.001 mm sides would result in a surface area of 6000 mm^2, or 1000 times that of the parent grain. (Reprinted by permission of McGraw-Hill Book Co. from *Soils and Soil Fertility* by Thompson and Troeh, 1973.)

1 mm in length, and the same volume of clay particles with sides 0.001 mm in length. The sand grain has a volume of 1 mm^3 and a surface area of 6 mm^2. If we slice this into bits with sides 0.1 mm in length, there will be 1000 bits, each one with 6×10^{-2} mm^2 surface area, or a total of 60 mm^2, which is an increase of ten times more than the surface area of the parent sand grain. Slice again, so that each soil particle has sides 0.01 mm in length, and the surface area increases to 600 mm^2. Slice again, so that clay particles have sides 0.001 in length, and surface area now equals 6000 mm^2, which is 1000 times that of the parent grain.

In summary, the sand fraction provides skeletal (structural) support, aeration and permeability for the soil. The clay fraction provides water holding and nutrient storage capacities (see also the section on soil chemistry). Silt also provides water holding capacity and, through weathering, provides plant nutrients and material to form clays, to replace that lost through leaching.

Soil Structure

Primary soil particles are arranged into secondary units, called **peds.*** The nature of the arrangement or **aggregation** of the peds is

---

*A ped is a unit of soil structure such as an aggregate, crumb, prism, block, or granule, formed by natural process.

called soil structure. Soil aeration is greatly enhanced by good soil structure, as is the movement of water and plant roots through the soil.

Aggregate stability, and the shapes and sizes of aggregates, are the most important characteristics of soil structure. **Stability** means the capacity of the peds to retain their structure, that is, to absorb water and not disintegrate. There are various agents which bind these aggregates. Fungal mycelia, microbial exudates, hydrogen and calcium ions, and certain clays promote stability; sodium ions, and expansion and contraction of certain clays promote instability.

The size and shape of peds are peculiarly significant to the success of plants and other soil-dwelling organisms. Table 16–4 presents illustrations and descriptions of several common soil aggregates and the classification of soil structures that possess them. Notice that granular soils are particularly suitable for plant growth, because both water and air can circulate between the peds. The presence of good structure—granular peds, high degree of aggregate stability—is especially desirable in the topsoil. If the texture of the B horizon is sandy, then structure is of little consequence, since drainage is assured. However, clay soils in the B horizon can create a "claypan" or hard pan, which has a deleterious effect on plant life. An excellent example of this is the Blacklock podzol soil (Typic Sideraquod) of the Pygmy Forest in Mendocino County, California. Jenny et al. (1969) note with some amusement the lack of agreement among naturalists as to the cause and effect:

> . . . . naturalists indulged in an apparent *circulus vitiosus*. On field trips the professors of botany would tell their students that the podsol soil is the cause of the unusual assortment of plant species, whereas the visiting professors of pedology (soil science) would attribute—in the light of classical podsol theory—the soil horizon features to the acid-producing vegetation. While it is true that a species individual responds to its soil niche, it is also true that it modifies that niche, which, in turn, reacts upon the individual.

This Blacklock soil is one of the most acid soils known, with a pH of 2.8–3.9 and a characteristic bleached white surface horizon underlain by an iron-cemented hardpan (Figure 16–11). Leachate from the acid conifer litter cannot drain through the hardpan, but continues to remove nutrients from the already depauperate surface soil.

SOIL CHEMISTRY

One of the most important aspects of soil chemistry is the state and the relationship between chemical elements in soil systems: as soluble ions in the soil solution, as adsorbed ions on charged particles, and as constituents of mineral and organic particles. Soil chemistry

Table 16-4. Classification, description, and illustration of various types of soil structure. (Reprinted by permission of McGraw-Hill Book Co. from *Soils and Soil Fertility* by Thompson and Troeh, 1973 (modified).)

Classification	Description	Illustration
Structureless:		
Single grain	Each soil particle independent of all others	
Massive	Entire soil mass clings together, no lines of weakness	
Structured:		
Granular	Primary soil particles grouped into roughly circular peds, such that there is space between them, as they do not fit tightly together. Enhances permeability.	Granular
Platy	Peds with horizontal dimensions greater than vertical. Common at soil surface, and often associated with lateral movement of water.	Platy
Blocky	Peds approximately equal in vertical, horizontal dimensions, but fit well together, unlike granular peds. May be angular or subangular.	Angular blocky Subangular blocky
Prismatic	Peds taller than wide, common in B horizons of well-developed soils. May be columnar, due to eluviation (old age or high sodium content will produce columnar peds of prismatic peds).	Prismatic
Structure destroyed:		
Puddled	Soils disturbed when wet, puddle or run together; structure is destroyed, pores collapse.	

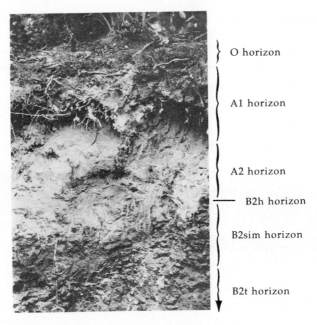

O horizon

A1 horizon

A2 horizon

B2h horizon

B2sim horizon

B2t horizon

Figure 16–11. Soil profile of Blacklock sand, a podzol soil in the Pygmy Forest, Mendocino County, California. The horizon boundaries are wavy rather than smooth. Note the highly bleached A2 horizon, beneath which is a B2h darkly stained with humus. The B2 SIM is a silica cemented hardpan, underlain by the B2t claypan. The entire profile encompasses only about one meter in depth.

deals primarily with reactive materials, which are the chemicals that are of significance to living things. Reactive materials tend toward equilibria in a changeable environment. There is virtually no correlation between the absolute and relative amounts of an element present in the soil and the significance of that element to life. Oxygen (49.5%), silicon (25.8%), and aluminum (7.5%) are the most abundant, the second most abundant, and the third most abundant elements, respectively, in the earth's crust. Yet, they are of little consequence to life and are relatively inaccessible due to the bound nature of their chemical compounds. By contrast, the most important elements in soil chemistry (Table 16–5) are, all together, less than 16% of the earth's crust materials. Soil pH, cation exchange capacity, and the nature of the interface between the solid phase and the liquid phase are all important aspects of soil chemistry.

Soil pH

Typical soils exhibit pH values that range from 4 to 8, although certain soils may have values higher or lower. For instance, saline soils commonly exhibit pH values ranging from 7.3 to 8.5. In southern Death Valley, California, saline soils support a saltbush scrub of desert holly *(Atriplex hymenelytra)*, Parry saltbush *(A. parryi)*, and honeysweet tidestromia *(Tidestromia oblongifolia)*. The distinctly alkaline soils of the broad desert valleys of the Great Basin region of western North America support a shadscale scrub vegetation dominated by shadscale saltbush *(A. confertifolia)* and bud sagebrush *(Artemisia spinescens)*. A pH of less than 6.6 is considered an acid soil, and vernal pools have pH values of perhaps 6.4–6.85. Very ancient soils, especially those covered by coniferous forest in humid regions, have very low pH values. The pygmy cypress *(Cupressus pygmaea)* grows in the Blacklock podzol soil of the Mendocino Forest, which we noted has an extremely acid soil (pH of 2.8–3.9). These highly weathered soils, perhaps half a million years old, are primarily very resistant quartz grains, and contain very small amounts of nutrients for plant growth. In general, leaching tends to remove basic cations, and thus lower the pH of the soil.

The influence of plants on the soil pH is very complex. Nutrients (both cations and anions) are brought up from subsurface soil by plants and then deposited with litter on the surface. Grasses tend to use more bases, and grass cover may act to keep the soil pH from dropping.

Table 16–5. Elements important in soil chemistry and their chemical symbols and principal forms. (Reprinted by permission of McGraw-Hill Book Co. from *Soils and Soil Fertility* by Thompson and Troeh, 1973.)

Element	Principal ions
Aluminum	Al^{+++}
Boron	$B_4O_7^{--}$
Calcium	Ca^{++}
Carbon	CO_3^{--}, HCO_3^-
Chlorine	Cl^-
Cobalt	Co^{++}
Copper	Cu^{++}
Hydrogen	H^+, OH^-
Iron	Fe^{++}, Fe^{+++}
Magnesium	Mg^{++}
Manganese	Mn^{++}, MnO_4^-
Molybdenum	MoO_4^{--}
Nitrogen	NH_4^+, NO_2^-, NO_3^-
Oxygen	with other elements
Phosphorus	$H_2PO_4^-$, HPO_4^{--}
Potassium	K^+
Sodium	Na^+
Sulfur	SO_4^{--}
Zinc	Zn^{++}

Coniferous litter, on the other hand, provides acid leachate, which effectively lowers the pH. Fewer cations are held on exchange sites, and more cations are released by acid weathering. Figure 16–12 shows the changing availability of plant nutrients in response to changing pH values.

Agricultural soils may develop a low pH due to fertilizer applications that add nitrogen and sulfur. Lime is often applied to acid soils to make them more basic. Very alkaline soils, on the other hand, may be treated with gypsum to lower pH. In most soils, the pH cannot be changed very much because of very effective soil pH buffering.

Although the direct effects of soil pH on plant growth are very limited, the indirect effects are numerous and significant. Toxicity of certain metals, such as aluminum and manganese (which are more soluble at lower pH) would be a direct effect of pH value. The most important indirect effect of soil pH is its influence on nutrient availability (Figure 16–12). Rates of weathering, availability of nutrients

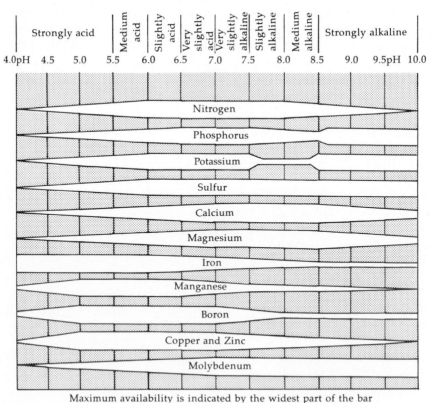

Maximum availability is indicated by the widest part of the bar

Figure 16–12. A schematic illustration of the relationship between pH and plant nutrient availability. (Reprinted by permission of McGraw-Hill Book Co. from *Soils and Soil Fertility* by Thompson and Troeh, 1973.)

such as nitrogen, phosphorus, and sulfur, and the leachability of nutrients such as potassium are all influenced by soil pH. The availability of nutrients with low solubilities, such as phosphorus, is dramatically influenced by pH changes. Calcium phosphates are less soluble as the pH climbs, and iron and aluminum phosphates are less soluble as the pH drops. A near neutral pH, between 6.5 and 7.5, is best for phosphorus availability, and indeed for the availability of most nutrients needed for plant growth.

Humus Formation

The action of microorganisms on organic materials eventually converts them into humus. The organic materials become very finely divided and nearly black in color. By virtue of their tiny size, these particles can coat soil particles to the point that the soil appears black, when it contains, for example, as little as 5% organic matter. Root exudates, earthworm excreta and the like may provide a significant contribution to this nonliving, finely divided organic matter, but probably 90% of humus is attributable to microorganisms, microbial tissue, and microbial activities on litter.

Humus, an intermediate product of decomposition, decomposes at an annual turnover rate of about 3% in temperate regions. It comprises not only the resistant fraction of the organic matter, but also the microbes that have accomplished the decomposition of the more decomposable fraction. In a climax community, or even a stable seral one, the amount of new humus formed each year approximately equals that which is released by decomposition. This may amount to 350–700 kg ha^{-1} yr^{-1}, and would be as much as 25% of the humus in a tropical soil, and as little as 1% in a very cool region. The significance of slowly decomposing humus is that it provides a secure supply of continually available nutrients for plant growth.

Although the composition of humus varies, 10–15% of its weight is made up of easily identifiable organic compounds: polysaccharides such as cellulose and its decomposition products; polyphenols such as tannins; proteins and their decomposition products; and a host of hydrocarbons, organic acids, alcohols, esters, and aldehydes. The rest of the humus is not so easily identified, but is characterized as humic materials, many of which contain reactive groups—carboxyl, amine, phenolic hydroxyl groups—combined with chains and rings, forming rather large and complex molecules.

Cation Exchange Capacity

Cation exchange capacity (CEC) is a measure of the number of negatively charged sites on soil particles that attract **exchangeable**

cations, that is, positively charged ions that can be replaced by other such ions in the soil solution. Three factors strongly influence the cation exchange capacity of a soil: its clay content, the kinds of clay minerals or amorphous colloids (allophane) that it contains, and the humus content. We have already commented on the significance of the surface-to-volume ratio to mineral storage. The negative charge, which provides for nutrient storage on clay mineral particles, is a result of, for instance, substitution of ions within clay mineral structures or ionization of hydrogen ions from OH groups, which also produces negatively charged surface sites. If a Mg^{++} or an Fe^{++} replaces an Al^{+++} within the octahedral sheet of a layer of silicate clay, there will be a net negative charge which must be balanced on or near that layer. A silicate clay particle with an expanding lattice (such as montmorillinite) will provide not only external surface sites but also internal exchange sites between adjacent layers. Within silicate clays with a nonexpanding lattice (for example, kaolinite) there are no interlayer exchangeable cations. Highway engineers value kaolinite clays because they have little tendency to swell and shrink. However, kaolinite clays provide low nutrient storage capacity for plant growth for the agriculturalists (see also Thompson and Troeh 1973).

Tropical Rainforest Soils

Let us consider the interaction of vegetation, soil texture, rainfall regime, temperature regime, and soil microorganisms, in reference to cation exchange in a particular ecosystem, the tropical rainforest. Soils of the tropical rainforest (TRF) are usually highly weathered and **laterized**: there is little or no horizon development in the reddish-brown loam, which is comprised of aluminum and ferric sesquioxides (Al_2O_3, Fe_2O_3). Lateritic soils (Oxisols) are very acidic and nutrient poor. The litter layer is extremely thin because mobilization of nutrients occurs at a very rapid pace in the warm, moist rainforest. Even wood is quickly recycled by subterranean termites.

Recall the reference (in Chapter 12) to constancy of concentration of chemicals in stream water as a measure of cation exchange capacity of soils drained by the stream in the Hubbard Brook Forest. Tropical watersheds may or may not show the same phenomenon. In the TRF of the Amazon Basin, soils are ancient, weathered, and nutrient poor, possessing low cation exchange capacity. As the volume of water draining these soils increases, the concentration of nutrients decreases, until runoff becomes very dilute. By contrast, the concentration of six major cations in a Costa Rican wet forest and in a Puerto Rican TRF shows very little sensitivity to even large fluctuations in soil water volume (Jordan and Kline 1972). The exchange capacity is very high: soils are saturated with nutrient elements. The contrast here with Amazonian soils is probably due to these soils being younger soils from rich volcanic rock, rather than typical lateritic soils.

The luxuriant growth of vegetation (52.5 t ha^{-1} yr^{-1} gross productivity) belies the extremely poor nutrient content of most TRF soils. Walter (1973) states that the entire nutrient reserve required by the forest is contained in the phytomass.

What mechanism allows for litter recycling so that mobile nutrients are not leached away by the inevitable rains? Work by Went and Stark (1968) showed that feeding roots of TRF trees are directly connected to litter. These rootlets exploit the hyphae of mycorrhizal fungi to obtain nutrients. As noted in Chapter 6, these are mutualistic fungi, interacting with the roots and increasing their "host's" ability to extract nutrients from the forest litter. Recall that many fungi are saprobic, obtaining nutrients by absorption. Due to this highly adaptive relationship, there is little chance that free, mobile ions released from litter biomass will be lost through leaching.

As long as the rainforest is not disturbed, the lack of a soil storage compartment is unimportant. The nearly complete recycling of materials allows for stability for hundreds or even thousands of years. Yet, deforestation through burning, clearing of the land for cultivation, or logging, removes the only significant storage facility, the phytomass, and quickly allows the mobilized ions to move out of the system. The luxuriant growth of the virgin forest is never (in human terms) renewed.

Soil Solution

There is no way that soil water can be withdrawn and studied as a separate entity; it must be understood *in situ*, in relationship to its function as the liquid medium which renders the soil a habitable environment. The readily available ions are present in the soil solution, and maintain an equilibrium with adsorbed ions. These, in turn, maintain an equilibrium with absorbed mineral ions (Figure 16–13). Plant removal of nutrients from the soil solution lowers the concentration of those nutrients in the rhizosphere; ions move toward the rhizosphere down a chemical gradient. This shift, due to solubility constants of the ions involved, enhances the release of adsorbed ions, and so forth.

For example, plant uptake of phosphorus—roots removing P from the soil solution—establishes a concentration gradient in the soil solution such that the movement of P is toward the root. Under optimum conditions, daily uptake of P in grassland may be as much as 50 times the quantity of P in the solution pool, the pool itself being replenished from the labile pool (that is, from adsorbed ions). When conditions of

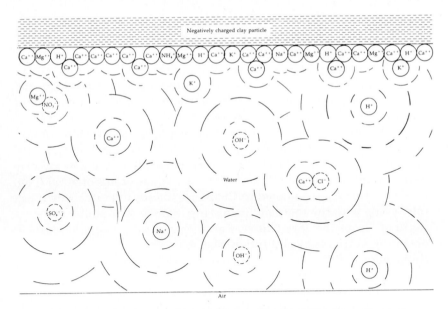

Figure 16–13. An illustration of how ions are distributed in the soil solution near a negatively charged clay particle. (Reprinted by permission of McGraw-Hill Book Co. from *Soils and Soil Fertility* by Thompson and Troeh, 1973.)

temperature and water are exactly right, the labile pool has the potential of replacing the solution pool 250 times daily. As we experience time, there is neither gain nor loss of phosphorus in grasslands, and the cycle itself is probably governed as much by biotic factors as by abiotic factors.

With continued uptake, the concentration of P in the cytoplasm of root cells may exceed concentrations (by up to two orders of magnitude) in ambient soil solution. It is well known that mycorrhizae greatly enhance uptake of P. This enhancement is attributed to the tremendously enlarged contact zone of root and soil, mediated through hyphal fungi.

A generalized phosphorus cycle is depicted in Figure 16–14. Phosphorus participates in a sedimentary cycle, which means that local deficits may occur. Sedimentary cycles are not as resilient to disruption as gaseous cycles, and perturbation may result in very wide oscillations, not easily damped. Note that no return is indicated from "loss to deep sediments."

SOIL TAXONOMY

The naming of soils allows one to impart a great deal of information simply by invoking the name of a certain soil. A name is applied to an individual kind of soil that has a definable existence on the land-scape. A unit of soil, large enough such that its profile and horizons can be studied and its properties defined, is called a **pedon**. This is the smallest practical volume that can be properly referred to as a soil. A pedon may range from 1 m^2 to 10 m^2 in area, depending on the conti-nuity or cyclic discontinuity on the soil's horizons within short linear distances (1 to 7 m). Pedons of soils with continuous horizons have the smallest area (1 m^2). Groups of similar adjacent pedons, termed **poly-pedons**, form the basis for soil survey and soil classification.

For names to be useful, a classification system must be scrupu-lously followed, and names must be assigned on the basis of certain specified properties. Characteristics that are used to define a soil are termed **differenting characteristics** or **differentiae**.

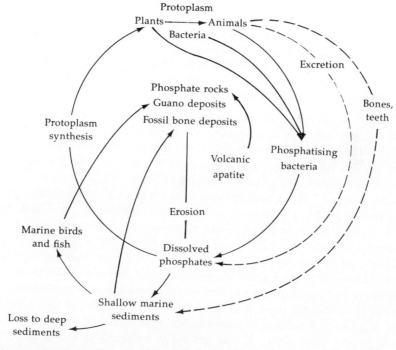

Figure 16-14. The phosphorus cycle. Phosphorus is a rare element compared with nitrogen. Its ratio to nitrogen in natural waters is about 1 to 23. Chemical erosion in the United States has been estimated at 34 t km^{-2} yr^{-1}. Fifty-year cultivation of virgin soils of the Middle West reduced the P$_2$O$_5$ content by 36%. As shown in the diagram, the evidence indicates that return of phosphorus to the land has not been keeping up with the loss to the ocean. (From Odum: *Fundamentals of Ecology*, copyright 1971 by the W. B. Saunders Co., Philadelphia, PA.)

Soil scientists strive for a classification scheme that reflects a natural relatedness, and therefore the scheme of naming is called **Soil Taxonomy**. Classification is the broader term, and may include schemes of ordering for specific practical purposes, whereas taxonomy is concerned with relatedness among groups. Nevertheless, any taxonomic system is arbitrary and is the creation of its author, reflecting certain biases and devised to fill a particular need.

History

The term soil is very old, and humans have, as agriculturalists, been concerned with soil as a medium for plant growth for millenia. In the 1870s, a Russian school led by Dokuchaiev developed and introduced a revolutionary new concept of soil and a system of naming soils (Soil Survey Staff 1975).

> Soils were . . . independent natural bodies, each with a unique morphology resulting from a unique combination of climate, living matter, earthy parent materials, relief, and age of land form. The morphology . . . reflected the combined effects of the particular set of genetic factors responsible for its development.

The Russian soil taxonomy was adopted first in Europe, then in this country early in this century. As more soils were described, and other information became available, the weaknesses of the Russian system became increasingly apparent. In 1938 a system was developed in the United States and published under the title "Classification of Soils on the Basis of Their Characteristics" (Baldwin et al. 1938). Then, starting in the 1950s, a taxonomic scheme was developed, modified, improved, and presented as a series of approximations, each one more refined and more workable than the last. Six of these approximations were tested by the Soil Survey Staff of the U.S. Department of Agriculture, various U.S. land grant universities, and interested European individuals and agencies. The seventh approximation was published and issued on a limited basis in 1960 at the 7th International Congress of Soil Science at Madison, Wisconsin, for wider testing and evaluation (Soil Survey Staff 1960). It was generally referred to as the Seventh Approximation or "the brown book." In 1965, a modified form was officially adopted for soil classification in this country by the National Cooperative Soil Survey. With additional modification, it was finally published for unrestricted use as a comprehensive system in 1975 by the Soil Survey Staff of the U.S. Department of Agriculture as Agricultural Handbook 436, Soil Taxonomy—A Basic System of Soil Classification for Making and Interpreting Soil Surveys. It is this system that we will discuss here.

To be useful, a soil taxonomy should have these attributes: (a) the definition of each taxon should have the same meaning for each user; (b) the taxonomy should be a multicategoric system, with many taxa in the lower categories; (c) the taxa should refer to real soils, known to occupy geographic areas; (d) differentiae should be soil properties that can be discerned in the field, inferred from other properties that are observable in the field, or taken from the combined data of soil science and other disciplines; (e) the taxonomy should be modifiable with a minimum of perturbation to the system; (f) the differentiae should maintain pristine soil and manipulated soil that is its equivalent in the same taxon, whenever possible; and (g) the taxonomy should provide taxa for all the soils of a landscape, or at least be capable of such provision (Soil Survey Staff 1975).

Differentiating Properties of Soils

The properties of a polypedon should serve as differentiae. Soil color and soil horizons have long been used for differentiae. Properties of soil that interact with other properties should be carefully considered when differentiae are selected to define soil taxa.

There are six defined diagnostic surface horizons or **epipedons**. An epipedon is a horizon, or horizons, that forms in the upper part of the soil profile. We will name these, and briefly identify them, but will not describe them in detail, due to limitations of space. A subsurface horizon (often B, but perhaps part of A) is also of diagnostic value, and the names, descriptions and genesis or mode of formation of these horizons are also important in soil classification. We will simply list these. The diagnostic epipedons are as follows:

1. mollic epipedon: from the Latin *mollis,* meaning soft; it is normally granular or easily workable, has appreciable organic matter and high base saturation (>50%).
2. anthropic epipedon: formed by human manipulation over a long time period under semiarid or arid conditions.
3. umbric epipedon: from the Latin *umbra,* meaning shade, thus dark soil; it has appreciable organic matter and a low base saturation (<50%).
4. histic epipedon: from the Greek *histos,* meaning tissue, thus a thin layer of muck or peat.
5. plaggen epipedon: from the German *plaggen,* meaning sod, thus a surface layer 50 cm or more thick, produced by long-continued manuring.
6. ochric epipedon: from the Greek *ochros,* meaning pale; a soil too dry, too thin, too high in chroma, etc., to qualify as a mollic or umbric epipedon.

The diagnostic subsurface horizons include argillic, agric, natric, spodic, placic, cambic, oxic, duripan, and the like.

Other properties used in defining a soil include: an abrupt textural change, mineralogical composition, amorphous material dominating the exchange complex, microrelief, lithic contact, organic soil materials, particle-size classes, and so on.

The Structure of Soil Taxonomy

The U.S.D.A. soil taxonomy system recognizes a hierarchy of categories. Each category comprises a set of taxa. The series is the unit of classification. One of the strengths of the system lies in its capacity for internal modification, as new information is gathered, without disruption of the rest of the system. The hierarchy is as follows:

Categories:	Example of taxa within categories:
Order	Entisol
Suborder	Orthent
Great Group	Cryorthent
Subgroup	Typic Cryorthent
Family	Loamy-skeletal, carbonatic Typic Cryorthent
Series	Swift Creek

Most of the formative elements in the names of soil orders (Table 16–6) are derived from Latin or Greek words appropriate to that soil order, and the soil order name is formed with the suffix -sol, from the Latin *solum* for soil. The formative elements *alf* and *ent* are not from Greek or Latin, but are meaningless syllables. Alfisols are variable, often found in continually wet, cold, or seasonally dry climates. Entisols are often young soils with little or no horizon development. The derivations of the other formative elements are listed in Table 16–6.

The names of suborders are a combination of the formative element of the parent order as the suffix with a prefix that suggests the diagnostic properties of the soil. For example, *alb* (from *albus*, white) with *oll* (from mollisol, soft) names a suborder of the Mollisol order with a white horizon—*Alboll*. Similarly, a great group name consists of the name of the suborder coupled with a prefix containing one or two formative elements to refer to definitive properties. For instance, *natralboll* indicates a Mollisol with an *albic* (white eluvial) horizon and a *natric* horizon (an illuvial subsoil with excess exchangeable sodium).

The great group name is combined with one or more modifiers to generate the subgroup name. *Typic* Natralboll would connote the subgroup thought to typify the great group. Families are named with polynomials consisting of the subgroup name modified by adjectives that follow a specified order and that are names of classes describing

Table 16–6. Formative elements in the names of soil orders. (From Soil Survey Staff 1975. Courtesy of the U.S. Dept. of Agriculture.)

Name of order	Formative element in name of order	Derivation of formative element	Pronunciation of formative element
Alfisol	Alf	Meaningless syllable	Ped*alf*er
Aridisol	Id	L. *aridus*, dry	A*rid*
Entisol	Ent	Meaningless syllable	Rec*ent*
Histosol	Ist	Gr. *histos*, tissue	*Hist*ology
Inceptisol	Ept	L. *inceptum*, beginning	Inc*ept*ion
Mollisol	Oll	L. *mollis*, soft	M*oll*ify
Oxisol	Ox	F. *oxide*, oxide	*Ox*ide
Spodosol	Od	Gr. *spodos*, wood ash	O*d*d
Ultisol	Ult	L. *ultimus*, last	*Ult*imate
Vertisol	Ert	L. *verto*, turn	Inv*ert*

the soil's particle size mineralogy, reaction, and temperature classes, among other factors. Series names are place names, usually taken from the region where the soil was first described. In the field, those profiles which do not match the descriptions of officially recognized series may be designated as **variants** of the existing, closely related series. With time and additional field information, variants may be defined as official soil series with their own unique name.

Table 16–7 provides the names of the suborders and great groups of four orders: Inceptisol, Mollisol, Oxisol, and Spodosol. The system is, by the very nature of its function, very complex. The words seem cryptic and very foreign. However, a minimal knowledge of Greek and Latin, and a little insight, allows even the novice to infer much about the properties of a named soil family or other taxon. Let us analyze the Henneke variant that we mentioned earlier in Figure 16–7. Remember that it was a loamy-skeletal, serpentinitic, thermic, Typic Xerochrept. The particle-size class, loamy-skeletal, means that rock fragments make up 35% or more by volume, and there is enough fine earth to fill interstices larger than 1 mm; the fine earth fraction fits the definition for a loamy soil texture. The serpentinitic mineralogy class tells us that the soil is more than 40 percent by weight serpentine minerals, and thermic tells us that the mean annual soil temperature, measured at 50 cm depth, is between 15°C and 22°C. A Typic Xerochrept is an Inceptisol of a mediterranean climate, with a xeric moisture regime. Many of these soils are on steep slopes and are shallow over rock, but others are on gentler slopes and have formed in geologically recent sediments.

In the 1938 soil classification system developed by Baldwin et al. (1938), these soils would have been grouped with Brown Forest and

Table 16–7. Examples of orders, suborders, and great groups of soils. (From Soil Survey Staff 1975. Courtesy of the U.S. Dept. of Agriculture.)

Order	Suborder	Great group	Order	Suborder	Great group
Inceptisols	Andepts	Cryandepts		Rendolls	Rendolls
		Durandepts		Udolls	Argiudolls
		Dystrandepts			Hapludolls
		Eutrandepts			Paleudolls
		Hydrandepts			Vermudolls
		Placandepts		Ustolls	Argiustolls
		Vitrandepts			Calciustolls
	Aquepts	Andaquepts			Durustolls
		Cryaquepts			Haplustolls
		Fragiaquepts			Natrustolls
		Halaquepts			Paleustolls
		Haplaquepts			Vermustolls
		Humaquepts		Xerolls	Argixerolls
		Placaquepts			Calcixerolls
		Plinthaquepts			Durixerolls
		Sulfaquepts			Haploxerolls
		Tropaquepts			Natrixerolls
	Ochrepts	Cryochrepts			Palexerolls
		Durochrepts	Oxisols	Aquox	Gibbsiaquox
		Dystrochrepts			Ochraquox
		Eutrochrepts			Plinthaquox
		Fragiochrepts			Umbraquox
		Ustochrepts		Humox	Acrohumox
		Xerochrepts			Gibbsihumox
	Plaggepts	Plaggepts			Haplohumox
	Tropepts	Dystropepts			Sombrihumox
		Eutropepts		Orthox	Acrorthox
		Humitropepts			Eutrorthox
		Sombritropepts			Gibbsiorthox
		Ustropepts			Haplorthox
	Umbrepts	Cryumbrepts			Sombriorthox
		Fragiumbrepts			Umbriorthox
		Haplumbrepts		Torrox	Torrox
		Xerumbrepts		Ustox	Acrustox
Mollisols	Albolls	Argialbolls			Eutrustox
		Natralbolls			Sombriustox
	Aquolls	Argiaquolls			Haplustox
		Calciaquolls	Spodosols	Aquods	Cryaquods
		Cryaquolls			Duraquods
		Duraquolls			Fragiaquods
		Haplaquolls			Haplaquods
		Natraquolls			Placaquods
	Borolls	Argiborolls			Sideraquods
		Calciborolls			Tropaquods
		Cryoborolls		Ferrods	Ferrods
		Haploborolls		Humods	Cryohumods
		Natriborolls			
		Paleborolls			
		Vermiborolls			

Noncalcic Brown soils. They have pale colored surface soils and weakly developed subsoils. The pristine vegetation was most likely shrubs and grasses in low-elevation regions and hardwoods and conifers at higher elevations. In California, these are the soils of parts of the Great Valley, of the foothills of the Sierra Nevada, and of parts of the Coast Ranges.

This variant differs from soils of the Henneke series in three important ways: (1) its epipedon is ochric rather than mollic, (2) it lacks an illuvial clay subsoil, and (3) it has a lithic contact below 50 cm in depth.

SUMMARY

The components of soil are mineral grains, organic matter, water, and air, providing, respectively, plant anchorage and nutrients, intra-system cycling, solvent medium, and oxygen and nitrogen. The upper layers of soil weather and change in response to abiotic and biotic factors. The first phase of soil formation is rock weathering and the second is biochemical weathering.

Soil formation relies on climate, organisms, topography, parent material, and time, as does soil profile development, which encompasses development of soil horizons. Soils can be classed as zonal, azonal, and intrazonal soils.

Soil texture refers to the content of the soil, by weight, of the sand, silt, and clay fractions. Skeletal support and permeability are provided by the largest particles; clay provides water and nutrient storage; silt aids in water storage and weathers to produce additional nutrients and clay.

Micelles (clay particles and bits of organic matter) provide the primary storage for nutrients in the soil. The nature of the arrangements of peds and aggregate stability are the most significant characteristics of soil structure.

Equilibrium among ions in the soil solution, adsorbed ions, and absorbed ions are concerns of soil chemistry. Most soils exhibit a pH range from about 4 to 8. Ancient soils may be very acid. Climate, vegetation, parent material, and relief determine soil pH, which in turn affects vegetation and rates of weathering.

Soil microorganisms convert organic matter to humus, a finely divided, nearly black fraction that decomposes at a stable, equilibrium rate in climax communities. Cation exchange capacity is influenced by the presence of and the kinds of clay minerals, allophane, and humus. Clays with expanding lattice provide more efficient cation exchange than do nonexpanding clays. Some tropical soils have little or no horizon development, and are ancient, nutrient poor soils with little exchange capacity. The phytomass associated with these soils provides the bulk of their nutrient storage.

Soils are classified by their differentiating characteristics. Soil taxonomy refers to a classification scheme designed to reflect a natural relatedness among soil groups. To be effective, a taxonomy must have definitions for each taxon that have the same meaning for each user. The system presently used in the United States was developed by the Soil Survey Staff of the U.S. Department of Agriculture, Soil Conservation Service. It relies on the properties of the polypedon for differentiating characters, and has a very complex but flexible multicategoric system. Names for the soil taxa are formed with syllables selected from terms that reflect the nature of the soil.

CHAPTER 17

WATER

Plant ecologists have expended a great deal of effort on considerations of water, since no other single environmental factor can be directly related to so many plant responses. We noted the significance of water in determining productivity in Chapter 11, in influencing biogeochemical cycles in Chapter 12, as an important factor in photosynthetic systems in Chapter 13, as an influence on leaf temperature in Chapter 14, and, if we were to look carefully, in almost every chapter in this text. In this chapter we hope to place the diverse effects of water on plants into perspective. This chapter includes the general aspects of water availability to plants, factors determining the distribution of water over the landscape, and the vegetational patterns which are correlated with the distribution of water.

THE HYDROLOGIC CYCLE: CONCEPTS AND COMPONENTS

Consider water availability in various environments, and the contrasts between a cypress swamp in the southeastern United States and the Sonoran Desert of Arizona are clear. Less apparent yet significant variations in water availability occur in most plant communities. It is the spatial and temporal variations in the hydrologic cycle which determine the amount and effectiveness of precipitation in plant communities. The cyclic nature of water is shown in Figure 17–1. Condensation of atmospheric water vapor in the form of precipitation, fog, or dew, places water into the biosphere where it is temporarily used or stored in bodies of water; regardless of its immediate fate, processes of evaporation and transpiration eventually cycle the water molecules back to the vapor state.

Water Potential Concept

Over the past 15 years the terminology relating to plant water, soil water, and water vapor in the atmosphere has changed dramatically. Prior to that nearly all such data were in relative values such as percent soil moisture. Percentage values convey information concerning the amounts of water present but do not give indications of the

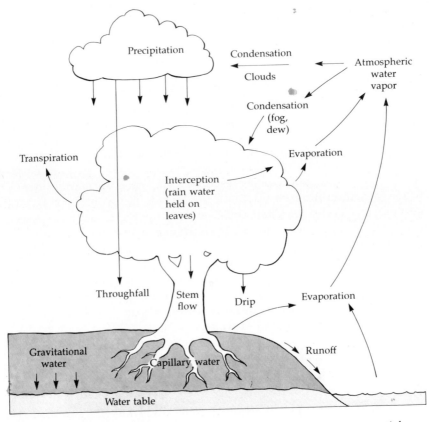

Figure 17–1. Schematic diagram of the hydrologic cycle showing potential pathways of water flux in an ecosystem.

direction of movement or the response of a plant to that amount of water. These problems led to the formulation of the water potential concept in which the status of water is expressed in thermodynamic terms. All matter tends to move from areas of high free energy (high capacity to do work) to areas of lower free energy. **Water potential (Ψ)** is a measure of the free energy of water in comparison to the free energy of pure water. The Ψ of pure water has been assigned a value of 0 bars. A bar is a pressure unit directly relatable to energy per unit mass (1 bar = 0.1 MPa = 0.987 atmospheres = 10^6 ergs g^{-1}). The chemical energy of water in the biosphere is lower than pure water and therefore values are expressed as negative numbers.

Water potential may be broken down into several components which influence the chemical energy of water. Water potential depends on three predominant forces:

$$\Psi = \Psi_m + \Psi_\pi + \Psi_p$$

Matric potential (Ψ_m) is a measure of the reduction in free energy caused by the attraction of the matrix for water molecules. Macromolecules, soil particles, cell walls, or other surfaces to which water adheres make up the matrix. Such attractions always reduce the free energy of water, and therefore, matric potential is always negative. **Osmotic potential** (Ψ_π) is a measure of the reduction in free energy (always negative) attributable to electrical attraction, reorientation, and increased entropy resulting from solutes in solution. In living systems a selectively permeable membrane must be present since no potential difference could exist without a restriction on the movement of solute molecules (Campbell 1977). **Pressure potential** (Ψ_p) is determined by the positive or negative hydrostatic pressure present in the system. Within living cells Ψ_p is usually positive; however, it is negative in xylem when transpiration is occurring. Water potential is a negative value except in fully turgid cells where Ψ_p may balance the negative Ψ_π and Ψ_m potentials (Kramer 1969). The terminology of plant water relations has changed very rapidly and the literature has been inundated with different symbols for the same quantities. Table 17–1 provides a comparison of the most commonly used terms and symbols.

Humidity and Vapor Pressure

Understanding evaporation and transpiration (**evapotranspiration**) processes is basic to determining the interchange of energy and water between the plant and the environment. Large amounts of energy are involved in any change in the state of water. For example, it takes most of the solar energy falling on 1 cm^2 on a clear summer day to evaporate a cubic centimeter of water. This rapid use of energy during the evaporation of water from plants (transpiration) causes an important cooling of the leaf surface. Plants grown in very high humidity conditions can develop an inhibitory heat load in the leaves because of a reduction in water loss (details and further discussion of this process may be found in Chapter 14). Court (1974) estimated that two-thirds of the precipitation falling on the conterminous United States evaporates—a massive energy sink. Conversely, when water condenses or freezes, large amounts of energy are released. This energy is in the form of heat and is called **latent heat**. When temperatures reach **dew point** (the temperature at which atmospheric water condenses), enough latent heat is released to significantly reduce the rate of cooling. Plants growing near bodies of water can escape frost damage because of this phenomenon (Rosenberg 1974).

Rates of evapotranspiration are indicators of water vapor transfer and energy exchange rates at leaf or soil surfaces. The rate of water loss is determined in part by the water vapor content of the atmosphere.

Table 17-1. Terminology for the osmotic quantities as used by several authors. (From Salisbury and Ross 1969. *Plant Physiology.* By permission of Wadsworth Publishing Co., Belmont, CA.)

This text and others[a]	Meyer et al. and others[b]	Bonner and Galston[c]	Levitt[d]	Steward[e]	Salisbury and Parke[f]	James[g]	Fogg[h]
Osmotic Potential Ψ_s (or π)	Osmotic Pressure OP	Osmotic Concentration OC	Osmotic Potential O	Osmotic Pressure P	Osmotic Potential ϕ	Osmotic Pressure O.P.	Osmotic Pressure P_i (internal) P_o (external)
Pressure Potential Ψ_p (or P)	Turgor Pressure (equal and opposite to Wall Pressure) TP	Turgor Pressure TP	Wall Pressure p	Wall Pressure W	Pressure P	Wall Pressure W.P.	Turgor Pressure T
Water Potential Ψ	Diffusion Pressure Deficit DPD	Diffusion Pressure Deficit DPD	Osmotic Equivalent E	Suction Pressure S	Enter Tendency E	Suction Pressure S.P.	Diffusion Pressure Deficit S
Water Potential Difference $\Delta\Psi$			Osmotic Potential Difference P				
$\Psi = \Psi_s + \Psi_p$	$DPD = OP - TP$	$DPD = OC - TP$	$E = O - p$	$S = P - W$	$E = \phi - P$	$S.P. = O.P. - W.P.$	$S = (P_i - P_o) - T$

[a]Gardner, W. R. 1965. Dynamic aspects of water availability to plants. *Annual Review of Plant Physiology.* 16:323–342.
Kramer, P. J., E. B. Knipling, and L. N. Miller. 1966. Terminology in cell water relations. *Science* 153:889–890.
Salisbury, F. B. and C. Ross. 1969. *Plant physiology.* Belmont, CA: Wadsworth.
Slatyer, R. O. 1967. *Plant water relationships.* New York: Academic Press.
Taylor, S. A. and R. O. Slatyer. 1962. Proposals for a unified terminology in studies of plant-soil-water relations. *UNESCO Arid Zone Research* 16:339–349.

[b]Devlin, R. M. 1966. *Plant physiology.* New York: Reinhold Publishing Corporation.
Ferry, J. R. and H. S. Ward. 1959. *Fundamentals of plant physiology.* New York: Macmillan.
Kramer, P. J. and T. T. Kozlowski. 1960. *Physiology of trees.* New York: McGraw-Hill.
Kozlowski, T. T. 1964. *Water metabolism in plants.* New York: Harper and Row.
Meyer, B. S., D. B. Anderson, and R. H. Bohning. 1960. *Introduction to plant physiology.* Princeton, NJ: D. Van Nostrand Co., Inc.
[c]Bonner, J. and A. W. Galston. 1952. *Principles of plant physiology.* San Francisco, CA: W. H. Freeman and Co.
[d]Levitt, J. 1954. *Plant physiology.* Englewood Cliffs, NJ: Prentice-Hall, Inc.
[e]Steward, F. C. 1964. *Plants at work.* Reading, MA: Addison-Wesley.
[f]Salisbury, F. B. and R. V. Parke. 1964. *Vascular plants: form and function.* Belmont, CA: Wadsworth.
[g]James, W. O. 1963. *An introduction to plant physiology.* Oxford: Clarendon Press.
[h]Fogg, G. E. 1963. *The growth of plants.* Baltimore, MD: Penguin Books Inc.

When, at a constant temperature, pure water evaporates into a closed space, an equilibrium will occur at saturation vapor pressure. **Saturation vapor pressure** is the maximum possible partial pressure of water that can be held in the air at a given temperature. Air is seldom saturated, so we need a means of expressing levels below saturation. One simple measurement of water vapor content of the atmosphere can be obtained with psychrometers which consist of two ventilated thermometers. One is dry and measures air temperature and the other is wet and measures the reduction in temperature caused by evaporative cooling. Figure 17–2 shows the relationship between vapor pressure, relative humidity, and wet and dry bulb temperatures. **Relative hu-**

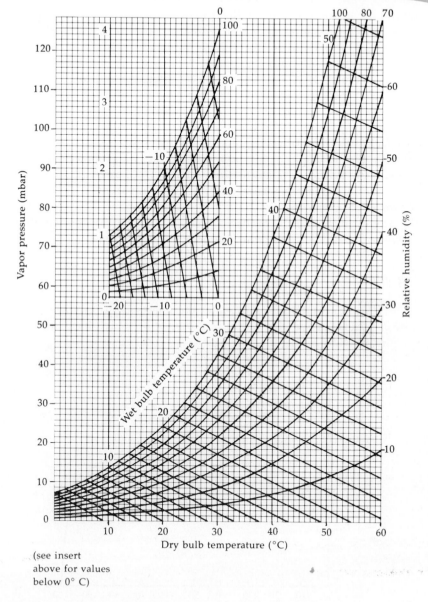

Figure 17–2. The relationship between vapor pressure, relative humidity, and wet and dry bulb temperature at sea level. (Modified from Campbell, G. S. 1977. *An Introduction to Environmental Biophysics.* By permission of Springer-Verlag, New York.) The inset expands the lower, left-hand corner of the graph, for cases where dry bulb temperatures are below 0°C.

midity is constant over a range of air temperatures and vapor pressures making it a meaningless term unless air temperature is specified. **Saturation deficit**, a more meaningful measure of the evaporative power of the air, is the difference between the actual vapor pressure and the saturation vapor pressure at the same temperature. For example, in Figure 17–2, the saturation vapor pressure (100% relative humidity) at 20°C is 22.5 mbar and at 50% relative humidity vapor pressure is 11.2 mbar, leaving a saturation deficit of 11.3 mbar. The saturation vapor pressure at 30°C is 40.5 mbar and at 50% relative humidity vapor pressure is 17.4 mbar, resulting in a 23.1 mbar saturation deficit. The potential evaporation is much greater at 30°C than at 20°C even though the relative humidity is, in both cases, 50%. It is important to understand that factors determining evaporation are not restricted to the physical state of the atmosphere but also reflect the water potential of the evaporating surface, a relationship discussed in Chapter 18. The importance of the evaporative power of the air is apparent in Figure 17–3, where at 25°C the relative humidity must be above 97% to approach the Ψ of even the most xerophytic plants. Consequently, there is almost always a Ψ drop between leaves and the atmosphere.

Fog and Dew

Dew, the condensation of water on objects in the environment, is a source of water for plants, the importance of which has been much

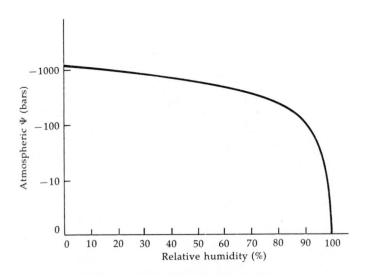

Figure 17–3. The relationship between atmospheric water potential (Ψ) and relative humidity at 25°C (1 bar = 0.1 MPa).

debated. In a lucid review of the literature on dew, Stone (1963) reported that there was no evidence that would suggest that the presence or absence of dew acted as a limiting factor, but that it was a contributing factor to competitive success. More recent studies support Stone's conclusions and document conditions where the presence of dew may promote survival.

Condensation of dew reaches the greatest levels under clear skies (clouds reduce reradiation), low wind speeds, high vapor pressures (which reduce the amount of cooling necessary), and beneath vegetation with open branching and leaf area spread over considerable heights. Lloyd (1961) found that no dew formed under a closed forest canopy, presumably because reradiation was reduced. Dew formation is also enhanced with cool surfaces, poleward facing slopes, in flats and valleys with cold air drainage (Chapter 14), and in areas subjected to cold ocean air when conditions are not such that fog will form. Even under optimal conditions the theoretical maximum dewfall during a single night is less than 1 mm (Slayter 1967; Stone 1963).

Soil water may be redistributed by vaporization and condensation, similar to dew formation, within the soil profile (referred to as **distillation** by Stone (1963)). Syvertsen et al. (1975) used this phenomenon to explain daily fluctuations in soil water potential in the 20–40 cm depth rooting zone of desert shrubs. As a result of a temperature lag in the soil profile (see Chapter 16), the soil below the rooting zone is warmer than the soil in the rooting zone. The result is vaporization below the rooting zone and condensation in the rooting zone, thus redistributing the soil water to areas where it may be absorbed.

Morphological differences in species may influence the relative amount of dew harvested when microenvironmental conditions are appropriate for condensation. Little quantitative data are available concerning dew harvest by various species. Shure and Lewis (1973) noted this difference in ability to harvest dew accidentally while introducing radioactive tracers into stems of old-field weeds. Wells placed on stems to hold tracers were found full of dew water on common ragweed (*Ambrosia artemisiafolia*) but empty on wild radish (*Raphanus raphanastrum*) and lamb's quarter (*Chenopodium album*). Ragweed concentrated 2.5–3.2 ml plant^{-1} day^{-1} for an estimated 1100 ml m^{-2} mo^{-1}. This consistent addition of soil water during the growing season may be the factor which gives ragweed the competitive edge during the first year of succession in the northeastern United States.

Stone also found that dew had a positive influence on *Pinus ponderosa* seedlings subjected to drought. Stressed seedlings lived an average of 3 months if sprayed with water while those not sprayed lived only 2 months. Unsprayed seedlings used twice the water of sprayed seedlings. Similar advantages were found for corn stressed to near wilting (Duvdevani 1964). Therefore, dew may be important in reducing water requirements by reducing the leaf-air Ψ gradient even when it never enters the soil.

Fog, especially along coastal areas (Figure 17–4), has long been recognized as an important ecological factor. Fog forms when warm, moist air: 1) passes over cold water or land as in the coastal communities of the Pacific Northwest; 2) cools as it ascends a mountain mass as in tropical cloud forests; or 3) is cooled by rapid reradiation from soil, as in inland valleys. Fog formation is greatly enhanced by the presence of condensation nuclei. Condensation nuclei are frequently salt particles released by the bursting of bubbles from surf, white caps, etc. (Boyce 1951), or dust, pollen, and other particulate matter in the atmosphere.

The importance of fog precipitation decreases dramatically with distance from the edge of a forest because the water droplets are rapidly removed by obstacles in the environment. This is why fog gauges are rain gauges with a cylindrical verticle screen extending above the opening, forming an obstacle causing water droplets to enter the rain gauge.

The giant coast redwood forests of northern California are frequently said to exist because of fog. This too has been the subject of controversy. On one hand, Byers (1953) indicated that during the critically dry summer months, fog drip is rare in the redwoods. He said that summer conditions are typically cloudy between 90 and 450 m eleva-

Figure 17–4. Early summer coastal fog in southern California.

tion, which is above most large redwood groves. Oberlander (1956), however, measured summer fog precipitation on the San Francisco peninsula and found that 4.8–150 cm of water was added to the soil during a single month without rainfall. Azevedo and Morgan (1974) measured 123–388 cm (depending on the species) of fog precipitation in coastal California forests during 28 summer days. Davis (1966) suggested a relationship between ocean fog and the distribution of spruce-fir forests on the coast of Maine, and Vogelmann et al. (1968) measured a 66.8% increase in precipitation due to fog at 1100 m elevation, but little or no increase below 850 m in the Green Mountains of Vermont. It is now established that fog is a potentially significant source of moisture, increasing precipitation 1.5–3 times in many habitats.

There may be benefits of dew and fog other than providing additional moisture. For example, the leaching of **foliar metabolites** from plants may result in rapid nutrient recycling, especially of K and Ca (Henderson et al. 1977; Tukey and Mecklenberg 1964; Azevedo and Morgan 1974). Tukey (1966) measured carbohydrates, amino acids, and organic acids in plant leachates and del Moral and Muller (1969) measured allelopathic substances in plant leachates.

The direct absorption of water from the surface of leaves into the plant does occur but is greatly restricted by a number of factors. The water potential at the surface of mesophyll cells is relatively high so that only a small energy gradient is established toward the leaf interior. Stomates are usually closed during periods of dew deposition thus severely restricting water entry, especially in xeric adapted plants (Vaadia and Waisel 1963). Slatyer (1967) concluded that absorption of surface water on leaves could never comprise more than a small portion of the water requirements of plants; at best it may speed nocturnal recovery of turgor pressure in wilted plants and may delay the onset of stress after sunrise.

Precipitation

Water droplets formed around condensation nuclei are called fog when in contact with the earth; they are called clouds when a layer with no condensation appears between them and the earth. As these water droplets become large enough to respond to gravity, precipitation will fall as rain. At temperatures below freezing and when conditions are such that the water arrives at the earth's surface as ice, the precipitation is called snow. Under certain circumstances sleet or hail may result from summer storms, but these forms of precipitation are of little consequence as sources of water for plants (although they may cause considerable physical damage on a local basis).

Condensation of atmospheric moisture sufficient to result in precipitation is stimulated by any of three sets of atmospheric conditions.

Temperatures generally decrease with altitude so that any condition where warm, moist air rises may cause precipitation. This type of precipitation is classified as either orographic precipitation (oreos = mountain) or convectional precipitation. **Orographic precipitation** occurs when an air mass is forced upward as it moves across a mountain mass. **Convectional precipitation** occurs during the summer as warm humid air rises abruptly from the earth's surface because of its low density. As this air reaches high altitudes, intense but usually brief summer showers or thunderstorms occur. Cool season precipitation in temperate areas is usually associated with low pressure centers moving easterly and along cold polar air masses. **Cyclonic precipitation** occurs as air masses move counterclockwise (in the Northern Hemisphere) around the low pressure center, become warmed on the advancing front and ascend over cool local air masses, cool, and cause large areas of sustained precipitation. Because of the seasonally predictable location of cool polar air, high pressure areas, and pathways of cyclonic storms, seasonal patterns and amounts of precipitation are quite predictable in those areas where the dominant cause of precipitation is cyclonic.

Rainfall

The spatial and temporal distribution of rainfall is very complex and is closely related to the productivity, distribution, and life forms of the major terrestrial biomes. The factors which determine the distribution and availability of rainfall and the response of plants to these variations have occupied plant ecologists for decades.

Vegetation may reflect patterns of orographic precipitation. In California, for example, the deciduous coastal sage scrub community near the coast gets less precipitation than the higher elevation chaparral. In very high mountains, such as the Sierra Nevada, air masses continue to move upward after the majority of water has precipitated so that subalpine and alpine communities get much less precipitation than middle altitude associations. Moisture-laden air masses frequently pass around high peaks through depressions and breaks in mountain ranges, leaving these peaks essentially devoid of orographic precipitation.

Once these air masses cross the summit of a mountain mass, the air expands and warms, greatly increasing its water holding capacity. The effect is to increase potential evapotranspiration and to decrease precipitation on the lee side. This low precipitation belt is referred to as a **rainshadow**. The pattern of precipitation in the Cascade Mountains of Washington is influenced primarily by topographic factors; it is illustrated in Figure 17–5. For the Sierra Nevada, see p. 497.

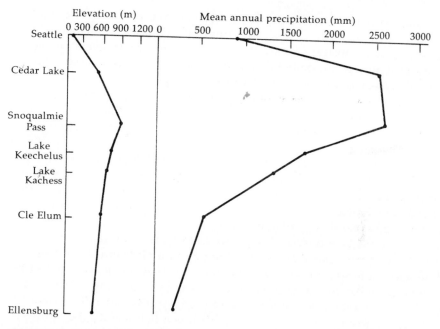

Figure 17-5. Precipitation along a cross section of the Cascade Range in the vicinity of Snoqualmie Pass, Washington (47° 25′ N. lat.); the distance from Seattle to Ellensburg is approximately 152 km. (From Franklin and Dyrness 1969, *Vegetation of Oregon and Washington,* U.S. Dept. of Agriculture Forest Service Research Paper PNW–80.)

Topography also influences the hydrological cycle by affecting potential evapotranspiration. Chapter 14 describes the variations in solar radiation incident upon slopes of various aspects. Significantly less total solar radiation is incident on poleward slopes, leading one to suspect that more water would be present on these slopes because of lowered evapotranspiration. However, Griffin (1973) and others showed that plants on poleward slopes could be under greater stress during the dry season than those on warmer slopes. Presumably these poleward slope plants have a reduced capacity to restrict water loss compared with those on other slopes, and therefore deplete the available water supply more quickly.

Humid air masses may have accumulated significant amounts of water vapor from several sources. The most important source is the oceans, where salt particles form condensation nuclei and cloud formation results. Ocean proximity then, is important in determining amounts of precipitation. Proximity to the ocean does not, however, always assure a mesic environment. In certain geographic situations such as in Baja California, Peru, and southwestern Africa, coastal deserts are formed when cold air masses (cooled by passing over cold ocean currents) contact a warm land mass; the air masses warm and cause high rates of evaporation. Conversely, warm ocean currents create warm air masses and are the source of very high levels of coastal precipitation when these heavily water-laden air masses contact cooler land masses, as in the Pacific Northwest.

Snow

Snowfall has significant ecological impact in many areas of the world as the prime source of water to replenish underground supplies, as a source of soil moisture, as an insulator in areas of extreme cold, and as a physical force important in controlling the distribution of vegetation. Snow varies considerably in water content per unit volume but typically 8–12 cm of snow is equivalent to 1 cm of rainfall. Increases in soil water from snow depend on the rate of melt. Sudden increases in temperature may cause rapid snowmelt so that much will be lost as runoff, similar to a very intense rainstorm. Slow melting will increase infiltration and water availability to plants in the immediate vicinity. Temperatures under snow seldom fall much below 0°C, thus protecting plants from extreme cold. Protection from wind also is important where temperatures are low and winds are desiccating. The physiognomic modification called **krummholz** (Figure 17–6) is due to physical and desiccation damage caused by wind and blowing ice crystals. Patterns of vegetation in subalpine forests are also dependent upon snow. Meadows within subalpine forests are often the result of deep snow accumulation and the resulting shorter growing season. Snow avalanches frequently clear large areas of subalpine forest, thereby modifying the vegetational mosaic of steep mountain slopes.

Snow has been more thoroughly studied in alpine tundra habitats than in other communities because of the very great importance of snow in determining the distribution and abundance of plants in that habitat. Summer rainfall is typically very low in alpine communities so that the snow cover provides the major source of soil moisture for these plants. In fact, most scientists agree (e.g., Billings and Bliss 1959; Johnson and Billings 1962; Buttrick 1977; Hrapka and LaRoi 1978; Komarkova and Webber 1978; Flock 1978) that the related factors of snow cover and soil moisture are the primary limiting factors for vascular and nonvascular plants in alpine communities. As shown in Figure 17–7, a constant supply of water below snowbeds results in meadow communities in drained areas and sedge-sphagnum communities in concave areas. The higher, steeper slopes above snow accumulations and melt are usually boulder fields with little vegetation because of the combined influence of maximum winter exposure and summer desiccation. The importance of snow as an ecological factor results from both its influence on water supply and its role in physical modification of the habitat (see also the discussion of alpine habitats in Chapter 19).

Soil Moisture

Soil water potential ultimately determines the availability of water for plant growth. It is important, then, to understand the relationship between the physical properties of soils discussed in Chapter 16 and the availability of water for plants. We will discuss the properties of soils which influence retention, infiltration, and movement of water, but first we need to define several terms basic to understanding soil water relations.

Moisture status in soils: basic definitions

Water entering the soil will fill most pore spaces and will drain downward through pores in response to the pull of gravity. We can add a term to our initial water potential formula for gravitational potential (Ψ_g):

$$\Psi_{soil} = \Psi_m + \Psi_p + \Psi_\pi + \Psi_g$$

Figure 17-6. Krummholz vegetation at timberline in the Rocky Mountains. Flagged appearance is caused by the abrasive and desiccating effects of wind and blowing ice crystals. (Photo courtesy of Harold Bradford.)

(a)

Figure 17-7. (a) Uneven patterns of snow accumulation in a Rocky Mountain alpine community. (b) Relationship between topography, moisture, and snow duration gradients in northern British Columbia alpine communities. ((a) Photo courtesy of Harold Bradford. (b) from Buttrick 1977. Reproduced by permission of the National Research Council of Canada from the *Canadian Journal of Botany*, Volume 55, pp. 1399–1409, 1977.)

(b)

Gravitational potential is only significant in saturated soils. Water moving through soil in response to gravitational forces is gravitational water. Once gravitational water has moved from the soil or is dispersed in the soil, the soil is left at **field capacity (FC)**. **Field capacity** is the amount of water that can be retained by a soil after gravitational water has been removed. Field capacity is a constant for a particular soil, depending on soil texture, soil structure, organic content and ≈⅓ bar.

The water available for plant use from soils at field capacity depends on the species, the physiological state of the individual, and environmental conditions. The lower limit of water availability is designated as the **permanent wilting point** (PWP). PWP is reached when Ψ_{soil} is equal to or below the minimum osmotic potential of the plant so that water cannot be removed from the soil and the plant wilts and remains wilted even if placed in a cool, dark, humid chamber. Addition of soil water may allow recovery. The amount of water between FC and PWP is called **available water** or **capillary water**, since it is retained against the pull of gravity by capillary action and matric forces. Plants are able to utilize capillary water to varying extents but cannot remove all water from the soil. Below the PWP, water is held so firmly to the soil matrix that it is removable only under conditions surpassing biological tolerance. The most important components of water potential in nonsaturated soils are matric potential and, in some cases such as saline soils, osmotic potential. PWP for mesophytes is often near −15 bars.

The relationship between FC, soil water content, and Ψ_{soil} depends to a large degree on soil texture (Figure 17–8). Clay soils hold large volumes of water at relatively lower water potentials than do coarse textured soils. This is because of the greater total pore space and greater particle-water contact space in fine textured soils.

Movement of soil water

Availability of water within a soil depends, in part, on how quickly and how far water will move upward from the water table

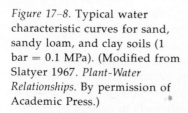

Figure 17–8. Typical water characteristic curves for sand, sandy loam, and clay soils (1 bar = 0.1 MPa). (Modified from Slatyer 1967. *Plant-Water Relationships*. By permission of Academic Press.)

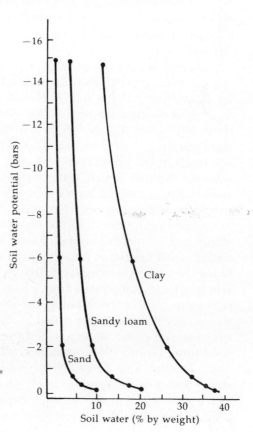

toward an absorbing root, or into the soil during a rain storm. **Infiltration** and movement of soil water depends on factors which determine soil water potential. Water will infiltrate into a dry soil faster than a wet soil if they have comparable texture and structure, because the gradient in water potential is greater in the dry soil. The rate of flow of water through a saturated soil (v) is:

$$v = -K \frac{\Delta \Psi_{soil}}{D}$$

where K is the **hydraulic conductivity** (the capacity of a soil to transport water); Ψ_{soil} is the water potential of the soil; and D is the depth in the soil. As Ψ_{soil} drops, there is a significant drop in hydraulic conductivity (Figure 17–9), especially in sandy soils where the continuity of the water is broken. In finer textured soils hydraulic conductivity also decreases but to a lesser extent because the continuity of water surfaces are maintained at a lower Ψ_{soil}. After gravitational water has been dispersed, the upper zone of the soil profile is at field capacity, and then there is a narrow zone of soil with a very steep drop in water content; the soil below that is essentially dry (Figure 17–9(b)). Water may rise from an underground water table in limited quantities. The rate of movement depends on the water potential gradient and the depth to the water table.

Lateral movement of water through the soil to zones of absorption by roots depends on the water potential gradient and the hydraulic conductivity of the soil. Root water potential must be lower than Ψ_{soil} for absorption to occur. This water potential difference can be quite low when Ψ_{soil} is -5 bars or higher, whereas at -15 bars, a significantly greater drop in potential is necessary for absorption. This is a result of the decrease in hydraulic conductivity as Ψ_{soil} drops. The distance through which water will move toward the dry zone surrounding a root depends on the plant's tolerance for stress and the soil properties which influence hydraulic conductivity.

Water infiltration into the soil is a critical factor in areas of steep slopes and/or intense rainfall. Sandy soils generally have rapid infiltration, with finer soils restricting the entrance of water in proportion to clay content. An exception exists for sandy soils in certain situations where organic compounds from plants form "skins" over soil particles so that, rather than penetrating, the soil water droplets bead up on the soil surface. Such soils are called hydrophobic or water repellent, and have been observed under a wide variety of species and environmental conditions (Bond 1964; Krammes and DeBano 1965; Adams et al. 1970; DeBano and Letey 1969). Water repellency increases surface runoff and post-fire erosion on chaparral soils in southern California. When chaparral burns, the water repellent zone moves into the soil (Figure 17–10), leaving a wettable zone at the surface. Post-fire rains

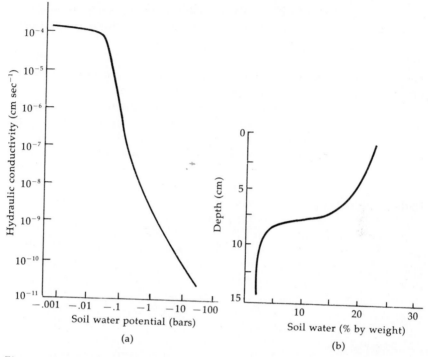

Figure 17-9. (a) The relationship between hydraulic conductivity and soil water potential in a loam soil (1 bar = 0.1 MPa). (b) The water content profile of an initially dry loam soil at one day following irrigation. (After Slatyer 1967. *Plant-Water Relationships.* By permission of Academic Press.)

penetrate and saturate the wettable zone, causing the surface soil to slip downslope.

Not all species in a habitat have the same influence on wettability and few communities have been characterized for wettability patterns. Soil wettability is another environmental factor which regulates water availability, and therefore, regulates the niche relations of plant species.

Precipitation Effectiveness

Knowledge of the total amount of precipitation falling on a community does not always give a clear picture of the availability of water to plants. The season, atmospheric condition, precipitation type, intensity, annual variation, soil condition, and vegetation physiognomy influence the availability of precipitation and its distribution within the habitat. Many attempts have been made to determine **site water balance** based on measurements of rainfall, runoff, percolation, and evapotranspiration (Rosenberg 1974). However, the accurate measurement of all the parameters necessary to estimate these properties is technologically difficult.

The conditions discussed earlier which influence atmospheric vapor pressure are important determinants of water loss through evaporation. In fact, measurements of temperature, saturation deficit, and wind have been combined to give a meaningful picture of biologically significant variations in climate (see Chapter 14). Significant evaporation occurs from free water surfaces, such as droplets intercepted by the vegetation canopy and slow draining depressions. Evaporation from exposed soils is limited to the upper 10–20 cm; there is, however, a certain amount of water vapor transfer from lower levels, especially in desert environments. A similar transfer to the vapor state may occur

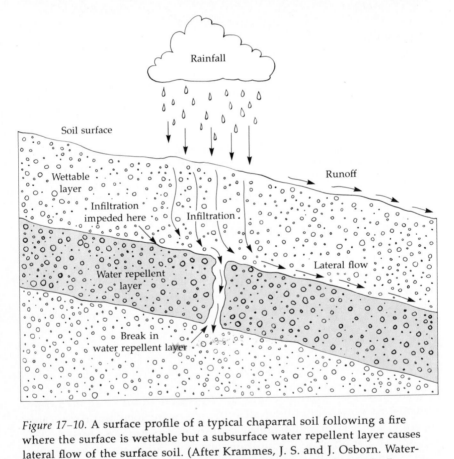

Figure 17–10. A surface profile of a typical chaparral soil following a fire where the surface is wettable but a subsurface water repellent layer causes lateral flow of the surface soil. (After Krammes, J. S. and J. Osborn. Water-repellent soils and wetting agents as factors influencing erosion. In *Proc. Symp. Water-repellent Soils.* L. F. DeBano and J. Letey, eds. 1969. By permission of the University of California Press, Riverside, CA.)

from the surface of snow or ice. This process, called **sublimation**, takes place as ice is transformed to the vapor state without going through a liquid phase. Sublimation, like evaporation, reduces the effectiveness of precipitation.

Seasonal distribution of precipitation determines, in part, how much of the water is available to the vegetation. An annual precipitation rate of 70–80 cm can support either certain deciduous forest communities, or the strongly drought adapted broad sclerophyll vegetation of mediterranean climates. The distinguishing factor is that almost all of the precipitation in mediterranean climates falls during the cool season so that chaparral plants must endure annual extended drought during the hottest months. If the precipitation were more evenly distributed or concentrated in the warm season, a more mesic vegetation could be expected. Where a large portion of the precipitation falls as snow, its effectiveness depends on the melt rate. When temperatures warm suddenly, a rapid melt will render most of the snow ineffective since much of the moisture will be lost as runoff, particularly if the soil is frozen or already saturated.

Intensity of rainfall is an important determinant of runoff and is related to seasonality. Typically, summer showers in temperate areas are of very short duration and may be so intense that most water will run off the surface before the slower soil infiltration process can take place.

Predictability of precipitation is also reflected in the flora and vegetation. Many deserts average enough annual precipitation to support more mesic plants, but rainfall is very unpredictable and periods between effective precipitation may extend up to several years. Plants are, therefore, adapted to respond quickly to precipitation and also to tolerate periods of drought. Where droughts are predictable, seasonality patterns in plant response are also predictable and little impact on the vegetation is apparent. However, in mesic environments periodic unpredictable droughts may have severe impacts on the vegetation.

Many attempts (e.g., Penman 1950; van Bavel 1966; Major 1977) to measure site water balance have been made. Recently, Grier and Running (1977) devised a site **water balance index** which correlates closely with leaf area in coniferous forest communities of western Oregon (Figure 17–11). The index was calculated by adding soil water storage to measured growing season precipitation and then subtracting open pan evaporation. This approach may be very instructive in those areas where low nutrient levels, severe temperature, physical damage, and high levels of runoff are not important factors. Major (1977) has compiled water balance information (based on the concepts of Thornthwaite; Chapter 14) for 12 sites on a transect across the Sierra Nevada (Figure 17–12). Water surpluses are greatest at mid-elevation where summer water deficits are less than at lower elevation sites in the Cen-

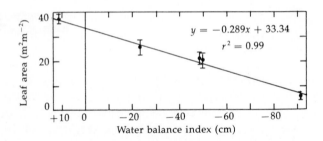

Figure 17-11. Relationship between water balance index and leaf area in five forest zones of western Oregon. Zones plotted are, from left to right: *Picia sitchensis, Tsuga heterophylla,* interior (Willamette) valley, east slope mixed conifer, and *Juniperus occidentalis.* Bars show range of observed leaf areas in each vegetation zone. (From Grier and Running 1977. Copyright 1977 by the Ecological Society of America.)

tral Valley or in the rainshadow of the Sierra Nevada where higher temperatures increase potential evapotranspiration.

Interception, Throughfall, and Stem Flow

The distribution of rainfall under a forest canopy depends on canopy structure and its influence on the amount of water entering the soil at the base of the trunk (stem flow) and the amount of rainfall coming through the canopy (throughfall). Lawson (1967), Leonard (1961), and Henderson et al. (1977) agreed that 10–20% of precipitation is intercepted by forest canopies (somewhat less during the dormant season). Henderson et al. (1977) compared throughfall in various forest types and found no significant differences (Table 17–2). Most of the intercepted water is lost to evaporation since little will be absorbed by the leaves. Stem flow causes a very uneven distribution of soil water by concentrating precipitation near the trunk of the tree. Geometric shapes which encourage stem flow may be a competitive advantage in water limiting environments.

WATER: GROWTH FORM RESPONSES AND HABITAT SELECTION

Many of the distribution patterns and even the growth forms of entire communities can be related to the spatial and temporal distribution of water. In this chapter we will consider ecological and morphological responses to water; in Chapter 18 we will discuss physiological and anatomical adaptations to water.

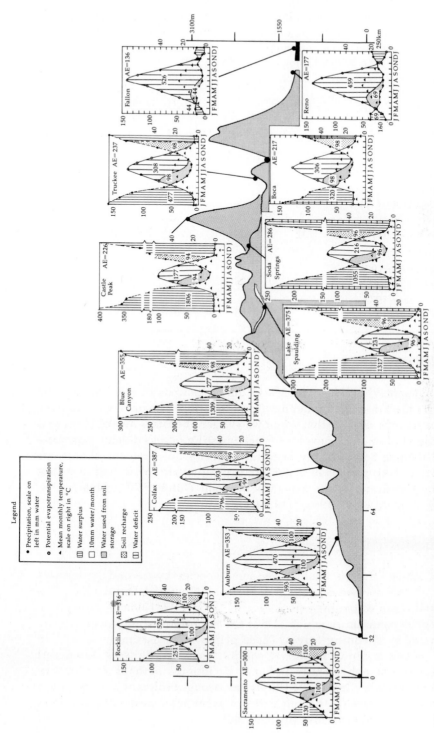

Figure 17–12. A transect over the Sierra Nevada from Sacramento (38.5°N), over Donner Summit (39.3°N), to Reno and Fallon, Nevada (39.5°N), showing water balances. Vertical exaggeration 14.5×. (From Major 1977. In *Terrestrial Vegetation of California*. Edited by Barbour and Major. Copyright 1977 John Wiley and Sons, Inc. Reprinted by permission of John Wiley and Sons, Inc.)

Table 17-2. Quantities of precipitation and throughfall (mm) in four forest types. Differences in throughfall between the forest types are not significant ($P = 0.05$) for any of the four periods. (From Henderson et al. 1977. By permission of the British Ecological Society.)

	Precipitation	Throughfall			
		Pinus	*Liriodendron*	*Quercus prinus*	*Quercus-Carya*
Growing Season:					
18 May 1971– 13 December 1971	710	596	615	591	588
18 May 1972– 7 December 1972	732	625	614	617	630
Dormant Season:					
13 December 1971– 18 May 1972	672	578	587	586	601
7 December 1972– 18 May 1973	942	805	806	793	801

Evergreen Plants

Plants which maintain leaves during periods of water stress must have a high degree of tolerance for desiccation. The obvious advantage of being evergreen is that when water again becomes available there is no lag while new tissues are formed; existing tissues quickly rehydrate and become active. Many evergreen species do, however, shed some leaves during periods of water stress; presumably this serves to reduce surface area and water loss (Mooney and Dunn 1970; Evanari et al. 1971; Oppenheimer 1960; Orshan 1972).

Drought-Deciduous Species

Water stress is avoided by some species by becoming dormant during the dry season (Mooney 1977; Mooney and Dunn 1970). The coastal scrub community of California (Figure 17-13) is an example of an entire community which avoids the hot, dry mediterranean climate summer by shedding leaves. These plants show maximum activity in the cool, wet winter months. Ocotillo (*Fouquieria splendens*) is a desert shrub which ephemerally becomes deciduous, perhaps four or five times a year. The leaves are simply not drought tolerant and they drop as stress increases, returning within a week after new rains. (Cannon 1905; Scott 1932) (see also pp. 539–540).

Paloverde *(Cercidium floridum)* is an example of a drought-deciduous plant which lives in desert areas in a leafless condition but has the advantage of green stem tissue. This gives paloverde and other similar desert plants the advantages of evergreenness and the advantage of increased photosynthetic area during cool, moist periods (Adams and Strain 1968, 1969; Adams et al. 1967).

Phreatophytes

Phreatophytes (literally, "well plants") are usually found in **riparian** (streamside) habitats and, even in arid zones, may develop into very large trees (Figure 17–14). They are restricted to habitats of permanent underground water supplies. Cottonwood *(Populus fremontii)*, willow *(Salix* spp.), sycamore *(Plantanus racemosa)*, fan palm *(Washingtonia filifera)*, and salt cedar *(Tamarix* spp.) are examples of phreatophyte trees which avoid many of the rigors of arid environments by having roots in constant contact with the fringe of capillary water above a water table. These and many associated shrubs are examples of obligate phreatophytes. Several species are apparently able to take advantage of ground water when present, but can tolerate periods of low water availability. These facultative phreatophytes (e.g., mesquite *(Prosopis glandulosa)* and saltbrush *(Atriplex polycarpa))* frequently occur in desert depressions where water and salts accumulate. Because of their high water requirement, many facultative and obligate phreatophytic

Figure 17–13. Drought-deciduous coastal scrub community of southern California.

Figure 17-14. Riparian community with large phreatophyte trees in the Chihuahuan Desert of the southwestern United States.

species must be tolerant of rather high levels of salinity. Horton (1977), Horton and Campbell (1974), McDonald and Hughes (1964), Campbell and Dick-Peddie (1964), and Vogl and McHargue (1966) have considered various aspects of phreatophyte ecology.

Not all plants associated with water courses are phreatophytic, but in a given location they are only able to survive in areas with more soil moisture. Riparian plants may also be those able to survive low oxygen conditions in flooded or highly saturated soils which eliminate upland species from flood plains. In mountain areas a species may be restricted to riparian habitats at lower elevations and extend on to slopes only at higher elevations where more water is available. In desert areas, some species (e.g., *Chilopsis linearis*) are winter deciduous and relatively intolerant of midsummer conditions, so they survive only near water courses where the additional moisture allows survival between spring and fall growth periods (Odening et al. 1974).

Ephemerals

Ephemerals are annual plants which germinate in response to periodic phenomena and are able to complete their life cycle during short periods of mesic conditions. Ephemerals are the most common life form in severe desert situations (Whittaker and Niering 1975), where they are able to survive because they grow only during periods of moderate temperatures and soil water availability. Germination and mortality of desert ephemerals are largely independent of photoperiod, since they are controlled by soil water and temperature (Went and Westergaard 1949; Went 1948; Tevis 1958a, b). Beatley (1974, 1969, 1967, 1966) has studied the relationship between the ephemerals of the Mojave Desert and the environmental factors which trigger germination and death (see Chapters 4 and 19). C_3 and C_4 plants occupy different temporal habitats, summer ephemerals being typically C_4, and winter ephemerals being typically C_3 in the southwestern desert areas (Mulroy and Rundel 1977; Syvertsen et al. 1976).

SUMMARY

The potential free energy of water in plants, soil, or atmosphere is expressed as water potential. Water potential is a measure of the combined influence of particle, surface, solute, and pressure forces as they influence the free energy of pure, free water. Water moves in response to decreasing gradients of water potential. Therefore, in order for water to move from soil through the plant and into the atmosphere, Ψ_{soil} must be greater (less negative) than Ψ_{plant}, and Ψ_{air} must be lower (more negative) than Ψ_{plant}.

The success of plants depends on a multitude of factors which, in terrestrial environments, always includes limitations related to the water factor. Environments differ in the relative amounts of rain, snow, dew, and fog precipitation they receive and in the seasonal distribution of available water. These factors interact with habitat and phenological development of plants to determine the effectiveness of precipitation. The distribution of 1000 mm of precipitation in an environment with a precipitation/evaporation ratio of 2 is presented in Figure 17-15. Large amounts of precipitation are stored in the soil or lost as runoff and drainage; minor amounts are intercepted by the vegetation canopy (most of which evaporates); and most of the soil water is vaporized through the vegetation as transpirational water loss. Evapotranspirational water flux depends on the amount of water present and on the evaporative power of the air.

The amount of precipitation received in a habitat depends on temperature, geography, and topography. Ocean proximity, temperature and moisture content of air masses, elevation, latitude, and relationship to seasonal changes in atmospheric pressure determine the amount, kind, and seasonal distribution of precipitation.

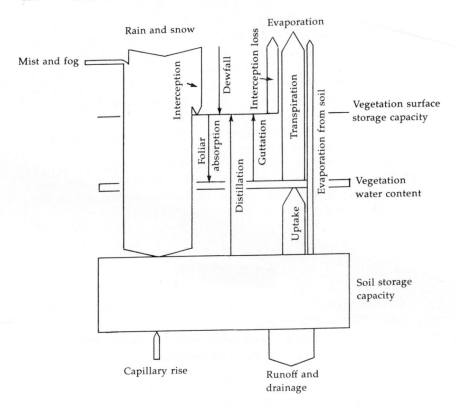

Figure 17–15. Diagram summarizing the hydrological cycle in a vegetation-soil system; the components are approximately to scale for an annual rainfall of 1000 mm and evaporation of 500 mm. (Guttation is the exudation of liquid water from leaves.) (From Rutter 1975. The hydrological cycle in vegetation. In *Vegetation and the Atmosphere,* vol. 1, ed. J. L. Monteith. By permission of Academic Press, London.)

Soil water exists in one of three categories: gravitational water moves down through the soil; capillary water is held by matric forces, and is the reserve from which plants obtain water; the remaining water is not available to plants because the soil water potential is equal to or below the minimum osmotic potential of the plant. Movement of soil water depends on the interaction of soil water potential and hydraulic conductivity.

Plants have adapted to variations in water availability by habitat selection and phenological adjustment as means of avoiding severe water stress. Drought-deciduousness (leaves present only during periods of low stress), ephemerality (existance as dormant seeds during periods of stress), riparian (growing near areas of greater water availability), and phreatophytism (growing where a perennial source of water is available) are mechanisms by which various plants avoid water stress. Evergreen plants are able to tolerate periods of drought and therefore endure, rather than avoid, water stress.

CHAPTER 18

PLANT WATER RELATIONS

Our objective in this chapter is to examine the nature of water movement through the plant and to examine the nature of water stress. Most terrestrial plants are periodically subjected to water stress. In this chapter we will discuss physiological and structural responses which render a plant resistant to water deficits, or postpone the onset of water stress. We will also discuss methods of measuring soil and plant water status.

THE SOIL-PLANT-ATMOSPHERE SYSTEM

A continuous path for water exists from the soil through the root, stem, leaf, and into the atmosphere. The factors which affect this pathway are shown in Figure 18-1. For water to move through the plant, it is necessary that $\Psi_{soil} > \Psi_{root} > \Psi_{stem} > \Psi_{leaf} > \Psi_{air}$. Even with a suitable Ψ gradient, there will be resistance to flow that is inherent in the system (Slatyer 1967; Meidner and Sheriff 1976). The pathway of water through the plant follows the scheme illustrated in Figure 18-2. Water enters the root primarily through root hairs, not only because they have greater permeability than older roots, but because they have a much greater surface area for absorption. Older roots lack root hairs and become suberized* and less permeable to water. From the root hairs, water passes along the course of least resistance through cell walls and between root cells to the endodermis. Here the water and dissolved material are forced to move through the cell membrane because the casparian strip makes the cell wall impermeable and attaches the membrane to the cell wall. Passage through the endodermis offers considerable resistance, but once the water (and minerals) enter the vascular cylinder and the xylem, resistance is significantly lower. Vas-

---

*Suberization is the formation of suberin in the walls of plant cells, which renders them less permeable to water.

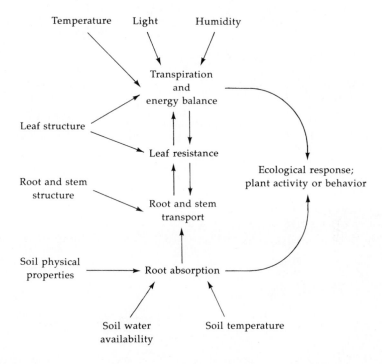

Figure 18–1. Diagram showing the interactions of the physical and biotic factors that are the most important in determining the ecological response of plants to water.

cular strands divide into very fine segments upon entering the leaves and are distributed throughout the mesophyll so that almost every cell is within 1 or 2 cells of vascular tissue (Esau 1965), but no vascular tissue is exposed directly to the intercellular spaces adjacent to stomates. From the vascular tissue, water continues to travel in the liquid phase through the cell walls and between mesophyll cells to substomatal cavities where it vaporizes and travels out into the atmosphere. This final loss of water from the plant (**transpiration**) is the most significant force in the movement of water because here the water potential gradient is the steepest. Slatyer (1967) indicated that a common set of water potential measurements along the gradient might be: soil −1 bar, stem −10 bars, leaf −15 bars, atmosphere −1000 bars. The most effective regulation of plant water status is associated with the stomates because of their location in the steepest portion of the gradient. Stomates, because of their ability to open and close, represent the major controlling factor of water flux from the plant to the air.

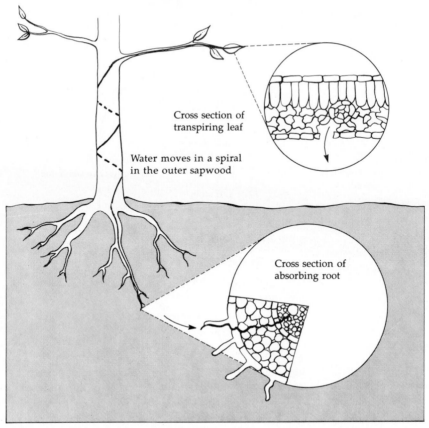

Figure 18–2. Diagrammatic representation of the course of water flow through the soil-plant-atmosphere continuum.

Energy Balance and Transpiration

The **energy balance** of the leaf depends on net radiation (R_n) (cal cm^{-2} sec^{-1}), sensible heat transfer* (H) (cal cm^{-2} sec^{-1}), and heat absorbed by the vaporization of water (LE):

$$R_n + H + LE = 0$$

(signs of any of the terms may be positive or negative) where L is latent heat of vaporization (590 cal g^{-1}) and E is transpiration (g cm^{-2} sec^{-1}). The transpiration rate is:

$$E = \frac{c_l - c_a}{\Sigma r}$$

*Sensible heat transfer is the energy transferred as heat by convection and conduction (combined terms C and G in the energy balance equation of Chapter 14).

which is the water vapor concentration gradient between the leaf (c_l) and the air (c_a) divided by the sum of the resistances to flow along the transfer path (Σr). You may recall that this is related to a form of Fick's law and that we used it to describe CO_2 flux in Chapter 13.

We can envision leaf energy balance by considering the fate of 1.4 cal cm^{-2} min^{-1} of energy absorbed by a hypothetical leaf (Figure 18–3). The diagram shows the relative importance to energy dissipation of net radiation, sensible heat transfer, and transpiration. When leaf and air temperatures are equal, the absorbed energy is effectively dissipated by transpiration and radiation emission (a). When transpiration is stopped due to stomatal closure (b), leaf temperature increases and energy is dissipated by sensible heat transfer and radiation emission. Situation (c) represents the usual case where absorbed energy is dissipated by the combined factors. Temperature and water loss are, therefore, intrinsically related. Their relative values, together with the photosynthetic capacity of a plant, which they directly influence, are a subject of profound importance in physiological ecology.

Plants evolve to maximize photosynthesis in all environments. However, where water is limited there is a trade off between CO_2 up-

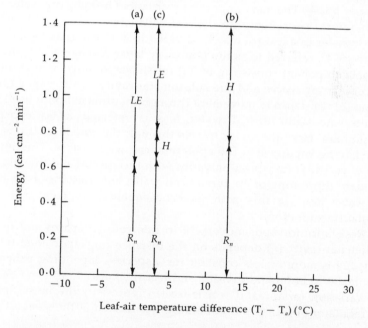

Figure 18–3. Energy exchange diagram for an idealized leaf of downwind width 10 cm, with air temperature 25°C and wind velocity 200 cm sec^{-1}. The total energy absorbed by the leaf is assumed to be 1.4 cal cm^{-2} min^{-1}. (Modified from Slatyer 1967. *Plant-Water Relationships.* By permission of Academic Press.)

take for photosynthesis and water loss by transpiration. Therefore, maximum **water use efficiency** (WUE) is selected for. WUE is the ratio between net photosynthesis (Mg CO_2) and the amount of water (g) consumed in transpiration. The nature and control of factors influencing WUE are the subject of the remainder of this chapter.

Water Transport and the Regulation of Plant Water Status

The rate of water loss is influenced by basic characteristics of the transport pathway and environmental conditions. We have discussed characteristics of the atmosphere and soil which influence plant water relations (Chapter 17), but we have not considered the plant characteristics and plant responses to the environment which form the remainder of the soil-plant-atmosphere continuum.

Leaf resistance

We noted earlier that the steepest water potential gradient in the soil-plant-atmosphere continuum occurs as water leaves the plant in the vapor phase. The most important resistances influencing water loss are, therefore, associated with the leaf. **Diffusive resistance*** to water vapor transfer in a leaf (r_l) may be divided into an external component (Figure 18–4), referred to as the **boundary layer resistance** (r_a), and an internal component consisting of: (a) resistance to movement through intercellular air spaces and the substomatal cavity in the vapor phase (r_i), and (b) resistance to movement through the stomatal pore itself (r_s). An alternate path of leaf water loss is referred to as cuticular transpiration. Here the water travels through the cuticle in the vapor phase having vaporized at the epidermal cell wall surface. The importance of cuticular transpiration varies with the species and the level of secondary thickening of epidermal cell walls, but resistance is so high that water loss via this path is inconsequential in comparison to stomatal transpiration.

Resistance to vapor movement in intercellular spaces and in the substomatal cavity (r_i) depends on the distance the vapor must travel. The value is more or less constant for a species, depending primarily on stomatal location and density, and leaf thickness.

Variation in **stomatal resistance** (r_s) represents the most significant regulator of transpiration because r_s is located at the steepest part

*Resistance units are $s\ cm^{-1}$ and express the same resistance to diffusive flow as equivalent path lengths of air of unit cross-sectional area divided by the diffusive coefficient of water in air ($2.57 \times 10^{15}\ m^2\ s^{-1}$ at 20°C). Conductance ($cm\ s^{-1}$), the inverse of resistance ($1/r$), is frequently used to evaluate barriers to flow.

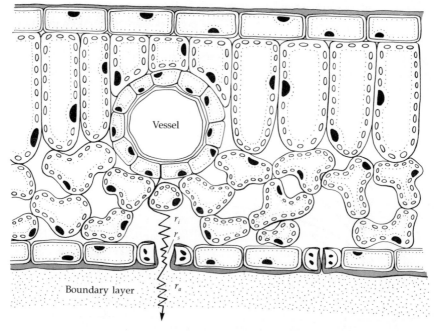

Figure 18–4. Cross section of a leaf showing the resistances associated with water loss. r_i is the resistance to vapor movement in the substomatal cavity, r_s is the resistance offered by the stomatal pore and guard cells, and r_a is the resistance offered by the boundary layer on the leaf surface.

of the Ψ gradient and the guard cells are sensitive to environmental factors. Stomates open and close in response to differences in turgidity stimulated by variations in light, leaf water potential, CO_2 concentrations, and humidity (Figure 18–5). Turgidity changes are related to osmotic potential variations in the guard cells caused by the movement of K^+ into and out of the cells. The mechanism which stimulates the transport of K^+ is not clearly understood even though several hypotheses have been proposed (for a review see Allaway and Milthorpe 1976 or Hsiao 1976). In any event, when potassium ions enter a guard cell, they cause increased guard cell turgidity and opening of stomates. Water stress will stimulate production of abscissic acid (ABA) which drives the K^+ pump in the other direction, causing stomatal closure. The level of water stress necessary to cause ABA production varies with the drought tolerance of the species. For any species there is a critical Ψ_{leaf} at which the stomatal aperture will decrease dramatically, increasing resistance to ∞. Resistance in fully open stomates may range as low as 0.4 s cm^{-1}.

The effect of atmospheric vapor pressure on stomatal aperture has only recently been satisfactorily documented (Lange et al. 1971;

Schulze et al. 1972, 1973; Camacho-B et al. 1974; Hall and Kaufmann 1975; Smith and Nobel 1977a; and others). Again, the mechanism of stomatal response has not been established. A possible explanation was proposed by Lange et al. (1971) based on the concept of **peristomatal transpiration**. Guard cells lose water more readily than other epidermal

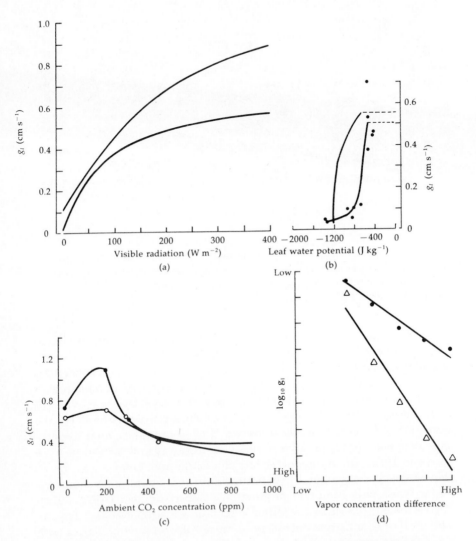

Figure 18–5. Leaf conductance (g_l) changes (primarily stomatal) in response to variations in (a) visible radiation, (b) leaf water potential, (c) ambient CO_2 concentration, and (d) vapor concentration difference between leaf and air for pairs of cultivated plants. (Modified from Burrows and Milthorpe 1976. Stomatal conductance in the control of gas exchange. In *Water Deficits and Plant Growth*, vol. IV, ed. T. T. Kozlowski. By permission of Academic Press.)

cells and, as a result, would respond directly to changes in vapor pressure. Confirmation of the peristomatal transpiration hypothesis remains a subject of scientific inquiry. Gradual closing of stomates, as evaporative demands of the atmosphere increase, may serve to maximize WUE in plants by increasing the time necessary to reach the critical Ψ_{leaf} where ABA stimulated stomatal closure would result.

Boundary layer resistance (r_a) is considered as part of leaf resistance because it is influenced by morphological characteristics of leaves. Resistance to diffusion is caused by a layer of air adjacent to the surface of the leaf which, as with laminar flow over any flat surface, does not readily mix with passing air. Vapor must pass through this potentially saturated layer by molecular diffusion before moving into the free atmosphere. The thickness of the boundary layer depends on leaf size and wind speed. Greater wind speeds cause turbulence closer to the leaf surface and reduce the depth of the boundary layer. Smaller leaves have thinner boundary layers because of increased convective exchange between the surface and the free air. Surface texture, leaf shape, and leaf orientation affect air turbulence near the surface and, therefore, the thickness of the boundary layer.

Optimal leaf form

Taylor (1975) and Campbell (1977) considered the interactions of physical factors such as temperature and light, and biological regulation of water loss (leaf resistance, r_l) on photosynthetic output and water use efficiency (WUE) of plants. Figure 18–6 shows the idealized photosynthetic and WUE responses of leaves of various sizes with different diffusive resistances when exposed to full sunlight. At high air temperatures (30°C; Figures 18–6(a) and 18–6(c)), larger leaves have lower net photosynthetic rates especially when diffusive resistance is high (e.g., $r_l = 32$ s cm^{-1}). This is a result of overheating because at low resistances (e.g., $r_l = 1$ s cm^{-1}), evaporative cooling maintains temperatures which support positive photosynthesis even in leaves 16 cm wide. This evaporative cooling, however, requires large amounts of water and, as a result, the water use efficiency is very low (<4 mg g^{-1}). There is a clear advantage for leaves less than 1 cm wide and with a diffusive resistance above 6 s cm^{-1} in a warm, dry environment. Larger leaves have higher photosynthetic rates than smaller leaves in cool air (10°C; Figure 18–6(b)) because they are at temperatures above ambient and therefore are closer to the optimum for photosynthesis. Figure 18–6(d) shows that water use efficiency drops as leaf size increases but, at comparable diffusive resistances, efficiency remains higher than at air temperatures of 30°C (Figure 18–6(c)). This theoretical model shows that the interactions of temperature, water, and photosynthesis should lead to plants with small leaves in warm, arid environments. This is, in

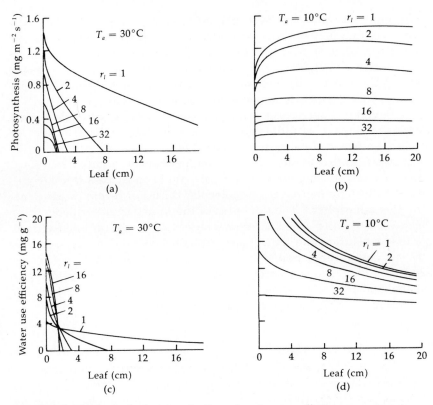

Figure 18–6. Theoretical photosynthetic and water use efficiency responses for leaves of different sizes and for different levels of resistance (r_l) when exposed to ambient temperatures (T_a) of 30°C and 10°C. (From Campbell, G. S. 1977. *An Introduction to Environmental Biophysics*. By permission of Springer-Verlag, New York.) WUE = MgCO$_2$ fixed per g H$_2$O transpired.

fact, the case, and may help explain the great success of leguminous trees in arid regions. Pinnately compound leaves with small leaflets are common in the desert leguminous trees and are apparently a significant advantage in desert habitats because of reduced heat load and greater WUE. In cooler environments, large leaves are warmer and, with low diffusive resistance, have high photosynthetic rates. There is a gradient of increasing leaf size from warm arid to cool moist habitats.

Theoretical photosynthetic and WUE responses to different light intensities and diffusive resistances are presented in Figure 18–7 for intermediate sized (3 cm) leaves. When ambient temperatures are cool (10°C; Figure 18–7(b)), maximum photosynthesis is attained at full sunlight (500 W m^{-2}), but when temperatures are higher (30°C; Figure 18–7(a)), intermediate light intensities are most effective. Full sunlight in conjunction with a 30°C air temperature results in low WUE. In fact, unless diffusive resistance is less than 4 s cm^{-1}, leaf temperature surpasses that tolerable in our hypothetical plant. High WUE is maintained across a range of light intensity when ambient temperatures are low (10°C; Figure 18-7(d)). Many leaves, especially larger leaves, of plants in arid or semiarid habitats are oriented vertically. Vertical orientation allows interception of maximum solar radiation in early morning and late afternoon when temperatures are lower, and reduces interception at midday when temperatures are high. Verticle orientation, then, would maximize both photosynthesis and WUE of our hypothetical plant.

While these relationships are for an idealized situation, they provide results which we can confirm, in general, by observing real

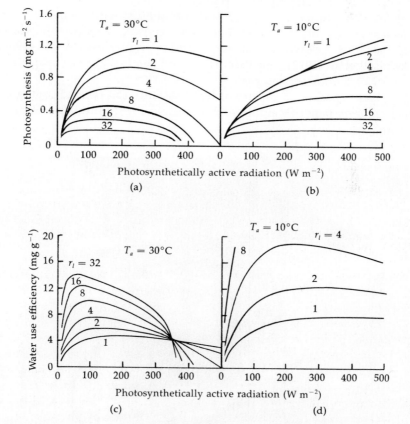

Figure 18-7. Theoretical photosynthetic and water use efficiency responses at different light intensities and for different levels of resistance (r_l) when exposed to ambient temperatures (T_a) of 30°C and 10°C. All leaves are intermediate sized (3 cm). (From Campbell, G. S. 1977. *An Introduction to Environmental Biophysics.* By permission of Springer-Verlag, New York.)

plants. Understanding adaptations of specific plants with regard to these variables will allow us to better interpret the habitat responses observed in nature.

Diffusive resistance and transpiration measurements

The most popular method of measuring leaf resistance is by the use of a **diffusion porometer** (Figure 18–8). A porometer measures the rate of increase in humidity in a small chamber attached to a leaf. Humidity is measured by hygrometers with lithium chloride or aluminum oxide sensors. Thermocouples are incorporated because of the interaction between vapor pressure and temperature (Figure 17–2). Some models have circulating air to reduce the effect of the boundary layer (Nobel 1974); others have plate humidity sensors which adjust close to the leaf surface to lessen r_a influences which may be inordinately large in the small, closed porometer chamber. Kanemasu et al. (1969), Beardsell et al. (1972), and Parkinson and Legg (1972) have proposed a variety of porometer designs.

Less expensive estimates of leaf resistance can be made using cobalt chloride paper (Meidner and Mansfield 1968; Milthorpe 1955), or

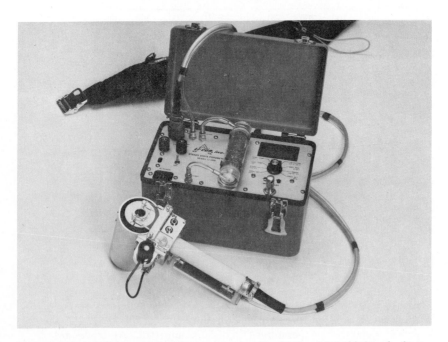

Figure 18–8. A diffusive resistance porometer, the Li-Cor LI–1600 Steady State Porometer. (Courtesy of Li-Cor, Inc.)

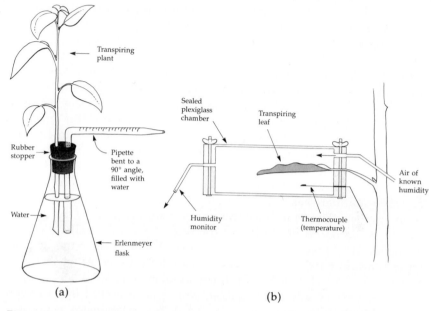

Figure 18-9. (a) Potometer for the measurement of transpiration of plant shoots or leaves. (b) A gas exchange chamber for the measurement of transpiration in continuously flowing air. See also Figure 13-17.

calculation from stomatal dimensions using infiltration techniques or direct measurements (for a review see Burrows and Milthorpe 1976).

Transpiration rates from detached leaves and potted plants are easily made by gravimetric analysis; repeated weighing of the parts over time gives an estimate of transpiration. For attached leaves and whole branches, a potometer is used. Figure 18-9 illustrates one form of potometer which manometrically measures water flux through an excised leaf, branch or shoot system of a plant. More accurate measurements are possible when transpiration measurements are incorporated into a gas exchange system (see Chapter 13).

Stem and root transport

Resistance to mass transport of water in vessels and tracheids is low in contrast to the resistances encountered in other phases of the continuum (i.e., root endodermis and mesophyll). Heine (1971) reviewed the development of concepts relating to conductivity in woody plants and stressed the importance of anatomy in determining water flux. Conductivity depends not only on the Ψ gradient but on lumen (central open portion of the cell) area and length of xylem elements. Carlquist (1975) has shown a relationship between element dimensions and the amount of available water. This close relationship between vascular element size and the environment is apparently related to the negative pressures which develop in xylem.

The Ψ gradient in xylem elements is an expression of a gradient in pressure potential because solute concentration and matric influences (see Chapter 17) are more or less constant throughout the xylem water column. As transpiration proceeds, resistance prevents the instantaneous replacement of transpired water, thereby establishing a negative hydrostatic pressure (tension) in the xylem. This negative pressure is a natural component of vascular water and is intensified when transpiration demand is high and replacement is slow because of low Ψ_{soil} or when high resistance to water flow increases the lag between inflow and outflow. Adhesion of water molecules to cell walls and cohesion between water molecules maintain the integrity of the water column. The more negative the pressure (lower Ψ) the greater the stress placed on the cells and the water column. Smaller vascular elements are, therefore, associated with tissues exposed to the greatest stress (Carlquist 1975). Rundel and Stecker (1977) report that this relationship is clearly evident in tracheid diameters of a 90 m nontranspiring Sierra redwood *(Sequoiadendron giganteum)* in Kings Canyon National Park, California (Figure 18–10(a)). This relationship is not surprising, since one would expect (because of the weight of water in the conducting tissues of tall trees) that upper xylem tissues would be subjected to higher levels of stress even in the absence of transpiration (Figure 18–10(b)).

Water uptake and transport in nonvascular tissues of roots is sensitive to the water potential gradient (as is vascular transport) but it is also sensitive to temperature changes. This affect is related to changes in endodermal cell membrane permeability with temperature. Water uptake in response to temperature is subject to modification by acclimation and varies with species and habitat (Anderson and McNaughton 1973). For example, soil temperature was identified by Fernandez and Caldwell (1975) as a potential stimulus for root activity in Great Basin Desert shrubs. Initiation of rapid growth and water absorption in *Atriplex confertifolia* occurred as the soil profile warmed to progressively deeper levels. Cessation of root activity was associated with soil water depletion so that activity in any profile zone was limited to one or two weeks of adequate water supply following the attainment of sufficiently warm temperatures. This mechanism prolongs the time that soil water is available by effectively partitioning the water for use during different periods.

SIGNIFICANCE OF PLANT WATER STRESS

Slatyer (1967) devised a hypothetical scheme (Figure 18–11) to show the relationship between Ψ_{soil}, Ψ_{root}, and Ψ_{leaf} as soil water is re-

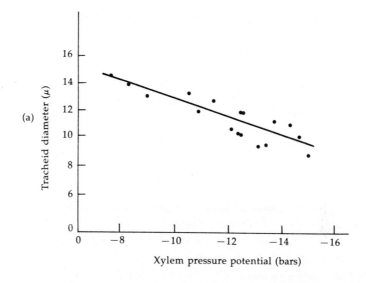

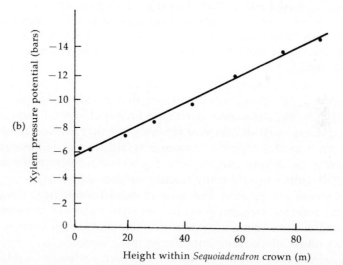

Figure 18-10. (a) The relationship between predawn xylem pressure potential and tracheid diameter in *Sequoiadendron giganteum*; (b) the relationship between height and predawn xylem pressure potential in *Sequoiadendron giganteum*. (From Rundel, P. W. and R. E. Stecker. 1977. Morphological adaptations of tracheid structure to water stress gradients in the crown of *Sequoiadendron giganteum*. *Oecologia* 27:135–139.)

duced from field capacity to below the permanent wilting point. Significant diurnal patterns of Ψ_{plant} are present even when Ψ_{soil} is near field capacity. The diurnal range must get larger as the soil becomes drier to maintain sufficient uptake from soils in which Ψ and hydraulic conductivity are decreasing. Following stomatal closure in response to

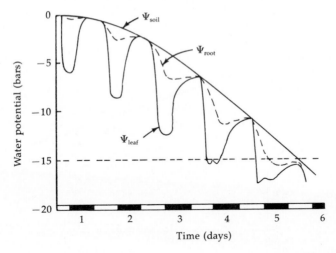

Figure 18–11. Schematic representation of changes in leaf water potential
(Ψ_{leaf}), root surface potential (Ψ_{root}), and soil mass water potential (Ψ_{soil}) as
transpiration proceeds from a plant rooted in initially wet ($\Psi_{root} \approx 0$)
soil. The same evaporative conditions are considered to prevail each day.
The horizontal dashed line indicates the value of Ψ_{leaf} at which wilting
occurs. (From Slatyer 1967. *Plant-Water Relationships.* By permission of
Academic Press.)

darkness, Ψ_{plant} gradually equilibrates with the soil so that just before
dawn, $\Psi_{soil} \sim \Psi_{leaf}$. The time necessary for equilibration increases as
Ψ_{soil} decreases, until $\Psi_{soil} \sim \Psi_{leaf}$ for only a brief period. Temporary
wilting and midday stomatal closure may reduce transpiration at the
point along the drying curve where Ψ_{leaf} = leaf osmotic potential, Ψ_{π}.
Mesophyll cells become permanently wilted when $\Psi_{soil} = \Psi_{\pi}$ (the
PWP), because a Ψ gradient sufficient to extract water from the soil is
no longer possible. Osmotic potential is not fixed within an individual
plant, and a wide range of values are found. We will discuss some
examples of osmotic adjustment later in this chapter.

Plant responses to water stress have been reviewed in detail recently by Hsaio (1976). We will briefly mention the physiological responses to water stress and later consider some specific adaptations to
avoid or deal with stress. Water is involved in and essential to all physiological processes. The photosynthetic decline as water stress increases
is, as we said in Chapter 13, primarily due to stomatal closure. However, when stress is great enough, photosynthesis, respiration, protein
synthesis, and most other processes involving chemical reactions may
be severely reduced because of protein (enzyme) denaturation. Most
physiological processes decrease gradually with increasing stress to the
point where all turgor is lost, and cease beyond that point. Under moderate stress biochemical processes may proceed but growth may be re-

duced because of lack of sufficient turgor pressure to cause cell en-largement. Lower overall cell size may increase the drought tolerance of the new tissues (Cutler et al. 1977). Newly formed cells may enlarge only when water becomes available, in which case, growth much ex-ceeds that expected by the rate of formation of new cells (Oechel et al. 1972). Water stress also increases the viscosity of phloem sap as well as reducing mass transfer in the xylem, thus limiting transport of nutri-ents, photosynthate, and hormones.

SPECIAL ADAPTATIONS

Solute Concentration

We learned earlier that the osmotic concentration of cell sap is the key quantity needed to understand variations in the level of stress necessary to induce wilting. **Osmotic concentration** depends on cell solute content and the amount of intracellular water. Only recently (Tyree and Hammel 1972) has the importance of such measurements been recognized and more recently (Cheung et al. 1975) interpreted in respect to ecologically meaningful adaptations of plants. Consequently, data on osmotic parameters of native plants are few (e.g., Osonubi and Davies 1978; Hellkvist et al. 1974; Roberts and Knoerr 1977).

Cheung et al. (1975) contrasted *Ginkgo* and *Salix* (Figure 18–12) as an example of plants showing very different osmotic qualities. *Ginkgo* has a lower (more negative) osmotic potential (point A for *Ginkgo*, Fig-ure 18–12) and would consequently remain turgid at much higher stress than would *Salix*. Plasmolysis (turgor pressure = 0 bars) occurs at point B, which is about −22 bars for *Ginkgo* and −12 bars for *Salix*. The steep slope for *Ginkgo* shows that very little water needs to be lost to reduce Ψ_{leaf} greatly. *Ginkgo*, the more xerophytic of the two species, can remove soil water to a much lower Ψ_{soil} than *Salix*, yet maintain a higher level of tissue hydration. This capacity is, in part, due to the more rigid cell walls of *Ginkgo*. Removal of small amounts of water from cells with rigid walls reduces turgor pressure rapidly. In *Salix*, the more flexible cell walls shrink, thus maintaining turgor pressure while creating only a minimal gradient for soil water extraction. *Salix*, there-fore, must grow in more mesic environments if it is to maintain a high level of tissue hydration. The ability to compare species water require-ments and responses at this fine scale offers exciting possibilities for expanding our knowledge of physiological ecology (see pressure cham-ber methods later in this chapter).

The capacity to reduce turgor pressure to negative values would give xerophytes further ability to extract soil water to very low water potentials. Negative turgor pressures are theoretically possible but

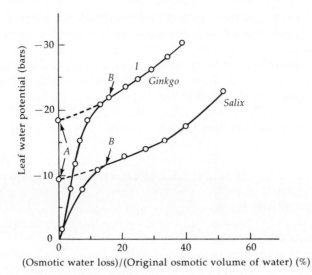

Figure 18-12. A plot of water potential versus osmotic water loss as predicted by the pressure-volume curves of single *Ginkgo biloba* and *Salix lasiandra* leaves. *A* is the estimate of original osmotic potential; *B* is the osmotic potential at incipient plasmolysis. (From Cheung, Tyree, and Dainty 1975. Reproduced by permission of the National Research Council of Canada from the *Canadian Journal of Botany*, Volume 53, pp. 1342–1346, 1975.)

have not been shown, beyond reasonable doubt, to exist, and have been the subject of considerable debate (Kyriakopoulos and Richter 1976; Tyree 1976). Such a phenomenon would help explain how some desert plants are able to reduce soil water potentials to less than −100 bars.

Osmotic adjustment is an effective **acclimation** device in response to water deficit (Osonubi and Davies 1978; Cutler and Rains 1978; Jones and Turner 1978; Cutler et al. 1977). Both stomatal response and wilting point depend on solute concentration, which increases in plants subjected to gradual drying or repeated periods of drought. The concentration increase may be due to increased solute content or to smaller solvent volumes in smaller cells formed under conditions of stress. Regardless of the mechanism operating in a particular species, many plants respond to previous conditions of water availability.

Halophytes are specifically adapted to widely varying external and internal concentrations of salt. Most saline habitats (Figure 18–13) contain NaCl (sodium chloride) as the primary salt, but sodium, magnesium, and calcium sulfates, magnesium and potassium chlorides, and sodium carbonate are also significant in many situations. Maintaining the chemical energy gradient necessary for water uptake requires that osmotic concentrations in halophytes must be higher than in glycophytes. Even though protoplasmic salt tolerance is higher in halophytes than in glycophytes, there is a limit to this tolerance and mechanisms which restrict cytoplasmic salt concentrations are necessary to survive.

Regulation of salt concentrations in the tissues of halophytes may be classified into three basic categories (Albert 1975). First, some halophyte species restrict the rate of salt uptake by increasing rates of ion accumulation only for brief periods or in certain habitats. Ungar (1977) has shown that saltbush *(Atriplex triangularis)* accumulates salt in direct proportion to the salt content of the soil. Salt uptake is restricted or excluded by root membranes of some species (e.g., mangrove) (Scholander 1968). When salts are excluded, cellular osmotic potential is adjusted by increasing the amounts of organic metabolites, in order to maintain the free energy gradient necessary for uptake. Various nitrogen compounds (amino acids and amines), carbohydrates, and organic acids accumulate in cells of halophytes. Such osmotic adjustments occur in ice plant *(Gasoul crystallinum = Mesembryanthemum crystallinum)*, a salt tolerant succulent which switches from C_3 photosynthesis to CAM and accumulates malate when growing in a saline medium (Lüttge et al. as cited in Flowers et al. 1977) (see Chapter 13). Brownell and Crossland (1972, 1974) have shown that some C_4 and CAM plants require or are stimulated by sodium ions in the soil. Such evidence has led to speculation (Laetsch 1974) that C_4 and CAM biochemical pathways were an adaptation to accumulate organic acids in order to counterbalance the osmotic influences of saline environments (Laetsch 1974; Flowers et al. 1977).

Salt dilution and salt extrusion are the other ways in which halophytes regulate salt concentrations in their tissues. **Succulence** caused by the accumulation of water in the tissues serves to dilute salts in cells, thereby reducing solute concentrations (Jennings 1968). Rapid growth of new tissues also has a diluting effect since salts are distributed to greater volumes of tissue (Greenway and Thomas 1965). **Salt extrusion** is another common mechanism by which excess salts may be

Figure 18–13. Halophytes growing in a desert depression. Note the salt accumulation at the soil surface.

removed from the plant body. Some species (e.g., *Atriplex* spp.) have salt glands which expel salt onto the surface of leaves where it can be washed from the plant by rain, dew, or fog condensation (Figure 18-14). Vesicular bladder hairs are another mechanism whereby salt is deposited on the leaf surface: the hair fills with salt water until it ruptures, releasing the salt onto the leaf. Salt may also be extruded by the shedding of organs which have accumulated salt. Salt evidently induces senility and, as the organ ages, nutrients and stored organics are mobilized to younger parts of the plant, followed by abscission of the organ containing the accumulation of salt.

Any one species usually employs a combination of these mechanisms to reduce the salt load. Further details may be obtained from the abundant literature, which was reviewed recently by Waisel (1972), Ranwell (1972), and Reimold and Queen (1974).

Figure 18-14. A scanning electronmicrograph of the leaf surface of saltbush (*Atriplex*) showing salt glands. (Reprinted by permission of McGraw-Hill Book Co. from *Probing Plant Structure* by Troughton and Donaldson, 1972.)

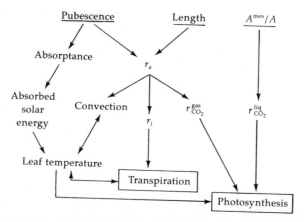

Figure 18-15. Schematic representation of the influences of pubescence, leaf length, and A^{mes}/A (ratio of mesophyll cell surface to outer leaf surface) on transpiration and photosynthesis. (Modified from Smith and Nobel 1977b. Copyright 1977 by the Ecological Society of America.)

Anatomical Adaptations

There is an abundant information base concerning anatomical responses to water deficits (e.g., Oppenheimer 1960; Esau 1965; Hanson 1917; Daubenmire 1974), much of which is now being critically re-evaluated in light of new technology and plant response information. **Xeromorphic leaves** are generally small, with reduced cell size, thicker blades, smaller, sunken, and more dense abaxial stomates, more pubescence (a covering of hairs), a thicker cuticle, more heavily lignified epidermal cells, better developed palisade mesophyll, and less intercellular space. It is true that these modifications are found in arid adapted plants, but we can no longer assume that all plants growing in hot and dry environments have similar anatomical adaptations. For example, creosote bush *(Larrea tridentata)*, the most common shrub in the warm North American deserts has, in the classical sense, **mesomorphic** leaf characteristics: no pubescence, stomates evenly distributed on both surfaces, stomates not in crypts, etc. (Barbour et al. 1974).

Pubescence, leaf size, and the ratio of mesophyll cell (internal leaf) surface to outer leaf surface (A^{mes}/A) interact to determine water use efficiency in desert plants (Cunningham and Strain 1969) (Figure 18-15). Leaf pubescence not only increases the depth of the boundary layer but is even more important as a regulator of light absorptance and, as a result, leaf temperature (Ehleringer et al. 1976; Smith and Nobel 1977b). We discussed leaf size earlier in this chapter and noted that small leaves are more efficient energy dissipators and lose less water per unit leaf area than large leaves. The ratio of mesophyll cell surface to outer leaf surface may be less important in regulating water loss than formerly assumed, being most important in the resistance to CO_2 uptake in the liquid phase (see Chapter 13). Smith and Nobel (1977*b*) studied the desert shrub *Encelia farinosa* which develops leaves which vary seasonally in size, pubescence, and A^{mes}/A. Pubescence and A^{mes}/A decreased as leaf size and absorptance increased in response to rainfall. Such morphological changes increase potential photosynthesis during a variety of environmental conditions. Water use efficiency was thus maximized in summer months and decreased during periods of cooler temperatures and more rainfall. Pubescence functioned to reduce absorptance, thus lowering leaf temperature in conjunction with reduced leaf size. Minimum r_l occurs during the wetter periods, maximizing photosynthesis when transpirational loss is not critical. *Mirabilis tenuiloba* coexists with *E. farinosa* but lacks the capacity to seasonally modify r_l and active photosynthesis is therefore restricted to a shorter period and the photosynthetic rate is lower during favorable periods (Figure 18–16).

Succulence was discussed in relation to halophytes and with respect to CAM but we have not yet addressed the water relations of desert succulents. Park Nobel (1977) studied the barrel cactus *(Ferocactus acanthodes)* in the Colorado Desert. Perhaps the most important aspect of the stored water in succulents lies in the ability of plants to continue a diurnal stomatal opening after Ψ_{soil} falls below Ψ_{plant}. In the barrel cactus, this occurred at a high (for desert plants) value of -4.5 to -4.7 bars; stomates continued to open for another 50–60 days, remaining closed only after Ψ_{soil} reached -90 bars. Following a 7 month period of no major rainfall, Ψ_{stem} fell to only -6.2 bars (close to the PWP, as osmotic potential was nearly the same). Such high levels of tissue hydration allowed stomatal response to 19 mm rainfall within 24 hours, with tissue rehydration and full stomatal opening in 48 hours. Cacti have the WUE advantages of CAM and store a volume of water large enough to continue CO_2 uptake for long periods after the Ψ_{soil} drops below the PWP. A question which deserves further study is: what keeps the water from diffusing out into the soil when Ψ_{stem} is over 80 bars higher than Ψ_{soil}?

PLANT AND SOIL WATER STATUS MEASUREMENTS

The technological advancements of the past two decades have dramatically improved our capabilities of measuring plant and soil water status. Most new techniques are developed around energy differences expressed by rates of evaporation as measured by heat flux from a wet thermocouple, or the negative xylem pressure created by transpiration. Nonelectronic methods based on the direction of water movement and total water content may also be used when more sophisticated equipment is not available.

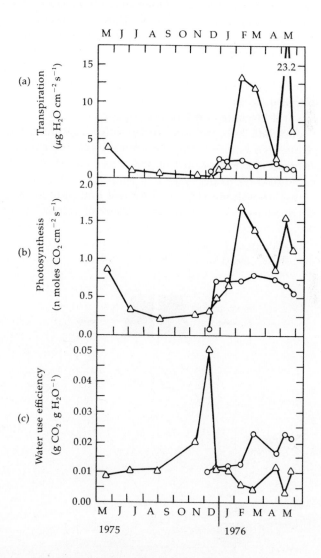

Figure 18–16. (a) Seasonal transpiration, (b) net photosynthesis, and (c) water use efficiency for *Encelia farinosa* (triangles) and *Mirabilis tenuiloba* (circles) based on field measurements. (Modified from Smith and Nobel 1977*b*. Copyright 1977 by the Ecological Society of America.)

Pressure Chamber

Scholander et al. (1964) developed the pressure chamber now widely used in ecophysiological research, and Waring and Cleary (1967) showed the great utility of **xylem pressure potentials** in solving ecological problems. This method is based on the fact that negative pressures (tensions) exist in the xylem and, when stems are severed, the pressure necessary to force water back to the cut surface is equivalent to the negative pressure in the xylem prior to cutting. A typical pressure chamber apparatus is diagrammed in Figure 18–17. An advantage of the pressure chamber lies in the large number of samples that can be measured in a short time. The apparatus has also been made portable so that data may be taken on plants in their native surroundings. Ritchie and Hinckley (1975) discussed in detail the applications, calibration, and data interpretation of pressure chambers. This review should be consulted before using the pressure chamber technique.

Recently, an application of the pressure chamber has been interpreted in ecological terms by Tyree and Hammel (1972) and Cheung et al. (1975). Pressure-volume relationships are determined by placing a fully hydrated, excised shoot into the pressure chamber, incrementally

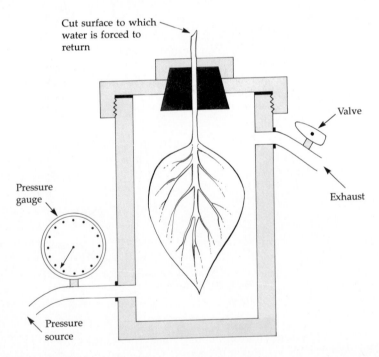

Figure 18–17. Schematic diagram of a pressure chamber used to measure plant water potential.

increasing the chamber pressure, and recording the volume of sap expressed at each pressure. When the upper, linear portion (*I* in Figure 18-12) of the curve is extended, it crosses the ordinate at a value equal to the original osmotic potential. Osmotic potential indicates the amount of solutes present per unit volume of cellular water, and sets the limit on the magnitude of turgor pressure which can develop. Stress tolerant plants would be expected to have a higher osmotic potential than more mesophytic species.

Pressure chamber values represent the total water status of the plant and can be used to determine Ψ_{plant} responses to environmental factors both diurnally and seasonally. These data, in combination with pressure chamber produced pressure-volume curves, provide an effective means of measuring plant water status and stress tolerance.

Psychrometry and Hygrometry

Psychrometers were developed to measure relative humidity (Chapter 17) of the air, and with the development of the water potential concept, it was found that Ψ could be determined from measurements of relative humidity. The relationship is:

$$\Psi = \frac{RT}{V} \ln wv$$

where *R* is the ideal gas constant, *T* is the absolute temperature, *V* is the volume of a mole of liquid water, and *wv* is the relative humidity. **Thermocouple psychrometers** have been developed which measure temperature depression of a wet surface (see Figure 17-2 to review the relationship between wet bulb temperature and humidity) and can detect small changes in relative humidity in a closed chamber in which the atmosphere is in equilibrium with a tissue or soil sample (Spanner 1951; Richards and Ogata 1958). Figure 18-18 is a diagrammatic sketch of a thermocouple psychrometer apparatus which can be used for soil or leaf disk measurements. Various chambers have been developed for specific purposes. Psychrometers with screen or ceramic covers may be inserted into the soil for Ψ_{soil} determinations. Thermocouple psychrometers can be inserted into a clamp and attached to leaves or stems as a nondestructive measure of water potential. Nondestructive methods are new and the influence of the rather extended time necessary for equilibration needs further testing (Campbell and Campbell 1974; Brown and McDonough 1977; Zanstra and Hagenzieker 1977).

It is possible to determine vapor pressure in sample chambers like those used in psychrometry by measuring the dew point. **Dew point hygrometers** contain a thermocouple or mirror with a film of water just as in a psychrometer, but the hygrometer maintains the film by staying

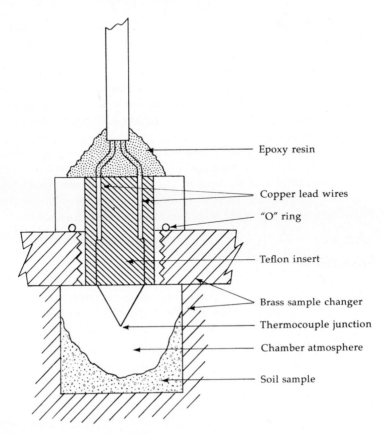

Figure 18–18. The Spanner (Peltier) thermocouple psychrometer and a psychrometer chamber shown with a soil sample. The entire assembly is immersed in a water bath to maintain constant temperature. (From Brown 1970. Measurement of water potential with thermocouple psychrometers: construction and applications. U.S. Dept. of Agriculture Forest Service Research Paper INT–80.)

at dew point temperature. Dew point hygrometry is a continuous measurement of humidity and is frequently utilized to measure water vapor in circulating air streams such as those in an infrared gas analyzer apparatus (see Chapter 13).

Dye Method

A simple and inexpensive method of measuring plant water potential is based on the change in density of a solution, sometimes referred to as the Shardokov dye method. Two series of replicate sugar solutions are prepared to span a range of osmotic potentials. Leaves or leaf discs are placed in one series of solutions. The second series is colored with methylene blue or other appropriate dye. After the tissue has been exposed to the solution for an hour or so, a drop of the colored solution is introduced into the center of the replicate solution which contains the tissue. If the drop moves up, the Ψ_{plant} was lower than the osmotic potential of the solution; if the drop moves down, the Ψ_{plant} was higher than the osmotic potential of the solution. The Ψ_{plant} is equal to the osmotic potential of the solution when the drop diffuses in all directions. Tables of osmotic potential versus molality of solutions are available. For details and precautions, see Knipling (1967), Knipling and Kramer (1967), and Barrs (1968).

Relative Water Content

Relative water content (RWC) provides a simple means by which information concerning the water status of plants can be obtained. RWC is an expression of the current plant water status relative to what it would be if the plant were fully hydrated.

$$RWC = \frac{fresh \; weight \; minus \; dry \; weight}{fully \; turgid \; weight \; minus \; dry \; weight} \times 100$$

Fully turgid weight is obtained by exposing the tissue to water until it no longer increases in weight. RWC is an accurate measure of water status but gives no information of the plant water status with respect to soil or atmosphere, and no indication of the plant's response to water stress.

Other Measures of Plant and Soil Water Status

Barrs (1968) has reviewed the extensive literature on water status methodology in plants and Rawlins (1976) has done the same for soils. Several more useful references and more specific information can be found in Brown (1970) and Brown and van Haveren (1972).

SUMMARY

The plant is part of a continuous system of water movement from the soil into the atmosphere. Water moves along a potential energy gradient which decreases from the soil to the leaf surface; the steepest decrease of the gradient occurs as water leaves the leaf surface and enters the atmosphere. The rate of transpiration depends on the steepness of the potential energy gradient and on the resistance of water flow through the soil, root, stem, mesophyll, substomatal cavities, stomates, and to the atmosphere. Leaf temperature and the rate of water loss depend on air temperature, evaporative cooling, leaf size, and solar radiation. Water loss and CO_2 uptake both occur through the stomates so that the balance between transpiration and photosynthesis (water use efficiency) determines the success of a plant in a water limiting environment.

Water stress develops when absorption by roots lags behind transpirational loss. Soil gradually dries as water is removed adjacent to the roots, thus restricting absorption until the plant can no longer extract water from the soil (plant osmotic potential = soil water potential); this is the point at which plants reach a 0 turgor pressure and wilt. As soil drying proceeds, photosynthesis gradually decreases as resistance to CO_2 uptake increases because of stomatal closure. This causes decreases in growth because turgor pressure is necessary to fully expand new cells. Respiration, protein synthesis, and most other processes involving chemical reactions are inhibited under severe stress because of protein denaturization. There is also a reduction in transport rates in both the xylem and phloem.

Plants adjust to water deficits by modifying resistances, by special anatomical characteristics, or by adjusting osmotic concentrations in the cells to overcome aridity caused by low soil water or salinity. Halophytes are able to regulate tissue salt concentrations by adjusting osmotic concentration, diluting salt with water (succulents), or extruding salt onto the leaf surface. Leaf characteristics thought to be important in water relations include size, pubescence, number and density of stomates, and heavily lignified and cutinized epidermal cells. Succulence provides a source of water, allowing a plant to reduce tissue water content after soil water is no longer available.

Methods of measuring soil and plant water include the pressure chamber, psychrometry, dew point hygrometry, the dye method, and relative water content. Many of these techniques depend on recently developed technology, so few studies have been conducted which utilize the full potential of these instruments. Much remains to be done if we are to understand the wide range of water related adaptations of terrestrial plants.

CHAPTER 19

MAJOR VEGETATION TYPES OF NORTH AMERICA

The North American continent has a moderately rich flora and a very rich diversity of habitats. The result is a complex mosaic of vegetation types and plant communities. One particular vegetation map for the conterminous United States shows over one hundred community types, but surveys of individual states or regions may show 50–100 community types because the level of detail is more refined than that used for the U.S. map. However, we choose to take a very broad survey view and so we will discuss only five vegetation types: **tundra, conifer forest, deciduous forest, grassland,** and **desert scrub,** although we will take account of regional differences in each case. Our objective is to present enough information on community structure, community dynamics, and autecology of dominants to underscore the uniqueness of each of the five types.

These five vegetation types are the major types covering the North American land mass. Every reader is likely to find one of these types nearby. At the same time, every reader is likely to find other types, not discussed here, equally nearby and possibly even dominating the local region. Such types might be chaparral, fresh water marsh, coastal scrub, woodland, or savanna. We apologize for these omissions, but as shown in Figure 19–1, most of North America can be mapped in one of the five categories we do discuss.

TUNDRA

Tundra vegetation occurs in regions beyond timberline which have cold, usually moist, climates with short growing seasons. The vegetation is low, often only 10 cm tall, and is dominated by perennial forbs, grasses, sedges, dwarf shrubs, mosses, and lichens (Figure 19–2). In the arctic, on the average, only the top 0.5 m of soil thaws in summer. The permanently frozen subsoil impedes drainage, root growth, and decomposer activity. In the Lapp and Russian languages, tundra means a marshy, unforested area, and these are its most prominent characteristics.

There are two major types of tundra in North America. **Arctic tundra** occurs at high latitudes in Canada and Alaska, but at low elevations. **Alpine tundra** occurs at higher elevations in mountain ranges further south. The vegetation is similar in both types of tundra, and

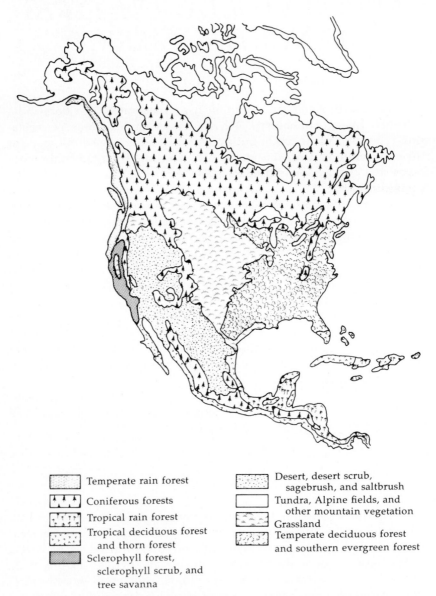

Temperate rain forest

Coniferous forests

Tropical rain forest

Tropical deciduous forest and thorn forest

Sclerophyll forest, sclerophyll scrub, and tree savanna

Desert, desert scrub, sagebrush, and saltbrush

Tundra, Alpine fields, and other mountain vegetation

Grassland

Temperate deciduous forest and southern evergreen forest

Figure 19–1. Major vegetation types of North America. (From *Environmental Conservation*, 3rd ed. Dasmann. Copyright 1972 John Wiley and Sons, Inc. Reprinted by permission of John Wiley and Sons, Inc.)

Figure 19–2. Tundra vegetation.

many species occur in both, but important environmental differences (such as maximum summer temperatures, extremes of daylength, and intensity of solar radiation) lead to important ecological differences. North of the tundra, or above it in mountains, is a barren region of permanent ice and snow called the **nival** zone. South of the arctic tundra, or below the alpine tundra, is the **boreal conifer forest** (also called the **taiga** or subalpine forest), which will be discussed later in this chapter.

Physical Features

The tundra's climate is rigorous and cold, and relatively few species of plants have evolved tolerances to withstand it. The short growing season is 50–90 days long, on the average, and the mean temperature of the warmest month is less than 10°C. Maximum air temperature, especially in the alpine tundra, can be much higher than 10°C, reaching above 25°C; in addition, leaves near the ground surface can be 10–20°C warmer than air temperature. Summer days in the alpine tundra exhibit greater fluctuations in temperature than in arctic tundra. The thin mountain air and lack of cloud cover allow rapid reradiation of heat at night, permitting frost to occur throughout the growing season.

Daylength may be shortened to zero hours in the arctic tundra, and during winter net radiation is negative. Winter conditions are very similar for both tundra types, with minimum temperatures below −50°C, a modest amount of snow cover, and strong winds of 15–30 m sec^{-1} (30–60 mph) being common. Annual precipitation varies greatly from region to region, but alpine tundra averages 100–200 cm, which is considerably more than the 10–50 cm which arctic areas receive. This annual precipitation may seem low, but evapotranspiration is quite low, so that the P/E ratio is far above 1.

Two soil orders dominate the tundra: **histosols** (bog soils, mucks, and organic soils with more than 20% organic matter) and **entisols**. The entisols have little soil profile development, and their traits reflect the character of the parent material more than they reflect the macro-environment. A typical entisol profile is shown in Figure 19–3. The top 15–60 cm of soil thaws in summer and freezes in winter; it is coarse textured, brown from undecomposed organic matter, and contains all the root mass of the vegetation above. Beneath this topsoil **(talik)** is permanently frozen parent material **(permafrost)**. If the vegetation is killed or removed, or if the talik is eroded or compacted, or if heat is supplied to the ground as from a buried pipeline or a house resting on the surface, then the upper part of the permafrost will melt, creating an unstable soil, further erosion or subsidence, and long-term vegetational change. Even one pass over tundra in summer with a normal wheeled vehicle will leave long-term scars. Bechtel Corporation developed a truck that ran on enormous, broad, air-filled tires to permit travel on tundra that accompanied construction of the famous oil pipeline (Figure 19–4) from Valdez to the Arctic Slope in Alaska. Whenever possible, the pipeline was built aboveground, and care was taken to prevent heat from "leaking" into the ground along the supports. Roadcuts made in alpine tundra in the Rocky Mountains 50 years ago still show only half the plant cover of nearby, undisturbed tundra.

Freeze/thaw cycles, over the course of thousands of years, produce a pattern to the landscape which is most visible from the air (Figure 19–5). The ground surface is broken into polygons, each polygon one to tens of meters across. The polygon edges may be marked by

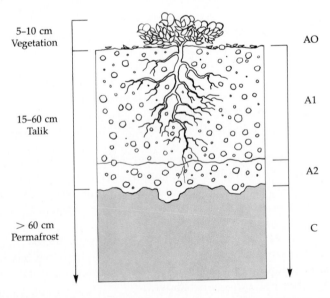

5–10 cm Vegetation		AO
15–60 cm Talik		A1
		A2
> 60 cm Permafrost		C

Figure 19–3. Profile of a typical tundra soil with permafrost. The plant is drawn to scale. AO = litter.

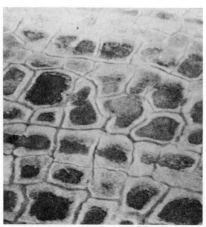

Figure 19-4. The Alaska pipeline, here above the ground and heavily insulated. Care has been taken in planning the supports so that heat is not transferred from the pipe to the ground.

Figure 19-5. Aerial photograph of frost polygons in the tundra. These polygons are up to 10 m across.

accumulations of rock, by ridges of soil, or by dense vegetation; the centers may be depressed and filled with standing water in summer, or they may be raised into hillocks several decimeters high. **Patterned ground** develops most extensively on arctic flood plains, and it is rare in the alpine tundra.

Microrelief affects tundra vegetation markedly, because (a) it will determine how late into summer snow will lie on the ground, first covering vegetation, then supplying it with water later on; (b) it will also determine the degree of protection plants receive from drying summer winds or ice-blasting winter winds; and (c) it will have a bearing on the depth of the talik, and hence on the amount of root space and the adequacy of drainage. Ridge tops support very little plant cover because in winter high winds remove protecting snow cover. Without the snow, abrasive ice crystals carried by the wind can prune plants back to the ground or, at least, winter winds can dessicate twig tissue. In summer, ridge soils dry rapidly. In contrast, as shown in Figure 19-6, gentle north-facing slopes may accumulate snow drifts that remain well into summer, providing melt water to **wet meadows** just below them. Wet meadows exhibit 100% ground cover and dominance by grasses and sedges. Rocky talus slopes and relatively level boulder-strewn areas **(fell fields)** support relatively little cover, scattered in patches where the microenvironment is favorable. Dwarf shrubs, for example, grow tallest in the lee of rocks or other projections. Productivity and biomass of fell fields can be about one-third that of wet meadows.

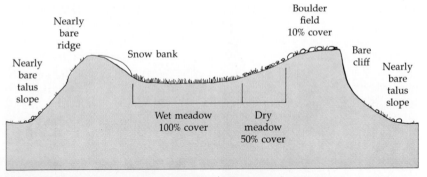

Figure 19-6. Major alpine tundra microenvironments, each with its own characteristic plant community.

Vegetation

The tundra flora is not rich. The arctic tundra of North America supports only about 600 species, which is only 3% of all angiosperms in the world. Tropical regions of smaller area support tens of thousands of species. The most common families in the tundra are: Caryophyllaceae, Asteraceae, Brassicaceae, Cyperaceae, Betulaceae, Ericaceae, Poaceae, Polygonaceae, Rosaceae, and Salicaceae. Only 1–5% of the species are annual; most are relatively long-lived (20–100 years) herbaceous perennials. Hemicryptophytes—herbaceous perennials with perennating buds just at the soil surface—make up more than 60% of the flora. Trees and tall shrubs are absent.

At its most luxurious state, as in wet meadows, the vegetation covers 100% of the ground, but it is quite short, 10–15 cm tall, with a leaf area index (LAI) of only 1–2. Most plant biomass is belowground, giving a root:shoot ratio of 4:1. Many of the perennial forbs are "cushion plants"—plants whose aboveground system is condensed into a compact canopy of small, xerophytic foliage, often with large, showy flowers incongruously scattered over the surface (Figure 19–7). The xerophytic foliage (small size, sometimes pubescent, sometimes succulent, often light colored, often vertically oriented) may be caused by low soil nitrogen levels more than by high light intensity or late summer drought. Although tundra soils are brown from organic matter, the nutrients are largely unavailable to the plants because decomposer activity is inhibited by the cold environment.

In spring, some plants resume growth while still covered by snow, and when root temperatures are barely above freezing. The hollow-stemmed forb *Mertensia ciliata* is an example of a subalpine snow bank plant able to conduct photosynthesis in the relatively warm, CO_2-rich air inside its stems even before the leaves have expanded. Seeds of tundra plants, on the other hand, require relatively high temperatures, close to 20°C, before they can germinate. Flower buds, formed one or more seasons ago, are induced to open by long days. About two-thirds of the flora is insect-pollinated, so the flowers are large and showy. Some flowers are dish-shaped and reflect solar radiation into their centers, creating a favorable thermal micro-environment for pollinators such as bumble bees. Pollinator density is low, however, and many species compensate by being facultatively or obligately self-fertilizing. For reasons that are still not clear, most tundra species are polyploids. Seed set is high, the seeds are light, and they have a long life.

During the short growing season, net primary productivity is low, amounting to only 100–200 g m^{-2} on a yearly basis. Much of this is channeled to roots and rhizomes and stored as lipid or protein. Part of the reason that NPP is low is that the ratio of net photosynthesis to respiration is low, only about 2:1 in contrast to a ratio of 10:1 for tropical plants. In addition, the maximum photosynthetic rate is low, typically 10 mg CO_2 dm^{-2} hr^{-1}, which is equivalent to that of a mesic, temperate zone sciophyte. A few tundra species have the C_4 photosynthetic pathway.

Dormancy, in late summer, is triggered by lowering temperatures, increasing soil moisture stress, a shortening daylength, or a changing quality in the solar spectrum that results from a more oblique solar angle (the latter is important only in the arctic tundra). Wet meadow species apparently have little control over stomatal aperture, and as the soil dries, plants lose water rapidly, finally becoming dormant at a tissue water potential of only −10 bars. Dry meadow and fell field plants can tolerate lower water potentials. Winter snow cover helps to maintain a modest tissue water potential. For example, needles of whitebark pine (Pinus albicaulis) in the timberline zone of the Sierra Nevada show a water potential of −18 bars in winter when covered with snow, but a water potential of −40 bars when exposed. In fact, the height of trees at timberline may correspond closely with the average depth of winter snowpack.

Figure 19–7. Cushion plant. Note the relatively large size of the flowers. The vegetative part of this plant is only 5 cm high.

Timberline

There is a broad ecotone in Canada and Alaska between the arctic tundra and the boreal forest to the south. In places, it may be 250 km wide. Within the ecotone, the forest first thins, no longer retaining 100% cover, then trees become restricted to patches, and finally trees in the patches become dwarfed, twisted, and shrublike (Figure 19–8). The German word **krummholz**, meaning twisted wood, has been applied often to both this growth form and to the ecotone "forest"; a synonym is **elfinwood**. Surrounding the patches of trees and krummholz is an increasingly continuous matrix of tundra. Finally, even the patches of krummholz disappear and the tundra proper is entered.

A similar ecotone exists in mountain ranges between alpine tundra and subalpine forest, but the ecotone is much narrower, extending upward through 150–300 m of elevational change. The exact elevation of timberline depends on the latitude and the distance from a maritime climate, as shown in Figure 19–9. Even within one mountain range at one point in latitude, timberline will be lower on the moister, ocean-facing slopes than on the drier, continental-facing slopes (Figure 19–9(b); compare the Olympic and Cascade Mountains).

The location of timberline is also a function of time, changing in response to changing climate. In some parts of Canada there is evi-

Figure 19–8. Krummholz at timberline in the Sierra Nevada (Sonora Pass, about 3400 m). The shrub-like thicket is a pure stand of white bark pine (*Pinus albicaulis*), about 1 m tall.

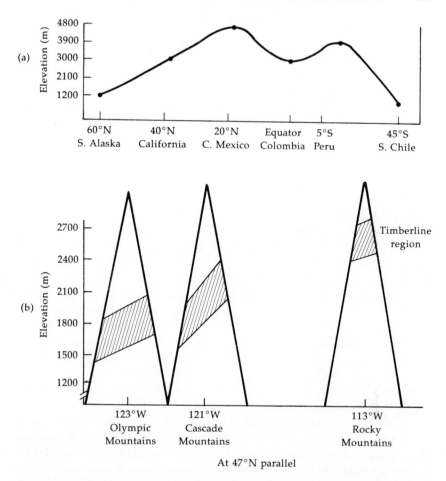

Figure 19–9. (a) The effect of latitude on the elevation of timberline, from the Coast Mountains of southern Alaska, through the Cascade Range, Sierra Nevada, Sierra Madre, and the Andes Mountains. (b) The effect of distance inland from oceanic influence, all at the same latitude, 47°N. The shaded region is the distance between the end of the closed forest and the end of the krummholz. ((a) from Daubenmire 1954. Alpine timberlines in the Americas and their interpretation. *Butler University Botanical Studies* 11:119–136.)

dence that timberline has retreated south in the past several hundred years, while in other parts there is evidence that it has been advancing. The ecotone is a dynamic tension zone within which the basic limitation to tree growth seems to be the amount of warmth received during the growing season. Below some critical level of seasonal heat, it becomes impossible to maintain a large amount of woody tissue.

CONIFER FORESTS

Several distinctive types of conifer forest exist in North America, and those which cover the most area will be included in this section.

The most extensive conifer forest type is the **taiga**, which stretches across most of Canada and Alaska, and dips down into the Great Lakes states and New England to about 45°N latitude. Taiga is a Russian word applied to Eurasian conifer forests described as damp, wild, and scarcely penetrable. A synonym is **boreal forest**. The taiga is a monotonous forest, with crowded trees of modest stature, and low species diversity in the overstory and understory. This low elevation forest extends over impressive distances, and it is interrupted often with lakes, ponds, and moss-covered bogs. It can be a quiet forest, not rich in vertebrate animal activity.

Conifer forests extend south of the taiga, at moderate to high elevations in mountain chains (Figure 19–10). Just below the alpine tundra is a **subalpine forest**. Different species dominate than in the taiga, but its physiognomy is the same. The **montane zones** below this are covered by a richer diversity of conifers, and the forests are often more open, the trees larger, more productive, and longer-lived than those of the taiga and subalpine forests. Below the montane zones lie vegetation types dominated by broadleaved trees, scrub, or grassland. One exception to this pattern is mountain ranges which penetrate north into the taiga; in these areas the surrounding low elevation vegetation is still conifer-dominated.

Another, far more magnificent, conifer forest dominates a wet, temperate, lowland coastal strip that reaches from northern California to Alaska: the **north coast temperate conifer forest**, sometimes called a **temperate rain forest**. Some of the largest trees in the world, and stands with the greatest biomass, exist in this forest.

Other low elevation conifer forests, all dominated by pines, are found in smaller areas of coastal California, in the Great Lakes area, in New Jersey, and in the southeastern coastal plain of the United States. Some of these are found on droughty, nutritionally poor soils, and many are subclimax types, maintained by fire. These will be discussed briefly at the end of this section.

Physical Features

The taiga is a cold, wet region, with a short growing season, only 1–3 months, and a mean temperature above 10°C. During more than 6 months of the year the mean temperature is below 0°C, and net radiation during that time is negative. Maximum summer temperatures reach into the low 20s and winter minimums are in the −50s. The temperature difference between annual extremes can be enormous in the interior, continental part of the taiga: from a minimum of −70°C to a maximum of +30°C, or a span of 100°C. Annual extremes are dampened near either the Atlantic or Pacific coast (Figure 19–11). Annual precipitation varies between 25 and 100 cm, and snow cover is not deep. Despite the modest amount of precipitation, the P/E ratio is still >2.

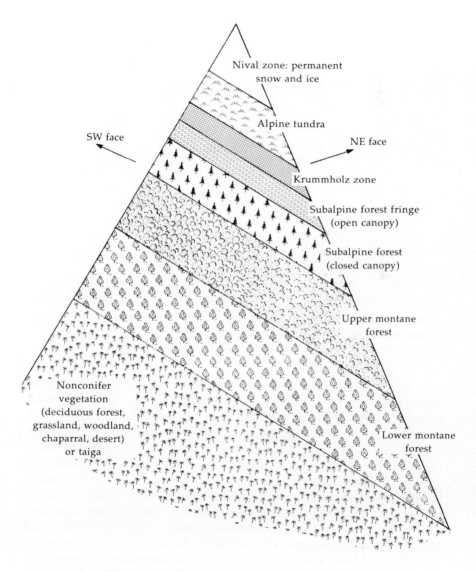

Figure 19–10. Elevational zones in a typical mountain range. The krummholz to lower montane zones are conifer-dominated.

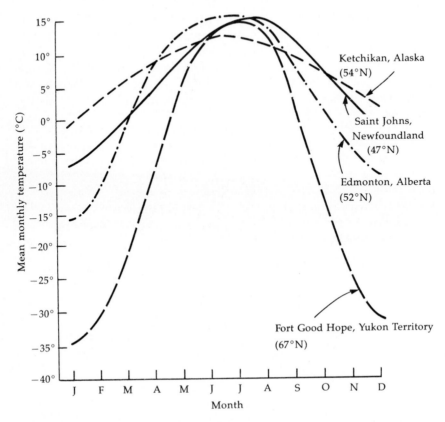

Figure 19–11. Mean monthly temperatures at four taiga locations. Two sites are coastal, and the oceanic influence moderates winter extremes: Ketchikan on the Pacific coast and Saint Johns on the Atlantic coast. Two are interior, with bitterly cold winters: Edmonton and Fort Good Hope.

The subalpine zone is generally 5°C warmer throughout the year and experiences 20–50% more precipitation than the taiga. There is a complex environmental gradient that corresponds to elevation in mountain ranges, and this gradient is responsible for the change in conifer species from one zone to the next. Two important environmental factors are temperature and precipitation. Figure 19–12 shows how average annual temperature and average annual precipitation change with elevation in the Sierra Nevada Mountains for west-facing (ocean-facing) and east-facing (desert-facing) slopes. Precipitation increases about 5 cm for every 100 m increase in elevation. The highest elevation shown, 2100 m, corresponds to the red fir forest in the upper montane zone. If higher elevations were shown, you would see that precipitation drops in the subalpine and alpine zones. The east slope is much drier and has a different **lapse rate** (rate of change) because a rain shadow has been created: winds generally come from the west, drop their moisture on the west slope as air rises, and descend on the east slope as dry winds. The west slope temperature lapse rate shows a drop of 4°C for every 1000 m rise in elevation. Neither lapse rate holds for all seasons or all mountain ranges. Considering many factors and locations, montane temperature lapse rates average about 3° per 1000 m.

The north coast forest lies in a wet, equable strip. Annual precipitation is generally between 200 and 300 cm, very little falling as snow. Mean January minimums are above freezing, at about 2°C, and mean July maximums are only about 20°C. Diurnal thermoperiods average only 10°C. A pronounced summer dry period exists from Washington to California, but it is moderated by frequent fog which lowers transpiration stress and also drips off the foliage onto the ground, adding perhaps 20 cm of "rain" in the Californian part of the range.

Taiga soils largely belong to the **spodosol** soil order, which has been described in Chapter 16. These are acidic soils (pH 3–5) whose upper layers have been leached of clay, organic matter, and many nu-

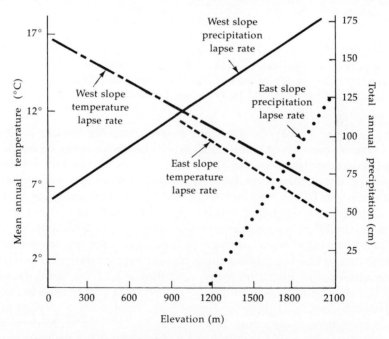

Figure 19–12. Temperature and precipitation lapse rates on west and east slopes of the Sierra Nevada at 38°N latitude. Temperature lapse rates are similar for both slopes, but precipitation lapse rates differ because of a rain shadow effect: the east side is drier. (From Major and Taylor 1977. In *Terrestrial Vegetation of California.* Edited by Barbour and Major. Copyright 1977 John Wiley and Sons, Inc. Reprinted by permission of John Wiley and Sons, Inc.)

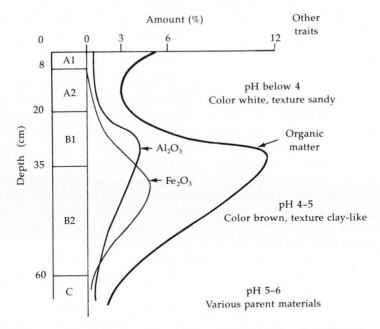

Figure 19–13. Some chemical characteristics of a spodosol soil profile. Note that depths of maximum accumulation of organic matter, aluminum, and iron are slightly different. (From the translation of *Plants, the Soil and Man* by M. G. Stålfelt. Copyright 1972 by Longman, London.)

trients (Figure 19–13). Conifers are characterized by having shallow root systems (root:shoot ratio = 0.2) which lie above zones of nutrient and clay accumulation in the B2 horizon. However, nutrient uptake is improved by ectomycorrhizae. All taiga conifers possess ectomycorrhizae (see Chapter 6) and there is experimental evidence which shows that tree growth and survival would be minimal without them. Soils are acidic in part because of leaching (P/E is >2), and in part because conifer foliage is acidic. The hydrogen ions replace nutrient cations on soil colloids, giving a low percent base saturation, they drive some cations into insoluble states, and they inhibit bacterial decomposers. Litter tends to accumulate, and as much as 60% of all ecosystem carbon is locked up in humus (Figure 7–8). In contrast, a tropical rain forest ecosystem retains only 15% of its carbon in humus. Mineral cycling is slow. The half-time for litter decay is about 3–5 yr. Permafrost may underlie taiga soils at the northern limit, but is not characteristic of the taiga proper.

Soils beneath montane and coast forests are generally either podzolic or are young, weakly developed soils (**Inceptisol** soil order). If podzolic, they are not as acidic as taiga soils and lack the ashy A2 horizon of taiga soils.

Taiga Vegetation

Throughout much of Canada the taiga is dominated by white spruce *(Picea glauca)* and balsam fir *(Abies balsamea)*. Black spruce *(P. mariana)* is a common associate and may dominate some stands. The trees are closely spaced, with dense canopies giving a LAI of 9–11 (40), but tree size (20 m tall, 30 cm dbh) and longevity (100–300 years) are not impressive. LAI can be 16–40 in some temperate conifer forests.

Shrubs and subdominant trees are not common, but a ground layer of dwarf shrubs, ferns, club mosses, true mosses, and lichens may reach 100% cover (Figure 19–14 and Table 19–1). The forest, then, is essentially a two-layered community.

Near the edge of bogs, the forest canopy thins and eventually only scattered trees of black spruce *(Picea mariana)* and larch *(Larix laricina,* also called tamarack) remain, underlain by a dense shrub canopy of *Vaccinium, Ledum,* and other ericads. Wetter parts of the bog are covered by *Sphagnum* moss. As discussed in Chapter 10, this moss can advance into more upland sites, creating a wetter, more sterile microenvironment which leads to retrogressive succession. Droughty, sandy soils often support stands of jack pine *(Pinus banksiana).*

Small areas are frequently disturbed by wind-throw from winter storms or by fire from summer dry lightning strikes. Succession often proceeds through meadow, then broadleaf tree *(Populus tremuloides, P. balsamifera, Betula papyrifera)* seral stages before reaching an even-aged spruce-fir climax in a span of about 300 years. These broadleaf trees also form an extensive parkland, up to 150 km wide, in an ecotone

Figure 19–14. The interior of a taiga forest. Note the high ground cover by ferns and other cryptogams.

Table 19–1. Vascular plants found in more than 40% of white spruce-balsam fir plots scattered through the taiga from Alaska to Newfoundland. A total of 34 plots, each approximately 300 × 300 m in area, were established. (From La Roi 1967. Copyright 1967 by the Ecological Society of America.)

Stratum	Common name	Scientific name
overstory tree	balsam fir	*Abies balsamea*
	paper birch	*Betula papyrifera*
	trembling aspen	*Populus tremuloides*
	balsam aspen	*P. balsamifera*
	white spruce	*Picea glauca*
	black spruce	*P. mariana*
low tree and tall shrub	alder	*Alnus crispa*
	chokeberry	*Pyrus decora*
medium and	gooseberry	*Ribes triste*
low shrub	rose	*Rosa acicularis*
	blackberry	*Rubus idaeus*
	viburnum	*Viburnum edule*
dwarf shrub and herb	wild sarsaparilla	*Aralia nudicaulis*
	clintonia	*Clintonia borealis*
	goldthread	*Coptis groenlandica*
	bunchberry	*Cornus canadensis*
	willow herb	*Epilobium angustifolium*
	rattlesnake plantain	*Goodyera repens*
	wood fern	*Dryopteris ausriaca*
	shield fern	*Gymnocarpium dryopteris*
	twin flower	*Linnaea borealis*
	club moss	*Lycopodium annotinum*
	false lily of the valley	*Maianthemum canadense*
	mitrewort	*Mitella nuda*
	lungwort	*Mertensia paniculata*
	one-flowered wintergreen	*Moneses uniflora*
	pink pyrola	*Pyrola asarifolia*
	one-sided pyrola	*P. secunda*
	blackberry	*Rubus pubescens*
	twisted stalk	*Streptopus roseus*
	woodland star	*Trientalis borealis*
	violet	*Viola renifolia*

where the boreal forest meets the central grasslands in Manitoba, Saskatchewan, and Alberta.

Net annual primary productivity is relatively low (600 g m^{-2}), but this is not surprising, given the short, cool growing season and the low nutrient status of the soil. The eastern deciduous forest shows twice that biomass gain, and the tropical rain forests of Central America show more than three times that gain.

Much of the taiga is still botanically and ecologically unexplored. Much of it is undisturbed by any human activities, in fact, and this may be due to small factors as well as large ones. The large factors have to do with climatic hardship and the absence of roads. The small factors are called black flies and mosquitoes. The Canadian writer Robert Stewart put it this way:

> The first insect assault of each year is launched by a tiny type of black fly nicknamed "No-see 'em." The sobriquet is a bit inaccurate, because you can see 'em all right—they sometimes travel in hordes so dense that in the distance they look like the smoke from a forest fire. . . . The little black flies provide only a prelude for their bigger cousins. By July the bush country whines like a power saw with black flies that are about a quarter of an inch long. . . . Black flies formidably influence human life in the wilder parts of the Canadian Shield country and deserve much of the credit for preserving Labrador as a wilderness. Because of the black flies, a Canadian woodsman in his summer clothes looks almost as if he were dressed for winter. . . . Yet Labrador's black flies are nearly matched by its mosquitoes. They, too, have helped protect Labrador from human incursion. . . . Who, then, is the lord of the wilderness after all?

Subalpine and Montane Vegetation

The Appalachian Mountains

The Appalachian Mountains extend from Maine south to Georgia. These mountains are unique in that the montane zones are largely dominated by hardwoods, and conifers are restricted to the subalpine zone. In the north, for example in the Adirondack Mountains of upstate New York, the upper limit of the montane zone is about 900 m; in the southern Smokey Mountains of North Carolina it is about 1600 m.

The taiga dominants, white spruce and balsam fir, do occur in the New England portion of the Appalachians, but they are not dominants. Red spruce (Picea rubens) dominates throughout the Appalachians, and it is joined by Fraser fir (Abies fraseri) south of Pennsylvania. Broad-leaved subdominant trees include mountain ash (Sorbus americana), yellow birch (Betula lutea), and maple (Acer spicatum).

The Rocky Mountains

Running obliquely from southeast to northwest, the Rocky Mountains cross 30° of latitude and 25° of longitude. There are regional differences in the tree species which associate at particular elevations, but several species range throughout the Rocky Mountains: Engelmann spruce *(Picea engelmanni)* and subalpine fir *(Abies lasiocarpa)* in the subalpine zone, Douglas fir *(Pseudotsuga menziesii)*, lodgepole pine *(Pinus contorta)*, and white fir *(Abies concolor)* in the upper montane zone, and ponderosa pine *(Pinus ponderosa)* in the lower montane zone.

The elevation of each zone increases to the south, and at any given latitude the zones are highest on southwest-facing slopes and lowest on northeast-facing slopes. In northern Alberta, the upper limit of the subalpine zone is about 2000 m, and the montane zone below begins at 1400 m. The lower montane further downslope is essentially the same as surrounding low elevation taiga. Far to the south, in Arizona, the upper limit of the subalpine zone is about 2700 m and the lower montane begins at 1800 m.

Regional associates of spruce-fir in the subalpine zone of the northern Rockies include white bark pine *(Pinus albicaulis)* and alpine larch *(Larix lyallii)*, the latter retaining a tree form higher into timberline than any other species in North America. Subalpine forest in the central Rockies may include white bark pine, lodgepole pine *(Pinus contorta)*, and mountain hemlock *(Tsuga mertensiana)*, in addition to spruce-fir. The southern Rockies have limber pine *(Pinus flexilis)*, foxtail pine *(P. balfouriana)*, and bristlecone pine *(P. longaeva*, which is famous for being the longest-lived tree in the world), in addition to spruce-fir.

The upper montane zone in the Rockies, of Montana and Idaho, is enriched in tree species because storm tracks coming east from Washington are able to penetrate that far. In this area, only 200 km separate the Rockies from the Cascade Mountains. Dense stands of grand fir *(Abies grandis)*, western red cedar *(Thuja plicata)*, and western hemlock *(Tsuga heterophylla)* are typical. Only in the drier, eastern part of this region—for example, in Yellowstone National Park—does this zone

Figure 19–15. The lower montane zone ponderosa pine forest of the Rocky Mountains. Note its relative openness. In the absence of ground fires every 8 years or so, other conifers will invade and the canopy will close.

support the Douglas fir and lodgepole pine forest more typical of the Rocky Mountains as a whole.

The lower montane zone is dominated by ponderosa pine, and it often has a rather open, parklike aspect (Figure 19–15). This open physiognomy and pure dominance has apparently been maintained by natural surface fires with a frequency of one fire every 3 to 10 years. Fire suppression policies over the past 50 years have resulted in denser stands, more brush, loss of grass, and the invasion of other conifers (see Chapter 15).

The Cascade and Sierra Nevada Mountains

The Cascade Range extends along 10° of latitude, from southern British Columbia to Mt. Lassen in northern California. The Sierra Nevada Mountains continue south from Mt. Lassen an additional 6° of latitude, to the edge of the Mojave Desert in southern California. There is considerable floristic shift along that distance—so much so that the northern and southern limits show a Sorensen-type community coefficient of only 30 (100 = identity, 0 = no species in common) (see Chapter 9).

The subalpine zone of the Cascade Range is dominated by mountain hemlock in the wettest, westernmost portions, and by subalpine fir. Associates are numerous and include Engelmann spruce, lodgepole pine, alpine larch, Douglas fir, western larch (*Larix occidentalis*), western white pine (*Pinus monticola*), and white bark pine.

Lodgepole pine, white bark pine, and mountain hemlock continue south into subalpine forests of the Sierra Nevada, and they are joined by red fir (*Abies magnifica*), limber pine, foxtail pine, and—in the extreme south—by bristlecone pine. The Sierran subalpine forest is often relatively open, contrasting strongly with the dense, closed forests in the upper montane zone below (Figure 19–16).

The upper montane zone in British Columbia and northern Washington is dominated by western hemlock, western red cedar, and (in the wettest, western portion) Pacific silver fir (*Abies amabilis*). In the southern Cascades, this zone is dominated by grand fir and is associated with white fir, ponderosa pine, lodgepole pine, Douglas fir, and western larch.

The upper montane zone of the Sierra Nevada is quite distinctive, with an upper portion dominated almost exclusively by red fir, and a lower portion called the **mixed conifer forest**. The red fir forest has been called a subalpine forest by some ecologists, and it does have some physiognomic similarities to subalpine spruce-fir. It is a dark, quiet, two-layered community, but the trees are relatively large and long-lived compared to typical spruce-fir. Below the red fir zone, be-

Figure 19-16. The subalpine forest of the central Sierra Nevada, about 3200 m. The open forest is dominated by white bark pine (Pinus albicaulis) and mountain hemlock (Tsuga mertensiana).

tween 1800 and 900 m, is a more open forest, richer in trees, shrub, and herbaceous species. Dominance is shared by five trees: ponderosa pine, Douglas fir, sugar pine (Pinus lambertiana), white fir, and incense cedar (Calocedrus decurrens, synonym is Libocedrus d).

A certain frequency of surface fires seems to maintain the importance of each species, and a fire suppression policy over the past 70 years has resulted in a major shift from the previous balance. Table 19-2 shows that the sapling community is quite different from the overstory community: white fir and incense cedar are 2–4 times as important in the sapling layer as in the overstory; sugar pine is holding its own; and Douglas fir and ponderosa pine are 2–9 times less important in the sapling layer than in the overstory. The mature forest 100 years from now will be a dense white fir forest, quite unlike the forest of the present and past. One of the reasons for the shift has to do with the seedling requirements of each species. Seedlings of ponderosa pine, for example, are known to do best when the little layer is thin or absent, and in relatively open sun. Seedlings of white fir can tolerate denser shade and a deeper litter layer.

Scattered within this mixed conifer forest are 75 groves of bigtree (Sequoiadendron giganteum), the most massive organism on earth (Figure 19-17). The largest groves cover more than 800 ha and contain 20,000 trees; the smallest and most northerly grove extends over only 0.4 ha and contains only 6 mature trees. The smaller stands appear to be losing mature trees to death by old age faster than they are gaining saplings. The reasons for the population decline are not known, but the decline has been going on for 500 or more years, so recent human activities are not necessarily the cause.

The lower montane zone is dominated by ponderosa pine throughout the Cascade and Sierra Nevada Mountains. It is a relatively open forest and closely resembles the Rocky Mountain lower montane. A well developed understory of shrubs (Symphoricarpos, Holodiscus, Chamaebatia, Rosa, Arctostaphylos, Physocarpus, Ceanothus spp.) or of grasses (Agropyron, Festuca, Stipa spp.) can exist. Subdominant oak trees, such as black oak (Quercus kelloggii), Oregon white oak (Q. garryana), and canyon live oak (Q. chrysolepis) are common.

As is true for the Rocky Mountains, many hectares are not covered by forest. In the Sierra Nevada, for example, more than half the land is covered by meadow or brushfield. **Brushfields (montane chaparral)** are often found on hot, rocky, dry exposures where tree growth is limited; they may also invade burned or logged areas on better sites that once supported conifer forest (Figure 19–18). In the latter case, montane chaparral is successional and will ultimately be replaced by a closed forest. The rate of succession is very slow, and management techniques are being developed which involve suppression of brush by selective herbicides that leave the conifers unaffected.

Temperate Low Elevation Conifer Forests

Pacific north coast conifer forest

This coastal forest extends from the Gulf of Alaska, at 62°N latitude, to the Mendocino coast of California, at 40°N latitude. It is un-

Table 19–2. Importance values (IV), average diameter at breast height (cm), and density per hectare of all conifer species in a virgin mixed conifer forest at Placer County Big Trees, California, 1700 m elevation. Data are means of about 80 points scattered over several hectares (point-centered quarter method; Chapter 8) sampled in 1974 and 1975 by plant ecology classes under the direction of Michael Barbour. Importance values (IV = relative basal area + relative density + relative frequency) are for the size classes shown, but average diameter at breast height and density are only for trees >3 cm dbh. Total basal area for trees >3 cm dbh was 49 m² ha⁻¹.

| | IV for three size classes | | | | |
Taxon	<3 cm dbh (saplings)	3–40 cm dbh	>40 cm dbh (overstory adults)	Average dbh (cm)	Density ha⁻¹
white fir (*Abies concolor*)	132	123	36	8.3	258
incense cedar (*Calocedrus decurrens*)	62	35	27	15.0	67
sugar pine (*Pinus lambertiana*)	59	38	69	23.1	87
ponderosa pine (*P. ponderosa*)	10	47	89	16.6	114
Douglas fir (*Pseudotsuga menziesii*)	37	57	79	19.1	122

Figure 19–17. Bigtree *(Sequoiadendron giganteum)* in the mixed conifer zone of the Sierra Nevada Mountains, about 1400 m.

usual for two reasons: first, hardwoods typically dominate temperate, mesic climates such as this, but here softwood volume is 1000 times hardwood volume; second, the size and longevity of the dominants is unprecedented. The largest trees of 10 genera are found here: *Abies, Chamaecyparis, Larix, Calocedrus, Picea, Pinus, Pseudotsuga, Sequoia, Thuja,* and *Tsuga.* Overstory trees are commonly 50–75 m tall and more than 2 m dbh. Aboveground standing crop averages 2000 t ha^{-1}. Longevity is often well beyond 500 years, and sometimes it is beyond 1000 years.

Figure 19–18. Montane chaparral on a disturbed area on the east side of the Sierra Nevada, about 1900 m. The dominant shrub is tobacco brush *(Ceanothus velutinus)*, and it is suppressing the growth of white fir *(Abies concolor)* coming up beneath it.

Except for the Californian part of the range, the most common overstory trees are Sitka spruce *(Picea sitchensis)*, western hemlock, and western red cedar. Common associates include Douglas fir, grand fir, silver fir, and two species of cedar *(Chamaecyparis)*. Douglas fir appears to be a successional species, despite its great size and long life; it does not reproduce well in its own shade and is succeeded by western hemlock. Broadleaf trees, such as red alder *(Alnus rubra)* may be scattered beneath the conifer canopy, and a lush ground layer of shrubs, perennial herbs, ferns, and cryptogams is typical (Figure 19–19).

In California, coast redwood *(Sequoia sempervirens)* becomes the dominant species. Up to 100 m tall and as old as 2000+ years, redwoods are nevertheless seral species, dependent on periodic disturbance by fire or flood in order to maintain their vigor, reproduction, and dominance.

Great Lakes pine forest

A mosaic of softwood and hardwood stands spreads over part of the Great Lakes region of the United States, in parts of Michigan, Wisconsin, and Minnesota. Three pines form nearly pure, even-aged stands: red pine *(Pinus resinosa)*, eastern white pine *(P. strobus)*, and jack pine *(Pinus banksiana)*. The pines are most often restricted to sandy soils, with beech, maple, yellow birch, and hemlock dominating the surrounding deciduous forest on more mesic soils. This mosaic lies in a tension zone, or ecotone, between the boreal forest and the deciduous forest, and there has been considerable debate about the climax nature of pine stands here (see also Chapter 5). The usual view is that pine is seral to hardwood, and at best the pine forms an edaphic climax on sites too dry to support the hemlock-hardwoods. These stands were extensively cut for lumber in the nineteenth century, and their area today does not reflect past dominance.

Figure 19–19. Interior of the north coast conifer forest at low elevations in Olympic National Park, Washington.

New Jersey pine barrens

A half million hectares of land in the coastal plain of New Jersey, and limited areas on Long Island and on Cape Cod, have very sandy, acidic soil of low fertility. They support a scrubby forest, about 6 m tall, of pitch pine (*Pinus rigida*) and several oak species. The tree canopy is somewhat open, and there is a well-developed understory of low shrubs, mainly oaks and ericads, and a ground layer of mosses, lichens and sedges. In part, the poor, droughty soils limit invasion of surrounding oak-hickory deciduous forest, but currently most ecologists believe that a high fire frequency is the major factor favoring the pines. In 1963, for example, 40,000 ha burned in one weekend alone. Where fire frequency is 10 years or less, the pines do not reach tree stature, but instead form a shrubby pygmy forest little more than 1 m high. Other parts of the barrens experience a fire frequency of about 20 years.

Despite a large surrounding human population, the pine barrens remains largely a wilderness, a precious resource in the most densely populated state in the United States.

Southeastern pine savanna

The coastal plain of North Carolina, South Carolina, Georgia, and Florida often supports an open stand of pines underlain with grass. The principal species of pine are: long leaf (*Pinus palustris*), short leaf, (*P. echinata*), slash (*P. eliottii*), and loblolly (*P. taeda*). The dependence of this forest type on frequent surface fires, once every several years, has been discussed in Chapter 15. In the absence of fire, pine is seral to hardwood forest.

DECIDUOUS FOREST

The deciduous forest offers many contrasts to the boreal forest. The deciduous forest is more open, permitting more light to reach the forest floor; it has more animal life; the overstory can be very rich in species diversity; and the structure of subdominant canopy layers can be more complex. Because of a longer, warmer growing season, net primary productivity and standing biomass are about twice what they are in the taiga. Litter decay is more rapid, litter half-life being on the order of one year instead of several years; as a result, mineral cycling is more rapid.

Deciduous forests cover about one-third of all land area in the conterminous United States. Some of this area is in the form of narrow fingers of riparian forest that extend west along river courses into non-forested regions, but the great bulk of the forest lies east of 95°W longitude and south of 45°N latitude. Only this eastern portion will be discussed here.

The eastern deciduous forest is not homogeneous throughout its extent. Certain groups of dominant tree species characterize particular regions. Much of what we know about the regional forests comes from the extensive field work done by Dr. Emma Lucy Braun (Chapter 2) prior to World War II. Her gender and diminutive size permitted her to gain the confidence of back-country people who helped her search for prime, virgin examples of each regional association. Many of the sites she described were logged during the war years, making her 1950 book an important record of what we once had.

Less technical descriptions, recorded by eighteenth century settlers, also give us a perspective. In Ohio and Pennsylvania, the forest overstory was dominated by 400-year-old oaks, sugar maples, and chestnuts, 1–2 m dbh, with straight boles rising 25 m before the first side branch. Black walnuts, shagbark hickories, and cottonwoods grew in flood plains near rivers, the latter being large enough for travelers such as Daniel Boone to make into dugout canoes 20 m long and more than 1 m across. The dense overstory canopy, together with grapevines which ran up many trunks, created a deep shade in the understory. There was little underbrush, so that it was easy for travelers to wind in and out among the great trunks, even though "one could not shoot an arrow in any direction for more than twenty feet without hitting a tree."

By day, the pristine forest was somber, dark, silent, and gloomy to some. "In the eternal woods it is impossible to keep off a particularly unpleasant, anxious feeling, which is excited irresistibly by the continuing shadow and the confined outlook," wrote one pioneer. The songbirds of our modern forest did not live in that dark forest. At night the forest became unnervingly vocal with calls from wolves, panthers, horned owls, and whip-poor-wills. "It is clear," wrote John Bakeless in his book on the accounts of early explorers, "that no one lamented the disappearance of the picturesque forest, since there were altogether too many trees for comfort."

Braun described nine regional forest associations (Figure 19–20). In this chapter, we will simplify the picture (and take into account some reinterpretations by other ecologists) by condensing those nine into five: **mixed mesophytic, oak-hickory, southern mixed hardwood, beech-maple**, and **hemlock northern hardwood**.

Physical Features

Annual precipitation ranges from 60 to 200 cm, generally increasing from the northwest towards the east and south. There is no pronounced seasonal wet or dry period. Except in the southern mixed hardwood forest, snow and hard frost in winter are common. Mini-

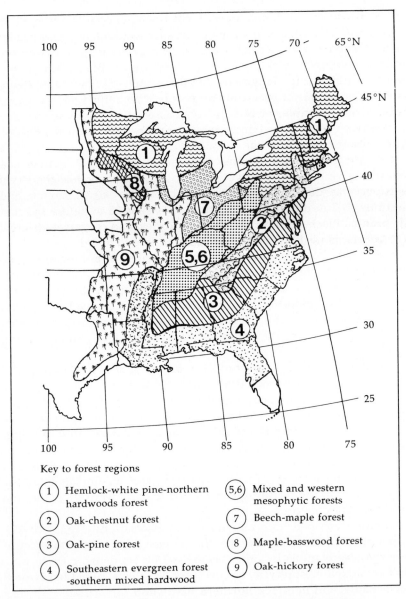

Figure 19-20. Nine regions of the eastern deciduous forest formation recognized by Braun (1950). (From Robichaud and Buell 1973. *Vegetation of New Jersey.* By permission of Rutgers University Press, New Brunswick.)

mum winter temperatures may drop to −30°C. The growing season is 4–6 months long (longer in the south) and maximum temperatures can reach 38°C. Even moderate temperatures of 30°C are uncomfortable because of the high humidity (as much as 10% of summer rain is intercepted by foliage and evaporates back to the air). The P/E ratio is greater than 1. Despite the wet summer, however, soil moisture deficits develop that are acute enough to affect herb growth and tree seedling survival.

Two soil orders characterize the eastern deciduous forest (Figure 19–21). The **alfisol** order extends from north of Virginia and Kentucky to well into an ecotone area with the taiga. Alfisols are also known as **gray-brown podzolics**; they are acidic (pH 5–6), but not as acidic as taiga podzols, partly because the pH of deciduous leaves is 5–7 rather than the pH 4 of spruce and pine. There is no bleached A2 horizon, and the fertility (cation exchange capacity and % base saturation) is relatively high. Diversity and activity of soil biota are high.

The **ultisol** order dominates the southern part of the forest. Ultisols have not been glaciated, hence they are older and more weathered than alfisols. Plowed fields show a B horizon which is bright yellow-orange as a result of oxidized iron. Ultisols are slightly more acidic than alfisols, and considerably lower in fertility.

Physiognomy and Phenology

The modern deciduous forest has been modified by 200 years of coexistence with farmers, harvesters, and industrialists. Community ar-

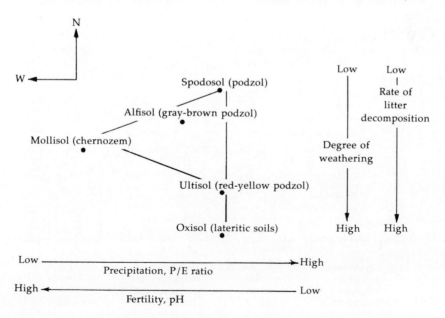

Figure 19–21. Some relationships between soil orders (and great soil groups) of the taiga, eastern deciduous forest, and grassland.

chitecture is no longer so strongly one-layered and impressive as it once was, except in locally protected areas. Today, much of the forest is four-layered (Figure 19–22). The overstory may reach up to 65 m in the best sites, but is more typically 35 m. It is a closed, interlocking canopy, with more than 100% cover. Beneath the overstory is a second tree canopy, that is much more open and consists of saplings and genetically smaller trees about 10 m tall, such as dogwood *(Cornus)*. A third layer is made up of scattered shrubs, often members of the Ericaceae, about 1–2 m tall. A fourth layer consists of seasonally active patches of herbs, mosses, and ferns. Vines such as the Virginia creeper *(Parthenocissus quinquefolia)* pass through several layers. Total LAI ranges between 4 and 8, and total biomass averages 30 kg m^{-2}. Annual net primary productivity is about 1200 g m^{-2}.

The aspect of the forest changes dramatically with the seasons because most species in each stratum are deciduous or die back. If we take New Jersey or Pennsylvania as a reference point, then snow melt in mid-March coincides with release of spring-flowering herbs from dormancy. At this time, the overstory tree canopy is leafless and more than half full sunlight reaches the ground. By late March to early April, the ground is green from renewed vegetative activity by herbs that have overwintered as bulbs or rhizomes. Geophytes (cryptophytes, see Figure 1–3) account for about 18% of all species in the forest, a proportion three times Raunkiaer's world average. By mid-April, the

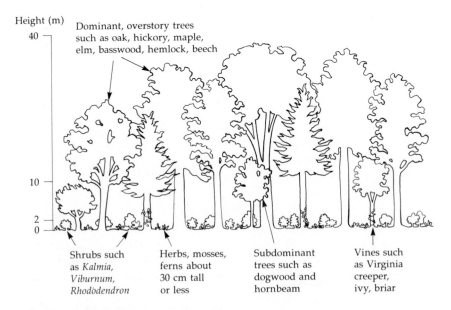

Figure 19–22. Physiognomy of a typical eastern deciduous forest. The dominant overstory trees may reach 60 m or more in height.

herbs are flowering and the overstory trees have resumed meristem activity and have begun to leaf out. Ring-porous trees resume radial growth well before diffuse porous species. As the canopy closes through May, light intensity on the forest floor is reduced to a minimum of 2–5% full sun. Less than 10 ly day^{-1} reach the forest floor at this time, in contrast to 450 ly day^{-1} above the canopy (Figure 13–5). Moving light flecks temporarily allow up to 70% full sun to reach ground herbs, and there is some evidence that this light can be biologically utilized. By July, many of the spring-flowering herbs have died back and become dormant. This dormancy may be induced by competition for moisture, rather than by competition for light. Other herbs, such as goldenrod (*Solidago* spp.), grow throughout the summer and flower in August and September. Leaves of shrubs and trees turn color in early October and are shed into November. Length of the growing season, from the time tree leaves are half expanded in early May to the time half their chlorophyll is lost in fall, averages 160 days. Radial tree growth averages 105 days in extent. Many phenological events may be triggered by environmental changes in the root zone, rather than in the air.

Major Ecotones

Taiga-deciduous forest ecotone: the hemlock-hardwoods forest

In the central part of North America, in Manitoba, Saskatchewan, and Alberta, the taiga passes into a spruce-aspen parkland, then an aspen parkland in a wide ecotone with the grasslands. East of this, in an equally wide ecotone, the taiga passes into the eastern deciduous forest. This ecotone centers on 45°N latitude (southern Maine, northern New York, and northern Michigan), then it bends north through northern Wisconsin and northeastern Minnesota. As balsam fir, white spruce, and red spruce become less abundant, a mixture of hardwoods and other conifers assume dominance.

From studies done in the Green Mountains of Vermont, the ecotone's location correlates closely with cloudiness and fog in spring and fall. Other factors are shortness of the frost-free period and soil depth. Above 800 m elevation, fog drip abruptly increases by 60% and the number of frost-free days per year decreases by 40%. Once spruce and fir are established, they modify the site to make it unfavorable for hardwoods: soil pH drops, litter depth increases, and soil moisture increases.

Ecotone hardwoods include sugar maple (*Acer saccharum*), yellow birch (*Betula allegheniensis*), beech (*Fagus grandifolia*), paper birch (*Betula papyrifera*), and basswood (*Tilia americana*). Conifers include three pines *(Pinus banksiana, P. resinosa, P. strobus)* and hemlock *(Tsuga canadensis)*. This **hemlock-hardwoods forest** is the type of forest Thoreau lived in at Walden Pond (Figure 19-23). To the south, it merges into beech-maple or oak-hickory forest. The pines are especially important in the Midwest, and this phase of the ecotone has been called the **Great Lakes Pine Forest**.

Prairie-deciduous forest ecotone: oak savanna and woodland

In an eastward arc, extending from Minnesota to Texas, the deciduous forest gives way rather abruptly to grassland. The ecotone projects east into Illinois, Indiana, Ohio, and Kentucky, permitting a **prairie peninsula** to cover almost a quarter of a million square kilometers east of the Mississippi River (Figure 19-24). The ecotone is very complex. Sometimes it is abrupt, with a transition from a closed oak forest to an open grassland in a matter of less than 10 m. In other places, the forest gradually thins to a woodland of more open trees with a grassy understory, then to a savanna of scattered trees (less than 30% canopy cover), and finally to a grassland in the distance of 50 km. The boundary is sinuous, with interdigitations of grassland and forest. It is also a mosaic, with numerous islands of prairie within the forest and outliers of forest in the grassland. Major tree species of the ecotone are bur oak *(Quercus macrocarpa)*, post oak *(Q. stellata)*, and blackjack oak *(Q. marilandica)*.

There are many physical and biotic factors which change across the ecotone, and it has not been possible for ecologists to determine which is most important. The major factors which correlate with change from forest to grassland are as follows:

Figure 19-23. The hemlock-hardwoods forest, an ecotone between the taiga and the deciduous forest.

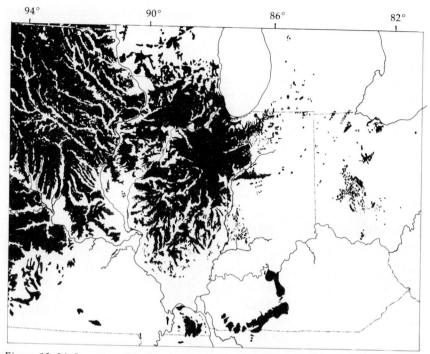

Figure 19–24. Location (black) of the prairie peninsula, as mapped by Transeau (1935). (By permission of the Ecological Society of America.)

1. P/E ratio drops below 1 (except in the prairie peninsula).
2. Annual precipitation drops below 60 cm. (Aridity increases to the west, as does prairie.)
3. Yearly variability in precipitation increases; the frequency of drought increases. Drought cannot be the total reason for forest to give way to grasslands, however, because grassland borders further west support trees under even more arid conditions.
4. Fire frequency increases, favoring grasses (see next paragraph).
5. Finer, deeper prairie soils favor fibrous roots of grasses.
6. Tree mycorrhizal fungi are absent in prairie soils, and tree saplings are at a competitive disadvantage without mycorrhizae.
7. Heavy grazing by bison until the late nineteenth century may have eliminated tree saplings.
8. The tree line may have been pushed east by the Xerothermic period, about 6000 years ago, and it has not yet regained its climatic limits.

In the pristine grassland, wildfires were common virtually every autumn. The perennial grasses could recover from rhizomes and root crowns below the surface, but the woody species were killed. Where forest cover was dense enough, ground vegetation was too thin to carry a fire, so in this way fire could maintain an abrupt forest-grassland ecotone. In the absence of fire, some Wisconsin studies show that woody species invade prairie at the average annual rate of 0.3 m. Burning also improves the vigor of the grasses, stimulating productivity and flowering two- to threefold due to removal of litter. In the absence of litter, soil temperatures in the root zone are 5°C higher, which stimulates plant growth and decomposer activity.

Mycorrhizal fungi of oaks and other forest trees appear to be absent from prairie soils, and without the mycorrhizal association, studies have shown trees to be at a competitive disadvantage with grasses. Thus, tree invasion may be slow because both seed and fungus must invade together.

Some Regional Forest Associations

The mixed **mesophytic forest association** of the Cumberland Mountains and Alleghany Mountains contains the richest collection of overstory dominants of any forest in North America. As many as 20–25 species of overstory and understory trees can be found in one hectare, and dominance fluctuates so greatly from stand to stand that the community cannot be named by even a handful of species. The most common, characteristic trees include sugar maple *(Acer saccharum)*, buckeye *(Aesculus octandra)*, beech *(Fagus grandifolia)*, tulip tree *(Liriodendron tulipifera)*, white oak *(Quercus alba)*, northern red oak *(Q. borealis var. maxima)*, and basswood *(Tilia heterophylla)*. Hemlock may sometimes be present, especially in mesic, protected cove forests of Appalachian foothills (Figure 19–25). Understory trees are also diverse and may produce a striking floral display in the spring. They include redbud *(Cercis canadensis)*, dogwood *(Cornus florida)*, witch hazel *(Hamamelis virginiana)*, ironwood *(Carpinus caroliniana)*, hornbeam *(Ostrya virginiana)*, hackberry *(Celtis occidentalis)*, and species of *Rhododendron* and *Magnolia*. Viburnum, sumac, huckleberry, blueberry, and blackberry shrubs are common, and midsummer is a berry-picker's delight.

Nearly surrounding the mixed mesophytic association is the **oak-hickory forest**, which has several species each of *Quercus* and *Carya*. A portion of this forest, just to the east of the mixed mesophytic association, was an area once dominated by chestnut *(Castanea dentata)*. As briefly described in Chapter 7, a fungal parasite of chestnut was accidentally introduced to the New York area from Europe early in the twentieth century. By the 1930s the fungus had spread throughout much of the range of chestnut and had killed all mature trees. Apart from an occasional stump sprout and a few trees, the only remains of chestnut today are "ghosts"—slowly decaying but sometimes still standing, skeletons of dead trees. Oaks and hickories now occupy the openings in the canopy vacated by chestnuts.

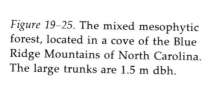

Figure 19-25. The mixed mesophytic forest, located in a cove of the Blue Ridge Mountains of North Carolina. The large trunks are 1.5 m dbh.

Along the Atlantic coast, just behind the beach and foredune, is a **live oak forest** dominated by short, twisted oaks with relatively small, sometimes evergreen, sclerophyllous leaves: *Quercus virginiana, Q. laurifolia,* and *Q. myrtifolia.* Below the dense canopy, only about 5 m tall, are many shrubs *(Ilex, Myrica, Vaccinium, Serenoa, Sabal)* and vines *(Vitis, Smilax, Rhus, Parthenocissus).* The pruning effect of wind, and of salt spray carried in the wind is quite evident (Figure 19-26). The maritime climate is more moderate than for other parts of the deciduous forest. Less than 1% of the hours each year have frost. The unique climate and the evergreen foliage make this community very different from the rest of the eastern deciduous forest.

A **beech-maple forest** extends through much of Ohio and Michigan. Major dominants are sugar maple *(Acer saccharum)* and beech *(Fagus grandifolia).* Common associates include yellow birch *(Betula alleghaniensis),* red maple *(Acer rubrum),* white ash *(Fraxinus americana),* and basswood *(Tilia americana).* The southern limit of this association corresponds to the limit of Wisconsin glaciation. The Hubbard Brook ecosystem study discussed in Chapter 12 is dominated by this forest. The **hemlock northern hardwood forest**, already discussed in the ecotone section, lies to the north, through much of Pennsylvania, New York, and New England.

The south-east is covered by the **southern mixed hardwood forest** (-southeastern evergreen forest in map, Figure 19-20). Dominants are many (up to 40 tree species), and include evergreens such as south-

Figure 19–26. A live oak forest (mainly *Quercus virginiana*) along the North Carolina coast.

ern magnolia *(Magnolia grandiflora)*, the semi-evergreen laurel oak *(Quercus laurifolia)*, live oak *(Q. virginiana)*, and several pines, and deciduous species such as beech *(Fagus grandifolia)*, sweet gum *(Liquidambar styraciflua)*, and some hickories. Epiphytes, such as Spanish moss *(Tillandsia usneoides)*, are common. A subdominant tree canopy may also be closed, but the shrub and herb strata are patchy.

Based on contribution to the importance value, evergreens account for about half of the canopy, but percentage evergreenness can vary from 0 to 100. In general, evergreen conifers are most abundant on sandy, sterile sites, where retention of leaves may have survival significance.

GRASSLANDS

Grasslands cover more area of North America than any other formation. The central grasslands are locked into the interior of the continent (Figure 19–27). They occupy a swath 1000 km wide between about 95°W and 105°W which tapers to a northern limit in Saskatchewan, at 52°N, and to a southern limit in central Texas, at 30°N.

This enormous strip of grassland is not homogenous. As precipitation decreases from east to west, the height, growth form, and identity of dominant grasses change. The easternmost portion, including the prairie peninsula, is the **tall grass prairie.** Flowering stalks of some sod-forming grasses here reach 2 m in height. Farther west is the **mixed** (sometimes called the **mid-) grass prairie,** where maximum plant height is about 1 m. Farther west, in the area of lowest rainfall (40 cm), is the **short grass prairie,** dominated by bunchgrasses less than 0.5 m tall. The spatial relationship of these grasslands to each other and to forests already discussed is summarized for latitude 37°N in Figure 19–28.

Outlying grasslands include: a) the **intermountain grassland,** extending through Wyoming, Idaho, Utah, Nevada, and eastern Oregon, and including the Palouse region of eastern Washington; b) the **desert grassland** in Texas, New Mexico, Arizona, and northern Mexico; and c) the **central valley grassland** of California.

All of the North American grasslands have been extensively modified, primarily through four activities: cultivation, fire suppression policies, grazing of domestic stock, and the accidental introduction of aggressive, weedy, alien plant species.

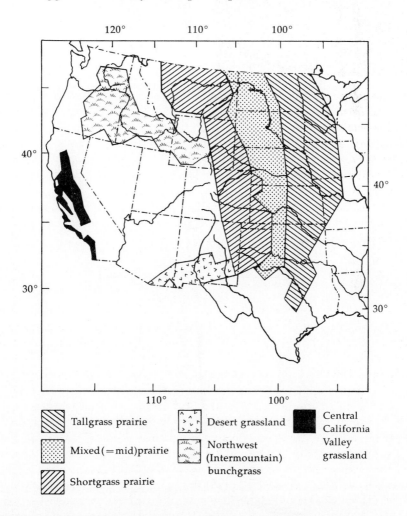

Figure 19-27. Diagrammatic location of the six grasslands discussed in the text. Extensions into Canada and Mexico are not shown. Prairie peninsula is not shown. (From Sims et al. 1978 (by permission of the British Ecological Society) and Heady 1977 (in *Terrestrial Vegetation of California.* Edited by Barbour and Major. Copyright 1977 John Wiley and Sons, Inc. Reprinted by permission of John Wiley and Sons, Inc.).)

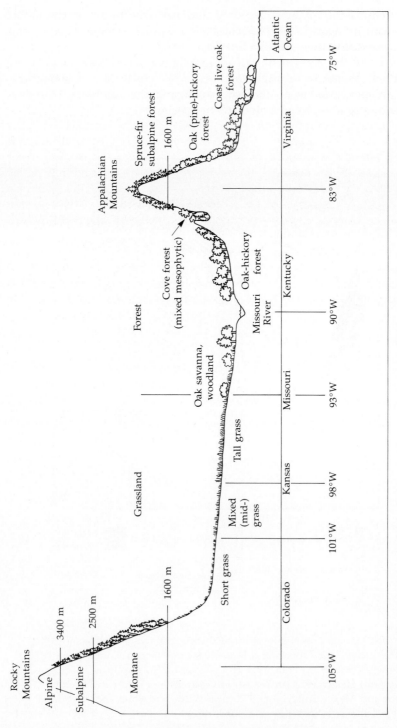

Figure 19–28. Diagrammatic representation of vegetation at 37°N latitude from the Atlantic coast to the Rocky Mountains.

Some Definitions

Grasslands are herbaceous communities dominated by **grami-noids** (grasses, sedges, rushes), but with **forbs** (non-graminoid herbs) present and sometimes seasonally dominant. Trees are absent except for local sites, such as along water courses.

Grasslands may be dominated by annual grasses, perennial **bunchgrasses**, or perennial **sod-forming** grasses. Annuals are most abundant on dry, overgrazed, or disturbed sites. Some annuals are cool season types, germinating in late fall and setting seed the following early summer. Cheatgrass *(Bromus tectorum)* in the intermountain area is such an annual. Other annuals are warm season types. They germinate in spring or summer and complete a much shorter life cycle in a matter of weeks, rather than months. Six-weeks grama *(Bouteloua barbata)*, in the desert grassland, is such an annual.

Annuals do not reproduce vegetatively with runners or rhizomes. Their lateral spread is limited to the production of **tillers** (stems) from buds near the root crown, giving the plant a bushy, clumped appearance (Figure 19–29). Perennial bunchgrasses, such as purple

Height (m)

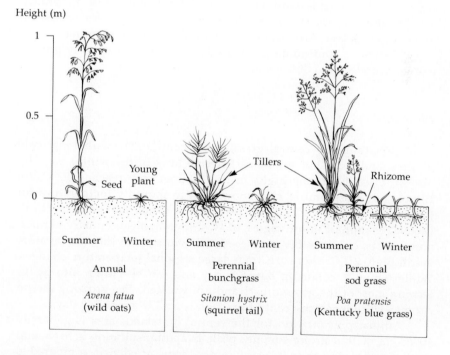

Figure 19–29. Summer and winter aspects of three major grass life forms: annual, bunchgrass, and sod-forming grass.

needlegrass (*Stipa pulchra*) in California, also produce tillers, but the continuation of that process for many years results in a large clump a decimeter or more in diameter. Bunchgrasses alone do not generally produce a community with 100% cover, but the spaces between clumps can be seasonally filled by forbs and other grasses.

Perennial sod-forming grasses, such as big bluestem (*Andropogon gerardii*) in the tall grass prairie, spread laterally by rhizomes. New shoots and roots arise from nodes on the rhizomes in such numbers that a **turf** results. The topsoil is thoroughly penetrated and held together by fibrous root systems, and a continuous sward of shoots covers the surface. Sod-formers are more resistant to grazing than bunchgrasses because they have so many more growing points. Some sod-formers are less aggressive than others, producing only short rhizomes, and some produce rhizomes only under certain environmental conditions. The growth forms of common grasses in North American grasslands are summarized in Table 19–3.

The term **prairie** is most often applied to a grassland dominated by tall, sod-forming grasses, with 100% cover. Steppe is often applied to a grassland dominated by short bunchgrasses, often interspersed with shrubs, with a ground cover that may be less than 100%. In this chapter, we will use **grassland** as a general term, applicable to prairie, steppe, and all intermediates. As precipitation increases, grassland gives way to forest. An ecotone vegetation type is the **savanna**. Savannas still retain a continuous grass cover, but trees are regularly present and contribute up to 30% cover.

Physical Features

Climate of the central grassland is continental, with long, cold winters, and long, hot summers. Extreme lows in the winter may dip to −40°C, and extreme highs in the summer may reach +45°C. An Illinois tall grass prairie may show a mean monthly temperature for January at −3°C, and 24°C for July. In the mixed grassland of North Dakota, or in the Palouse region of Washington, mean January temperature can be considerably lower, −14°C, while mean July temperature is still high, 21°C. There are, then, large seasonal temperature changes. Winter snow is common, but it is dry and can be blown into drifts, leaving most of the ground bare and uninsulated. The growing season for mid-latitudes in the central grasslands area is about 240 days.

Annual precipitation for the central grasslands area is generally 40–60 cm, with a strong summer peak. Evapotranspiration is high, and the P/E ratio is less than 1. A pronounced moisture deficit, compared to a minor one in the deciduous forest, characterizes grassland soils in summer. Summer thunderstorms may be violent, accompanied by hail

Table 19-3. Common dominant grasses in each of the six major grasslands of North America. A = annual, B = bunchgrass, S = sod-forming grass, * = introduced, not native, ‡ = known to be C_4 (some of those not marked may be C_4).

		Tall	Mixed	Short	Inter-mountain/ Palouse	Desert	California
switch grass (*Panicum virgatum*)	S	x					
Kentucky bluegrass (*Poa pratensis*)	S*	x					
Indian grass (*Sorghastrum nutans*)	B‡	x					
quack grass (*Agropyron repens*)	S*	x	x				
western wheatgrass (*A. smithii*)	S	x	x	x	x		
big bluestem (*Andropogon gerardii*)	S‡	x	x	x			
little bluestem (*A. scoparius*)	B‡	x	x	x			
hairy grama (*Bouteloua hirsuta*)	B or S‡	x	x	x			
sideoats grama (*B. curtipendula*)	B‡	x	x	x	x		
blue grama (*B. gracilis*)	B‡	x	x	x	x		
needlegrass (*Stipa spartea*)	B	x	x	x			
needle and thread (*S. comata*)	B		x	x	x		
June grass (*Koeleria cristata*)	B		x	x	x		x
dropseed (*Sporobolus cryptandrus*)	B		x	x			
sand bluestem (*Andropogon hallii*)	S			x		x	

Table 19-3 continued.

	Tall	Mixed	Short	Inter-mountain/ Palouse	Desert	California
buffalo grass (*Buchloe dactyloides*)	S[‡]		x			
vine mesquite (*Panicum obtusum*)	S		x			
tobosa (*Hilaria mutica*)	B[‡]				x	
galleta (*H. jamesii*)	S				x	
black grama (*Bouteloua eriopoda*)	S				x	
six-weeks grama (*B. barbata*)	A				x	
three awn (*Aristida* spp.)	B			x		
Indian rice grass (*Oryzopsis hymenoides*)	B	x	x	x		
muhly (*Muhlenbergia* spp.)	S or B		x	x		
sacaton (*Sporobolus wrightii*)	B			x		
alkali sacaton (*S. airoides*)	B	x	x	x		
burro grass (*Scleropogon brevifolius*)	S			x		
Arizona fescue (*Festuca arizonica*)	B			x		
bluebunch fescue (*F. idahoensis*)	B			x		
red fescue (*F. rubra*)	B			x		
greenleaf fescue (*F. viridula*)	B			x		
California oat grass (*Danthonia californica*)	B			x		

Table 19–3 continued.

	Tall	Mixed	Short	Inter-mountain/ Palouse	Desert	California
bluebunch wheatgrass (*Agropyron spicatum*)	B			x		
slender wheatgrass (*A. trachycaulum*)	B			x		
cheatgrass (*Bromus tectorum*)	A*			x		
California brome (*B. carinatus*)	A or B			x		x
brome (*B. marginatus*)	B			x		
pinegrass (*Calamagrostis rubescens*)	B or S			x		
wild rye (*Elymus* spp.)	B or S			x		x
melic (*Melica* spp.)	B					x
needlegrass (*Stipa* spp.)	B			x	x	x
soft chess (*Bromus mollis*)	A*					x
rip gut (*B. diandrus*)	A*					x
fescue (*Festuca* spp.)	A					x
wild oats (*Avena* spp.)	A*					x

and high winds. Total rainfall can fluctuate two- to threefold in successive years, and periodic droughts (as in the Dust Bowl years of the 1930s) are to be expected. It has been suggested that the distribution limits of some grassland species correlate with occasional extreme lows in rainfall, rather than with long-term means.

Precipitation in the desert, intermountain, and Californian grasslands is lower, about 25–45 cm, and the seasonality also differs from that in central grasslands. The desert area has both summer and winter peaks, while the intermountain and Californian areas have winter peaks. Climographs for three grasslands are shown in Figure 19–30. Notice how closely the tall and short grass areas overlap.

Grassland soils are quite diverse. In the prairie peninsula and in part of the tall grass area, the prevailing soils are in the ultisol order. A P/E ratio of 1 permits weak podzolization to occur. Soil pH is close to neutral, however, and decay of fibrous root systems gives the soil a dark brown color. There is no bleached A horizon. Where precipitation is less than evapotranspiration, **mollisol** soils become prominent. A layer of calcium carbonate accumulates in the subsoil, deposited at shallower and shallower levels as the P/E ratio declines. The most mesic mollisol is a **chernozem**, with up to 16% organic matter in the

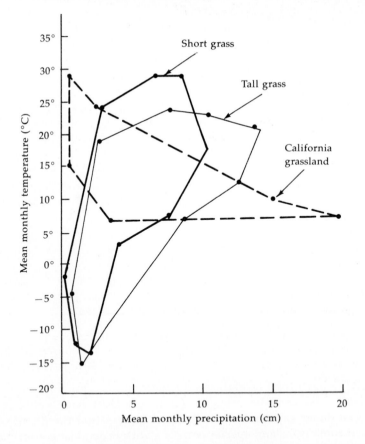

Figure 19–30. Climographs (hythergraphs, pp. 358–9) for three grasslands. Note the similarity between tall and short grassland regions, and the uniqueness of the California central valley grassland.

nearly black A horizon, a clay loam texture, high cation exchange capacity, very high percent base saturation, pH slightly basic, and carbonate accumulation only at 100 cm depth. Chernozem is a Russian word, meaning black earth. The color is a result of richness in organic matter. Chernozems give way to **chestnut**, then to **brown** soils as one moves through mixed and short grass areas. Intermountain grasslands are also on mollisols, but desert grasslands can occur on **aridisols**, which have little profile development, and contain less than 1% organic matter in the root zone.

Central Grassland Vegetation

The **tall grass prairie** has a nearly equal mix of bunchgrasses and sod-forming grasses, which together provide a dense canopy of more than 100% cover and a standing aboveground biomass of 200 g m^{-2}. The LAI is 5–8, about the same as that in the deciduous forest. Most biomass is belowground, and the root:shoot ratio averages 6–7. Annual aboveground net productivity can be more than 1000 g m^{-2}, the great bulk of which is contributed by warm season grasses, some of which are C$_4$ species (e.g., little bluestem and blue grama).

Many species are listed in Table 19–3, but the most typical dominants are big bluestem and Indian grass, both with flowering shoots up to 2 m tall. Associated with them are shorter grasses, such as little bluestem and gramas, and a variety of perennial forbs. As in all the grasslands, forb species outnumber graminoids by 3 or 4 to 1, but their productivity only accounts for 10–20% of total community productivity. There is considerable latitudinal variation in the associated species, and even at one particular location there are major shifts that reflect the microenvironment. Prairie cordgrass (*Spartina pectinata*), for example, can be prominent in wet depressions, and Kentucky bluegrass can be abundant where the range is heavily grazed.

Most of the tall grass prairie is gone, replaced by today's corn belt, but the records of early travelers can help us reconstruct it. This is from a journal dated 1837:

> The view from this mound . . . begs all description. An ocean of prairie surrounds the spectator whose vision is not limited to less than 30 or 40 miles. This great sea of verdure is interspersed with delightfully varying undulations, like the vast swells of ocean, and every here and there, sinking in the hollows or cresting the swells, appears spots of trees. . . .

The **mixed grass prairie** is perhaps the most floristically complex of the central grasslands. The tall sod-formers of the east generally become restricted to locally heavy soils, and the usual dominants are shorter bunchgrasses such as little bluestem, hairy grama, blue grama, and needlegrass. To those who grow up around forests, a grassland such as this may appear homogeneous and monotonous (Figure 19–31), but careful observation reveals a complex mosaic of communities. Table 19–4 shows the importance value of major grass and forb species in 10 stands, all in a relatively local region of western North Dakota. There are enormous differences, fully as significant as tree differences among the five or so associations of the deciduous forest. Stands 1 and 2 are dominated by sedge and needlegrass; stand 3 by blue grama; stand 5 is reminiscent of tall grass prairie, with big bluestem dominant; stands 6 and 7 are dominated by prairie dropseed; stands 4 and 9 have unusually high forb components; only stands 8 and 10 are "typical" mixed grass prairie dominated by little bluestem. Sorensen community coefficients, for pairs of stands based on these importance values, show a maximum similarity of only 63% and a minimum of 8%.

Cover in the mixed grassland is still 100%, but standing, live aboveground biomass is only half that of tall grass prairie, and only half the annual productivity derives from warm season grasses.

Short grass communities are dominated by bunchgrasses such as buffalo grass, sand bluestem, Indian ricegrass, blue grama, and species of *Muhlenbergia*. A few species are sod-formers. Buffalo grass and blue grama, among others, are C_4 species, but about half the annual productivity derives from cool season C_3 species. Aboveground standing crop is only 50 g m^{-2}. When mixed grass prairie is stressed by heavy grazing or drought, short grass species tend to increase in abundance. All three grasslands are on a moisture continuum, with tall grass prairie in the wettest region and short grass in the driest.

Figure 19–31. Mixed grassland on rolling hills in North Dakota.

Table 19-4. Importance values of major species in 10 mixed grassland stands in North Dakota. Importance value = relative frequency + relative cover. Asterisks indicate dominants of each stand. (From Redmann 1975. Copyright 1975 by the Ecological Society of America.)

	Little Shell Creek Area					Packinaw Hills Area				
	1	2	3	4	5	6	7	8	9	10
western wheatgrass (Agropyron smithii)	25	15	13							
big bluestem (Andropogon gerardii)					34*					
little bluestem (A. scoparius)					53*		6	73*	65*	86
sagebrush (Artemisia frigida)	2	7	23	1	2					
sagebrush (A. ludoviciana)						16		8	1	2
aster (Aster ericoides)	4	16	2	13	6					
Brauneria angustifolia	1	4	2	12	13					
blue grama (Bouteloua gracilis)	2	22	40*		2					
sedge (Carex pennsylvanica)	41*	32*	13	3						
sunflower (Helianthus rigida)				34	22	24		7	8	6
panic grass (Panicum wilcoxianum)								6	7	14
knot weed (Polygonum alba)		3	28		2					
rose (Rosa arkansana)				16	8	1	3		14	1
prairie dropseed (Sporobolus heterolepis)						69*	104*			
needle and thread (Stipa comata)	6	32*	20				1	8	6	21
needlgrasss (S. spartea)				58*	26	32*	32*	34*	16	2
needlegrass (S. viridula)	60*	15	3							
all graminoids	138	120	116	90	128	112	140	134	103	146
all forbs	62	80	84	110*	72	88	60	66	97*	54
No. of species	21	32	21	25	21	27	16	32	34	27

Desert Grassland

Plateaus above 1000 m elevation at the edge of the Sonoran and Chihuahan deserts were covered until recently by short bunchgrasses, with desert scrub restricted to ravines, knolls, or other locally poor sites. Important grasses were Indian ricegrass, *Muhlenbergia*, black grama, tobosa, and galleta.

A significant fraction of this grassland's acreage has disappeared over the past 100 years. **Desert scrub** has invaded from the locally poor sites and has come to dominate many hectares (Figure 19–32). An 1858 vegetation survey of the Jornada Experimental Range in southcentral New Mexico showed 33,800 ha to have 100% relative cover by grasses. Desert shrubs were absent from those hectares at that time. A resurvey in 1915 showed only 14,400 ha could still be so classified. In 1963, no hectares had 100% relative grass cover (Buffington and Herbel 1965). At the same time, cover by such desert shrubs as creosote bush, mesquite, and tarbush increased twentyfold. A 50+ year study of desert grassland on the Santa Rita Experiment Station in southeastern Arizona showed a similar trend, creosote bush alone increasing thirteenfold between 1904 and 1958 (Humphrey and Mehrhoff 1958).

Some ecologists have hypothesized that the shrub explosion was due to overgrazing of grasses by cattle. Even after cattle exclosures were erected in 1931 at Jornada and Santa Rita, shrubs continued to spread in both. This may be due to erosion of the thin top soil formerly held in place by the grasses. Other causes suggested for shrub encroachment include fire suppression policies and a slight warming-drying trend in the climate of the southwest region.

Intermountain-Palouse-Willamette Grassland

The northern part of the Great Basin, between the Rocky Mountains and the Cascade Range, is a **shrub steppe** (Figure 19–33). Short bunchgrasses share dominance with cold desert shrubs, especially species of sagebrush *(Artemisia)*. Many dominants of this region are also found in the loess-covered Columbia River Basin of eastern Washington (the **Palouse** area), and in another grassland that used to clothe the Willamette Valley of Oregon. Like the central valley of California (next section), cultivation has eliminated the Willamette grassland and we must rely on incomplete records to reconstruct the pristine vegetation.

Dominants of these three regions include the following cool season bunchgrasses: bluebunch wheatgrass, bluebunch fescue, several other fescues, wild rye, sacaton, and California oat grass (see Table 19–3 for scientific names). It is this intermountain grassland which has been invaded and changed by annual cheatgrass, as discussed in Chapter 5.

Figure 19–32. This stand of desert scrub, dominated by creosote bush in New Mexico, was desert grassland a century ago, according to historical land descriptions.

Figure 19–33. Artemisia shrub steppe, Wind River area of Wyoming.

A recent analysis of C_4 grasses shows that the intermountain and Californian grasslands are much lower in C_4 species than the central grasslands. Tall and mixed grassland floras are 40% C_4 (grass species only), while the intermountain-Palouse-Willamette-California grassland floras are only 15% C_4.

California Central Valley Grassland

This grassland is located mainly in the Sacramento and San Joaquin Valleys, although historically it was also found in the Los Angeles basin and in several other coastal valleys of central and southern California. (An interrupted grassland along the north coast of California is more closely related to the Palouse and Willamette grassland than it is to the central valley grassland.)

Two hundred years ago, when first viewed by Europeans, the vegetation may have been dominated by cool season bunchgrasses such as purple needlegrass (*Stipa pulchra*), nodding needlegrass (*S. cernua*), wild rye (*Elymus* spp.), pine bluegrass (*Poa scabrella*), three awn (*Aristida* spp.), June grass (*Koeleria cristata*) and deer grass (*Muhlenbergia rigens*). These grasses contributed about 50% cover and the spaces between were filled with great masses of annual and perennial grasses and forbs which flowered in late spring. During the June-October dry period, forbs and grasses alike died back, turning the region golden brown. This grassland occupied over 5 million hectares, about 13% of the modern state's land area.

Overgrazing, drought, and the introduction of hundreds of weedy annuals have changed this grassland into an annual type, with forage of lower nutritional value. Modern dominants are wild oats, bromes, ryegrasses (*Lolium* spp.), foxtails (*Hordeum* spp.), and forbs such as filaree (*Erodium* spp.). Cattle exclosure studies indicate that the native bunchgrasses are very slow to reinvade or to increase in cover, given the presence of these aggressive annuals. It appears that the California grassland has been permanently changed, and the weeds are "new natives," to use the expression of one ecologist.

DESERT SCRUB

Desert scrub is a vegetation type dominated by shrubs with less than 100% cover and generally restricted to semiarid regions receiving 5–25 cm precipitation a year. In North America, desert scrub occupies more than 1 million square kilometers. Based on climatic, vegetational, and floristic differences, four desert regions can be recognized (Figure 19–34).

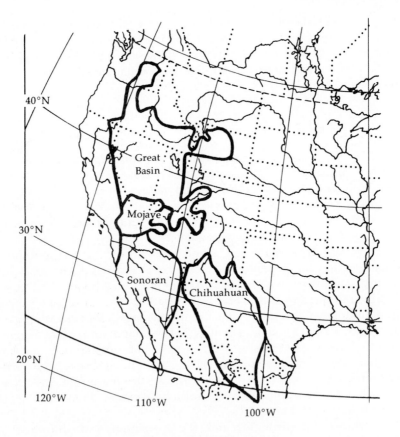

Figure 19–34. Desert regions of North America.

Figure 19–35. Great Basin desert, eastern Oregon, dominated by big sagebrush *(Artemisia tridentata).*

Figure 19–36. Close-up of creosote bush *(Larrea tridentata)* with mature fruit.

The **Great Basin desert** (also called the **cold desert**) has a considerable amount of snow and hard frost in winter. The intermountain grassland, discussed in the last section, extends through the northern part of this desert, but most of the region is thoroughly dominated by big sagebrush (*Artemisia tridentata*, Figure 19–35). The other three deserts are collectively called the **warm** or **hot deserts** because winters are mild. One shrub dominates all three: *Larrea tridentata* (Figure 19–36), variously called creosote bush, greasewood, gobernadora, or hediondilla. Creosote bush is slow growing but persistent. It is able to spread vegetatively by a splitting of the root crown followed by adventitious rooting at the base of lateral branches beneath the surface. Gradually, clonal circles of shrubs develop which may be hundreds to thousands of years old (Figure 19–37).

Some parts of the desert have been misused by people for grazing, recreation, and even residential development. Off-road vehicles such as motorcycles and dune buggies have altered or denuded the landscape. Like the tundra, even one pass with a heavy vehicle can cause damage, in this case to the superficial roots of some plants, to the food plants of herbivores, and to the burrows of many animals. There are more than 5 million off-road vehicles in the United States and many of these are in or adjacent to desert areas.

Productivity

When viewed from a low, oblique angle, plant cover seems high, but when the canopies are projected vertically down, only 10–25% of the ground typically lies beneath perennial cover (extremes are 5–50%).

Aboveground biomass is low, averaging 700 g m^{-2}, which is nearly the same as tundra standing crop. LAI is less than 1, and during the dry part of the year, plants are either absent, leafless, dormant, or functioning at a low level. Over the course of a year, then, the fraction of solar radiation which is trapped in photosynthesis is quite low—as low as 0.003%, which is an order or two in magnitude lower than the efficiency of other vegetation types. Because of all these reasons, net primary productivity is the lowest of all vegetation types examined in this chapter, about 100 g m^{-2} yr^{-1}, which is 50% lower than tundra productivity.

It is a mistake to think that the low community productivity means low rates of photosynthesis for each species. If productivity is expressed in terms of grams fixed per gram of leaf area instead of per unit ground area, then desert shrubs are fully as productive as shrubs in the mesic eastern deciduous forest (e.g., *Larrea* = 0.28 g CO_2 g leaf^{-1} yr^{-1} versus 0.16–0.37 for mesic shrubs). Some desert subshrubs and annuals have net photosynthesis rates per unit leaf area comparable to or above those of crop plants under irrigated cultivation, with rates in the range of 60–90 mg CO_2 dm^{-2} hr^{-1}.

Figure 19–37. Creosote bush scrub, showing a clonal ring of creosote bushes. The diameter of the ring is several meters. (Courtesy of Frank Vasek.)

Table 19–5. Some climatic features of the four desert regions of North America.

	Great Basin	Mojave	Sonoran	Chihuahuan
annual precipitation (mm)	100–300	100–200	50–300	150–300
precipitation falling in summer (% of total)	30	35	45	65
snowfall (cm; 10 cm snow = 1 cm rain)	150–300	25–75	trace	trace
winter mean max/min temperatures (°C)	+8/−8	+15/0	+18/+4	+16/0
hours of frost (% of total)	5–20	2–5	0–1	2–5
summer mean max/min temperatures (°C)	34/10	39/20	40/26	34/19

Physical Features

Some temperature and precipitation data for each desert are summarized in Table 19–5. Cold winter temperatures in the Great Basin and Chihuahuan deserts reflect their relatively high elevation, on plateaus generally above 1000 m. The Mojave desert covers a wide amplitude of elevation, from +1300 to −86 m (the lowest spot in Death Valley). The Sonoran desert generally lies below 700 m. Annual precipitation does not vary much from desert to desert, but its seasonality and form do. The Great Basin and Mojave have winter peaks, and experience moderate to considerable snowfall. Portions of the Sonoran have winter and summer peaks of equal intensity, but no snow, and the Chihuahuan has a pronounced summer peak only. As in the grassland, fluctuation in rainfall from year to year can be large. The P/E ratio averages 0.3, but it may drop below 0.1 at the head of the Gulf of California. Summer temperatures are almost equally high in all four deserts.

Some important topographic features common to all deserts are diagrammed in Figure 19–38. The lower slopes of desert mountains are long and gentle and they are given the Spanish name **bajada**. Their average rise is only 1–5%, which is so gradual that one does not realize the elevation gain until one looks back downslope. Bajadas are composed of coarse alluvium eroded from steeper slopes in the montane zone above. They are dissected with channels, called **arroyos, washes**, or **wadis**, cut by intermittent streams that only flow after heavy rains.

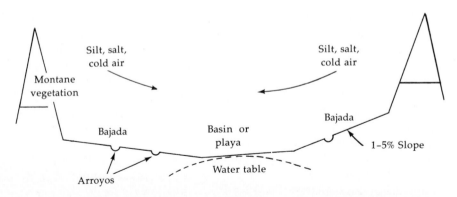

Figure 19–38. Major topographic features of the desert landscape. Typical desert vegetation is restricted to the bajadas.

Bajadas support typical desert vegetation, with the highest species diversity of any desert habitat. Evidently the coarseness of the soil provides a wide range of microhabitats. On a typical Sonoran bajada near Tucson, Arizona, the upper bajada community consisted of 18 perennial species with 35% cover and a species diversity index (H) of 1.2. Near the base of the bajada, on finer soil, the community consisted of only 6 perennial species with 20% cover and a diversity index of 0.7. Along the same gradient, rock and gravel fell from 45% to 15%. Throughout the desert, species diversity correlates better with soil texture than with rainfall.

Arroyos have low plant cover because of frequent disturbance, but the larger ones may have scattered trees. These riparian trees are able to survive because their root systems are in contact with greater supplies of water than are available on the surrounding bajadas. It is only in areas of significant summer rainfall as in the Arizona Sonoran desert where these trees occur on the bajadas. Common examples include smoke tree *(Dalea spinosa)*, desert ironwood *(Olneya tesota)*, mesquite *(Prosopis* spp.), and palo verde *(Cercidium* spp.). All of these are legumes with large seeds which require scarification for germination.

Bajada soils are typical **aridisols**, with little profile development, less than 1% organic matter, sandy loams to loamy sands, pH 7–8.5, and often with an indurated layer of calcium carbonate (called a **caliche** layer) 25–50 cm below the surface. Such a layer impedes root growth, and the shallowness of the caliche affects the species composition of the community above. In some areas, erosional processes produce a **desert pavement**—a surface layer of close-fitting rocks, polished and burnished on their exposed faces. Plant establishment is very difficult here, and plant cover is unusually low. Another surface phenomenon can be a thin but persistent water-repellent layer beneath the canopy of certain shrubs, such as creosote bush. Evidently, material released from living or decaying leaves combines with the soil particles to produce the layer, which may remain for years after the plant is gone.

Playas are undrained basins at the base of bajadas. Runoff from the bajadas carries fine textured soil and dissolved salts. This material accumulates in the playa in such a way that there are circular zones of increasing salinity from the edge to the lowest part of the basin. Each zone is characterized by a different community of plants, and the innermost zone may be devoid of plants, showing only a crust of salts on the surface. Soil aeration is low because of the fine texture and because a water table may be close to or at the surface of the playa bottom. Some research indicates that bajada species are prevented from colonizing playas mainly because of low soil oxygen, rather than because of the high soil salinity (sometimes more than 35,000 ppm salt, pH 9–11).

An additional microenvironmental trait of playas is that they are colder than surrounding slopes. Cold air at night sinks and collects in the basin, and frost can be more common there. One 10 year study in the Mojave desert of southern Nevada showed that a 60 m drop in elevation from bajada slope to playa bottom correlated with a drop in mean minimum temperatures from several degrees above freezing to several degrees below freezing (Figure 19–39). Thus, there are several factors which account for major community differences between playas and bajadas.

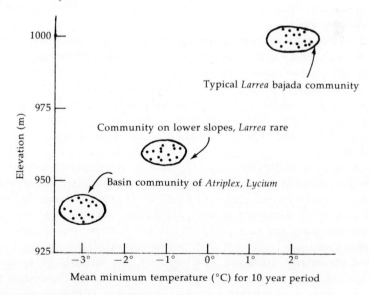

Figure 19–39. Relationship between elevation, vegetation, and mean minimum temperatures over a 10 year period for bajadas and basins in southern Nevada. (From Beatley 1975. Reprinted by permission of *American Midland Naturalist*.) Each dot represents one stand.

Not all community changes can be accounted for by elevational or soil changes. In some level parts of the Great Basin a mosaic of communities exists with quite narrow ecotones. One community may be strongly dominated by shadscale *(Atriplex confertifolia)*, another by winterfat *(Eurotia lanata)*, a third by salt sage *(Atriplex nuttallii)*, and a fourth by sagebrush. Detailed studies of soil texture, chemistry, and water retention have failed to reveal any changes significant enough to account for the mosaic.

Autecology of Major Life Forms

Annuals

Annuals account for a larger fraction of the flora than any other life form: over 40% of the flora is annual, in contrast to Raunkiaer's world normal of 13%. There are two categories of desert annuals. **Winter annuals** germinate in fall or winter and flower in late spring; **summer annuals** germinate in midsummer and flower in late summer or fall. Rarely does the same species belong to both groups, partly because germination requirements of winter and summer annuals usually differ. If soil temperature is below 18°C, for example, Mohave desert winter annuals will germinate, but if soil temperature is above 26°C, Mojave summer annuals will germinate. Germination of summer annuals is also enhanced, in some cases, by dry storage for several weeks at 50°C, a requirement which prevents seed germination between the time seeds are shed in fall and a rainy period the following summer (see Chapter 14).

The diversity of annual species at one season and in one region may be high, and the species may group into distinctive communities that reflect unique microenvironments. Some species grow on bajadas, some at the edge of playas, some in playas; some grow beneath certain canopies and avoid others; some grow only in the open. These patterns may be due to nurse plant requirements (for shade, organic matter), to allelopathy, to water-repellent layers beneath some shrubs, or to salinity tolerances.

Long term studies by Dr. Janice Beatley in southern Nevada show that winter annuals of the Mojave desert germinate after fall or winter rains in excess of 15 mm. Below this critical limit germination is absent. The rain may be necessary to leach inhibitors from the seed coat or to provide moisture for some critical inhibitional period of time. The density of annuals correlates with increasing rainfall between 15 and 25 mm. As briefly discussed in Chapter 4, the annuals grow slowly through winter, are resistant to frost down to −18°C, then grow rapidly in spring as temperatures rise. Flowering, fruiting, and death occurs in late May, for a total life cycle of 5–8 months. Herbivory accounts for some mortality, but most seedlings succumb to drought and high temperature. Most do not survive to maturity. Under the best conditions, density may be 1000 m^{-2}, cover 30%, and biomass 60 g m^{-2}, but typically density is 100 m^{-2} and biomass 10 g m^{-2} (Figure 19–40).

Mojave summer annuals germinate in August or September after heavy rains, generally are C$_4$ in metabolism, remain small, and mature by the time of autumn frosts. Their life span is measured in terms of weeks rather than months.

Drought-deciduous species

Some **drought-deciduous** species, such as ocotillo or desert coach-whip *(Fouquieria splendens)* (Figure 19–41), are able to produce several crops of leaves a year, losing each one as a dry period follows a wet period. Most drought-deciduous species, however, produce only one crop of leaves a year and they enter a long dormancy following leaf drop. Their leaves are mesophytic in terms of size, anatomy, and shape (see Chapter 17), making them energetically inexpensive for the plant to manufacture, compared to evergreen leaves. Their large intercellular spaces and thin blades permit rapid gas exchange (and water loss), and hence their photosynthetic rate is relatively high, 20–30 mg CO$_2$ g^{-1} hr^{-1}, compared to 10 mg for evergreens. The costs and benefits of this deciduous mode of survival were discussed in Chapter 4.

Winter-deciduous riparian species

Most desert riparian species are winter-deciduous, a trait common to riparian species throughout all vegetation types in North America. Onset of stem growth, leaf expansion, and blooming is photoperiodically controlled, in contrast to the moisture controls for other desert life forms. Leaves appear in early summer and they exhibit some

 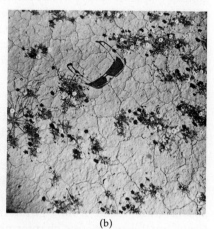

(a)   (b)

Figure 19–40. Summer annuals (mainly *Pectis*) in the Sonoran desert. (a) Aspect. (b) Close-up, showing that only 10% of the surface is actually covered by the plants.

Figure 19–41. Ocotillo *(Fouquieria splendens),* a drought-deciduous species, (a) in full leaf and (b) leafless.

xeromorphic features, possibly because most of these species maintain leaves during the hot summer, thus the water potentials developed within the plant are more negative than those of drought-deciduous species. A considerable amount of photosynthesis is accomplished by young, chlorenchymatous twigs in some species.

As already mentioned, many riparian desert species are legumi-nous trees (Figure 19–42) which produce large seeds with tough seed coats. The scarification they require is easily provided when the seeds are carried for some distance by raging water down an arroyo. Some riparian species are phreatophytes and seedlings must be able to produce a root that can grow fast enough to keep ahead of soil drying down from the surface. One of the fastest reported root growth rates is for mesquite: 102 mm day^{-1}, which is an order of magnitude greater than for typical xerophytes or mesophytes.

Succulents

There are many species of **stem** and **leaf succulents**, in several families, but the cacti are the best known. Recall from Chapter 13 that they have the CAM photosynthetic pathway, with stomates open at night. Consequently their water use efficiency is very high, but their net photosynthesis rate is quite low. Internal water stress rarely ex-ceeds −5 bars. They possess a shallow root system which is able to absorb water even from light rains (the other perennial desert life forms we discuss in this chapter generally have no roots in the upper 10 cm of soil), and in dry periods much of this root system dies. In wet periods, water is stored in large parenchyma cells, swelling the stem; in dry periods this water is used and the stem shrinks.

Most succulents are not frost resistant, and some ecologists have correlated their distribution limits to isolines of average consecutive hours of frost. Nevertheless, two species of beavertail cactus (*Opuntia* spp.) do extend throughout the Great Basin desert.

Although succulents have a slow growth rate, the largest plants in the desert are cacti: the cardón cactus of Baja California and Sonora, Mexico *(Pachycereus pringlei)* can reach 18 m and (presumably) hundreds of years in age. The more widely known saguaro cactus of Arizona and Sonora *(Carnegiea gigantea = Cereus giganteus)* can reach 10 m and an age of at least 200 years. Some populations of saguaro are clearly senescent, dominated by middle-aged and old plants (Figure 19–43), possibly because of increasing cattle and rodent populations, both of which graze on young plants.

Evergreens

These are true xerophytes because they grow and transpire throughout even the driest part of the year, at most shedding only a fraction of their leaves. Creosote bush *(Larrea tridentata)* and jojoba *(Simmondsia chinensis)* are good examples. For reasons not yet understood, their cytoplasm is able to resist unusual desiccation. Net photosynthesis is possible even when leaf water potential drops to −50 bars. When moisture is not limiting, their transpiration rates are equivalent to those of mesophytes, but when moisture is limiting their transpiration rates are very low. Their leaves often possess stereotyped morphological features which ecologists presume are responsible for the low rate of transpiration: for example, thick cuticle, lack of intercellular spaces, sunken stomates, palisade parenchyma beneath both surfaces, small leaves, and vertical leaf orientation. Hairs on the surface may also serve to reflect light, hence lowering the heat load on the leaf. Many evergreens invest considerable metabolic energy into the production and accumulation of resins and other complex molecules that may have anti-herbivore significance.

Figure 19–42. Foothill palo verde *(Cercidium)*, a leguminous tree of the Sonoran desert.

These evergreens give each desert its characteristic appearance, as they dominate the bajadas in terms of cover and biomass. In the warm deserts, CAM succulents and drought-deciduous species may also be important on bajadas, but C_4 plants are significant only in playas. A recent study of bajada and playa vegetation in the Chihuahuan desert of New Mexico very dramatically documents this pattern (Table 19–6): 90% of playa biomass was contributed by C_4 species, primarily summer annuals, but only 1% of bajada biomass was contributed by C_4 taxa. Evidently, some aspects of C_4 metabolism are ecologically disadvantageous out on the bajadas, regardless of life form.

Great Basin Vegetation

The cold desert is covered by a one-layered shrub community, about 1 m tall and very low in species diversity. The overwhelming dominant is sagebrush, *Artemisia tridentata* (see Figure 19–35). At 1400–2100 m, just below the lower montane **pinion-juniper woodland**, sagebrush steppe becomes dominant.

In pristine times, this shrub steppe contained sagebrush, bitter-brush *(Purshia tridentata)*, desert peach *(Prunus andersonii)*, and horse-

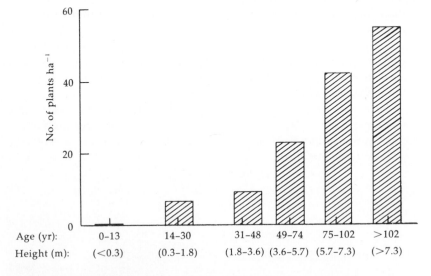

Figure 19–43. Age distribution of a saguaro *(Carnegiea gigantea = Cereus giganteus)* population in Saguaro National Monument, Arizona. The distribution is just the opposite of what one would expect for a vigorous, stable population. (From "The Saguaro: A Population in Relation to its Environment," by Niering, W. A., Whittaker, R. H., and Lowe, C. L., *Science*, Vol. 142, pp. 15–23, 5 September 1963. Copyright 1963 by the American Association for the Advancement of Science.)

Table 19-6. Percent C_3, C_4, and CAM of the total species and of the total biomass in three adjacent communities in the Chihuahuan desert of southern New Mexico. Data were accumulated over the course of several years. (From Syvertsen et al. 1976. By permission of *Southwestern Naturalist.*)

Community	Attribute	C_3	C_4	CAM	Absolute totals	Dominants
playa bottom	species	52	48	0	21 species	*Panicum*
	biomass	10	90	0	1243 kg ha^{-1}	*obtusum* (C$_4$)
playa fringe	species	54	44	1	77 species	*Hilaria mutica*
	biomass	42	50	8	11,200 kg ha^{-1}	(C$_4$), *Prosopis glandulosa* (C$_3$)
bajada	species	66	24	10	144 species	*Larrea tridentata*
	biomass	51	1	48	2088 kg ha^{-1}	(C$_3$), *Yucca elata* (CAM), *Prosopis glandulosa* (C$_3$), *Flourensia cernua* (C$_3$)

brush *(Tetradymia glabrata)*. Scattered abundantly with and between the shrubs were perennial (mainly bunch) grasses, such as western wheatgrass, needlegrass, and bluebunch wheatgrass (Table 19-3). Under grazing pressure by domestic stock, the grasses and palatable shrubs lost cover and the following introduced weedy annuals gained in cover: cheatgrass, medusa head *(Elymus caput-medusae)*, *Halogeton glomeratus*, Russian thistle *(Salsola iberica)*, tumble mustard *(Sisymbrium altissimum)*, and filaree *(Erodium* spp.). Sagebrush, which is not palatable, has been reduced in cover by fire (it is not a sprouter). This degraded range will not recover without active management.

At lower elevations, common associates of sagebrush are shadscale *(Atriplex confertifolia)*, spiny hopsage *(Grayia spinosa)*, winterfat *(Eurotia lanata)*, and Mormon tea *(Ephedra* spp.). Total cover is 15–40%. In parts of southern Nevada and eastern Utah, shadscale becomes the dominant, associated with budsage *(Artemisia spinescens)*, greasewood *(Sarcobatus baileyi)*, and *Ephedra nevadensis*. This has been called the **shadscale zone** of the Great Basin, and its climate is significantly warmer and drier than sagebrush-dominated areas. Ground cover is only 10%.

Warm Desert Vegetation

The **Mojave** is dominated by a two-layered community, an overstory shrub layer mostly of creosote bush, and an understory subshrub layer mostly of burro bush (also called bur sage, *Ambrosia dumosa*) (Figure 19–44). It is a monotonous, open community, with only 5–10% cover. At higher elevations (1200–1800 m) creosote bush is displaced by blackbrush *(Coleogyne ramosissima)*, Joshua tree *(Yucca brevifolia)*, spiny hopsage, matchweed *(Gutierrezia* spp.), and winterfat. The **blackbrush community** is an ecotone between **creosote bush scrub** below and pinon-juniper woodland above, sharing half of its flora with each.

The **Sonoran desert** is so complex that it has been divided into as many as seven vegetational regions by some ecologists. We will only consider two extremes here. The **Colorado desert** of southern California, southwestern Arizona, and around the northern part of the Gulf of California is very arid and the vegetation is reminiscent of the Mojave. Creosote bush and burro bush dominate. But there are many differences in annuals and other life forms between this region and the Mojave, for of 545 Mojave plant species, only half are also found in the Sonoran. Many hectares of the Coachella and Imperial Valleys in California are now dominated by citrus, date palms, and sugar beets, irrigated by water diverted from the Colorado River.

The other extreme, typical of central Baja California, the **Arizona uplands**, and parts of Sonora, is a complex four-layered community with about 25% ground cover (Figure 19–45). The overstory, 3–5 m tall, is of arborescent cacti and trees. A tall shrub-cactus layer below this is 2–3 m tall and has creosote bush, *Acacia* species, coachwhip, and chollas. A third layer is comprised of subshrubs 1 m or less in height, such as two species of burro bush *(Ambrosia dumosa* and *A. deltoidea)*,

Figure 19–44. Typical *Larrea-Ambrosia* community on a Mojave bajada.

Figure 19–45. Arizona uplands community of the Sonoran desert, with saguaro *(Carnegiea gigantea)* and mesquite *(Prosopis juliflora)* prominent in the overstory.

Figure 19–46. Typical Chihuahuan desert vegetation at Big Bend National Park, Texas, with creosote bush *(Larrea tridentata), Agave* species, and cacti prominent.

brittle bush *(Encelia farinosa)*, and a great variety of cacti (barrel, fish hook, cholla, beavertail). There are approximately five times as many species of cacti in the Sonoran desert as in the Mojave. A fourth layer is dominated by some perennial grasses and many species of summer and winter annuals. Because there are two nearly equal peaks of rainfall in winter and summer in these more mesic parts of the Sonoran, the diversity of species for each annual category is about the same. In contrast, the winter-wet Mojave desert has about six times as many winter as summer annuals.

The **Chihuahuan desert** lacks an arboreal element, but it has great species richness in two shrub layers. Beneath an overstory of creosote bush, coachwhip, mesquite, and *Acacia* are: tarbush *(Flourensia cernua)*, guayule *(Parthenium incanum)*, leather plant *(Jatropha* spp.), crucifixion thorn *(Koeberlinia spinosa)*, many cacti, and several distinctive members of the Agavaceae with spinescent, basal leaves (Figure 19–46). The latter includes: century plant *(Agave lechuguilla* and others), sotol *(Dasylerion wheeleri)*, Spanish bayonet *(Yucca elata, Y. baccata)*, and *Nolina*. Ground cover is relatively high, about 20–25%. Summer annuals can be abundant.

CHAPTER 19 REFERENCES

This is a very personal list of references. Works cited cover the factual material in the text, and they give a sampling of both current and classical research. Some relatively narrow papers are included, while some broad references are omitted, purely because of our own biases. In short, this list is neither consistent nor exhaustive.

TUNDRA

Billings, W. D. 1974a. Adaptations and origins of alpine plants. *Arctic and Alpine Research* 6:129–142.

——. 1974b. Arctic and alpine vegetation: plant adaptations to cold summer climates. In *Arctic and alpine environments*, ed. J. D. Ives and R. G. Barry, pp. 403–443. London: Methuen.

Billings, W. D. and P. J. Godfrey. 1967. Photosynthetic utilization of internal carbon dioxide by hollow-stemmed plants. *Science* 158:121–123.

Billings, W. D. and H. A. Mooney. 1968. The ecology of arctic and alpine plants. *Biological Review* 43:481–529.

Bliss, L. C. 1956. Comparison of plant development in microenvironments of arctic and alpine tundras. *Ecological Monographs* 26:303–337.

——. 1962. Adaptations of arctic and alpine plants to environmental conditions. *Arctic* 15:117–144.

——. 1963. Alpine plant communities of the Presidential Range, New Hampshire. *Ecology* 44:687–697.

——. 1966. Plant productivity in alpine microenvironments on Mt. Washington, New Hampshire. *Ecological Monographs* 36:125–155.

Britton, M. E. 1966. *Vegetation of the arctic tundra*. Corvallis, OR: Oregon State University Press.

Caldwell, M. M. 1968. Solar ultraviolet radiation as an ecological factor for alpine plants. *Ecological Monographs* 38:243–268.

Canaday, B. B. and R. W. Fonda. 1974. The influence of subalpine snowbanks on vegetation pattern, production, and phenology. *Bulletin of the Torrey Botanical Club* 101:340–350.

Chabot, B. F. and W. D. Billings. 1972. Origins and ecology of the Sierran alpine flora and vegetation. *Ecological Monographs* 42:143–161.

Clausen, J. 1965. Population studies of alpine and subalpine races of conifers and willows in the California high Sierra Nevada. *Evolution* 19:56–68.

Daubenmire, R. F. 1954. Alpine timberlines in the Americas and their interpretation. *Butler University Botanical Studies* 11:119–136.

Dennis, J. G. and P. L. Johnson. 1970. Shoot and rhizome-root standing crops of tundra vegetation at Barrow, Alaska. *Arctic and Alpine Research* 2:253–266.

Douglas, G. W. and L. C. Bliss. 1977. Alpine and high subalpine plant communities of the North Cascades Range, Washington and British Columbia. *Ecological Monographs* 47:113–150.

Greller, A. M. 1974. Vegetation of roadcut slopes in the tundra of Rocky Mountain National Park, Colorado. *Biological Conservation* 6(2):84–93.

Griggs, R. F. 1938. Timberlines in the northern Rocky Mountains. *Ecology* 19:548–564.

———. 1946. The timberlines of North America and their interpretation. *Ecology* 27:257–289.

Hoffman, R. S. 1958. The meaning of the word "taiga." *Ecology* 39: 540–541.

Hustich, I. 1953. The boreal limits of conifers. *Arctic* 6:149–162.

Johnson, D. A. and M. M. Caldwell. 1976. Water potential components, stomatal function, and liquid phase water transport resistances of four arctic and alpine species in relation to moisture stress. *Physiologia Plantarum* 36:271–278.

Kevan, P. G. 1975. Sun-tracking solar furnaces in high arctic flowers: significance for pollination and insects. *Science* 189:723–726.

Klikoff, L. G. 1965. Microenvironmental influence on vegetational pattern near timberline in the Sierra Nevada. *Ecological Monographs* 35:187–211.

LaMarche, V. C., Jr. and H. A. Mooney. 1972. Recent climatic change and development of the bristlecone pine (*P. longaeva* Bailey) krummholz zone, Mt. Washington, Nevada. *Arctic and Alpine Research* 4:61–72.

Larsen, J. A. 1965. The vegetation of the Ennadai Lake Area, N.W.T. *Ecological Monographs* 35:37–59.

Major, J. and S. A. Bamberg. 1967. Comparison of some North American and Eurasian alpine ecosystems. In *Arctic and alpine environments*, ed. H. E. Wright, Jr. and W. H. Osburn, pp. 89–118. Bloomington, IN: Indiana Univ. Press.

Major, J. and D. W. Taylor. 1977. Alpine. In *Terrestrial vegetation of California*, ed. M. G. Barbour and J. Major, pp. 601–675. New York: Wiley-Interscience.

Marr, J. W. 1948. Ecology of the forest-tundra acotone on the east coast of Hudson Bay. *Ecological Monographs* 18:117–144.

Moldenke, A. R. 1976. California pollination ecology and vegetation types. *Phytologia* 34:305–361.

Pewe, T. L. 1966. *Permafrost and its effect on life in the north*. Corvallis, OR: Oregon State University Press.

Polunin, N. 1948. *Botany of the Canadian eastern arctic, III. Vegetation and ecology*. National Museum of Canada Bulletin 104. Ottawa, Canada: National Museum of Canada.

Rickard, W. E., Jr. and J. Brown. 1974. Effects of vehicles on arctic tundra. *Environmental Conservation* 1:55–62.

Shaver, G. R. and W. D. Billings. 1977. Effects of daylength and temperature on root elongation in tundra graminoids. *Oecologia* 28:57–65.

Stålfelt, M. G. 1960. *Plants, the soil and man.* 1972 translation by M. S. Jarvis and P. G. Jarvis. New York: Wiley.

Teeri, J. A. 1976. Phytotron analysis of a photoperiodic response in a high arctic plant species. *Ecology* 57:374–379.

CONIFER FORESTS

Arno, S. F. and J. R. Habeck. 1972. Ecology of alpine larch (*Larix lyallii* Parl.) in the Pacific northwest. *Ecological Monographs* 42:417–450.

Azevedo, J. and D. L. Morgan. 1974. Fog precipitation in coastal California forests. *Ecology* 55:1135–1141.

Buell, M. F. and W. A. Niering. 1957. Fir-spruce-birch forest in northern Minnesota. *Ecology* 38:602–610.

Chapman, H. H. 1932. Is the longleaf type a climax? *Ecology* 13:328–334.

Cooper, C. F. 1960. Changes in vegetation, structure, and growth of southwestern pine forests since white settlement. *Ecological Monographs* 30:129–164.

Cooper, W. S. 1957. Vegetation of the northwest American province. *Proc., 8th Pacific Science Congress* 4:133–138.

Curtis, J. T. 1959. *The vegetation of Wisconsin*. Madison, WI: University of Wisconsin Press.

Dansereau, P. and F. Segadas-Vianna. 1952. Ecological study of the peat bogs of eastern North America. *Canadian J. of Botany* 30:490–520.

Daubenmire, R. F. 1943. Soil temperature versus drought as a factor in determining lower altitudinal limits of trees in the Rocky Mountains. *Botanical Gazette* 105:1–13.

———. 1943. Vegetational zonation in the Rocky Mountains. *Botanical Review* 9:325–393.

Davis, R. B. 1966. Spruce-fir forests of the coast of Maine. *Ecological Monographs* 36:79–94.

Fonda, R. W. and J. A. Bernardi. 1976. Vegetation of Sucia Island in Puget Sound, Washington. *Bulletin of the Torrey Botanical Club* 103:99–109.

Franklin, J. F. and C. T. Dyrness. 1973. *Natural vegetation of Oregon and Washington.* U.S. Dept. of Agriculture Forest Service General Technical Report PNW–8, Portland, OR. U.S. Dept. of Agriculture Forest Service.

Garren, K. H. 1943. Effects of fire on vegetation of the southeastern United States. *Botanical Review* 9:617–654.

Goff, F. G. and P. H. Fedler. 1968. Structural gradient analysis of upland forests in the western Great Lakes area. *Ecological Monographs* 38:65–86.

Habeck, J. R. and R. W. Mutch. 1973. Fire-dependent forests in the northern Rocky Mountains. *J. of Quarternary Research* 3:408–424.

Hartesveldt, R. J., H. T. Harvey, H. S. Shellhammer, and R. E. Stecker. 1975. *The giant sequoia of the Sierra Nevada.* U.S. Dept. of the Interior National Park Service Publ. no. 120. Washington, D.C.: U.S. Dept. of the Interior, National Park Service.

Heinselman, M. L. 1963. Forest sites, bog processes, and peatland types in the glacial Lake Aggasiz region, Minnesota. *Ecological Monographs* 33:327–374.

Jones, E. W. 1945. The structure and reproduction of the virgin forests of the north temperate zone. *The New Phytologist* 44:130–148.

LaRoi, G. H. 1967. Ecological studies in the boreal spruce-fir forests of the North American taiga. *Ecological Monographs* 37:229–253.

Larsen, J. A. 1930. Forest types of the northern Rocky Mountains and their climatic controls. *Ecology* 11:631–672.

Lowe, C. H. and D. E. Brown. 1973. *The natural vegetation of Arizona.* Arizona Resources Information System Cooperative Publ. no. 2. Phoenix, AZ: Arizona Resources Information System.

Lutz, H. J. 1956. *Ecological effects of forest fires in the interior of Alaska.* U.S. Dept. of Agriculture Technical Bulletin 1133. Washington, D.C.: U.S. Dept. of Agriculture.

Maycock, P. F. 1961. The spruce-fir forests of the Keneenaw Peninsula, northern Michigan. *Ecology* 42:357–365.

McCormick, J. 1970. *The pine barrens.* New Jersey State Museum Report no. 2. Trenton, NJ: New Jersey State Museum.

McIntosh, R. P. and R. T. Hurley. 1964. The spruce-fir forests of the Catskill Mountains. *Ecology* 45:314–326.

Moss, E. H. 1955. The vegetation of Alberta. *Botanical Review* 21:493–567.

Oosting, H. J. and W. D. Billings. 1951. A comparison of virgin spruce-fir forest in the northern and southern Appalachian system. *Ecology* 32:84–103.

Oosting, H. J. and J. F. Reed. 1944. Ecological composition of pulpwood forests in northwestern Maine. *American Midland Naturalist* 31:181–210.

Oosting, H. J. and J. F. Reed. 1952. Virgin spruce-fir of the Medicine Bow Mountains, Wyoming. *Ecological Monographs* 22:69–91.

Potzger, J. E. 1946. Phytosociology of the primeval forest in central and northern Wisconsin and upper Michigan and a brief postglacial history of the Lake Forest Formation. *Ecological Monographs* 16:211–250.

Ritchie, J. C. 1956. The vegetation of northern Manitoba. *Canadian J. of Botany* 34:523–561.

Robichaud, B. and M. F. Buell. 1973. *Vegetation of New Jersey.* New Brunswick: Rutgers University Press.

Rundel, P. W. 1971. Community structure and stability in the giant sequoia groves of the Sierra Nevada. *American Midland Naturalist* 85:478–492.

Rundel, P. W., D. J. Parsons, and D. T. Gordon. 1977. Montane and subalpine vegetation of the Sierra Nevada and Cascade Ranges. In *Terrestrial vegetation of California,* ed. M. G. Barbour and J. Major, pp. 559–599. New York: Wiley-Interscience.

Shirley, H. L. 1945. Reproduction of upland conifers in the Lake States as affected by root competition and light. *American Midland Naturalist* 33:537–612.

Stallard, H. 1929. Secondary succession in the climax forest formation of Northern Minnesota. *Ecology* 10:476–548.

Stewart, R. 1977. *Labrador.* Amsterdam: Time-Life.

Taylor, D. W. 1977. Floristic relationships along the Cascade-Sierran axis. *American Midland Naturalist* 97:333–349.

Zinke, P. J. 1977. The redwood forest and associated north coast forests. In *Terrestrial vegetation of California*, ed. M. G. Barbour and J. Major, pp. 679–698. New York: Wiley-Interscience.

DECIDUOUS FOREST

Bakeless, J. 1961. *The eyes of discovery.* New York: Dover.

Bormann, F. H., T. G. Siccama, G. E. Likens, and R. H. Whittaker. 1970. The Hubbard Brook ecosystem study: composition and dynamics of the tree stratum. *Ecological Monographs* 40:373–388.

Boyce, S. G. 1954. The salt spray community. *Ecological Monographs* 24:29–67.

Braun, E. L. 1957. Development of the deciduous forests of eastern North America. *Ecological Monographs* 17:211–219.

———. 1950. *Deciduous forests of eastern North America.* Philadelphia, PA: Blakiston.

———. 1955. The phytogeography of the eastern United States and its interpretation. *Botanical Review* 21:297–375.

Buell, M. F. and W. E. Martin. 1961. Competition between maple-basswood and spruce-fir communities in Itasca Park, Minnesota. *Ecology* 42:428–429.

Cain, S. A. 1943. The tertiary character of the cove hardwood forests of the Great Smoky Mountains. *Bulletin of the Torrey Botanical Club* 70:213–245.

Core, E. L. 1966. *Vegetation of West Virginia.* Charleston, WV: McClain, Parsons.

Curtis, J. T. 1959. *The vegetation of Wisconsin.* Madison, WI: University of Wisconsin Press.

Daubenmire, R. F. 1936. The "Big Woods" of Minnesota. *Ecological Monographs* 6:233–268.

Dyksterhuis, E. J. 1948. The vegetation of western Cross Timbers. *Ecological Monographs* 18:325–376.

Eyre, S. R. 1963. *Vegetation and soils.* Chicago: Aldine.

Forcier, L. K. 1975. Reproductive strategies and the co-occurrence of climax tree species. *Science* 189:808–810.

Graham, S. A. 1941. Climax forests of the upper peninsula of Michigan. *Ecology* 22:355–362.

Holt, P. C., ed. 1970. *The distributional history of the biota of the southern Appalachians. Part II: Flora.* Blacksburg, VA: Virginia Polytechnic Institute and State University.

Hutchison, B. A. and D. R. Matt. 1977. The distribution of solar radiation within a deciduous forest. *Ecological Monographs* 47:185–207.

Keever, C. 1953. Present composition of some stands of the former oak-chestnut forest in the southern Blue Ridge Mountains. *Ecology* 34:44–54.

———. 1973. Distribution of major forest species in southeastern Pennsylvania. *Ecological Monographs* 43:303–327.

Koyama, H. and S. Kawano. 1973. Biosystematic studies on *Maianthemum* (Liliaceae-Polygonatae). VII. Photosynthetic behavior of *M. dialatatum. Botanical Magazine of Tokyo* 86:89–101.

Küchler, A. W. 1964. *Potential natural vegetation of the conterminous United States.* American Geographical Society Special Publ. no. 36. New York: American Geographical Society.

Lutz, H. J. 1930. The vegetation of Heart's Content: a virgin forest in northwestern Pennsylvania. *Ecology* 11:1–29.

Monk, C. D. 1965. Southern mixed hardwood forest of north-central Florida. *Ecological Monographs* 35:335–354.

———. 1966. An ecological significance of evergreenness. *Ecology* 47:504–505.

Mowbray, T. B. and H. J. Oosting. 1968. Vegetation gradients in relation to environment and phenology in a southern Blue Ridge gorge. *Ecological Monographs* 38:309–344.

Muller, R. N. and F. H. Bormann. 1976. Role of *Erythronium americanum* Ker. in energy flow and nutrient dynamics of a northern hardwood forest ecosystem. *Science* 193:1126–1128.

Oosting, H. J. 1942. An ecological analysis of the plant communities of the Piedmont, North Carolina. *American Midland Naturalist* 28:1–126.

Oosting, H. J. and P. F. Bourdeau. 1955. Virgin hemlock forest segregates in the Joyce Kilmer Memorial Forest of western North Carolina. *Botanical Gazette* 116:340–359.

Oosting, H. J. and P. F. Bourdeau. 1959. The maritime live oak forest in North Carolina. *Ecology* 40:148–152.

Potzger, J. E. 1946. Phytosociology of the primeval forest of central-northern Wisconsin and upper Michigan. *Ecological Monographs* 16:211–250.

Potzger, J. E., M. E. Potzger, and J. McCormick. 1956. The forest primeval of Indiana as recorded in the original land surveys and an evaluation of previous interpretations of Indiana vegetation. *Butler University Botanical Studies* 13:95–111.

Quarterman, E. and C. Keever. 1962. Southern mixed hardwood forest: climax in the southeastern coastal plain, USA. *Ecological Monographs* 32:167–185.

Rice, E. L. and W. T. Penfound. 1959. The upland forests of Oklahoma. *Ecology* 40:593–607.

Robichaud, B. and M. F. Buell. 1973. *Vegetation of New Jersey.* New Brunswick: Rutgers University Press.

Siccama, T. G. 1974. Vegetation, soil, and climate on the Green Mountains of Vermont. *Ecological Monographs* 44:325–349.

Siccama, T. G., F. H. Bormann, and G. E. Likens. 1970. The Hubbard Brook ecosystem study: productivity, nutrients, and phytosociology of the herbaceous layer. *Ecological Monographs* 40:389–402.

Transeau, E. N. 1935. The prairie peninsula. *Ecology* 16:423–437.

Vogelmann, H. W., T. G. Siccama, D. Leedy, and D. C. Ovitt. 1968. Precipitation from fog moisture in the Green Mountains of Vermont. *Ecology* 49:1205–1207.

Whittaker, R. H. 1956. Vegetation of the Great Smoky Mountains. *Ecological Monographs* 26:1–80.

Williams, A. B. 1936. The composition and dynamics of a beech-maple climax forest (Ohio). *Ecological Monographs* 6:319–408.

Woods, F. W. and R. E. Shanks. 1959. Natural replacement of chestnut by other species in the Great Smoky Mountains National Park. *Ecology* 40:349–361.

GRASSLANDS

Albertson, F. W. 1937. Ecology of mixed prairie in west central Kansas. *Ecological Monographs* 7:481–547.

Albertson, F. W. and J. E. Weaver. 1947. Reduction of ungrazed mixed prairie to short grass as a result of drought and dust. *Ecological Monographs* 16:449–463.

Ayyad, M. A. G. and R. L. Dix. 1964. An analysis of a vegetation-microenvironmental complex on prairie slopes in Saskatchewan. *Ecological Monographs* 34:421–442.

Barry, W. J. 1972. *The central valley prairie.* Sacramento, CA: Dept. of Parks and Recreation.

Beetle, A. A. 1974. Distribution of the native grasses of California. *Hilgardia* 9:309–357.

Bogusch, E. R. 1952. Brush invasion in the Rio Grande Plain of Texas. *Texas J. of Science* 4:85–91.

Borchert, J. R. 1950. The climate of the central North American grassland. *Annals of the Association of American Geographers* 40:1–39.

Brown, A. L. 1950. Shrub invasion of southern Arizona desert grassland. *J. of Range Management* 3:172–177.

Buffington, L. C. and C. H. Herbel. 1965. Vegetational changes on a semidesert grassland range from 1858 to 1963. *Ecological Monographs* 35:139–164.

Carpenter, J. R. 1940. The grassland biome. *Ecological Monographs* 10:617–684.

Coupland, R. T. 1961. A reconsideration of grassland classification in the northern Great Plains of North America. *J. of Ecology* 49:135–167.

Daubenmire, R. F. 1968. Ecology of fire in grasslands. *Advances in Ecological Research* 5:209–266.

Dyksterhuis, E. J. 1946. The vegetation of the Ft. Worth prairie. *Ecological Monographs* 16:1–29.

Ellison, L. 1960. Influence of grazing on plant succession of rangelands. *Botanical Review* 26:1–78.

Franklin, J. F. and C. T. Dyrness. 1973. *Natural vegetation of Oregon and Washington.* U.S. Dept. of Agriculture Forest Service General Technical Report PNW-8. Portland, OR: U.S. Dept. of Agriculture Forest Service.

Gardner, J. L. 1951. Vegetation of the creosote bush area of the Rio Grande Valley in New Mexico. *Ecological Monographs* 21:379–403.

Gay, C. W. Jr. and D. D. Dwyer. 1965. *New Mexico range plants.* New Mexico State University Cooperative Extension Service Circular 374. Las Cruces, NM: Cooperative Extension, New Mexico State University.

Gleason, H. A. 1923. The vegetational history of the middle west. *Annals of the Association of American Geographers* 12:39–85.

Hadley, E. B. and R. P. Buccos. 1967. Plant community composition and net primary production within a native eastern North Dakota prairie. *American Midland Naturalist* 77:116–127.

Harris, G. A. 1967. Some competitive relationships between *Agropyron spicatum* and *Bromus tectorum. Ecological Monographs* 37:89–111.

Hastings, J. R. and R. M. Turner. 1965. *The changing mile.* Tucson, AZ: University of Arizona Press.

Heady, H. F. 1977. Valley grassland. In *Terrestrial vegetation of California,* ed. M. G. Barbour and J. Major, pp. 491–514. New York: Wiley-Interscience.

Hitchcock, A. S. 1950. *Manual of the grasses of the United States.* 2nd ed. U.S. Dept. of Agriculture Misc. Publ. 200. Washington, D.C.: U.S. Dept. of Agriculture.

Humphrey, R. R. 1953. The desert grassland: a history of vegetational change and an analysis of causes. *Botanical Review* 24:193–252.

Humphrey, R. R. and L. A. Mehrhoff. 1958. Vegetation changes on a southern Arizona grassland range. *Ecology* 39:720–726.

Johnston, M. C. 1963. Past and present grasslands of southern Texas and N.E. Mexico. *Ecology* 44:456–466.

Jones, C. H. 1944. Vegetation of Ohio prairies. *Bulletin of the Torrey Botanical Club* 71:536–548.

Kucera, C. L., R. C. Dahlman, and M. R. Koelling. 1967. Total net productivity and turnover on an energy basis for tallgrass prairie. *Ecology* 48:536–541.

Larson, F. 1940. The role of the bison in maintaining the short grass plains. *Ecology* 21:113–121.

McMillan, D. 1959. The role of ecotypic variation in the distribution of the central grassland of North America. *Ecological Monographs* 29:285–308.

Old, S. M. 1969. Microclimates, fire, and plant production in an Illinois prairie. *Ecological Monographs* 39:355–384.

Redmann, R. E. 1975. Production ecology of grassland plant communities in western North Dakota. *Ecological Monographs* 45:83–106.

Rice, E. L. and R. L Parenti. 1978. Causes of decreases in productivity in undisturbed tall grass prairie. *American J. of Botany* 65:1091–1097.

Sims, P. L., J. S. Singh, and W. K. Lauenroth. 1978. The structure and function of ten western North American grasslands. *J. of Ecology* 66:251–285.

Sprague, H. B., ed. 1959. *Grasslands.* Washington, D.C.: American Association for the Advancement of Science.

Teeri, J. A. and L. G. Stowe. 1976. Climatic patterns and the distribution of C_4 grasses in North America. *Oecologia* 23:1–12.

Weaver, J. E. 1954. *The North American prairie.* Lincoln, NB: Johnsen.

Weaver, J. E. and F. W. Albertson. 1956. *Grasslands of the Great Plains.* Lincoln, NB: Johnsen.

Weaver, J. E. and T. J. Fitzpatrick. 1932. Ecology and relative importance of the dominants of tall-grass prairie. *Botanical Gazette* 93:113–150.

Weaver, J. E. and T. J. Fitzpatrick. 1934. The prairie. *Ecological Monographs* 4:109–295.

White, D. 1941. Prairie soil as a medium for tree growth. *Ecology* 22:399–407.

Williams, W. A. 1966. Range improvements as related to net productivity, energy flow, and foilage configuration. *J. of Range Management* 19:29–34.

Young, J. A., R. A. Evans, and J. Major. 1977. Sagebrush steppe. In *Terrestrial vegetation of California,* ed. M. G. Barbour and J. Major, pp. 762–796. New York: Wiley-Interscience.

Young, J. A., R. A. Evans, and P. T. Tueller. 1975. Great Basin plant communities— pristine and grazed. In *Holocene climates in the Great Basin,* Occasional Paper, Nevada Archeological Survey, ed. R. Elston, pp. 186–215. Reno, NV: Nevada Archeological Survey.

DESERT SCRUB

Adams, S., B. R. Strain, and M. S. Adams. 1970. Water-repellent soils, fire, and annual plant cover in a desert scrub community of southeastern California. *Ecology* 51: 696–700.

Anderson, D. J. 1971. Pattern in desert perennials. *J. of Ecology* 59:555–560.

Axelrod, D. I. 1950. Evolution of desert vegetation. In Carnegie Institution of Washington Publ. no. 590, pp. 215–306. Washington, D.C.: Carnegie Institution of Washington.

———. 1959. Evolution of the Madro-Tertiary geoflora. *Botanical Review* 24:433–509.

———. 1967. Drought, diastrophism, and quantum evolution. *Evolution* 21:201–209.

———. 1972. Edaphic aridity as a factor in angiosperm evolution. *American Naturalist* 106:311–320.

Barbour, M. G. 1969. Age and space distribution of the desert shrub *Larrea divaricata. Ecology* 50:679–685.

———. 1973. Desert dogma re-examined: root/shoot productivity and plant spacing. *American Midland Naturalist* 89:41–57.

Beatley, J. C. 1969. Biomass of desert winter annual plant populations in southern Nevada. *Oikos* 20:261–273.

———. 1974. Phenological events and their environmental triggers in Mojave Desert ecosystems. *Ecology* 55:856–863.

———. 1975. Climates and vegetation pattern across the Mojave/Great Basin Desert transition of southern Nevada. *American Midland Naturalist* 93:53–70.

Benson, L. and R. A. Darrow. 1954. *The trees and shrubs of the southwestern deserts.* Tucson, AZ: University of Arizona Press.

Billings, W. D. 1949. The shadscale vegetation zone of Nevada and eastern California in relation to climate and soils. *American Midland Naturalist* 42:87–109.

Björkman, O., R. W. Pearcy, A. T. Harrison, and H. A. Mooney. 1972. Photosynthetic adaptation to high temperatures: a field study in Death Valley, California. *Science* 175:786–789.

Burk, J. H. 1977. Sonoran desert. In *Terrestrial vegetation of California*, ed. M. G. Barbour and J. Major, pp. 869–889. New York: Wiley-Interscience.

Cannon, W. A. 1919. Relation of the rate of root growth in seedlings of *Prosopis velutina* to the temperature of the soil. *Carnegie Institution of Washington yearbook* no. 18.

Capon, B. and W. Van Asdall. 1967. Heat pretreatment as a means of increasing germination of desert annual seeds. *Ecology* 48:305–306.

Chew, R. M. and A. E. Chew. 1965. The primary productivity of a desert shrub *(Larrea tridentata)* community. *Ecological Monographs* 35:353–375.

Cloudsley-Thompson, J. L. and M. J. Chadwick. 1964. *Life in deserts.* Philadelphia, PA: Dufour.

Cooke, R. U. and A. Warren. 1973. *Geomorphology in deserts.* Berkeley, CA: University of California Press.

Flowers, S. 1934. Vegetation of the Great Salt Lake region. *Botanical Gazette* 95:353–418.

Gardner, J. L. 1951. Vegetation of the creosote bush area of the Rio Grande Valley in New Mexico. *Ecological Monographs* 21:379–403.

Gates, D. H., L. A. Stoddart, and C. W. Cook. 1956. Soil as a factor influencing plant distribution on salt deserts of Utah. *Ecological Monographs* 26:155–175.

Hastings, J. R., R. M. Turner, and D. K. Warren. 1972. *An atlas of some plant distributions in the Sonoran desert.* University of Arizona Institute of Atmospheric Physics Technical Report on the Meterology and Climatology of Arid Regions no. 21. Tucson, AZ: University of Arizona Institute of Atmospheric Physics.

Hunt, C. B. 1966. *Plant ecology of Death Valley, California.* U.S. Geological Survey Professional paper no. 509. Washington, D.C.: U.S. Geological Survey.

Jaeger, E. C. 1957. *The North American deserts.* Stanford, CA: Stanford University Press.

Lowe, C. H. and D. E. Brown. 1973. *The natural vegetation of Arizona.* Arizona Resources Information System Cooperative Publ. no. 2. Phoenix, AZ: Arizona Resources Information System.

Lunt, O. R., J. Letey, and S. B. Clark. 1973. Oxygen requirements for root growth in three species of desert shrubs. *Ecology* 54:1356–1362.

Mabry, T. J., J. H. Hunziker, and D. R. Difeo, Jr. 1977. *Creosote bush.* Stroudsburg, PA: Dowden, Hutchinson, and Ross.

Marks, J. B. 1950. Vegetation and soil relations in the lower Colorado desert. *Ecology* 31:176–193.

McGinnies, W. G., B. J. Goldman, and P. Paylore, eds. 1968. *Deserts of the world.* Tucson, AZ: University of Arizona Press.

Mitchell, J. E., N. E. West, and R. W. Miller. 1966. Soil physical properties in relation to plant community patterns in the shadscale zone of northwestern Utah. *Ecology* 47:627–630.

Muller, W. H. and C. H. Muller. 1956. Association patterns involving desert plants that contain toxic products. *American J. of Botany* 43:354–361.

Mulroy, T. W. and P. W. Rundel. 1977. Annual plants: adaptations to desert environments. *BioScience* 27:109–114.

Niering, W. A., R. H. Whittaker, and C. H. Lowe. 1963. The saguaro: a population in relation to its environment. *Science* 142:15–23.

Odening, W. R., B. R. Strain, and W. C. Oechel. 1974. The effect of decreasing water potential on net CO_2 exchange of intact desert shrubs. *Ecology* 55:1086–1095.

Oppenheimer, H. R. 1960. Adaptation to drought: xerophytism. In *Plant-water relationships in arid and semi-arid conditions, reviews of research,* pp. 105–138. Paris: UNESCO.

Orians, G. H. and O. T. Solbrig. 1977. *Convergent evolution in warm deserts.* Stroudsburg, PA: Dowden, Hutchinson, and Ross.

Pearson, L. C. 1966. Primary productivity in a northern desert area. *Oikos* 15:211–228.

Rickard, W. H. and J. R. Murdock. 1963. Soil moisture and temperature survey of a desert vegetation mosaic. *Ecology* 44:821–824.

Shantz, H. L. and R. L. Piemeisel. 1924. Indicator significance of the natural vegetation of the southwestern desert region. *J. of Agricultural Research* 28:721–802.

Shreve, F. 1942. The desert vegetation of North America. *Botanical Review* 8:195–246.

Shreve, F. and A. L. Hinkley. 1937. Thirty years of change in desert vegetation. *Ecology* 18:463–478.

Shreve, F. and I. L. Wiggins. 1964. *Vegetation and flora of the Sonoran desert.* Stanford, CA: Stanford University Press.

Stebbins, R. C. 1974. Off-road vehicles and the fragile desert. *The American Biology Teacher* 36:203–208, 294–304.

Steenbergh, W. F. and C. H. Lowe. 1969. Critical factors during the first years of life of the saguaro *(Cereus giganteus)* at Saguaro National Monument, Arizona. Ecology 50:825–834.

Stocker, O. 1960. Physiological and morphological changes in plants due to water deficiency. In *Plant-water relationships in arid and semi-arid conditions, reviews of research,* pp. 63–104. Paris: UNESCO.

Syvertsen, J. P., G. L. Nickell, R. W. Spellenberg, and G. L. Cunningham. 1976. Carbon reduction pathways and standing crop in three Chihuahuan desert plant communities. *Southwestern Naturalist* 21:311–320.

Turner, R. W. 1963. Growth in four species of Sonoran desert trees. *Ecology* 44:760–765.

Valentine, K. A. and J. J. Norris. 1964. A comparative study of soils of selected creosote bush sites in southern New Mexico. *J. of Range Management* 17:23–32.

Vasek, F. C. and M. G. Barbour. 1977. Mojave desert scrub vegetation. In *Terrestrial vegetation of California,* ed. M. G. Barbour and J. Major, pp. 835–867. New York: Wiley-Interscience.

Vasek, F. C., H. B. Johnson, and G. D. Brum. 1975. Effects of power transmission lines on vegetation of the Mojave desert. *Madroño* 23:114–130.

Vasek, F. C., H. B. Johnson, and D. H. Eslinger. 1975. Effects of pipeline construction on creosote bush scrub vegetation of the Mojave desert. *Madroño* 23:1–13.

Vogl, R. J. and L. T. McHargue. 1966. Vegetation of California fan palm oases on the San Andreas fault. *Ecology* 47:532–540.

Wallace, A. and E. M. Romney. 1972. *Radioecology and ecophysiology of desert plants at the Nevada Test Site.* U.S. Atomic Energy Commission TID-25954. Washington, D.C.: U.S. Atomic Energy Commission.

Went, F. W. 1942. The dependence of certain annual plants on shrubs in southern California deserts. *Bulletin of the Torrey Botanical Club* 69:100–114.

———. 1949. Ecology of desert plants. II. The effect of rain and temperature on germination and growth. *Ecology* 30:1–13.

Went, F. W. and M. Westergaard. 1949. Ecology of desert plants. III. Development of plants in the Death Valley National Monument, California. *Ecology* 30:26–38.

West, N. E. and K. I. Ibrahim. 1968. Soil-vegetation relationships in the shadscale zone of southeastern Utah. *Ecology* 49:445–456.

LITERATURE CITED

Abrahamson, W. G. and M. Gadgil. 1973. Growth form and reproductive effort in goldenrods (*Solidago*, Compositae). *American Naturalist* 107: 651–661.

Ackerman, E. A. 1941. The Köppen classification of climates in North America. *Geographical Review* 31:105–111.

Adams, M. S. and B. R. Strain. 1968. Photosynthesis in stems and leaves of *Cercidium floridum*: spring and summer diurnal field response and relation to temperature. *Oecologia Plantarum* 3:285–297.

Adams, M. S., and B. R. Strain. 1969. Seasonal photosynthetic rates in stems of *Cercidium floridum* Benth. *Photosynthetica* 3:55–62.

Adams, M. S., B. R. Strain, and M. S. Adams. 1970. Water-repellent soil, fire, and annual plant cover in a desert shrub community of southeastern California. *Ecology* 51:696–700.

Adams, S., B. R. Strain, and J. P. Ting. 1967. Photosynthesis in chlorophyllous stem tissue and leaves of *Cercidium floridum*: accumulation and distribution of ^{14}C from $^{14}CO_2$. *Plant Physiology* 42:1797–1799.

Agee, J. K. 1973. *Prescribed fire effects on physical and hydrologic properties of mixed-conifer forest floor and soil.* Water Resources Center Contribution Report no. 143. Davis, CA: University of California.

Ahlgren, I. F. 1974. The effect of fire on soil organisms. In *Fire and ecosystems*, ed. T. T. Kozlowski and C. E. Ahlgren, ch. 3. New York: Academic Press.

Al-Ani, H. A., B. R. Strain, and H. A. Mooney. 1972. The physiological ecology of diverse populations of the desert shrub *Simmondsia chinesis*. *J. Ecology* 60:41–57.

Albert, R. 1975. Salt regulation in halophytes. *Oecologia* 21:57–71.

Allaway, W. G. and F. L. Milthorpe. 1976. Structure and functioning of stomata. In *Water deficits and plant growth*, vol. IV, ed. T. T. Kozlowski, pp. 57–102. New York: Academic Press.

Allen, E. A. and R. T. T. Forman. 1976. Plant species removals and old-field community structure and stability. *Ecology* 57:1233–1243.

Anderson, D. J. 1967. Studies on structure in plant communities. III. Data on pattern in colonizing species. *J. of Ecology* 55:397–404.

Anderson, J. E. and S. J. McNaughton. 1973. Effects of low soil temperature on transpiration, photosynthesis, leaf relative water content, and growth among elevationally diverse plant populations. *Ecology* 54:1220–1233.

Anderson, J. M. 1973. The breakdown and decomposition of sweet chestnut (*Castanea sativa* Mill.) and beech (*Fagus sylvatica* L.) leaf litter in two deciduous woodland soils. I. Breakdown, leaching, and decomposition. *Oecologia* 12:251–274.

Anderson, K. L. 1965. Fire ecology—some Kansas prairie forbs. In *Proc. Tall Timbers Fire Ecol. Conf.*, no. 4, pp. 153–159. Tallahassee, FL: Tall Timbers Research Station.

Anderson, M. C. 1973. Solar radiation and carbon dioxide in plant communities—conclusions. In *Photosynthesis and productivity in different environments*, ed. J. P. Cooper, pp. 345–354. London: Cambridge University Press.

Andrews, F., D. C. Coleman, J. E. Ellis, and J. S. Singh. 1974. Energy flow relationships in a shortgrass prairie ecosystem. In *Proc. First Int. Congr. Ecol.*, pp. 22–28. Wageningen, The Netherlands: Center for Agricultural Publishing and Documentation.

Armstrong, W. P. 1977. Fire followers in San Diego County. *Fremontia* 4:3–9.

Art, H. W., F. H. Bormann, G. K. Voigt, and G. M. Woodwell. 1974. Barrier island forest ecosystem: role of meteorologic nutrient inputs. *Science* 184:60–62.

Ashton, D. H. 1975. Studies of litter in *Eucalyptus regnans* forest. *Australian J. of Botany* 23:413–433.

Atkinson, I. A. E. 1970. Successional trends in the coastal and lowland forest of Mauna Loa and Kilauea Volcanoes, Hawaii. *Pacific Science* 24:387–400.

Axelrod, D. I. 1966. A method for determining the altitudes of Tertiary floras. *The Paleobotanist* 14:144–171.

——. 1977. Outline history of California vegetation. In *Terrestrial vegetation of California*, ed. M. G. Barbour and J. Major, pp. 139–193. New York: Wiley-Interscience.

Axelrod, D. I. and H. P. Bailey. 1969. Paleotemperature analysis of Tertiary floras. *Palaeogeography, Palaeoclimatology, Palaeoecology* 6:163–195.

Ayala, F. J. 1969. Experimental invalidation of the principle of competitive exclusion. *Nature* 224:176–179.

Ayyad, M. A. G. and R. L. Dix. 1964. An analysis of a vegetation-microenvironmental complex on prairie slopes in Saskatchewan. *Ecological Monographs* 34:421–442.

Azevedo, J. and D. L. Morgan. 1974. Fog precipitation in coastal California forests. *Ecology* 55:1135–1141.

Bakeless, J. 1961. *The eyes of discovery.* New York: Dover.

Baldwin, M., C. E. Kellogg, and J. Thorp. 1938. Classification of soils on the basis of their characteristics. In *Soils and man: 1938 yearbook of agriculture*, pp. 979–1001. Washington, D.C.: U.S. Dept. of Agriculture.

Bamberg, S. A., G. E. Kleinkopf, A. Wallace, and A. T. Vollmer. 1975. Comparative photosynthetic production of Mojave Desert shrubs. *Ecology* 56:732–736.

Bamberg, S. A., A. T. Vollmer, G. E. Kleinkopf, and T. L. Ackerman. 1976. A comparison of seasonal primary production of Mojave Desert shrubs during wet and dry years. *American Midland Naturalist* 95:398–405.

Bannister, P. 1976. *Introduction to physiological plant ecology.* New York: Halsted Press.

Barbour, M. G. 1970a. Seedling ecology of *Cakile maritima* along the California coast. *Bulletin of the Torrey Botanical Club* 97:280–289.

——. 1970b. Is any angiosperm an obligate halophyte? *American Midland Naturalist* 84:106–119.

——. 1973a. Chemistry and community composition. In *Air pollution damage to vegetation*, Advances in Chemistry no. 122, ed. M. G. Barbour, pp. 85–100. Washington, D.C.: American Chemical Society.

——. 1973b. Desert dogma re-examined: root/shoot productivity and plant spacing. *American Midland Naturalist* 89:41–57.

Barbour, M. G., R. B. Craig, F. R. Drysdale, and M. T. Ghiselin. 1973. *Coastal ecology: Bodega Head.* Berkeley, CA: University of California Press.

Barbour, M. G., G. L. Cunningham, W. C. Oechel, and S. A. Bamberg. 1977. Growth and development, form and function. In *Creosote bush*, ed. T. J. Mabry, J. H. Hunziker, and D. R. Difeo, Jr., ch. 4. Stroudsberg, PA: Dowden, Hutchinson, and Ross.

Barbour, M. G., D. V. Diaz, and R. W. Breidenbach. 1974. Contributions to the biology of *Larrea* species. *Ecology* 55:1199–1215.

Barbour, M. G., J. A. MacMahon, S. A. Bamberg, and J. A. Ludwig. 1977. The structure and function of *Larrea* communities. In *Creosote bush*, ed. T. J. Mabry, J. H. Hunziker, and D. R. Difeo, Jr., ch. 7. Stroudsberg, PA: Dowden, Hutchinson, and Ross.

Barclay-Estrup, P. and C. H. Gingham. 1969. The description and interpretation of cyclic processes in a heath community. I. Vegetational change in relation to the *Calluna* cycle. *J. of Ecology* 57:737–758.

Bard, G. 1952. Secondary succession on the Piedmont of New Jersey. *Ecological Monographs* 22:195–215.

Barrs, H. D. 1968. Determination of water deficits in plant tissues. In *Water deficits and plant growth*, vol. I, ed. T. T. Kozlowski, pp. 235–268. New York: Academic Press.

Bartholomew, B. 1970. Bare zone between California shrub and grassland communities: the role of animals. *Science* 170:1210–1212.

Bates, H. W. 1962. *The naturalist on the River Amazons.* Berkeley, CA: University of California Press.

Bauer, H. L. 1943. The statistical analysis of chaparral and other plant communities by means of transect samples. *Ecology* 24:45–60.

Beard, J. S. 1946. The mora forests in Trinidad, British West Indies. *J. of Ecology* 33:173–192.

Beardsell, M. F., P. G. Jarvis, and B. Davidson. 1972. A null-balance diffusion porometer suitable for use with leaves of many shapes. *J. of Applied Ecology* 9:677–690.

Beatley, J. C. 1966. Winter annual vegetation following a nuclear detonation in the northern Mojave Desert (Nevada Test Site). *Radiation Botany* 6:69–82.

———. 1967. Survival of winter annuals in the northern Mojave Desert. *Ecology* 48:745–750.

———. 1969. Biomass of desert winter annual plant populations in southern Nevada. *Oikos* 20:261–273.

———. 1974. Phenological events and their environmental triggers in Mojave Desert ecosystems. *Ecology* 55:856–863.

Beauchamp, J. J. and J. S. Olson. 1973. Corrections for bias in regression estimates after logarithmic transformation. *Ecology* 54:1403–1407.

Becking, R. W. 1957. The Zurich-Montpellier school of phytosociology. *Botanical Review* 23:411–488.

Bender, M. M. 1971. Variation in the $^{13}C/^{12}C$ ratios of plants in relation to pathway of photosynthetic carbon fixation. *Phytochemistry* 10:1239–1244.

Bender, M. M., I. Rougani, H. M. Vines, and C. C. Black, Jr. 1973. $^{13}C/^{12}C$ ratio changes in crassulacean acid metabolism plants. *Plant Physiology* 52:427–430.

Berthet, P. 1960. La mesure de la temperature par determination de la vitesse d'inversion du saccharose. *Vegetatio* 9:197–207.

Bidwell, R. G. S. 1974. *Plant physiology.* New York: Macmillan.

Billings, W. D. 1938. The structure and development of old field short-leaf pine stands and certain associated physical properties of the soil. *Ecological Monographs* 8:437–499.

———. 1949. The shadscale vegetation zone of Nevada and eastern California in relation to climate and soils. *American Midland Naturalist* 42:87–109.

———. 1952. The environmental complex in relation to plant growth and distribution. *Quarterly Review of Biology* 27:251–265.

———. 1970. *Plants, man, and the ecosystem.* 2nd ed. Belmont, CA: Wadsworth.

Billings, W. D. and L. C. Bliss. 1959. An alpine snowbank environment and its effects on vegetation, plant development, and productivity. *Ecology* 40:388–397.

Billings, W. D., P. J. Godfrey, B. F. Chabot, and D. P. Bourque. 1971. Metabolic acclimation to temperature in arctic and alpine ecotypes of *Oxyria digyna. Arctic and Alpine Research* 3:277–289.

Biswell, H. H. 1958. Prescribed burning in Georgia and California compared. *J. of Range Management* 11:293–298.

———. 1963. Research in wildland fire ecology in California. In *Proc. Tall Timbers Fire Ecol. Conf.,* no. 2, pp. 63–97. Tallahassee, FL: Tall Timbers Research Station.

———. 1967. Forest fire in perspective. In *Proc. Tall Timbers Fire Ecol. Conf.,* no. 7, pp. 43–63. Tallahassee, FL: Tall Timbers Research Station.

———. 1974. Effects of fire on chaparral. In *Fire and ecosystems,* ed. T. T. Kozlowski and C. E. Ahlgren, ch. 10. New York: Academic Press.

———. 1977. *Giant sequoia fire ecology.* University Extension Course X417.4. Davis, CA: University of California.

Björkman, O. 1968a. Carbosydismutase activity in shade-adapted species of higher plants. *Physiologia Plantarum* 21:1–10.

———. 1968b. Further studies on differentiation of photosynthetic properties of sun and shade ecotypes of *Solidago virgaurea. Physiologia Plantarum* 21:84–99.

Björkman, O., H. A. Mooney, and J. Ehleringer. 1975. Photosynthetic characteristics of plants from habitats with contrasting thermal regimes: comparisons of photosynthetic responses of intact plants. *Carnegie Institution of Washington Yearbook* 74:743–748.

Black, C. C. 1973. Photosynthetic carbon fixation in relation to net CO_2 uptake. *Annual Review of Plant Physiology* 24:253–286.

Blackman, G. E. 1935. A study by statistical methods of the distribution of species in grassland associations. *Annals of Botany* 49:749–778.

Bleak, A. T. 1970. Disappearance of plant material under a winter snow cover. *Ecology* 51:915–917.

Blum, U. and E. L. Rice. 1969. Inhibition of symbiotic nitrogen-fixation by gallic and tannic acid, and possible roles in old-field succession. *Bulletin of the Torrey Botanical Club* 96:531–544.

Bocock, K. L., O. J. W. Gilbert, C. K. Capstick, D. C. Twinn, J. S. Waid, and M. J. Woodman. 1960. Changes in the leaf litter when placed on the surface of soils with contrasting humus types. I. Losses in dry weight of oak and ash leaf litters. *J. of Soil Science* 11:1–9.

Bond, R. D. 1964. The influence of the microflora on the physical properties of soils. II. Field studies on water-repellent sands. *Australian J. of Soil Research* 2:123–131.

Bonner, J. 1940. On the growth factor requirements of isolated roots. *American J. of Botany* 27:692–701.

Booth, W. E. 1941. Revegetation of abandoned fields in Kansas and Oklahoma. *American J. of Botany* 28:415–422.

Bormann, F. H. 1953. The statistical efficiency of sample plot size and shape in forest ecology. *Ecology* 34:474–487.

Bormann, F. H. and G. E. Likens. 1967. Nutrient cycling. *Science* 155:424–429.

Bormann, F. H., G. E. Likens, and J. M. Melillo. 1977. Nitrogen budget for an aggrading northern hardwood forest ecosystem. *Science* 196:981–983.

Bormann, F. H., G. E. Likens, T. G. Siccama, R. S. Pierce, and J. S. Eaton. 1974. The export of nutrients and recovery of stable conditions following deforestation at Hubbard Brook. *Ecological Monographs* 44:255–277.

Bormann, F. H., T. G. Siccama, G. E. Likens, and R. H. Whittaker. 1970. The Hubbard Brook ecosystem study: composition and dynamics of the tree stratum. *Ecological Monographs* 40:373–388.

Botkin, D. B., J. F. Janak, and J. R. Wallis. 1972. Some ecological consequences of a computer model of forest growth. *J. of Ecology* 60:849–872.

Bourdeau, P. F. 1953. A test of random vs. systematic ecological sampling. *Ecology* 34:499–512.

Bowers, W. S., T. Ohta, J. S. Cleere, and P. A. Marsella. 1976. Discovery of insect anti-juvenile hormones in plants. *Science* 193:542–547.

Box, T. W. 1967. Brush, fire and west Texas rangeland. In *Proc. Tall Timbers Fire Ecol. Conf.*, no. 6, pp. 7–19. Tallahassee, FL: Tall Timbers Research Station.

Boyce, S. G. 1951. Source of atmospheric salts. *Science* 113:620–621.

———. 1954. The salt spray community. *Ecological Monographs* 24:29–67.

Boyer, J. S. 1976. Water deficits and photosynthesis. In *Water deficits and plant growth*, vol. IV, ed. T. T. Kozlowski, pp. 153–190. New York: Academic Press.

Bradshaw, A. D. 1965. Evolutionary significance of phenotypic plasticity in plants. *Advances in Genetics* 13:115–155.

———. 1969. An ecologist's viewpoint. In *Ecological aspects of the mineral nutrition of plants*, ed. I. H. Rorison, pp. 415–427. Oxford: Blackwell Scientific Publications.

Brady, N. C. 1974. *The nature and properties of soils.* 8th Ed. New York: Macmillan.

Braun, E. L. 1916. *The physiographic ecology of the Cincinnati region.* Ohio Biological Survey Bulletin 7.

———. 1950. *Deciduous forests of eastern North America.* Philadelphia, PA: Blakiston.

Braun-Blanquet, J. 1932. *Plant sociology: the study of plant communities.* New York: McGraw-Hill.

Bray, J. R. and J. T. Curtis. 1957. An ordination of the upland forest communities of southern Wisconsin. *Ecological Monographs* 27:325–349.

Bray, J. R. and E. Gorhman. 1964. Litter production in forests of the world. *Advances in Ecological Research* 2:101–157.

Brian, P. W. 1957. Effects of antibiotics on plants. *Annual Review of Plant Physiology* 8:413–426.

Briggs, D. and S. M. Walters. 1969. *Plant variation and evolution.* New York: McGraw-Hill.

Brinson, M. B. 1977. Decomposition and nutrient exchange of litter in an alluvial swamp forest. *Ecology* 58:601–609.

Brower, L. P. and J. V. Z. Brower. 1964. Birds, butterflies, and plant poisons: a study in ecological chemistry. *Zoologica* 49:137.

Brown, R. T. 1967. Influence of naturally occurring compounds on germination and growth of jack pine. *Ecology* 48:542–546.

Brown, R. W. 1970. *Measurement of water potential with thermocouple psychrometers: construction and applications.* U.S. Dept. of Agriculture Forest Service Research Paper INT–80. Washington, D.C.: U.S. Dept. of Agriculture.

Brown, R. W. and W. T. McDonough. 1977. Thermocouple psychrometer for *in situ* leaf water potential determinations. *Plant and Soil* 48:5–10.

Brown, R. W. and B. P. van Haveren, eds. 1972. *Psychrometry in water relations research.* Utah Agricultural Experiment Station. Logan, UT: Utah State University.

Brownell, P. F. and C. J. Crossland. 1972. The requirement of sodium as a micronutrient by species having the C_4 dicarboxylic acid photosynthetic pathway. *Plant Physiology* 49:794–797.

Brownell, P. F. and C. J. Crossland. 1974. Growth responses to sodium by *Bryophyllum tubiflorum* under conditions inducing crassulacean acid metabolism. *Plant Physiology* 54:416–417.

Bryson, R. A. 1974. A perspective on climatic change. *Science* 184:753–759.

Bunce, J. A. 1977. Nonstomatal inhibition of photosynthesis at low water potentials in intact leaves of species from a variety of habitats. *Plant Physiology* 59:348–350.

Burbank, M. P. and R. B. Platt. 1964. Granite outcrop communities on the Piedmont Plateau in Georgia. *Ecology* 45:292–305.

Burk, J. H. and W. A. Dick-Peddie. 1973. Comparative production of *Larrea divaricata* Cav. on three geomorphic surfaces in southern New Mexico. *Ecology* 54:1094–1102.

Burkholder, P. R. 1952. Cooperation and conflict among primitive organisms. *American Scientist* 40:601–631.

Burris, R. H. and C. C. Black. 1976. *CO_2 metabolism and plant productivity.* Baltimore, MD: University Park Press.

Burrows, F. J. and F. L. Milthorpe. 1976. Stomatal conductance in the control of gas exchange. In *Water deficits and plant growth*, vol. IV, ed. T. T. Kozlowski, pp. 103–152. New York: Academic Press.

Busby, J. R., L. C. Bliss, and C. D. Hamilton. 1978. Microclimate control of growth rates and habitats of the boreal forest mosses, *Tomenthypnum nitens* and *Hylocomium splendens. Ecological Monographs* 48:95–110.

Buttrick, S. C. 1977. The alpine flora of Teresa Island, Atlin Lake, British Columbia, with notes on its distribution. *Canadian J. of Botany* 55:1399–1409.

Byers, H. R. 1953. Coast redwoods and fog drip. *Ecology* 34:192–193.

Cable, D. R. 1975. Influence of precipitation on perennial grass production in the semi-desert southwest. *Ecology* 56:981–986.

Cain, S. A. and G. M. De O. Castro. 1959. *Manual of vegetation analysis.* New York: Harper.

Caldwell, M. M., R. S. White, R. T. Moore, and L. B. Camp. 1977. Carbon balance, productivity, and water use of cold-winter desert shrub communities dominated by C_3 and C_4 species. *Oecologia* 29:275–300.

Camacho-B, S. E., A. E. Hall, and M. R. Kaufmann. 1974. Efficiency and regulation of water transport in some woody and herbaceous species. *Plant Physiology* 54:169–172.

Campbell, C. J. and W. A. Dick-Peddie. 1964. Comparison of phreatophyte communities on the Rio Grande in New Mexico. *Ecology* 45:492–502.

Campbell, G. S. 1977. *An introduction to environmental biophysics.* New York: Springer-Verlag.

Campbell, G. S. and M. D. Campbell. 1974. Evaluation of a thermocouple hygrometer for measuring leaf water potential *in situ. Agronomy J.* 66:24–27.

Cannon, W. A. 1905a. On the transpiration of *Fouquieria splendens. Bulletin of the Torrey Botanical Club* 32:397–414.

———. 1905b. On the water-conducting systems of some desert plants. *Botanical Gazette* 39:397–408.

———. 1911. *The root habits of desert plants.* Carnegie Institution of Washington Publ. 131. Washington, D.C.: Carnegie Institution of Washington.

Capon, B. and W. Van Asdall. 1966. Heat pretreatment as a means of increasing germination of desert annual seeds. *Ecology* 48:305–306.

Carlquist, S. 1975. *Ecological strategies of xylem evolution.* Berkeley, CA: University of California Press.

Carlisle, A., A. H. F. Brown, and E. J. White. 1966. The organic matter and nutrient elements in the precipitation beneath a sessile oak *(Quercus petraea)* canopy. *J. of Ecology* 54:87–98.

Caswell, H. 1976. Community structure: a neutral model analysis. *Ecological Monographs* 46:327–354.

Cates, R. G. 1975. The interface between slugs and wild ginger: some evolutionary aspects. *Ecology* 56:391–400.

Cates, R. G. and G. H. Orians. 1975. Successional status and the palatability of plants to generalized herbivores. *Ecology* 56:410–418.

Cavers, P. B. and J. L. Harper. 1967a. Germination polymorphism in *Rumex crispus* and *R. obtusifolius. J. of Ecology* 54:367–382.

——. 1967b. The comparative biology of closely related species living in the same area. IX. *Rumex*: the nature of adaptation to a sea-shore habitat. *J. of Ecology* 55:73–82.

Center, T. D. and C. D. Johnson. 1974. Coevolution of some seed beetles (Coleoptera: Bruchidae) and their hosts. *Ecology* 55:1096–1103.

Ceska, A. and H. Roemer. 1971. A computer program for identifying species-relevé groups in vegetation studies. *Vegetatio* 23:255–277.

Chadwick, H. W. and P. D. Dalke. 1965. Plant succession in dune stands in Fremont County, Idaho. *Ecology* 46:765–780.

Chaney, R. W. 1948. *The ancient forests of Oregon.* Condon Lectures. Eugene, OR: Oregon State System of Higher Education.

Chapman, S. B. 1976. Production ecology and nutrient budgets. In *Methods in plant ecology,* ed. S. B. Chapman, pp. 157–228. New York: Halsted Press.

Chapman, V. J. 1960. *Salt marshes and salt deserts of the world.* New York: Interscience.

Chatterton, N. J. 1970. *Physiological ecology of Atriplex polycarpa: growth, salt tolerance, ion accumulation and soil-plant-water relations.* Ph.D. dissertation. Riverside, CA: University of California.

Chepurko, N.I. 1972. The biological productivity and the cycle of N and ash elements in the dwarf shrub tundra ecosystems of the Khibini Mountains (Kola Peninsula). In *Int. Biol. Programme, Tundra Biome, Proc. IV Int. Meet. on the Biol. Productivity of Tundra,* ed. F. L. Weilgolaski and T. Rosswall, pp. 236–247. Stockholm, Sweden.

Cheung, Y. N. S., M. T. Tyree, and J. Dainty. 1975. Water relations parameters on single leaves obtained in a pressure bomb and some ecological interpretations. *Canadian J. of Botany* 53:1342–1346.

Chew, R. M. 1974. Consumers as regulators of ecosystems: an alternative to energetics. *Ohio J. of Science* 74:359–370.

Chew, R. M. and A. E. Chew. 1965. The primary productivity of a desert shrub *(Larrea tridentata)* community. *Ecological Monographs* 35:353–375.

Christensen, N. and C. H. Muller. 1975. Effects of fire on factors controlling plant growth in *Adenostoma* chaparral. *Ecological Monographs* 45:29–55.

Churchill, G. B., H. H. John, D. P. Duncan, and A. C. Hodson. 1964. Long term effects of defoliation of aspen by the forest tent caterpillar. *Ecology* 45:630–633.

Clapham, A. R. 1932. The form of the observational unit in quantitative ecology. *J. of Ecology* 20:192–197.

Clark, P. J. and F. C. Evans. 1954. Distance to nearest neighbor as a measure of spatial relationships in populations. *Ecology* 35:445–453.

Clausen, J. 1962. *Stages in the evolution of plant species.* New York: Hafner.

Clausen, J., D. D. Keck, and W. M. Hiesey. 1940. *Experimental studies on the nature of species. I. The effect of varied environments on western North American plants.* Carnegie Institution of Washington Publ. 520. Washington, D.C.: Carnegie Institution of Washington.

Clements, E. S. 1960. *Adventures in ecology.* New York: Pageant Press.

Clements, F. E. 1916. *Plant succession: an analysis of the development of vegetation.* Carnegie Institution of Washington Publ. 242. Washington, D.C.: Carnegie Institution of Washington.

——. 1920. *Plant indicators: the relation of plant communities to process and practice.* Carnegie Institution of Washington Publ. 290. Washington, D.C.: Carnegie Institution of Washington.

——. 1936. Nature and structure of the climax. *J. of Ecology* 24:252–284.

Clements, F. E. and G. W. Goldsmith. 1924. *The phytometer method in ecology.* Carnegie Institution of Washington Publ. 356. Washington, D.C.: Carnegie Institution of Washington.

Cloudsley-Thompson, J. L. and M. J. Chadwick. 1964. *Life in deserts.* Philadelphia, PA: Dufour.

Coats, R. N., R. L. Leonard, and C. R. Goldman. 1976. Nitrogen uptake and release in a forested watershed, Lake Tahoe Basin. *Ecology* 57:995–1004.

Cody, M. L. 1974. Optimization in ecology. *Science* 183:1156–1164.

Cole, C. V., R. S. Innis, and J. W. B. Stewart. 1977. Simulation of phosphorus cycling in semiarid grasslands. *Ecology* 58:1–15.

Cole, D. W., S. P. Gessel, and S. F. Dice. 1967. Distribution and cycling of nitrogen, phosphorus, potassium and calcium in a secondary growth Douglas-fir ecosystem. In *Symp. primary productivity and mineral cycling in natural ecosystems,* pp. 196–232. Orono, ME: University of Maine Press.

Colinvaux, P. A. 1973. *Introduction to ecology.* New York: Wiley.

Collier, B. D., G. W. Cox, A. W. Johnson, and P. C. Miller. 1973. *Dynamic ecology.* Englewood Cliffs, NJ: Prentice-Hall.

Conard, H. S. 1935. The plant associations on central Long Island: a study in descriptive sociology. *American Midland Naturalist* 16:433–516.

Connell, J. H. and R. O. Slatyer. 1977. Mechanism of succession in natural communities and their role in community stability and organization. *American Naturalist* 111:1119–1144.

Cooper, C. F. 1961. The ecology of fire. *Scientific American* 204:150–160.

Cooper, W. S. 1913. The climax forest of Isle Royale, Lake Superior, and its development. *Botanical Gazette* 55:1–44, 115–140, 189–235.

——. 1919. Ecology of the strand vegetation of the Pacific coast of North America. *Carnegie Institution of Washington Yearbook* 18:96–99.

——. 1922. *The broad-sclerophyll vegetation of California.* Carnegie Institution of Washington Publ. 319. Washington, D.C.: Carnegie Institution of Washington.

——. 1923. The recent ecological history of Glacier Bay, Alaska. *Ecology* 4:93–128, 223–246, 355–365.

——. 1931. A third expedition to Glacier Bay, Alaska. *Ecology* 12:61–95.

——. 1936. The strand and dune flora of the Pacific coast of North America: a geographic study. In *Essays in geobotany,* ed. T. H. Goodspeed, pp. 141–187. Berkeley, CA: University of California Press.

Cornforth, I. S. 1970. Leaf fall in a tropical rain forest. *J. Applied Ecology* 7:603–608.

Cottam, G. and J. T. Curtis. 1956. Use of distance measures in phytosociological sampling. *Ecology* 37:451–460.

Cottam, W. P., J. M. Tucker, and R. Drobnick. 1959. Some clues to Great Basin postpluvial climates provided by oak distributions. *Ecology* 40:361–377.

Court, A. 1974. Water balance estimates for the United States. *Weatherwise* 27:252–256.

Cowles, H. C. 1901. The physiographic ecology of Chicago and vicinity: a study of the origin, development, and classification of plant societies. *Botanical Gazette* 31:73–108, 145–182.

——. 1909. Present problems in plant ecology: the trend of ecological philosophy. *American Naturalist* 43:356–368.

——. 1911. The causes of vegetative cycles. *Botanical Gazette* 51:161–183.

Cox, G. W. 1976. *Laboratory manual of general ecology.* 3rd ed. Dubuque, IA: W. C. Brown.

Critchfield, W. B. 1971. *Profiles of California vegetation.* U.S. Dept. of Agriculture Forest Service Research Paper PSW-76, Berkeley, CA. Washington, D.C.: U.S. Dept. of Agriculture.

Crocker, R. L. and J. Major. 1955. Soil development in relation to vegetation and surface age at Glacier Bay, Alaska. *J. of Ecology* 43:427–448.

Crosby, J. S. 1961. *Litter and duff fuel in shortleaf pine stands in southeast Missouri.* Central States Forest Experiment Station Technical Paper 178. Columbus, OH: U. S. Forest Service.

Cunningham, G. L. and B. R. Strain. 1969. Irradiance and productivity in a desert shrub. *Photosynthetica* 3:69–71.

Currey, D. B. 1965. An ancient bristlecone pine stand in eastern Nevada. *Ecology* 46:564–566.

Curtis, J. T. and G. Cottam. 1962. *Plant ecology workbook*. Minneapolis, MN: Burgess.

Curtis, J. T. and R. P. McIntosh. 1950. The interrelations of certain analytic and synthetic phytosociological characteristics. *Ecology* 31: 434–455.

Curtis, J. T. and R. P. McIntosh. 1951. An upland forest continuum in the prairie-forest border region of Wisconsin. *Ecology* 32:476–498.

Cutler, J. M. and D. W. Rains. 1978. Effects of water stress and hardening on the internal water relations and osmotic constituents of cotton leaves. *Physiologia Plantarum* 42:261–268.

Cutler, J. M., D. W. Rains, and R. S. Loomis. 1977. The importance of cell size in the water relations of plants. *Physiologia Plantarum* 40: 255–260.

Dahlman, R. C. and C. L. Kucera. 1965. Root productivity and turnover in native prairie. *Ecology* 46:84–89.

Dansereau, P. 1951. Description and recording of vegetation upon a structural basis. *Ecology* 32:172–229.

——. 1961. Essai de representation cartographique des elements structuraux de la vegetation. In *Methods de la cartographie de la vegetation*, 97th Colloquium, ed. H. Gaussen, pp. 233–255. Paris: Centre National de la Recherche Scientifique.

Darwin, C. R. 1859. *On the origin of species by means of natural selection, or the preservation of favoured races in the struggle for life*. London: John Murray.

Dasmann, R. F. 1972. *Environmental Conservation*, 3rd ed. New York: Wiley.

Daubenmire, R. F. 1938. Merriam's life zones of North America. *Quarterly Review of Biology* 13:327–332.

——. 1952. Forest vegetation of northern Idaho and adjacent Washington and its bearing on concepts of vegetation classification. *Ecological Monographs* 22:301–330.

——. 1959. A canopy-coverage method of vegetational analysis. *Northwest Science* 33:43–66.

——. 1968a. *Plant communities*. New York: Harper and Row.

——. 1968b. Ecology of fire in grasslands. *Advances in Ecological Research* 5:209–266.

——. 1974. *Plants and environment*. 3rd ed. New York: Wiley.

Davis, P. H. and V. H. Heywood. 1963. *Principles of angiosperm taxonomy*. London: Oliver and Boyd.

Davis, R. B. 1966. Spruce-fir forests of the coast of Maine. *Ecological Monographs* 36:79–94.

Davis, T. A. W. and P. W. Richards. 1933. The vegetation of Moraballi Creek, British Guiana; an ecological study of a limited area of tropical rain forest. *J. of Ecology* 21:350–385.

DeBano, L. F., P. H. Dunn, and C. E. Conrad. 1977. Fire's effect on physical and chemical properties of chaparral soils. In *Proc. Symp. Env. Cons. Fire and Fuel Mangmt. in Medit. Ecosyst.*, pp. 65–74. Washington, D.C.: U.S. Dept. of Agriculture Forest Service.

DeBano, L. F. and J. Letey, eds. 1969. *Proc. Symp. on Water-repellent Soils*. Riverside, CA: University of California.

De Candolle, A. L. 1855. *Geôgraphie botanique raisoné*. Paris: V. Masson.

——. 1874. Constitution dans le régne végétale de groupes physiologiques applicables á la geógraphie botanique ancienne et moderne. *Arch. Sci. Phys. Nat.*

Deevey, E. S., Jr. 1947. Life tables for natural populations. *The Quarterly Review of Biology* 22:283–314.

——. 1970. Mineral cycles. *Scientific American* 223(3):148–160.

DeJong, T. M. 1975. A comparison of three diversity indices based on their components of richness and evenness. *Oikos* 26:222–227.

de la Cruz, A. A. and B. C. Gabriel. 1974. Caloric, elemental, and nutritive changes in decomposing *Juncus roemerianus* leaves. *Ecology* 55:882–886.

del Moral, R. and C. H. Muller. 1969. Fog drip: a mechanism of toxin transport from *Eucalyptus globulus*. *Bulletin of the Torrey Botanical Club* 96:467–475.

Dennis, G. J. and P. L. Johnson. 1970. Shoot and rhizome-root standing crops of tundra vegetation at Barrow, Alaska. *Arctic and Alpine Research* 2:253–266.

de Wit, C. D. 1960. *On competition*. Versl. Landbouwkl. Onderzoek. no. 66.8. Wageningen, The Netherlands: Centre for Agricultural Publications and Documentation.

——. 1961. Space relationships within populations of one or more species. In *Mechanisms in biological competition*, ed. F. L. Milthorpe, pp. 314–429. New York: Cambridge University Press.

Dix, R. L. 1957. Sugar maple in forest succession at Washington, D.C. *Ecology* 38:663–665.

Doman, E. R. 1967. Prescribed burning and brush type conversion in California National Forests. In *Proc. Tall Timbers Fire Ecol. Conf.*, no. 7, pp. 225–243. Tallahassee, FL: Tall Timbers Research Station.

Dorf, I. 1960. Climate changes of past and present. *American Scientist* 48:341–364.

Douglas, L. A. and J. C. F. Tedrow. 1959. Organic matter decomposition rates in arctic soils. *Soil Science* 88:305–312.

Downes, R. W. and J. D. Hesketh. 1968. Enhanced photosynthesis at low O_2 concentrations: differential response of temperate and tropical grasses. *Planta* 78:79–84.

Downton, W. J. S. 1975. Checklist of C_4 species. *Photosynthetica* 9:96–105.

Downton, W. J. S. and E. B. Tregunna. 1968. Carbon dioxide compensation—its relation to photosynthetic carboxylation reaction, systematics of the Graminae, and leaf anatomy. *Canadian J. Botany* 46:207–215.

Drude, O. 1890. *Handbuch der Pflanzengeographie*. Stuttgart, Germany: Engelmann.

Drury, W. H., Jr. 1956. Bog flats and physiographic processes in the upper Kuskokwin River region, Alaska. *Gray Herbarium Contributions* 178:1–30.

Drury, W. H., Jr. and I. C. T. Nisbet. 1973. Succession. *J. of the Arnold Arboretum* 54:31–368.

Dunn, P. H. and L. F. DeBano. 1977. Fire's effect on biological and chemical properties of chaparral soils. In *Proc. Symp. Env. Cons. Fire and Fuel Mangmt. in Medit. Ecosyst.*, pp. 75–84. Washington, D.C.: U.S. Dept. of Agriculture Forest Service.

DuRietz, G. E. 1931. Life forms of terrestrial flowering plants. *Acta Phytogeographica Suecica*, vol. III. Uppsala.

Duvduvani, S. 1964. Dew in Israel and its effect on plants. *Soil Science* 98:14–21.

Dyer, M. I. 1975. The effects of red-winged blackbirds (*Agelaius phoeniceus* L.) on biomass production of corn grain (*Zea mayes* L.) *J. of Applied Ecology* 12:719–726.

Dyer, M. I. and U. G. Bokhari. 1976. Plant-animal interactions: studies on the effects of grasshopper grazing on blue grama grass. *Ecology* 57:762–772.

Edwards, C. A. and G. W. Heath. 1963. The role of soil animals in break-down of leaf material. In *Soil organisms*, ed. J. Docksen and J. Van Der Drift, pp. 76–84. Amsterdam: North Holland Publishing Co.

Egerton, F. N. 1976. Ecological studies and observations before 1900. In *Evolution of issues, ideas and events in America: 1776–1976*, ed. B. J. Taylor, pp. 311–351. Norman, OK: University of Oklahoma Press.

Egler, F. E. 1954. Vegetation science concepts. I. Initial floristic composition, a factor in old-field vegetation development. *Vegetatio* 4:412–417.

Ehleringer, J., O. Björkman, and H. A. Mooney. 1976. Leaf pubescence: effects on absorptance and photosynthesis in a desert shrub. *Science* 192:376–377.

Ehrlich, P. R. and P. H. Raven. 1965. Butterflies and plants: a study in coevolution. *Evolution* 18:568–608.

Eickmeier, W. G. 1978. Photosynthetic pathway distributions along an aridity gradient in Big Bend National Park, and implications for enhanced resource partitioning. *Photosynthetica* 12:290–297.

Ellenberg, H. 1958. Bodenreaktion (einschlieplich Kalfrage). In *Handbuch der Pflanzenphysiologie*, vol. 4, ed. W. Ruhland, pp. 638–708. Berlin: Springer-Verlag.

England, C. B. and E. H. Lesesne. 1962. Evapotranspiration research in western North Carolina. *Agricultural Engineering* 43:526–528.

Epstein, E. 1972. *Mineral nutrition of plants: principles and perspectives*. New York: Wiley.

Esau, K. 1965. *Plant Anatomy*. New York: Wiley.

Eshel, Y. and Y. Waisel. 1965. The salt relations of *Prosopis farcta* (Banks et Sol.) Eig. *Israel J. of Botany* 14:50–51 (abstract).

Etherington, J. R. 1975. *Environment and plant ecology*. New York: Wiley.

Evanari, M., L. Shanan, and N. H. Tadmor. 1971. *The Negev, the challenge of a desert*. Cambridge, MA: Harvard University Press.

Evans, G. C. 1976. A sack of uncut diamonds. *J. of Ecology* 64:1–39.

Eyre, S. R. 1963. *Vegetation and soils*. Chicago: Aldine.

Faegri, K. and L. van der Pijl. 1971. *The principles of pollination ecology*. New York: Pergamon Press.

Feeny, P. P. and H. Bostock. 1968. Seasonal changes in the tannin content of oak leaves. *Phytochemistry* 7:871–880.

Ferguson, C. W. 1968. Bristlecone pine: science and esthetics. *Science* 159:839–846.

Fernandez, O. A. and M. M. Caldwell. 1975. Phenology and dynamics of root growth of three cool semi-desert shrubs under field conditions. *J. of Ecology* 63:703–714.

Finck, A. 1969. *Pflanzenernahrung in Stichworten*. Kiel, Germany: F. Hirt.

Flint, R. F. 1957. Glacial and pleistocene geology. New York: Wiley.

Flock, J. A. W. 1978. Lichen-Bryophyte distributions along a snow-cover-soil-moisture gradient, Niwot Ridge, Colorado. *Arctic and Alpine Research* 10:31–47.

Flowers, T. J., P. F. Troke, and A. R. Yeo. 1977. The mechanism of salt tolerance in halophytes. *Annual Review of Plant Physiology* 28:89–121.

Fonda, R. W. 1974. Forest succession in relation to river terrace development in Olympic National Park, Washington. *Ecology* 55: 927–942.

Forcier, L. K. 1975. Reproductive strategies and the co-occurrence of climax tree species. *Science* 189:808–810.

Ford, J. M. and J. E. Monroe. 1971. *Living systems*. San Francisco, CA: Canfield Press.

Forman, R. T. T. 1975. Canopy lichens with blue-green algae: a nitrogen source in a Columbian rain forest. *Ecology* 56:1176–1184.

Foster, N. W. and I. K. Morrison. 1976. Distribution and cycling of nutrients in a natural *Pinus banksiana* ecosystem. *Ecology* 57:110–120.

Fortescue, J. A. C. and G. G. Marten. 1970. Analysis of temperate forest ecosystems. *Ecological studies*, vol. 1, ed. D. E. Reichle, pp. 173–198. New York: Springer.

Fowells, H. A., ed. 1965. *Silvics of forest trees of the United States*. Agriculture Handbook no. 271. Washington, D.C.: U.S. Dept. of Agriculture Forest Service.

Franklin, J. F. and C. T. Dyrness. 1969. *Natural vegetation of Oregon and Washington*. U.S. Dept. of Agriculture Forest Service General Technical Report PNW–8. Portland, OR: U.S. Dept. of Agriculture Forest Service.

Fryer, J. H. and F. T. Ledig. 1972. Microevolution of the photosynthetic temperature optimum in relation to an elevational complex gradient. *Canadian J. of Botany* 50: 1231–1235.

Gadgil, M. and O. T. Solbrig. 1972. The concept of r- and K-selection: evidence from wildflowers and some theoretical considerations. *The American Naturalist* 106:14–31.

Gambell, A. W. and D. W. Fisher. 1966. *Chemical composition of rainfall, eastern North Carolina and southeastern Virginia*. U.S. Geological Survey Water Supply Paper 1535-K.

Garrison, G. A. 1949. Uses and modifications for the "moosehorn" crown closure estimator. *J. of Forestry* 47:733–735.

Garth, R. E. 1964. The ecology of Spanish moss (*Tillandsia usneoides*). *Ecology* 45:470–481.

Gates, D. M. 1962. *Energy exchange in the biosphere*. New York: Harper and Row.

———. 1965a. Heat transfer in plants. *Scientific American* 213:76–83.

———. 1965b. Radiant energy, its receipt and disposal. *Meterological Monographs* 6:1–26.

———. 1972. *Man and his environment: climate*. New York: Harper and Row.

Gause, G. F. 1934. *The struggle for existence*. Baltimore, MD: Williams and Wilkins.

Gerdemann, J. W. 1968. Vesicular-arbuscular mycorrhizae and plant growth. *Annual Review of Phytophatology* 6:397–418.

Gill, A. M. 1977. Plant traits adaptive to fires in Mediterranean land ecosystems. In *Proc. Symp. Env. Cons. Fire and Fuel Mangmt. in Medit. Ecosyst.*, pp. 17–26. Washington, D.C.: U.S. Dept. of Agriculture Forest Service.

Gilmour, J. S. L. and J. Heslop-Harrison. 1954. The -deme terminology and the units of micro-evolutionary change. *Genetica* 27:146–161.

Gleason, H. A. 1926. The individualistic concept of the plant association. *Bulletin of the Torrey Botanical Club* 53:1–20.

———. 1953. Dr. H. A. Gleason, Distinguished Ecologist. *Bulletin of the Ecological Society of America* 34(2):40–42.

Goldsmith, F. B. and C. M. Harrison. 1976. Description and analysis of vegetation. In *Methods in plant ecology*, ed. S. B. Chapman, pp. 85–155. New York: Halsted Press.

Goldsworthy, A. 1976. *Photorespiration*. Carolina Biology Readers, no. 80. Burlington, NC: Carolina Biological Suppliers.

Golley, F. B. 1961. Energy values of ecological materials. *Ecology* 42: 581–584.

———. ed. 1977. *Ecological succession*. Stroudsberg, PA: Downden, Hutchinson, and Ross.

Good, R. E. 1931. A theory of plant geography. *The New Phytologist* 30:149–203.

———. 1953. *The geography of the flowering plants*. 2nd ed. New York: Longmans, Green, and Co.

Good, R. E. and N. F. Good. 1975. Growth characteristics of two populations of *Pinus ridgida* Mill, from the pine barrens of New Jersey. *Ecology* 56:1215–1220.

Goodall, D. W. 1953. Objective methods for the classification of vegetation. I. The use of positive interspecific correlation. *Australian J. of Botany* 1:39–63.

———. 1957. Some considerations in the use of point quadrat methods for the analysis of vegetation. *Australian J. of Biological Science* 5:1–41.

———. 1973. Sample similarity and species correlation. In *Handbook of vegetation science. V. Ordination and classification of communities*, ed. R. H. Whittaker, pp. 107–156. The Hague: Junk.

Goodland, R. J. 1975. The tropical origin of ecology: Eugen Warming's jubilee. *Oikos* 26:240–245.

Goss, R. W. 1960. Mycorrhizae of ponderosa pine in Nebraska grassland soils. *Nebraska Agricultural Experiment Station Research Bulletin* 192:1–47.

Gosz, J. R., G. E. Likens, and F. H. Bormann. 1973. Nutrient release from decomposing leaf and branch litter in the Hubbard Brook Forest, New Hampshire. *Ecological Monographs* 43:173–191.

Graham, B. F., Jr. and F. H. Bormann. 1966. Natural root grafts. *Botanical Review* 32:255–292.

Gray, A. 1889. *Scientific papers*, selected by C. S. Sargent. New York: Houghton Mifflin.

Gray, W. D. 1959. *The relation of fungi to human affairs.* New York: Holt-Dryden.

Greenway, H. and D. A. Thomas. 1965. Plant response to saline substrates. V. Chloride regulation in the individual organs of *Hordeum vulgare* during treatment with sodium chloride. *Australian J. of Biological Science* 18:505–524.

Gregor, J. W. 1946. Ecotypic differentiation. *The New Phytologist* 45:254–270.

Greig-Smith, P. 1964. *Quantitative plant ecology.* London: Butterworths.

Greig-Smith, P. and K. A. Kershaw. 1958. The significance of pattern in vegetation. *Vegetatio* 8:189–192.

Grier, C. C. and S. W. Running. 1977. Leaf area of mature northwestern coniferous forests: relation to site water balance. *Ecology* 58:893–899.

Griffin, J. R. 1965. Digger pine seedling response to serpentinite and non-serpentinite soil. *Ecology* 46:801–807.

Griffin, J. R. 1973. Xylem sap tension in three woodland oaks of central California. *Ecology* 54:152–159.

Griffin, J. R. and W. B. Critchfield. 1972. *The distribution of forest trees in California.* U.S. Dept. of Agriculture Forest Service Research Paper PSW–82/1972. Washington, D.C.: U.S. Dept. of Agriculture Forest Service.

Griffin, K. O. 1975. Vegetation studies and modern pollen spectra from the Red Lake peatland, northern Minnesota. *Ecology* 56: 531–546.

Grime, J. P. 1977. Evidence for the existence of three primary strategies in plants and its relevance to ecological and evolutionary theory. *American Naturalist* 111:1169–1194.

Grime, J. P. 1979. *Plant strategies and vegetation processes.* New York: Wiley.

Grime, J. P. and K. Thompson. 1976. An apparatus for measurement of the effect of amplitude of temperature fluctuation upon the germination of seeds. *Annals of Botany* 40:795–799.

Grisebach, A. H. R. 1872. *Die Vegetation der Erde nach ihrer klimatischen Anordnung.* Leipzig, Germany: Engelmann.

Grosenbaugh, L. R. 1952. Plotless timber estimates—new, fast, easy. *J. of Forestry* 50:32–37.

Grubb, P. J., H. E. Green, and R. C. J. Merrifield. 1969. The ecology of chalk heath: its relevance to the calcicole-calcifuge and soil acidification problems. *J. of Ecology* 57:175–212.

Gupta, S. R. and J. S. Singh. 1977. Decomposition of litter in a tropical grassland. *Pedobiologia* 17:330–333.

Gutterman, Y. 1972. Delayed seed dispersal and rapid germination as survival mechanisms of the desert plant *Blepharis persica* (Burm.) Kuntze. *Oecologia* 10:145–149.

Gutterman, Y., T. H. Thomas, and W. Heydecker. 1975. Effect on the progeny of applying different day length and hormone treatments to parent plants of *Lactuca scariola*. *Physiologia Plantarum* 34:30–38.

Hacskaylo, E., ed. 1971. *Proc. of the First North American Conf. on Mycorrhizae, 1969.* U.S.

Dept. of Agriculture Forest Service Misc. Publ. 1189. Washington, D.C.: U.S. Dept. of Agriculture Forest Service.

Hadley, E. B. 1970. Net productivity and burning responses of native eastern North Dakota prairie communities. *American Midland Naturalist* 84:121–135.

Haeckel, E. H. P. A. 1866. *Generelle Morphologie der Organismen.* Berlin: Reimer.

Hagan, R. M., H. R. Haise, and T. W. Edminster, eds. 1967. *Irrigation of agricultural lands.* Madison, WI: American Society of Agronomy.

Hall, A. E. and M. R. Kaufmann. 1975. The regulation of water transport in the soil-plant-atmosphere continuum. In *Perspectives in biophysical ecology,* ed. D. M. Gates and R. B. Schmerl, pp. 187–202. Berlin: Springer-Verlag.

Hall, A. E., E. D. Schulze, and O. L. Lange. 1976. Current perspectives of steady-state stomatal responses to environment. In *Water and plant life: problems and modern approaches,* ed. O. L. Lange, L. Kappen, and E. D. Schulze, pp. 169–188. Berlin: Springer-Verlag.

Halligan, J. P. 1973. Bare areas associated with shrub stands in grassland: the case of *Artemesia californica. BioScience* 23:429–432.

Halm, G. J. 1972. *The phosphorus cycle in a grassland ecosystem.* Ph.D. dissertation. Saskatoon, Canada: University of Saskatchewan.

Hanes, T. L. 1977. California chaparral. In *Terrestrial vegetation of California,* ed. M. G. Barbour and J. Major, pp. 417–469. New York: Wiley-Interscience.

Hannapel, R. J., W. H. Fuller, S. Bosma, and J. S. Bullock. Phosphorus movement in a calcareous soil: I. Predominance of organic forms of phosphorus in phosphorus movement. *Soil Science* 97:350–357.

Hanson, H. C. 1917. Leaf-structure as related to environment. *American J. of Botany* 4:553–560.

———. 1962. *Dictionary of ecology.* New York: Philosophical Library.

Harley, J. L. 1969. *The biology of mycorrhizae.* 2nd ed. London: Leonard Hill.

Harper, J. L. 1961. Approaches to the study of plant competition. *Symposium of the Society for Experimental Biology* 15:1–39.

———. 1964. The individual in the population. *J. of Ecology* 52 (Supplement):149–158.

———. 1967. A Darwinian approach to plant ecology. *J. of Ecology* 55:247–270.

———. 1977. *Population biology of plants.* New York: Academic Press.

Harper, J. L. and J. Ogden. 1970. The reproductive strategy of higher plants. I. The concept of strategy with special reference to *Senecio vulgaris. J. of Ecology* 58:681–698.

Harper, J. L. and J. White. 1974. The demography of plants. *Annual Review of Ecology and Systematics* 5:419–463.

Harris, G. A. 1967. Some competitive relationships between *Agropyron spicatum* and *Bromus tectorum. Ecological Monographs* 37:89–111.

Harris, P. 1974. A possible explanation of plant yield increases following insect damage. *Agro-Ecosystems* 1:219–225.

Harrison, A. T., E. Small, and H. A. Mooney. 1971. Drought relationships and distribution of two Mediterranean-climate California plant communities. *Ecology* 52:869–875.

Harshberger, J. W. 1911. *Phytogeographic survey of North America.* New York: Stechert.

———. 1911. An hydrometric investigation of the influence of sea water on the distribution of salt water and estuarine plants. *Proc. of the American Philosophical Society* 50:457–497.

———. 1916. *The vegetation of the New Jersey pine-barrens.* Philadelphia, PA: Sower.

Hartesveldt, R. J. 1964. The fire ecology of the giant sequoias. *Natural History* 73:12–19.

Hartesveldt, R. J. and H. T. Harvey. 1967. The fire ecology of sequoia regeneration. In *Proc. Tall Timbers Fire Ecol. Conf.,* no. 7, pp. 65–78. Tallahassee, FL: Tall Timbers Research Station.

Hartesveldt, R. J., H. T. Harvey, H. S. Shellhammer, and R. E. Stecker. 1975. *The giant sequoia of the Sierra Nevada.* U.S. Dept. of the Interior National Park Service Publ. no. 120. Washington, D.C.: U.S. Dept of the Interior, National Park Service.

Hartsock, T. L. and P. S. Nobel. 1976. Watering converts a CAM plant to daytime CO_2 uptake. *Nature* 262:574–576.

Harvey, H. T., H. S. Shellhammer, and R. E. Stecker. 1977. *Giant sequoia ecology: fire and reproduction.* U.S. Dept. of the Interior National Park Service Sci. Mono. 12. Washington, D.C.: U.S. Dept. of the Interior, National Park Service.

Hastings, J. R. and R. M. Turner. 1965. *The changing mile*. Tucson, AZ: University of Arizona Press.

Hatch, A. B. 1937. The physical basis of mycotrophy in *Pinus. Black Rock Forest Bulletin* 6.

Hatch, M. D. and C. R. Slack. 1966. Photosynthesis by sugar-cane leaves: a new carboxylation reaction and the pathway of sugar formation. *Biochemical J.*, 101:103–111.

Heine, R. W. 1971. Hydraulic conductivity in trees. *J. of Experimental Botany* 22:503–511.

Hellkvist, J., G. P. Richards, and P. G. Jarvis. 1974. Verticle gradients of water potential and tissue water relations in Sitka spruce trees measured with the pressure chamber. *J. of Applied Ecology* 11:637–667.

Hellmers, H. 1964. An evaluation of the photosynthetic efficiency of forests. *Quarterly Review of Biology* 39:249–257.

———. 1966. Growth response of redwood seedlings to thermoperiodism. *Forest Science* 12:276–283.

Henderson, G. S., W. F. Harris, D. E. Todd, Jr., and T. Grizzard. 1977. Quantity and chemistry of throughfall as influenced by forest type and season. *J. of Ecology* 65:365–374.

Heslop-Harrison, J. 1964. Forty years of genecology. *Advances in Ecological Research* 2:159–247.

Heyward, F. 1939. The relation of fire to stand composition of longleaf pine forests. *Ecology* 20:287–304.

Heywood, V. H. 1967. *Plant taxonomy*. New York: St. Martin's Press.

Hickman, J. C. 1975. Environmental unpredictability and plastic energy allocation strategies in the annual *Polygonum cascadense. J. of Ecology* 63:689–701.

Hickman, J. C. 1977. Energy allocation and niche differentiation in four coexisting species of *Polygonum* in western North America. *J. of Ecology* 65:317–326.

Hickman, J. C. and L. F. Pitelka. 1975. Dry weight indicates energy allocation in ecological strategy analyses of plants. *Oecologia* 21:117–121.

Highkin, H. R. 1958. Transmission of phenotypic variability within a pure line. *Nature* 182:1460.

Holdridge, L. R. 1967. *Life zone ecology*, rev. ed. San José, Costa Rica: Tropical Science Center.

Hopkins, A. D. 1938. *Bioclimatics: a science of life and climatic relations*. U.S. Dept. of Agriculture Misc. Publ. 280. Washington, D.C.: U.S. Dept. of Agriculture.

Hopkins, B. 1957. Pattern in the plant community. *J. of Ecology* 45:451–463.

Horn, H. S. 1971. *The adaptive geometry of trees*. Princeton, NJ: Princeton University Press.

———. 1974. The ecology of secondary succession. *Annual Review of Ecology and Systematics* 5:25–37.

———. 1975. Forest succession. *Scientific American* 232(5): 90–98.

Horton, J.S. 1977. The development and perpetuation of the permanent Tamarisk type in the phreatophyte zone of the southwest. In *Importance, preservation, and management of riparian habitat: symp. proc.*, pp. 124–127. U.S. Dept. of Agriculture Forest Service General Technical Report RM–43. Washington, D.C.: U.S. Dept. of Agriculture Forest Service.

Hough, A. F. 1936. A climax forest community on East Tionesta Creek in northwestern Pennsylvania. *Ecology* 17:9–28.

Hrapko, J. O. and G. A. La Roi. 1978. The alpine tundra vegetation of Signal Mountain, Jasper National Park. *Canadian J. of Botany* 56:309–332.

Hsiao, T. C. 1973. Plant responses to water stress. *Annual Review of Plant Physiology* 24:519–570.

———. 1976. Stomatal ion transport. In *Encyclopaedia of plant physiology*, vol. 2, part B, ed. A. Pirson and M. H. Zimmerman, pp. 195–217. Berlin: Springer-Verlag.

Huey, R. B. and E. R. Pianka. 1977. Patterns of niche overlap among broadly sympatric versus narrowly sympatric Kalahari lizards (Scinidae: Maybuya). *Ecology* 58:119–128.

Hull, R. J. and O. A. Leonard. 1964. Physiological aspects of parasitism in mistletoes (*Arceuthobium* and *Phoradendron*), parts I and II. *Plant Physiology* 39:996–1017.

Humboldt, A. von and A. Bonpland. 1807. *Essai sur la geographie des plantes*. Paris.

———. 1807–1834. *Voyage aux régions equinoxiales du nouveau continent*. 30 vols.

Humphrey, R. R. 1974. Fire in the deserts and desert grassland of North America. In *Fire and ecosystems*, ed. T. T. Kozlowski and C. E. Ahlgren, ch. 11. New York: Academic Press.

Hunt, W. H. 1977. A simulation model for decomposition in grasslands. *Ecology* 58: 469–484.

Hurlbert, S. H. 1971. The nonconcept of species diversity: a critique and alternative parameters. *Ecology* 52:577–586.

Hutchings, M. J. and J. P. Barkham. 1976. An investigation of shoot interactions in *Mercurialis perennis* L., a rhizomatous perennial herb. *J. of Ecology* 64:723–743.

Hutchison, B. A. and D. R. Matt. 1977. The distribution of solar radiation within a deciduous forest. *Ecological Monographs* 47:185–207.

Ingram, M. 1957. Microorganisms resisting high concentrations of sugars and salts. In *Seventh Symposium of the Society for General Microbiology*, pp. 90–133. Cambridge, England: Cambridge University Press.

Ivimey-Cook, R. B. and M. C. F. Proctor. 1966. The application of association-analysis to phytosociology. *J. of Ecology* 54:179–192.

Iwaki, H., M. Monsi, and B. Midorikawa. 1966. Dry matter production of some herb communities in Japan. *The Eleventh Pacific Science Congress*, pp. 1–15. Tokyo, Japan: Science Council of Japan.

Jackson, J. R. and R. W. Willemsen. 1976. Allelopathy in the first stages of secondary succession on the piedmont of New Jersey. *American J. of Botany* 63:1015–1023.

Jackson, M. T. 1966. Effects of microclimate on spring flowering phenology. *Ecology* 47:407–415.

Janzen, D. H. 1970. Herbivores and the number of tree species in tropical forests. *American Naturalist* 104:501–528.

Janzen, D. H. 1973. Community structure of secondary compounds in plants. *Pure and Applied Chemistry* 34:529–538.

———. 1975. *Ecology of plants in the tropics.* London: Edward Arnold.

Jeffree, E. P. 1960. Some long term means from the phenological reports (1891–1948) of the Royal Meteorological Society. *Quarterly J. of the Royal Meteorological Society* 86:95–103.

Jennings, D. H. 1968. Halophytes, succulence, and sodium in plants—a unified theory. *New Phytologist* 67:899–911.

Jenny, H. 1941. *The factors of soil formation.* New York: McGraw-Hill.

Jenny, H., R. J. Arkley, and A. M. Schultz. 1969. The pygmy forest-podsol ecosystem and its dune associates of the Mendocino Coast. *Madroño* 20:60–74.

Jenny, H., S. P. Gessel, and F. T. Bingham. 1949. Comparative study of decomposition rates of organic matter in temperate and tropical regions. *Soil Science* 68:419–432.

Johnson, D. A. and M. M. Caldwell. 1975. Gas exchange of four arctic and alpine tundra plant species in relation to atmospheric and soil moisture stress. *Oecologia* 21:93–108.

Johnson, H. B. 1976. Vegetation and plant communities of southern California deserts—a functional view. In *Symposium proc., plant communities of southern California*, Special Publ. no. 2 of the California Native Plant Society, ed. J. Latting, pp. 125–164. Berkeley, CA: California Native Plant Society.

Johnson, P. L., ed. 1977. *An ecosystem paradigm for ecology.* Oak Ridge, TN: Oak Ridge Associated Universities.

Johnson, P. L. and W. D. Billings. 1962. The alpine vegetation of the Beartooth Plateau in relation to cryopedogenic processes and patterns. *Ecological Monographs* 32:105–135.

Jones, C. E. and S. L. Buchmann. 1974. Ultra-violet floral patterns as functional orientation cues in Hymenopteran pollination systems. *Animal Behavior* 22:481–485.

Jones, E. W. 1945. The structure and reproduction of the virgin forests of the north temperate zone. *The New Phytologist* 44:130–148.

Jones, M. M. and N. C. Turner. 1978. Osmotic adjustment in leaves of sorghum in response to water deficits. *Plant Physiology* 61:122–126.

Jordan, C. F. 1971. A world pattern in plant energetics. *American Scientist* 59:425–433.

Jordan, D. F. and J. R. Kline. 1972. Mineral cycling: some basic concepts and their application in a tropical rain forest. *Annual Review of Ecology and Systematics* 3:33–50.

Kalle, K. 1958. In *Handbuch der Pflanzenphysiologie*, Bd. IV, ed. W. Ruhland, pp. 170–178. Berlin–Gottingen–Heidelberg: Springer-Verlag.

Kanemasu, E. T., G. W. Thurtell, and C. B. Tanner. 1969. Design, calibration and field use of a stomatal diffusion porometer. *Plant Physiology* 44:881–885.

Kaplanis, J. N., M. J. Thompson, W. E. Robbins, and B. M. Bryce. 1967. Insect hormones: alpha ecdysone and 20-hydroxyecdysone in bracken fern. *Science* 157:1436–1438.

Keeley, J. E. 1977. Fire-dependent strategies in *Arctostaphylos* and *Ceanothus*. In *Proc. Symp. Env. Cons. Fire and Fuel Mangmt. in Medit. Ecosyst.*, pp. 391–396. Washington, D.C.: U.S. Dept. of Agriculture Forest Service.

Keeley, S. 1977. The relationship of precipitation to post-fire succession in southern California chaparral. In *Proc. Symp. Env. Cons. Fire and Fuel Mangmt. in Medit. Ecosyst.*, pp. 387–390. Washington, D.C.: U.S. Dept. of Agriculture Forest Service.

Keever, C. 1950. Causes of succession on old fields of the piedmont, North Carolina. *Ecological Monographs* 20:229–250.

Kelly, G. J., E. Latzko, and M. Gibbs. 1976. Regulatory aspects of photosynthetic carbon metabolism. *Annual Review of Plant Physiology* 27:181–205.

Kelly, J. M. 1975. Dynamics of root biomass in two eastern Tennessee old-field communities. *American Midland Naturalist* 94:54–61.

Kenoyer, L. A. 1927. A study of Raunkiaer's Law of Frequency. *Ecology* 8:341–349.

Kerner, A. 1863. *The plant life of the Danube Basin*. Translated by H. S. Conard. Republished. Ames, IA: Iowa State College Press. 1951.

———. 1895. *The natural history of plants, their forms, growth, reproduction and distribution*. Translated by F. W. Oliver. London: Blackie.

Kershaw, K. A. 1973. *Quantitative and dynamic plant ecology*. 2nd ed. New York: American Elsevier.

Kilgore, B. M. 1972. The role of fire in giant sequoia-mixed conifer forest. In *Proc. American Association for the Advancement of Science Symp., 1971*, pp. 1–38.

Kilgore, B. M. and H. H. Biswell. 1971. Seedling germination following fire in a giant sequoia forest. *California Agriculture Magazine* 25:8–10.

Kira, T. 1975. Primary production of forests. In *Photosynthesis and productivity in different environments*, ed. J. P. Cooper, pp. 5–40. Cambridge, England: Cambridge University Press.

Kira, T., H. Ogawa, K. Yoda, and K. Ogina. 1967. Comparative ecological studies on three main types of forest vegetation in Thailand. IV. Dry matter production, with special reference to the Khao Chong rain forest. *Nature and Life in Southeast Asia* 5:149–174.

Kittredge, J. 1944. Estimation of the amount of foliage on trees and stands. *J. of Forestry* 42:905–912.

Kluge, M. 1974. Metabolism of carbohydrates and organic acids. In *Progress in botany*, ed. H. Ellenberg, et al., vol. 36, pp. 90–98. Berlin: Springer-Verlag.

Knapp, R. 1957. Über den Einfluss der Temperatur während der Keimung auf die spatere Entweicklung einiger annueller Pflanzenarten. *Zeit. Naturforsch.* 126:564–568.

Knipling, E. B. 1967. Measurement of leaf water potential by the dye method. *Ecology* 48:1038–1041.

Knipling, E. B. and P. J. Kramer. 1967. Comparison of the dye method with the thermocouple psychrometer for measuring leaf water potentials. *Plant Physiology* 42:1315–1320.

Koelling, M. R. and C. L. Kucera. 1965. Production and turnover relationships in native tall-grass prairie. *Iowa State J. of Science* 39:387–392.

Komarek, E. V., Sr. 1963. Fire, research, and education. In *Proc. Tall Timbers Fire Ecol. Conf.*, no. 2, pp. 181–187. Tallahassee, FL: Tall Timbers Research Station.

———. 1964. The natural history of lightning. In *Proc. Tall Timbers Fire Ecol. Conf.*, no. 3, pp. 139–184. Tallahassee, FL: Tall Timbers Research Station.

———. 1965. Fire ecology—grasslands and man. In *Proc. Tall Timbers Fire Ecol. Conf.*, no. 4, pp. 169–220. Tallahassee, FL: Tall Timbers Research Station.

———. 1967. Fire—and the ecology of man. In *Proc. Tall Timbers Fire Ecol. Conf.*, no. 6, pp. 143–170. Tallahassee, FL: Tall Timbers Research Station.

———. 1968. Lightning and lightning fires as ecological forces. In *Proc. Tall Timbers Fire Ecol. Conf.*, no. 8, pp. 169–197. Tallahassee, FL: Tall Timbers Research Station.

———. 1969. Fire and animal behavior. In *Proc. Tall Timbers Fire Ecol. Conf.*, no. 9, pp. 161–207. Tallahassee, FL: Tall Timbers Research Station.

———. 1974. Effects of fire on temperate forests and related ecosystems: southeastern United States. In *Fire and ecosystems*, ed. T. T. Kozlowski and C. E. Ahlgren, ch. 8. New York: Academic Press.

Komarkova, V. and P. J. Webber. 1978. An alpine vegetation map of Niwot Ridge, Colorado. *Arctic and Alpine Research* 10:1–30.

Kortschak, H. P., C. E. Hartt, and G. O. Burr. 1965. Carbon dioxide fixation in sugar cane leaves. *Plant physiology* 40:209–213.

Kozlovsky, D. G. 1968. A critical evaluation of the trophic level concept. *Ecology* 49:48–60.

Kramer, P. J. 1969. *Plant and soil water relationships: a modern synthesis.* New York: McGraw-Hill.

Krammes, J.S. and L. F. DeBano. 1965. Soil wettability: a neglected factor in watershed management. *Water Resources Research* 1:283–286.

Krammes, J. S. and J. Osborn. 1969. Water-repellent soils and wetting agents as factors influencing erosion. In *Proc. Symp. on Water-repellent Soils,* ed. L. F. DeBano and J. Letey, pp. 117–187. Riverside, CA: University of California.

Krebs, C. J. 1972. *Ecology: the experimental analysis of distribution and abundance.* New York: Harper and Row.

Krenzer, E. G., Jr., D. N. Moss, and R. K. Crookston. 1975. Carbon dioxide compensation points of flowering plants. *Plant Physiology* 56:194–206.

Krieger, R. I., P. P. Feeny, and C. F. Wilkinson. 1971. Detoxication enzymes in the guts of caterpillars: an evolutionary answer to plant defenses? *Science* 172:579–581.

Kruckeberg, A. R. 1954. The ecology of serpentine soils. III. Plant species in relation to serpentine soils. *Ecology* 35:267–274.

Kucera, C. L., R. C. Dahlman, and M. R. Koelling. 1967. Total net productivity and turnover on an energy basis for tallgrass prairie. *Ecology* 48:536–541.

Küchler, A. W. 1964. *The potential natural vegetation of the conterminous United States.* American Geographical Society Special Publ. no. 36. New York: American Geographical Society.

——. 1967. *Vegetation mapping.* New York: Ronald Press.

——. 1974. A new vegetation map of Kansas. *Ecology* 55:586–604.

——. 1977. Potential natural vegetation of California. In *Terrestrial vegetation of California,* ed. M. G. Barbour and J. Major, pp. 909–938 and map. New York: Wiley-Interscience.

Küchler, A. W. and J. McCormick. 1965. *International bibliography of vegetation maps. Volume I: North America.* Lawrence, KS: University of Kansas Press.

Kuhlman, H. 1977. *Regeneration of a knobcone pine forest after a fire.* San Jose State University Biology 160 (General Ecology). Unpublished.

Kuijt, J. 1969. *The biology of parasitic flowering plants.* Berkeley, CA: University of California Press.

Kyriakopoulos, E. and H. Richter. 1976. A comparison of methods for the determination of water status in *Quercus ilex* L. *Zhurnal Pflanzenphysiologie* 82:14–27.

Laetsch, W. M. 1974. The C_4 syndrome: a structural analysis. *Annual Review of Plant Physiology* 25:1974.

LaMarche, V. C., Jr. 1974. Paleoclimatic inferences from long tree-ring records. *Science* 183:1043–1048.

Landsberg, H. E., H. Lippmann, K. H. Paffen, and C. Troll. 1966. *World maps of climatology.* Berlin: Springer-Verlag.

Lang, G. E. 1974. Litter dynamics in a mixed oak forest on the New Jersey Piedmont. *Bulletin of the Torrey Botanical Club* 101:277–286.

Lange, O. L. and L. Kappen. 1972. Photosynthesis of lichens from Antarctica. p. 83–95 In *Antarctic terrestrial biology,* ed. G. A. Llano, pp. 83–95. Antarctic Research Series, vol. 20. Washington, D.C.: American Geophysical Union.

Lange, O. L., R. Lösch, E. D. Schulze, and L. Kappen. 1971. Responses of stomata to changes in humidity. *Planta* 100:76–86.

Langlet, O. 1959. A cline or not a cline—a question of scots pine. *Sylvae Genetica* 8:13–22.

Larcher, W. 1975. *Physiological plant ecology.* Translated by M. A. Bierdman-Thorson. New York: Springer-Verlag.

La Roi, G. H. 1967. Ecological studies in the boreal spruce-fir forests of the North American taiga. I. Analysis of the vascular flora. *Ecological Monographs* 37:229–253.

Larsen, J. A. 1922. Effect of removal of the virgin white pine stand upon the physical factors of site. *Ecology* 3:302–305.

Laudermilk, J. D. and P. A. Munz. 1938. *Plants in the dung of Nothrotherium from Gypsum Cave, Nevada.* In Carnegie Institution of Washington Publ. 453, pp. 29–37. Washington, D.C.: Carnegie Institution of Washington.

Lawson, E. R. 1967. Throughfall and stemflow in a pine-hardwood stand in the Ouachita Mountains of Arkansas. *Water Resources Research* 3:731–735.

Lee, J. J. and D. L. Inman. 1975. The ecological role of consumers—an aggregated systems view. *Ecology* 56:1455–1458.

Lee, R. 1969. Chemical temperature determination. *J. of Applied Meteorology* 8:423–430.

Lemon, E. R. 1960. Photosynthesis under field conditions. II. An aerodynamic method for determining the turbulent carbon dioxide exchange between the atmosphere and a corn field. *Agronomy J.* 52:697–703.

Lemon, P. C. 1968. Fire and wildlife grazing on an African plateau. In *Proc. Tall Timbers Fire Ecol. Conf.*, no. 8, pp. 71–88. Tallahassee, FL: Tall Timbers Research Station.

Leonard, R. E. 1961. Net precipitation in a northern hardwood forest. *J. of Geophysical Research* 66:2417–2421.

Leopold, A. 1966. *A sand county almanac.* San Francisco/New York: Sierra Club/Ballantine Books.

Leopold, A. C. and P. E. Kriedemann. 1975. *Plant growth and development.* 2nd ed. New York: McGraw-Hill.

Levin, S. A. 1976. Alkaloid-bearing plants: an ecological perspective. *The American Naturalist* 110:261–284.

Levitt, J. 1974. The mechanism of stomatal movement—once more. *Protoplasma* 82:1–17.

———. 1976. Physiological basis of stomatal response. In *Water and plant life: problems and modern approaches*, ed. O. L. Lange, L. Kappen, and E. D. Schulze, pp. 160–168. Ecological Studies, vol. 19. Berlin: Springer-Verlag.

Liebig, J. 1840. *Chemistry in its agriculture and physiology.* London: Taylor and Walton.

Lieth, H. 1968. The determination of plant dry-matter production with special emphasis on the underground parts. In *Proc. Copenhagen Symp.: Functioning of terrestrial ecosystems at the primary production level*, ed. F. E. Eckard. *Natural Resources Research* 5:179–186. Paris: UNESCO.

———. 1973. Primary production: terrestrial ecosystems. *Human Ecology* 1:303–332.

Lieth, H. and R. H. Whittaker, eds. 1975. *The primary production of the biosphere.* New York: Springer-Verlag.

Likens, G. E. and F. H. Bormann. 1972. Nutrient cycling in ecosystems. In *Ecosystem structure and function*, ed. J. A. Weins, pp. 25–67. Corvallis, OR: Oregon State University Press.

Likens, G. E., F. H. Bormann, N. M. Johnson, W. D. Fisher, and R. S. Pierce. 1970. Effects of forest cutting and herbicide treatment on nutrient budgets in the Hubbard Brook watershed-ecosystem. *Ecological Monographs* 40:23–47.

Likens, G. E., F. H. Bormann, N. M. Johnson, and R. S. Pierce. 1967. The calcium, magnesium, potassium, and sodium budgets for a small forested ecosystem. *Ecology* 48:772–785.

Likens, G. E., F. H. Bormann, R. S. Pierce, J. S. Eaton and N. M. Johnson. 1977. *Biogeochemistry of a forested ecosystem.* New York: Springer-Verlag.

Lindsey, A. A. 1955. Testing and line-strip method against full tallies in diverse forest types. *Ecology* 36:485–495.

———. 1956. Sampling methods and community attributes in forest ecology. *Forest Science* 2:287–296.

Lindsey, A. A., J. D. Barton, and S. R. Miles. 1958. Field efficiencies of forest sampling methods. *Ecology* 39:428–444.

Lindsey, A. A. and J. O. Sawyer, Jr. 1971. Vegetation-climate relationships in the Eastern United States. *Proc. Indiana Academy of Science* 80:210–214.

Livingstone, D. A. 1968. Some interstadial and postglacial pollen diagrams from eastern Canada. *Ecological Monographs* 38:87–125.

Lloyd, M. G. 1961. The contribution of dew to the summer water budget of northern Idaho. *Bulletin of the American Meteorological Society* 42:572–580.

Loach, K. 1967. Shade tolerance in tree seedlings. I. Leaf photosynthesis and respiration in plants raised under artificial shade. *New Phytologist* 66:607–621.

Logan, K. T. 1970. Adaptations of the photosynthetic apparatus of sun- and shade-grown yellow birch (*Betula alleghaniensis* Britt.). *Canadian J. of Botany* 48:1681–1688.

Longman, K. A. and J. Jenik. 1974. *Tropical forest and its environment.* London: Longman.

Lorimer, C. G. 1977. The presettlement forest and natural disturbance cycle of northeastern Maine. *Ecology* 58:139–148.

Lotan, J. E. 1974. Cone serotiny-fire relationships in lodgepole pine. In *Proc. Tall Timbers Fire Ecol. Conf.*, no. 14, pp. 267–278. Tallahassee, FL: Tall Timbers Research Station.

Lotka, A. J. 1925. *Elements of physical biology.* Baltimore, MD: Williams and Wilkins.

Loucks, O. L. 1970. Evolution of diversity, efficiency, and community stability. *American Zoologist* 10:17–25.

Lowry, W. P. 1969. *Weather and Life.* New York: Academic Press.

Lutz, H. J. 1930. The vegetation of Heart's Content, a virgin forest in northwestern Pennsylvania. *Ecology* 11:1–29.

MacArthur, R. H. 1958. Population ecology of some warblers of northeastern coniferous forests. *Ecology* 39:599–619.

———. 1962. Some generalized theorems of natural selection. *Proc. National Academy of Sciences* 48:1893–1897.

———. 1972. *Geographical ecology: patterns in the distribution of species.* New York: Harper and Row.

Machta, L. 1972. Mauna Loa and global trends in air quality. *American Meteorological Society Bulletin* 53:402–420.

Macior, L. W. 1973. The pollination ecology of *Pedicularis* on Mount Ranier. *American J. of Botany* 60:863–871.

Mack, R. N. and J. L. Harper. 1977. Interference in dune annuals: spatial pattern and neighborhood effects. *J. of Ecology* 65:345–363.

Madgwick, H. A. I. 1968. Seasonal changes in biomass and annual production of an old-field *Pinus virginiana* stand. *Ecology* 49:149–152.

Madgwick, H. A. I. and T. Satoo. 1975. On estimating the aboveground weights of tree stands. *Ecology* 56:1446–1450.

Major, J. 1969. Historical development of the ecosystem concept. In *The ecosystem concept in natural resource management*, ed. G. M. Van Dyne, pp. 9–22. New York: Academic Press.

———. 1974. Kinds and rates of changes in vegetation and chronofunctions. In *Handbook of vegetation science, part VIII, vegetation dynamics*, ed. R. Knapp, pp. 9–18. The Hague: Junk.

———. 1977. California climate in relation to vegetation. In *Terrestrial vegetation of California*, ed. M. G. Barbour and J. Major, pp. 11–74. New York: Wiley-Interscience.

Malcolm, W. M. 1966. Biological interactions. *Botanical Review* 32:243–254.

Margalef, R. 1968. *Perspectives in ecological theory.* Chicago: University of Chicago Press.

Marks, G. C. and T. T. Kozlowski, eds. 1973. *Ectomycorrhizae.* New York: Academic Press.

Marsh, G. P. 1864. *Man and nature.* Reprint 1965. Ed. D. Lowenthal. Cambridge, MA: Harvard University Press.

Marshall, D. R. and S. K. Jain. 1969. Interference in pure and mixed populations of *Avena fatua* and *A. barbata*. *J. of Ecology* 57:251–270.

Matthaei, G. L. C. 1905. Experimental researches on vegetable assimilation and respiration. III. On the effect of temperature on carbon-dioxide assimilation. *Philosophical Transactions of the Royal Society of London, Series B* 197:47–105.

Mattson, W. J. and N. D. Addy. 1975. Phytophagous insects as regulators of forest primary production. *Science* 190:515–522.

Maycock, P. F. 1967. Jozef Paczoski: founder of the science of phytosociology. *Ecology* 48:1031–1034.

Mayr, E. 1970. *Populations, species, and evolution.* Cambridge, MA: Belknap Press of Harvard University Press.

McBride, J. and E. C. Stone. 1976. Plant succession on the sand dunes of the Monterey Peninsula, California. *American Midland Naturalist* 96:118–132.

McCormick, J. 1968. Succession. *Via* 1:22–35, 131–132. Philadelphia, PA: Graduate School of Fine Arts, University of PA.

McCormick, J. and M. F. Buell. 1968. The plains: pygmy forests of the New Jersey pine barrens, a review and annotated bibliography. *Bulletin New Jersey Academy of Science* 13:20–34.

McCormick, J. and P. A. Harcombe. 1968. Phytograph: useful tool or decorative doodle? *Ecology* 49:13–20.

McCutchan, M. H. 1977. Climatic features as a fire determinant. In *Proc. Symp. Env. Cons. Fire and Fuel Mangmt. in Medit. Ecosyst.*, pp. 1–11. Washington, D.C.: U.S. Dept. of Agriculture Forest Service.

McDonald, C. D. and G. H. Hughes. 1964. *Studies of consumptive use of water by phreatophytes and hydrophytes near Yuma, Arizona.* Geological Survey Professional Paper 486-F. Washington, D.C.: U.S. Geological Survey.

McDougall, W. B. 1927. *Plant ecology.* Philadelphia: Lea and Febiger.

———. 1922. Symbiosis in a deciduous forest, part I. *Botanical Gazette* 73:200–212.

———. 1925. Symbiosis in a deciduous forest, part II. *Botanical Gazette* 79:95–102.

McIntosh, R. P. 1962. Raunkiaer's "Law of Frequency." *Ecology* 43:533–555.

———. 1974. Plant ecology, 1947–1972. *Annals of the Missouri Botanical Garden* 61:132–165.

———. 1976. Ecology since 1900. In *Evolution of issues, ideas and events in America: 1776–1976,* ed. B. J. Taylor, pp. 353–372. Norman, OK: University of Oklahoma Press.

McMillan, C. 1956. The edaphic restriction of *Cupressus* and *Pinus* in the Coast Ranges of central California. *Ecological Monographs* 26:177–212.

McNaughton, S. J. 1966. Thermal inactivation properties of enzymes from *Typha latifolia* L. ecotypes. *Plant Physiology* 41:1736–1738.

———. 1967. Photosynthetic system II: racial differentiation in *Typha latifolia. Science* 156:1363.

———. 1972. Enzymatic thermal adaptations: the evolution of homeostasis in plants. *The American Naturalist* 106:165–172.

———. 1973. Comparative photosynthesis of Quebec and California ecotypes of *Typha latifolia. Ecology* 54:1260–1270.

McPherson, J. K. and C. H. Muller. 1969. Allelopathic effects of *Adenostoma fasciculatum,* "chamise," in the California chaparral. *Ecological Monographs* 39:177–198.

Meidner, H. and T. A. Mansfield. 1968. *Physiology of stomata.* New York: McGraw-Hill.

Meidner, H. and D. W. Sheriff. 1976. *Water and plants.* New York: Wiley.

Melin, E. 1930. Biological decomposition of some types of litter from North American forests. *Ecology* 11:72–101.

Merriam, C. H. 1890. Results of a biological survey of the San Francisco Mountain region and desert of the Little Colorado, Arizona. *North American Fauna* 3:1–136.

———. 1894. Laws of temperature control of the geographic distribution of animals and plants. *National Geographic* 6:229–238.

———. 1898. Life zones and crop zones of the United States. *U.S. Dept. of Agriculture Biological Survey Division Bulletin* 10:9–79.

Meyer, F. H. 1973. Ectomycorrhizae. In *Ectomycorrhizae,* eds. G. C. Marks and T. T. Kozlowski. New York: Academic Press.

Miller, A. and J. C. Thompson. 1975. *Elements of meteorology.* 2nd ed. Columbus, OH: Merrill.

Miller, C. E. 1938. *Plant physiology.* New York: McGraw-Hill.

Miller, P. C. and L. Tieszen. 1972. A preliminary model of processes affecting primary production in the arctic tundra. *Arctic and Alpine Research* 4:1–18.

Miller, P. R. 1969. Air pollution and the forests of California. *California Air Environment* 1:1–3.

Milthorpe, F. L. 1955. The significance of the measurements made by the cobalt chloride paper method. *J. of Experimental Botany* 6:17–19.

Moldenke, A. R. 1975. Niche specialization and species diversity along a California transect. *Oecologia* 21:219–242.

Monk, C. D. 1966. An ecological significance of evergreenness. *Ecology* 47:504–505.

Monteith, J. L. 1973. *Principles of environmental physics.* London: Arnold.

Mooney, H. A. 1972. The carbon balance of plants. *Annual Review of Ecology and Systematics* 3:315–346.

———. 1977a. Southern coastal scrub. In *Terrestrial vegetation of California.* ed. M. G. Barbour and J. Major, pp. 471–489. New York: Wiley-Interscience.

———. ed. 1977b. *Convergent evolution in Chile and California: Mediterranean climate ecosystems.* Stroudsburg, PA: Dowden, Hutchinson, and Ross.

Mooney, H. A. and W. D. Billings. 1961. Comparative physiological ecology of arctic and alpine populations of *Oxyria digyna. Ecological Monographs* 31:1–29.

Mooney, H. A. and E. L. Dunn. 1970a. Convergent evolution of Mediterranean-climate evergreen sclerophyll shrubs. *Evolution* 24:292–303.

Mooney, H. A. and E. L. Dunn. 1970b. Photosynthetic systems of Mediterranean-climate shrubs and trees of California and Chile. *The American Naturalist* 104:447–453.

Mooney, H. A., J. Ehleringer, and J. Berry. 1976. High photosynthetic capacity of a winter annual in Death Valley. *Science* 194:322–323.

Mooney, H. A., J. Ehleringer, and O. Björkman. 1977. The energy balance of leaves of the evergreen desert shrub *Atriplex hymenelytra. Oecologia* 29:301–310.

Mooney, H. A. and A. T. Harrison. 1970. The influence of conditioning temperature on subsequent temperature-related photosynthetic capacity in higher plants. In *Prediction and measurement of photosynthetic productivity*, ed. C. T. de Wit, pp. 411–417. Wageningen, The Netherlands: Center for Agricultural Publishing and Documentation.

Mooney, H. A. and M. West. 1964. Photosynthetic acclimation of plants of diverse origin. *American J. of Botany* 51:825–827.

Moore, R. M., ed. 1970. *Australian grasslands.* Canberra, Australia: Australian National University Press.

Mueller-Dombois, D. and H. Ellenberg. 1974. *Aims and methods of vegetation ecology.* New York: Wiley.

Muir, J. 1878. The new sequoia forest of California. *Harpers* 57:813–827.

———. 1894. *The mountains of California.* Garden City, NY: Doubleday.

Muller, C. H. 1953. The association of desert annuals with shrubs. *American J. of Botany* 40:53–60.

———. 1966. The role of chemical inhibition (allelopathy) in vegetational composition. *Bulletin of the Torrey Botanical Club* 93:332–351.

Muller, C. H., R. B. Hanawalt, and J. K. McPherson. 1968. Allelopathic control of herb growth in the fire cycle of California chaparral. *Bulletin of the Torrey Botanical Club* 95:225–231.

Muller, W. H. and C. H. Muller. 1956. Association patterns involving desert plants that contain toxic products. *American J. of Botany* 43:354–361.

Mulroy, T. W. and P. W. Rundel. 1977. Annual plants: adaptations to desert environments. *BioScience* 27:109–114.

Mutch, R. W. 1970. Wildland fires and ecosystems—a hypothesis. *Ecology* 51:1046–1051.

Nash, R. 1973. *Wilderness and the American mind.* Revised ed. New Haven, CT: Yale University Press.

Neales, T. F. 1973. The effect of night temperatures on CO_2 assimilation, transpiration, and water use efficiency in *Agave americana* L. *Australian J. of Biological Science* 26:705–714.

Newbould, P. J. 1967. *Methods for estimating the primary production of forests.* IBP Handbook no. 2. Oxford: Blackwell Scientific Publ.

———. 1968. Methods of estimating root production. In *Proc. Copenhagen Symp.: Functioning of terrestrial ecosystems at the primary production level*, ed. F. E. Eckhard. *Natural Resources Research* 5:187–190. Paris: UNESCO.

Newell, S. J. and E. J. Tramer. 1978. Reproductive strategies in herbaceous plant communities during succession. *Ecology* 59:228–234.

Niering, W. A., R. H. Whittaker, and C. H. Lowe. 1963. The saguaro: a population in relation to its environment. *Science* 142:15–23.

Nixon, S. W., C. A. Oviatt, J. Garber, and V. Lee. 1976. Diel metabolism and nutrient dynamics in a salt marsh embayment. *Ecology* 57:740–750.

Nobel, P. S. 1974a. *Introduction to biophysical plant physiology.* San Francisco: W. H. Freeman.

———. 1974b. Boundary layers of air adjacent to cylinders: estimation of effective thickness and measurements on plant material. *Plant Physiology* 54:177–181.

———. 1976. Water relations and photosynthesis of a desert CAM plant, *Agave deserti. Plant Physiology* 58:576–582.

———. 1977. Water relations and photosynthesis of barrel cactus, *Ferocactus acanthodes*, in the Colorado Desert. *Oecologia* 27:117–133.

Noy-Meir, I. 1975. Stability of grazing systems: an application of predator-prey graphs. *J. of Ecology* 63:459–481.

Nye, P. H. 1961. Organic matter and nutrient cycles under moist tropical forests. *Plant and Soil* 13:333–346.

Oberlander, G. T. 1956. Summer fog precipitation on the San Francisco peninsula. *Ecology* 37:851–852.

Odening, W. R., B. R. Strain, and W. C. Oechel. 1974. The effect of decreasing water potential on net CO_2 exchange of intact desert shrubs. *Ecology* 55:1086–1095.

Odum, E. P. 1963. Limits of remote ecosystems containing man. *American Biology Teacher* 25:429–443.

———. 1969. The strategy of ecosystem development. *Science* 164:262–270.

———. 1971. *Fundamentals of ecology.* 3rd ed. Philadelphia, PA: W. B. Saunders.

Ødum, S. 1965. Germination of ancient seeds: floristic observations and experiments with archeologically dated soil samples. *Dansk Botanik Arkiven* 24(2):1–70.

Oechel, W. C., B. R. Strain, and W. R. Odening. 1972. Tissue water potential, photosynthesis, ^{14}C-labeled photosynthetic utilization and growth in the desert shrub *Larrea divaricata. Ecological Monographs* 42:127–141.

Olmsted, C. E. III and E. L. Rice. 1970. Relative effects of known plant inhibitors on species from the first two stages of old-field succession. *Southwestern Naturalist* 15:165–173.

Olson, J. S. 1958. Rates of succession and soil changes on southern Lake Michigan sand dunes. *Botanical Gazette* 119: 125–170.

———. 1963. Energy storage and balance of producers and decomposers in ecological systems. *Ecology* 44:322–331.

Oosting, H. J. 1942. An ecological analysis of the plant communities of Piedmont, North Carolina. *American Midland Naturalist* 28:1–126.

———. 1956. *The study of plant communities.* 2nd ed. San Francisco, CA: Freeman.

Oosting, H. J. and W. D. Billings. 1951. A comparison of virgin spruce-fir forest in the northern and southern Appalachian system. *Ecology* 32:84–103.

Oppenheimer, H. R. 1960. Adaptation to drought: xerophytism. In *Plant-water relationships in arid and semi-arid conditions,* pp. 105–138. Paris: UNESCO.

Orloci, L. 1966. Geometric models in ecology. I. The theory and application of some ordination methods. *J. of Ecology* 54:193–215.

Orshan, G. 1972. Morphological and physiological plasticity in relation to drought. In *Wildland shrubs—their biology and utilization,* ed. C. M. McKell, J. P. Blaisdell, and J. R. Goodin, pp. 245–254. U.S. Dept. of Agriculture Forest Service General Technical Report INT-1. Ogden, UT: U.S. Dept. of Agriculture Forest Service.

Osmond, C. B. 1976. CO_2 assimilation and dissimilation in the light and dark in CAM plants. In *CO_2 metabolism and plant productivity,* ed. R. H. Burris and C. C. Black, pp. 217–233. Baltimore, MD: University Park Press.

Osonubi, O. and W. J. Davies. 1978. Solute accumulation in leaves and roots of woody plants subjected to water stress. *Oecologia* 32:323–332.

Ovington, J. D., D. Heitkamp, and D. B. Lawrence. 1963. Plant biomass and productivity of prairie, savanna, oakwood and maize field ecosystems in central Minnesota. *Ecology* 44:52–63.

Paczoski, J. 1921. *Osnowy fitosocjologji.* Cherson, Izd. Stud. Comitet Tech.

Parkinson, K. J. and B. J. Legg. 1972. A continuous flow porometer. *J. of Applied Ecology* 9:669–675.

Parmeter, J. R. Jr. 1977. Effects of fire on pathogens. In *Proc. Symp. Env. Cons. Fire and Fuel Mangmt. in Medit. Ecosyst.,* pp. 58–64. Washington, D.C.: U.S. Dept. of Agriculture Forest Service.

Parrish, J. A. D. and F. A. Bazzaz. 1976. Underground niche separation in successional plants. *Ecology* 57:1281–1288.

Parsons, D. J. 1976. Vegetation structure in the Mediterranean scrub communities of California and Chile. *J. of Ecology* 64:435–447.

Parsons, D. J. and A. R. Moldenke. 1975. Convergence in vegetation structure along analogous climatic gradients in California and Chile. *Ecology* 56:950–957.

Pearcy, R. W. 1976. Temperature responses of growth and photosynthetic CO_2 exchange rates in coastal and desert races of *Atriplex lentiformis. Oecologia* 26:245–255.

Pearcy, R. W. and A. T. Harrison. 1974. Comparative photosynthetic and respiratory gas exchange characteristics of *Atriplex lentiformis* (Torr.) Wats. in coastal and desert habitats. *Ecology* 55:1104–1111.

Pearson, G. A. 1942. Herbaceous vegetation a factor in natural regeneration of ponderosa pine in the southwest. *Ecological Monographs* 12:315–338.

Peet, R. K. 1974. The measurement of species diversity. *Annual Review of Ecology and Systematics* 5:285–307.

———. 1975. Relative diversity indices. *Ecology* 56:496–498.

Penfound, W. T. 1964. Effects of denudation on the productivity of grassland. *Ecology* 45:838–845.

Penman, H. L. 1950. Evaporation over the British Isles. *Quarterly J. of the Royal Meteorological Society* 76:372–383.

Perino, J. V. and P. G. Risser. 1972. Some aspects of structure and function in Oklahoma old-field succession. *Bulletin of the Torrey Botanical Club* 99:233–239.

Peters, G. A. 1978. Blue-green algae and algal associations. *BioScience* 28:580–585.

Petrusewicz, K. and W. L. Grodzinski. 1975. The role of herbivore consumers in various ecosystems. In *Productivity of world ecosystems*, pp. 64–70. Washington, D.C.: National Academy of Sciences.

Phillips, E. A. 1959. *Methods of vegetation study.* New York: Holt, Rinehart, and Winston.

Phillips, H. 1965. Fire—as master and servant: its influence in the bioclimatic regions of Trans-Saharan Africa. In *Proc. Tall Timbers Fire Ecol. Conf.*, no. 4, pp. 7–109. Tallahassee, FL: Tall Timbers Research Station.

———. 1974. Effects of fire in forest and savanna ecosystems of Sub-Saharan Africa. In *Fire and ecosystems*, ed. T. T. Kozlowski and C. E. Ahlgren, ch. 13. New York: Academic Press.

Philpot, C. W. 1977. Vegetative features as determinants of fire frequency and intensity. In *Proc. Symp. Env. Cons. Fire and Fuel Mangmt. in Medit. Ecosyst.*, pp. 12–16. Washington, D.C.: U.S. Dept. of Agriculture Forest Service.

Pianka, E. R. 1970. On r- and K-selection. *American Naturalist* 104:592–597.

Pickett, S. T. A. and F. A. Bazzaz. 1976. Divergence of two co-occurring successional annuals on a soil moisture gradient. *Ecology* 57:169–176.

Pielou, E. C. 1961. Segregation and symmetry in two-species populations as studied by nearest neighbor relations. *J. of Ecology* 49:255–269.

Pielou, E. C. 1969. *An introduction to mathematical ecology.* New York: Wiley-Interscience.

Pike, L. H., D. M. Tracy, M. A. Sherwood, and D. Nielsen. 1972. Estimates of biomass and fixed nitrogen of epiphytes from old-growth Douglas-fir. In *Proc.—research on coniferous forest ecosystems—a symp.*, ed. J. F. Franklin, L. J. Dempster, and R. H. Waring, pp. 177–187. Portland, OR: Pacific Northwest Forest and Range Experiment Station. Washington, D.C.: U.S. Dept. of Agriculture Forest Service.

Pinder, J. E. III. 1975. Effects of species removal on an old-field plant community. *Ecology* 56:747–751.

Pomeroy, L. R. 1970. The strategy of mineral cycling. *Annual Review of Ecology and Systematics* 1:171–190.

Poore, M. E. D. 1955a. The use of phytosociological methods in ecological investigations. I. The Braun-Blanquet system. *J. of Ecology* 43:226–244.

———. 1955b. The use of phytosociological methods in ecological investigations. II. Practical issues involved in an attempt to apply the Braun-Blanquet system. *J. of Ecology* 43:245–269.

Porsild, A. E., C. R. Harington, and G. A. Mulligan. 1967. *Lupinus arcticus* Wats. grown from seeds of Pleistocene age. *Science* 158:113–114.

Post, L. J. 1970. Dry matter production of mountain maple and balsam fir in northwestern New Brunswick. *Ecology* 51:548–550.

Potzger, J. E., M. E. Potzger, and J. McCormick. 1956. The forest primeval of Indiana as recorded in the original land surveys and an evaluation of previous interpretations of Indiana vegetation. *Butler University Botanical Studies* 13:95–111.

Preston, F. W. 1948. The commonness and rarity of species. *Ecology* 29:254–283.

Quick, C. R. 1959. Ceanothus seeds and seedlings on burns. *Madroño* 15:79–81.

Rabotnov, T. A. 1969. On coenopopulations of perennial herbaceous plants in natural coenoses. *Vegetatio* 19:87–95.

Ranwell, D. S. 1972. *Ecology of salt marshes and sand dunes.* London: Chapman and Hall.

Raschke, K. 1976. How stomata resolve the dilemma of opposing priorities. *Philosophical Transactions of the Royal Society of London, Series B* 273:551–560.

Raunkiaer, C. 1918. Récherches statistiques sur les formations végétales. *Kgl. Danske Videnskabernes Selskab Biologiske Meddelelser* 1:1–47.

———. 1934. *The life forms of plants and statistical plant geography.* Oxford: Clarendon Press.

Raven, P. H. and H. Curtis. 1970. *Biology of plants.* New York: Worth.

Rawlins, S. L. 1976. Measurement of water content and the state of water in soils. In *Water deficits and plant growth,* vol. IV, ed. T. T. Kozlowski, pp. 1–55. New York: Academic Press.

Reardon, P. O., C. L. Leinweber, and L. B. Merrill. 1972. The effect of bovine saliva on grasses. *J. of Animal Science* 34:897–898.

Reardon, P. O., C. L. Leinweber, and L. B. Merrill. 1974. Response of sideoats grama to animal saliva and thiamine. *J. of Range Management* 27:400–401.

Redmann, R. E. 1975. Production ecology of grassland plant communities in western North Dakota. *Ecological Monographs* 45:83–106.

Reed, H. S. 1942. *A short history of the plant sciences.* New York: Ronald Press.

Reichle, D. E., R. A. Goldstein, R. I. Van Hook, Jr., and G. J. Dodson. 1973. Analysis of insect consumption in a forest canopy. *Ecology* 54:1076–1084.

Reimold, R. J. and W. H. Queen. 1974. *Ecology of halophytes.* New York: Academic Press.

Reiners, W. A. 1972. Structure and energetics of three Minnesota forests. *Ecological Monographs* 42:71–94.

Reiners, W. A. and N. M. Reiners. 1970. Energy and nutrient dynamics of forest floors in three Minnesota forests. *J. of Ecology* 58:497–519.

Reuss, J. O. and G. S. Innis. 1977. A grassland nitrogen flow simulation model. *Ecology* 58:379–388.

Rhoades, D. H. and R. G. Cates. 1976. Toward a general theory of plant antiherbivore chemistry. *Recent Advances in Phytochemistry* 10:168–213.

Rice, E. L. 1964. Inhibition of nitrogen-fixing and nitrifying bacteria by seed plants, part I. *Ecology* 45:824–837.

———. 1965. Inhibition of symbiotic nitrogen-fixing and nitrifying bacteria by seed plants, part II. *Physiologia Plantarum* 18:255–268.

———. 1974. *Allelopathy.* New York: Academic Press.

Rice, E. L. and R. W. Kelting. 1955. The species-area curve. *Ecology* 36:7–11.

Rice, E. L. and R. L. Parenti. 1978. Causes of decreases in productivity in undisturbed tall grass prairie. *American J. of Botany* 65:1091–1097.

Richards, L. A. and G. Ogata. 1958. Thermocouple for vapor pressure measurements in biological and soil systems at high humidity. *Science* 128:1089–1090.

Richards, P. W. 1936. Ecological observations on the rain forest of Mount Dulit, Sarawak. *J. of Ecology* 24:1–37 and 340–360.

———. 1973. The tropical rain forest. *Scientific American* 229(6):58–67.

Richardson, J. L. 1977. *Dimensions of ecology.* Baltimore, MD: Williams and Wilkins.

Ricklefs, R. E. 1973. *Ecology.* Newton, MA: Chiron Press.

Riley, D. and A. Young. 1974. *World vegetation.* London: Cambridge University Press.

Ritchie, G. A. and T. M. Hinckley. 1975. The pressure chamber as an instrument for ecological research. *Advances in Ecological Research* 9:166–254.

Roberts, S. W. and K. R. Knoerr. 1977. Components of water potential estimated from xylem pressure measurements in five tree species. *Oecologia* 28:191–202.

Robinson, T. 1967. *The organic constituents of higher plants.* 2nd ed. Minneapolis, MN: Burgess.

Rochow, J. J. 1974. Litter fall relations in a Missouri forest. *Oikos* 25:80–85.

Rockwood, L. L. 1974. Seasonal changes in the susceptibility of *Crescentia alata* leaves to the flea beetle, *Oedionyehus* sp. *Ecology* 55:142–148.

———. 1976. Plant selection and foraging patterns in two species of leaf cutting ants (Atta). *Ecology* 57:48–61.

Rodin, L. E. and N. I. Bazilevich. 1967. *Production and mineral cycling in terrestrial vegetation.* Edinbergh, Scotland: Oliver and Boyd.

Rodin, L. E., and N. I. Bazilevich. 1968. World distribution of plant biomass. In *Proc. Copenhagen Symp.: functioning of terrestrial ecosystems at the primary production level,* ed. F. E. Eckhard. *Natural Resources Research* 5:45–52. Paris: UNESCO.

Rook, D. A. 1969. The influence of growing temperature on photosynthesis and respiration of *Pinus radiata* seedlings. *New Zealand J. of Botany* 7:43–55.

Rorison, I. H., ed. 1969. *Ecological aspects of the mineral nutrition of plants.* Oxford: Blackwell.

Rosenberg, N. J. 1974. *Micro-climate: the biological environment.* New York: Wiley.

Rosenzweig, M. L. 1968. Net primary productivity of terrestrial communities: prediction from climatological data. *American Naturalist* 102:67–74.

Rowe, J. S. 1964. Environmental preconditions with special reference to forestry. *Ecology* 45:399-403.

Rozema, J. 1975. The influence of salinity, inundation, and temperature on the germination of some halophytes and nonhalophytes. *Oecologia Plantarum* 10:341-353.

Rübel, E. 1930. *Pflanzengesellschaften der Erde.* Berlin: H. Huber.

Rundel, P. W. 1971. Community structure and stability in the giant Sequoia groves of the Sierra Nevada, California. *American Midland Naturalist* 85:478-492.

Rundel, P. W. 1974. Water relations and morphological variation in *Ramalina menziesii* Tayl. *The Bryologist* 77:23-32.

Rundel, P. W., D. J. Parsons, and D. T. Gordon. 1977. Montane and subalpine vegetation of the Sierra Nevada and Cascade Ranges. In *Terrestrial vegetation of California,* ed. M. G. Barbour and J. Major, ch. 17. New York: Wiley-Interscience.

Rundel, P. W. and R. E. Stecker. 1977. Morphological adaptations of tracheid structure to water stress gradients in the crown of *Sequoiadendron giganteum. Oecologia* 27:135-139.

Rutter, A. J. 1975. The hydrological cycle in vegetation. In *Vegetation and the atmosphere,* vol. 1, ed. J. L. Monteith, pp. 111-154. New York: Academic Press.

Saeki, T. 1973. Distribution of radiant energy and CO_2 in terrestrial communities. In *Photosynthesis and productivity in different environments,* ed. J. P. Cooper, pp. 297-322. London: Cambridge University Press.

Salisbury, E. 1952. *Downs and dunes.* London: G. Bell and Sons.

Salisbury, F. B. and C. Ross. 1969. *Plant physiology.* Belmont, CA: Wadsworth.

Sampson, A. W. 1944. Plant succession on burned chaparral lands in northern California. *California Agriculture Experiment Station Bulletin* 685:1-144.

Sanders, F. E., B. Mosse, and P. B. Tinker, eds. 1976. *Endomycorrhizas.* New York: Academic Press.

Sarukhan, J. and J. L. Harper. 1973. Studies on plant demography: *Ranunculus repens* L., *R. bulbosa* L., and *R. acris* L. *J. of Ecology* 61:675-716.

Saunier, R. E. and R. F. Wagle. 1965. Root grafting in *Quercus turbinella* Greene. *Ecology* 46:749-750.

Savage, J. M. 1963. *Evolution.* New York: Holt, Rinehart and Winston.

Schimper, A. F. W. 1898. *Pflanzengeographie auf physiologischer Grundlage.* Jena, Germany: Fischer.

———. 1903. Plant geography upon a physiological basis. Translated by W. R. Fisher. Oxford: Clarendon Press.

Schlesinger, W. H. and P. L. Marks. 1977. Mineral cycling and the niche of Spanish moss, *Tillandsia usneoides* L. *American J. of Botany* 64: 1254-1262.

Scholander, P. F. 1968. How mangroves desalinate seawater. *Physiologia Plantarum* 21:251-261.

Scholander, P. F., H. T. Hammel, E. A. Hemmingsen, and E. D. Bradstreet. 1964. Hydrostatic pressure and osmotic potential in leaves of mangroves and some other plants. *National Academy of Sciences* 52:119-125.

Schouw, J. F. 1822. *Grundtraek til en almindelig plantegeographie.* Copenhagen, Denmark: Glydendal.

Schultz, A. M. 1969. A study of an ecosystem: the arctic tundra. In *The ecosystem concept in natural resource management,* ed. G. M. Van Dyne, pp. 77-93. New York: Academic Press.

Schultz, A. M., R. P. Gibbens, and L. DeBano. 1961. Artificial populations for teaching and testing range techniques. *J. of Range Management* 14:237-242.

Schulze, E. D., O. L. Lange, U. Buschbom, L. Kappen, and M. Evenari. 1972. Stomatal responses to changes in humidity in plants growing in the desert. *Planta* 108:259-270.

Schulze, E. D., O. L. Lange, L. Kappen, U. Buschbom, and M. Evenari. 1973. Stomatal responses to changes in temperature at increasing water stress. *Planta* 110:29-42.

Scott, D. and W. D. Billings. 1964. Effects of environmental factors on standing crop and productivity of an alpine tundra. *Ecological Monographs* 34:243-270.

Scott, F. M. 1932. Some features of the anatomy of *Fouquieria splendens. American J. of Botany* 19:673-678.

Scott, G. D. 1969. *Plant symbiosis.* New York: St. Martin's Press.

Seigler, D. and P. W. Price. 1976. Secondary compounds in plants: primary functions. *The American Naturalist* 110:101-105.

Semikhatova, O. A. 1960. The after-effect of temperature on photosynthesis. *Botanisheskii Zhurnal* 45:1488–1501.

Sestak, Z., J. Catsky and P. G. Jarvis. 1971. *Plant photosynthetic production: manual of methods.* The Hague: Junk.

Shanks, R. E. and J. S. Olson. 1961. First year breakdown of leaf litter in southern Appalachian forests. *Science* 134:194–195.

Shannon, C. E. and W. Weaver. 1949. *The mathematical theory of communication.* Urbana, IL: University of Illinois Press.

Shantz, H. L. 1945. Frederick Edward Clements (1874–1945). *Ecology* 26:317–319.

Sharitz, R. R. and J. F. McCormick. 1973. Population dynamics of two competing annual plant species. *Ecology* 54:723–739.

Shelford, V. E. 1913. *Animal communities in temperate America.* Chicago: University of Chicago Press.

Shelton, J. S. 1966. *Geology illustrated.* San Francisco, CA: W. H. Freeman.

Shimshi, D. 1969. A rapid field method for measuring photosynthesis with labeled carbon dioxide. *J. Experimental Botany* 20:381–401.

Shimwell, D. W. 1971. *The description and classification of vegetation.* Seattle, WA: University of Washington Press.

Shirley, H. L. 1945. Reproduction of upland conifers in the Lake States as affected by root competition and light. *American Midland Naturalist* 33:537–612.

Shreve, E. B. 1923. Seasonal changes in the water relations of desert plants. *Botanical Gazette* 39:397–408.

Shreve, F. 1914. The role of winter temperatures in determining the distribution of plants. *American J. of Botany* 1:194–202.

———. 1917. The establishment of desert perennials. *J. of Ecology* 5:210–216.

———. 1942. The desert vegetation of North America. *Botanical Review* 8:195–246.

Shugart, H. H., Jr., D. E. Reichle, N. T. Edwards, and J. R. Kercher. 1976. A model of calcium-cycling in an east Tennessee *Liriodendron* forest. *Ecology* 57:99–109.

Shure, D. J. and A. J. Lewis. 1973. Dew formation and stem flow on common ragweed (*Ambrosia artemisiafolia*). *Ecology* 54:1152–1155.

Siccama, T. G., F. H. Bormann, and G. E. Likens. 1970. The Hubbard Brook ecosystem study: productivity, nutrients, and phytosociology of the herbaceous layer. *Ecological Monographs* 40:389–402.

Simpson, E. H. 1949. Measurement of diversity. *Nature* 163:688.

Singer, S. F., ed. 1970. *Global effects of environmental pollution.* New York: Springer-Verlag.

Singh, J. S. and S. R. Gupta. 1977. Plant decomposition and soil respiration in terrestrial ecosystems. *Botanical Review* 43:449–528.

Singh, J. S., W. K. Lauenroth, and R. K. Steinhorst. 1975. Review and assessment of various techniques for estimating net aerial primary productivity in grasslands from harvest data. *Botanical Review* 41:181–232.

Slatyer, R. O. 1967. *Plant-water relationships.* New York: Academic Press.

Slobodkin, L. B. 1974. Comments from a biologist to a mathematician. In *Proc. SIAM-SIMS Conference,* Alta, UT, ed. S. A. Leven, pp. 318–329.

Slobodkin, L. B. and A. Rapoport. 1974. An optimal strategy of evolution. *The Quarterly Review of Biology* 49:181–200.

Slobodkin, L. B., F. E. Smith, and N. G. Hairston. 1967. Regulation in terrestrial ecosystems, and the implied balance of nature. *American Naturalist* 101:109–124.

Smith, A. D. 1940. A discussion of the application of a climatological diagram, the hythergraph, to the distribution of natural vegetation types. *Ecology* 21:184–191.

Smith, B. N. and S. Epstein. 1971. Two categories of $^{13}C/^{12}C$ ratios for higher plants. *Plant Physiology* 47:380–384.

Smith, P. F. 1966. Leaf analysis of citrus. In *Fruit Nutrition,* ed. N. F. Childers. New Brunswick, NJ: Horticultural Publications.

Smith, W. H. 1976. Character and significance of forest tree root exudates. *Ecology* 57:324–331.

Smith, W. K. and P. S. Nobel. 1977a. Temperature and water relations for sun and shade leaves of a desert broadleaf, *Hyptis emoryi*. *J. of Experimental Botany* 28:169–183.

Smith, W. K. and P. S. Nobel. 1977b. Influences of seasonal changes in leaf morphology on water use efficiency for three desert broadleaf shrubs. *Ecology* 58:1033–1043.

Soil Survey Staff. 1960. *Soil classification, a comprehensive system (7th Approximation).* Washington, D.C.: U.S. Dept. of Agriculture Soil Conservation Service.

———. 1975. *Soil taxonomy.* Agriculture Handbook no. 436. Washington, D.C.: U.S. Dept. of Agriculture Forest Service.

Sondheimer, E. and J. B. Simeone, eds. 1970. *Chemical ecology.* New York: Academic Press.

Soo Hoo, C. F. and G. Fraenkel. 1964. The resistance of ferns to the feeding of *Prodenia eridania* larvae. *Annuals of the Entomological Society of America* 57:788–790.

Spanner, P. C. 1951. The Peltier effect and its use in the measurement of suction pressure. *J. of Experimental Botany* 2:145–168.

Spurr, S. H. and B. V. Barnes. 1973. *Forest ecology.* New York: Ronald Press Co.

Stalter, R. and W. T. Batson. 1969. Transplantation of salt marsh vegetation, Georgetown, South Carolina. *Ecology* 50:1087–1089.

Stearn, W. T., ed. 1968. *Humboldt, Bonplant, Kunth and tropical American botany, a miscellany on the Nova Genera et Species Plantarum.* Cramer.

Steenbergh, W. F. and C. H. Lowe. 1969. Critical factors during the first years of life of the saguaro *(Cereus giganteus)* at Saguaro National Monument. *Ecology* 50:825–834.

Stewart, O. C. 1963. Barriers to understanding the influence of fire by aborigines on vegetation. In *Proc. Tall Timbers Ecol. Conf.*, no. 2, pp. 117–126. Tallahassee, FL: Tall Timbers Research Station.

Stone, E. C. 1963. The ecological importance of dew. *Quarterly Review of Biology* 38:328–341.

Stowe, L. G. and J. A. Teeri. 1978. The geographic distribution of C_4 species of the Dicotyledonae in relation to climate. *American Naturalist* 112:609–623.

Strahler, A. N. and A. H. Strahler. 1973. *Environmental geosciences: interaction between natural systems and man.* Santa Barbara, CA: Hamilton Publishing Co.

Strahler, A. N. and A. H. Strahler. 1974. *Introduction to environmental science.* Santa Barbara, CA: Hamilton Publishing Co.

Strain, B. R. and V. C. Chase. 1966. Effect of past and prevailing temperatures on the carbon dioxide exchange capacities of some woody desert perennials. *Ecology* 47:1043–1045.

Sweeney, J. R. 1956. Responses of vegetation to fire. *University of California Publications in Botany* 28:143–206.

———. 1967. Ecology of some "fire type" vegetations in Northern California. *Proc. Tall Timbers Fire Ecol. Conf.*, no. 7, pp. 111–125. Tallahassee, FL: Tall Timbers Research Station.

———. 1977. *Fire ecology.* Paper given to California Botanical Society, 17 March 1977 at University of California, Berkeley.

Syvertsen, J. P., G. L. Cunningham, and T. V. Feather. 1975. Anomalous diurnal patterns of stem xylem water potentials in *Larrea tridentata. Ecology* 56:1423–1428.

Syvertsen, J. P., G. L. Nickell, R. W. Spellenberg, and G. L. Cunningham. 1976. Carbon reduction pathways and standing crop in three Chihuahuan desert plant communities. *Southwestern Naturalist* 21:311–320.

Tadros, T. M. 1957. Evidence of the presence of an edaphobiotic factor in the problem of serpentine tolerance. *Ecology* 38:14–23.

Tall Timbers Fire Ecol. Confs. 1962–1979. *Proc. Tall Timbers Fire Ecol. Conf.* Tallahassee, FL: Tall Timbers Research Station.

Tansley, A. G. 1914. Presidential address to the first annual general meeting of the British Ecological Society. *J. of Ecology* 2:194–202.

———. 1935. The use and abuse of vegetation concepts and terms. *Ecology* 16:284–307.

———. 1939. The plant community and the ecosystem. *J. of Ecology* 27:513–530.

———. 1940. Henry Chandler Cowles, 1869–1939. *J. of Ecology* 28:450–452.

———. 1947a. Frederic Edward Clements, 1874–1945. *J. of Ecology* 34:194–196.

———. 1947b. The early history of modern plant ecology in Britain. *J. of Ecology* 35:130–137.

Tappeiner, J. C. and A. A. Alm. 1975. Undergrowth vegetation effects on the nutrient content of litterfall and soils in red pine and birch stands in northern Minnesota. *Ecology* 56:1193–1200.

Taylor, N. 1938. A preliminary report on the salt marsh vegetation of Long Island, New York. *New York State Museum Bulletin* 316:21–84.

———. 1939. Salt tolerance of Long Island salt marsh plants. *New York State Museum Circular* 23:1–42.

Taylor, S. E. 1975. Optimal leaf form. In *Perspectives in biophysical ecology*, ed. D. M. Gates and R. B. Schmerl, pp. 73–86. New York: Springer-Verlag.

Teeri, J. A. and L. G. Stowe. 1976. Climatic patterns and the distribution of C_4 grasses in North America. *Oecologia* 23:1–12.

Tevis, L., Jr. 1958a. Germination and growth of ephemerals induced by sprinkling a sandy desert. *Ecology* 39:681–687.

——. 1958b. A population of desert ephemerals germinated by less than one inch of rain. *Ecology* 39:688–695.

Thompson, D. Q. and R. H. Smith. 1970. The forest primeval in the Northeast—A great myth? *Proc. Tall Timbers Ecol. Conf.*, no. 10, pp. 255–265. Tallahassee, FL: Tall Timbers Research Station.

Thompson, L. M. and F. R. Troeh. 1973. *Soils and soil fertility.* New York: McGraw-Hill.

Thomson, J. W. 1974. Lichenology in North America 1947–1972. *Annals of the Missouri Botanical Garden* 61:45–55.

Thornthwaite, C. W. 1931. The climates of North America; according to a new classification. *Geographical Review* 21:633–655.

Tikhomirov, B. A. 1963. *Contributions to the biology of Arctic plants.* Leningrad, USSR: Acad. Nauk.

Torrey, J. G. 1978. Nitrogen fixation by actinomycete-nodulated angiosperms. *BioScience* 28:586–592.

Transeau, E. N. 1926. The accumulation of energy by plants. *Ohio J. of Science* 26:1–10.

Triska, F. J. and J. R. Sedell. 1976. Decomposition of four species of leaf litter in response to nitrate manipulation. *Ecology* 57:783–792.

Troughton, J. and L. A. Donaldson. 1972. *Probing plant structure.* New York: McGraw-Hill.

Tukey, H. B., Jr. 1966. Leaching of metabolites from aboveground plant parts and its implications. *Bulletin of the Torrey Botanical Club* 93:385–401.

Tukey, H. B., Jr. and R. A. Mecklenberg. 1964. Leaching of metabolites from foliage and subsequent reabsorption and redistribution of the leachate in plants. *American J. of Botany* 51:737–742.

Turesson, G. 1922a. The species and variety as ecological units. *Hereditas* 3:100–113.

——. 1922b. The genotypical response of the plant species to the habitat. *Hereditas* 3:211–350.

——. 1925. The plant species in relation to habitat and climate. *Hereditas* 6:147–236.

——. 1930. The selective effect of climate upon the plant species. *Hereditas* 14:99–152.

Turner, R. M., S. M. Alcorn, G. Olin, and J. A. Booth. 1966. The influence of shade, soil, and water on saguaro seedling establishment. *Botanical Gazette* 127:95–102.

Turrill, W. B. 1946. The ecotype concept, a consideration with appreciation and criticism especially of recent trends. *The New Phytologist* 45:34–43.

Tyler, G. 1971. Distribution and turnover of organic matter and minerals in a shore meadow ecosystem. *Oikos* 22:265–291.

Tyree, M. T. 1976. Negative turgor pressure in plant cells: fact or fallacy? *Canadian J. of Botany* 54:2738–2746.

Tyree, M. T. and H. T. Hammel. 1972. The measurement of the turgor pressure and the water relations of plants by the pressure-bomb technique. *J. of Experimental Botany* 23:267–282.

UNESCO. 1973. *International classification and mapping of vegetation.* Paris: UNESCO.

Ungar, I. A. 1962. Influence of salinity on seed germination in succulent halophytes. *Ecology* 43:763–764.

——. 1966. Salt tolerance of plants growing in saline areas of Kansas and Oklahoma. *Ecology* 47:154–155.

——. 1977. The relationship between soil water potential and plant water potential in two inland halophytes under field conditions. *Botanical Gazette* 138:498–501.

Vaadia, Y. and Y. Waisel. 1963. Water absorption by the aerial organs of plants. *Physiologia Plantarum* 16:44–51.

van Bavel, C. H. M. 1966. Potential evaporation: the combination concept and its experimental verification. *Water Resources Research* 2:445–467.

van der Valk, A. G. 1974. Mineral cycling in coastal foredune plant communities in Cape Hatteras National Seashore, *Ecology* 55:1349–1358.

van Dobben, W. H. and R. R. Lowe-McConnell, eds. 1972. *Unifying concepts in ecology.*

Wageningen, The Netherlands: Centre for Agricultural Publishing and Documentation.

Van Dyne, G. M., W. G. Vogel, and H. G. Fisser. 1963. Influence of small plot size and shape on range herbage production estimates. *Ecology* 44:746–759.

Van Valen, L. 1975. Life, death, and energy of a tree. *Biotropica* 7:260–269.

Varley, G. C. 1967. The effects of grazing by animals on plant productivity. In *Secondary productivity of terrestrial ecosystems (principles and methods)*, vol. II, ed. K. Petrusewicz, pp. 773–777. Warsaw: Polish Academy of Sciences.

Vasek, F. C., H. B. Johnson, and D. H. Eslinger. 1975. Effects of pipeline construction on creosote bush scrub vegetation of the Mojave Desert. *Madroño* 23:1–13.

Viereck, L. A. 1966. Plant succession and soil development on gravel outwash of the Muldrow Glacier, Alaska. *Ecological Monographs* 36:181–199.

Viro, P. J. 1974. Effects of forest fire on soil. In *Fire and ecosystems*, ed. T. T. Kozlowski and C. E. Ahlgren, ch. 2. New York: Academic Press.

Vogelmann, H. W., T. G. Siccama, D. Leedy, and D. C. Ovitt. 1968. Precipitation from fog moisture in the Green Mountains of Vermont. *Ecology* 49:1205–1207.

Vogl, R. J. 1972. Effects of fire on southeastern grasslands. *Proc. Tall Timbers Fire Ecol. Conf.*, no. 12, pp. 175–198. Tallahassee, FL: Tall Timbers Research Station.

———. 1973. Ecology of knobcone pine in the Santa Ana Mountains, California. *Ecological Monographs* 43:125–143.

———. 1974. Effects of fire on grasslands. In *Fire and ecosystems*, ed. T. T. Kozlowski and C. E. Ahlgren, ch. 5. New York: Academic Press.

Vogl, R. J., W. P. Armstrong, K. L. White, and K. L. Cole. 1977. The closed-cone pines and cypress. In *Terrestrial vegetation of California*, ed. M. G. Barbour and J. Major, pp. 295–358. New York: Wiley-Interscience.

Vogl, R. J. and L. T. McHargue. 1966. Vegetation of California fan palm oases on the San Andreas Fault. *Ecology* 47:532–540.

Voigt, G. K. 1965. Nitrogen recovery from decomposing tree leaf tissue and forest humus. *Soil Science Society of America, Proc.* 29:756–759.

Volterra, V. 1926. Fluctuations in the abundance of a species considered mathematically. *Nature* 118:558–560.

Waisel, Y. 1972. *Biology of halophytes.* New York: Academic Press.

Wakimoto, R. H. 1977a. Chaparral growth and fuel assessment in southern California. In *Proc. Symp. Env. Cons. Fire and Fuel Mangmt. in Medit. Ecosyst.*, pp. 412–418. Washington, D.C.: U.S. Dept. of Agriculture Forest Service.

———. 1977b. Presentation at giant sequoia fire ecology extension course, October 1977, University of California, Davis.

Walker, D. 1970. Direction and rate in some British postglacial hydroseres. In *The vegetational history of the British Isles*, ed. D. Walker and R. West, pp. 117–139. Cambridge, England: Cambridge University Press.

Walker, R. B. 1954. The ecology of serpentine soils. II. Factors affecting plant growth on serpentine soils. *Ecology* 35:259–266.

Wallace, J. W. and R. L. Mansell, eds. 1976. *Biochemical interaction between plants and insects.* New York: Plenum Press.

Walter, H. 1973. *Vegetation of the earth in relation to climate and the eco-physiological conditions.* New York: Springer-Verlag.

Wang, J. Y. 1979. *Instruments for physical environmental measurement*, vol. 1. San Jose, CA: Milieu Information Service.

Wanner, H. 1970. Soil respiration, litter fall, and productivity of tropical rain forest. *J. of Ecology* 58:543–547.

Waring, R. H. and B. D. Cleary. 1967. Plant moisture stress: evaluation by pressure bomb. *Science* 155:1248–1254.

Warming, J. E. B. 1892. *Lagoa Santa.* Copenhagen, Denmark: Lun.

———. 1895. *Plantesamfund, grundtraek af den okologiske plantegeografi.* Copenhagen, Denmark: Philipsen.

———. 1909. *Oecology of plants.* London: Oxford University Press.

Watt, A. S. 1947. Pattern and process in the plant community. *J. of Ecology* 35:1–22.

———. 1955. Bracken versus heather, a study in plant sociology. *J. of Ecology* 43:490–506.

Weaver, J. E. 1920. *Root development in the grassland formation.* Carnegie Institution of Washington Publ. 286. Washington, D.C.: Carnegie Institution of Washington.

Weaver, J. E. and F. E. Clements. 1929. *Plant ecology.* New York: McGraw-Hill.

Weaver, J. E. and T. J. Fitzpatrick. 1934. The prairie. *Ecological Monographs* 4:109–295.

Webley, D. M., D. J. Eastwood, and C. H. Gimingham. 1952. Development of a soil microflora in relation to plant succession on sand dunes, including the "rhizosphere" flora associated with colonising species. *J. of Ecology* 40:168–178.

Weier, T. E., C. R. Stocking, and M. G. Barbour. 1974. *Botany—an introduction to plant biology,* 5th ed. New York: Wiley.

Weigert, R. G. and F. C. Evans. 1964. Primary production and disappearance of dead vegetation on an old field in southeastern Michigan. *Ecology* 45:49–63.

Weigert, R. G. and D. F. Owen. 1971. Trophic structure, available resources and population density in terrestrial vs. aquatic ecosystems. *J. of Theoretical Biology* 30:69–81.

Wells, P. V. and R. Berger. 1967. Late Pleistocene history of coniferous woodland in the Mojave Desert. *Science* 155:1640–1647.

Went, F. W. 1942. The dependence of certain annual plants on shrubs in southern California deserts. *Bulletin of the Torrey Botanical Club* 69:100–114.

——. 1948. Ecology of desert plants. I. Observations on germination in Joshua Tree National Monument. *Ecology* 29:242–253.

Went, F. W. and N. Stark. 1968. Mycorrhiza. *BioScience* 18:1035–1039.

Went, F. W. and M. Westergaard. 1949. Ecology of desert plants. III. Development of plants in the Death Valley National Monument, California. *Ecology* 30:26–38.

Werner, P. A. 1976. Ecology of plant populations in successional environments. *Systematic Botany* 1:246–268.

West, N. E. and P. T. Tueller. 1971. Special approaches to studies of competition and succession in shrub communities. In *Wildland shrubs,* U.S. Dept. of Agriculture Forest Service General Technical Report INT-1, ed. C. M. McKell, J. P. Blaisdell, and J. R. Goodin, pp. 165–171. Logan, UT: U.S. Dept. of Agriculture Forest Service.

Westman, W. E. 1975. Edaphic climax pattern of the pygmy forest region of California. *Ecological Monographs* 45:109–135.

Westman, W. E. and R. H. Whittaker. 1975. The pygmy forest region of northern California: studies on biomass and primary productivity. *J. of Ecology* 63:493–520.

White, D. 1941. Prairie soil as a medium for tree growth. *Ecology* 22:399–407.

Whittaker, R. H. 1953. A consideration of climax theory: the climax as a population and pattern. *Ecological Monographs* 23:41–78.

——. 1954. The ecology of serpentine soils. I. Introduction. *Ecology* 35:258–259.

——. 1956. Vegetation of the Great Smoky Mountains. *Ecological Monographs* 26:1–80.

——. 1960. Vegetation of the Siskiyou Mountains of California. *Ecological Monographs* 30:279–338.

——. 1962. Classification of natural communities. *Botanical Review* 28:1–239.

——. 1966. Forest dimensions and production in the Great Smoky Mountains. *Ecology* 47:103–121.

——. 1967. Gradient analysis of vegetation. *Biological Review* 42:207–264.

——. 1973a. Climax concepts and recognition. *Handbook of Vegetation Science* 8:137–154.

——. ed. 1973b. *Ordination and classification of communities.* Gotingen, The Netherlands: Junk.

——. 1975. *Communities and ecosystems.* 2nd ed. New York: Macmillan.

Whittaker, R. H. and P. P. Feeny. 1971. Allelochemics: chemical interactions between species. *Science* 171:1757–1770.

Whittaker, R. H., S. A. Levin, and R. B. Root. 1973. Niche, habitat, and ecotype. *The American Naturalist* 107:321–338.

Whittaker, R. H. and G. E. Likens. 1975. The biosphere and man. In *The primary production of the biosphere,* ed. H. Lieth and R. H. Whittaker, pp. 305–328. New York: Springer-Verlag.

Whittaker, R. H. and W. A. Niering. 1975. Vegetation of the Santa Catalina Mountains, Arizona: V. Biomass, production and diversity along the elevation gradient. *Ecology* 56:771–790.

Whittaker, R. H. and G. M. Woodwell. 1968. Dimension and production relations of trees and shrubs in the Brookhaven Forest, New York. *J. of Ecology* 56:1–25.

Whittaker, R. H. and G. M. Woodwell. 1969. Structure, production, and diversity of the oak-pine forest at Brookhaven, New York. *J. of Ecology* 57:155–174.

Whittaker, R. H. and G. M. Woodwell. 1972. Evolution of natural communities. In *Ecosystem structure and function,* Oregon State University Annual Biology Colloquia, volume 31, ed. J. A. Wiens, pp. 137–156. Corvallis, OR: Oregon State University.

Wiegert, R. G. and J. T. McGinnis. 1975. Annual production and disappearance of detritus on three South Carolina old fields. *Ecology* 56:129–140.

Wieland, N. K. and F. A. Bazzaz. 1975. Physiological ecology of three codominant successional annuals. *Ecology* 56:681–688.

Williams, C. B. 1964. *Patterns in the balance of nature.* New York: Academic Press.

Williams, C. M. 1970. Hormonal interactions between plants and insects. In *Chemical ecology,* ed. E. Sondheimer and J. B. Simeone, pp. 103–132. New York: Academic Press.

Williams, W. T. and J. M. Lambert. 1959. Multivariate methods in plant ecology. I. Association analysis in plant communities. *J. of Ecology* 47:83–101.

Williams, W. T. and J. M. Lambert. 1961. Multivariate methods in plant ecology. III. Inverse association-analysis. *J. of Ecology* 49:717–729.

Willis, A. J. 1963. Braunton burrows: the effects on the vegetation of the addition of mineral nutrients to the dune soils. *J. of Ecology* 51:353–374.

Willmot, A. and P. D. Moore. 1973. Adaptation to light intensity in *Silene alba* and *S. dioica. Oikos* 24:458–464.

Wilson, R. E. and E. L. Rice. 1968. Allelopathy as expressed by *Helianthus annuus* and its role in old-field succession. *Bulletin of the Torrey Botanical Club* 95:432–448.

Winn, C. 1977. *Decomposition in southern California chaparral.* Masters Thesis. Fullerton, CA: California State University.

Winter, K. 1975. *Die Rolle des Crassulacean-säurestoffwechsels als Biochemishe grundlage zur anpassung von Halophyten an Standorte hoher Salinität.* Dissertation Darmstadt, Germany.

Winter, K. and U. Lüttge. 1976. Balance between C_3 and CAM pathway of photosynthesis. *Ecological Studies,* 19:323–334.

Woods, F. W. and K. Brock. 1964. Interspecific transfer of Ca_{45} and P_{32} by root systems. *Ecology* 45:886–889.

Woods, F. W. and R. E. Shanks. 1959. Natural replacement of chestnut by other species in the Great Smoky Mountains National Park. *Ecology* 40:349–361.

Woodwell, G. M. and H. H. Smith, eds. 1969. *Diversity and stability in ecological systems.* Brookhaven Symposia in Biology no. 22. Upton, NY: Brookhaven National Laboratories.

Woodwell, G. M., R. H. Whittaker, and R. S. Houghton. 1975. Nutrient concentrations in plants in the Brookhaven oak-pine forest. *Ecology* 56(2):318–332.

Workman, J. P. and N. E. West. 1967. Germination of *Eurotia lanata* in relation to temperature and salinity. *Ecology* 48:659–661.

Wright, E. and R. F. Tarrant. 1957. *Microbial soil properties after logging and slash burning.* U.S. Forest Service Pacific Northwest Forest Range Experimental Station Research Notes 157. Washington, D.C.: U.S. Dept. of Agriculture Forest Service.

Wright, R. D. 1974. Rising atmospheric CO_2 and photosynthesis of San Bernardino Mountain plants. *American Midland Naturalist* 91:360–370.

Yeaton, R. I. 1978. A cyclic relationship between *Larrea tridentata* and *Opuntia leptocaulis* in the northern Chihuahuan Desert. *J. of Ecology* 66:651–656.

Yemm, E. W. and R. G. S. Bidwell. 1969. Carbon dioxide exchanges in leaves. I. Discrimination between $^{14}CO_2$ and $^{12}CO_2$ in photosynthesis. *Plant Physiology* 44:1328–1334.

Young, J. A., R. A. Evans, and P. T. Tueller. 1975. Great Basin plant communities— pristine and grazed. In *Holocene climates in the Great Basin,* Occasional Paper, Nevada Archeological Survey, ed. R. Elston, pp. 186–215. Reno, NV: Nevada Archeological Survey.

Zanstra, P. E. and F. Hagenzieker. 1977. Comments on the psychrometric determination of leaf water potentials *in situ. Plant and Soil* 48:347–367.

Zavitkovski, J. and M. Newton. 1968. Ecological importance of snowbrush, *Ceanothus velutinus,* in the Oregon Cascades agricultural ecology. *Ecology* 49:1134–1145.

Zinke, P. J. 1977. The redwood forest and associated north coast forests. In *Terrestrial vegetation of California,* ed. M. G. Barbour and J. Major, pp. 679–698. New York: Wiley-Interscience.

INDEX

*The page on which the term is defined is indicated by (def.).